Signaling and Communication in Plants

Series Editor
František Baluška
Department of Plant Cell Biology, IZMB
University of Bonn
Bonn, Germany

More information about this series at http://www.springer.com/series/8094

François Chaumont • Stephen D. Tyerman
Editors

Plant Aquaporins

From Transport to Signaling

Springer

Editors
François Chaumont
Institut des Sciences de la Vie
Université catholique de Louvain
Louvain-la-Neuve
Belgium

Stephen D. Tyerman
School of Agriculture, Food and Wine
University of Adelaide
Adelaide
Australia

ISSN 1867-9048 ISSN 1867-9056 (electronic)
Signaling and Communication in Plants
ISBN 978-3-319-84153-3 ISBN 978-3-319-49395-4 (eBook)
DOI 10.1007/978-3-319-49395-4

Printed on acid-free paper

This Springer imprint is published by Springer Nature
The registered company is Springer International Publishing AG
The registered company address is: Gewerbestrasse 11, 6330 Cham, Switzerland

Preface

The discovery of the water channel activity of the first plant aquaporin, γ-TIP or TIP1;1, in 1993 has significantly challenged the concepts by which plants control cell water homeostasis but also the water relations of the whole organism. In addition, it appeared rapidly that plant aquaporins or MIPs (membrane intrinsic proteins) facilitate also the membrane diffusion of an increasing amount of small solutes, such as urea, CO_2, H_2O_2, ammonium, metalloids, etc. This diversity of substrates probably evolves from the high number of aquaporin genes identified in plant genomes. Higher plant aquaporins cluster into five phylogenetic subfamilies (PIPs, plasma membrane intrinsic proteins; TIPs, tonoplast intrinsic proteins; NIPs, NOD26-like intrinsic protein; SIPs, small basic intrinsic proteins; and XIPs, X intrinsic proteins) and are present in different cell membranes.

This book integrates exciting data illustrating the various regulation mechanisms leading to active aquaporins in their target membranes and addresses the involvement of different aquaporins in many physiological processes at different cell, organ, and tissue levels and in several environmental conditions. It includes the roles and regulation of aquaporins in plant water homeostasis, but also in plant distribution of other small solutes including nitrogen, CO_2, and metalloids. There are still many more discoveries to be made in how aquaporins are regulated and how their selectivity to different solutes are controlled, particularly those that appear to have dual permeation properties. Their interaction with plant mycorrhizae and their contribution in signaling processes are also discussed. This volume, by the diversity of the aspects developed in the different chapters, illustrates the importance of the aquaporins and their regulation in controlling plant physiology and development.

Louvain-la-Neuve, Belgium — François Chaumont
Adelaide, Australia — Stephen D. Tyerman

August 2016

Contents

Structural Basis of the Permeation Function of Plant Aquaporins

Sukanya Luang and Maria Hrmova

Abstract Aquaporins facilitate rapid and selective bidirectional water and uncharged low-molecular-mass solute or ion movements in response to osmotic gradients. The term 'aquaporin' was coined by Peter Agre and colleagues, who in 1993 suggested that major intrinsic proteins (MIPs) that facilitate rapid and selective movement of water in the direction of an osmotic gradient be named 'aquaporins (AQPs)' (Agre et al. 1993). Aquaporins are spread across all kingdoms of life including archaea, bacteria, protozoa, yeasts, plants and mammals. Plant aquaporins are classified within the ancient superfamily of MIPs, and based on sequence homology and subcellular localisation, they constitute several subfamilies. Genome-wide identifications of aquaporin genes are now available from around 15 plant species, and this information provides a rich source of sequence data for molecular studies through structural bioinformatics, three-dimensional (3D) modelling and molecular dynamics simulations. These studies have capacity to reveal new information, unavailable to X-ray diffraction studies of time- and space-averaged molecules confined in crystal lattices.

1 Summary

Aquaporins facilitate rapid and selective bidirectional water and uncharged low-molecular-mass solute or ion movements in response to osmotic gradients. The term 'aquaporin' was coined by Peter Agre and colleagues, who in 1993 suggested that major intrinsic proteins (MIPs) that facilitate rapid and selective movement of water in the direction of an osmotic gradient be named 'aquaporins (AQPs)' (Agre et al. 1993). Aquaporins are spread across all kingdoms of life including archaea, bacteria, protozoa, yeasts, plants and mammals. Plant aquaporins are classified within

S. Luang • M. Hrmova (✉)
School of Agriculture, Food and Wine, University of Adelaide, Glen Osmond, South Australia 5064, Australia
e-mail: maria.hrmova@adelaide.edu.au

F. Chaumont, S.D. Tyerman (eds.), *Plant Aquaporins*, Signaling and Communication in Plants, DOI 10.1007/978-3-319-49395-4_1

the ancient superfamily of MIPs, and based on sequence homology and subcellular localisation, they constitute several subfamilies. Genome-wide identifications of aquaporin genes are now available from around 15 plant species, and this information provides a rich source of sequence data for molecular studies through structural bioinformatics, three-dimensional (3D) modelling and molecular dynamics simulations. These studies have capacity to reveal new information, unavailable to X-ray diffraction studies of time- and space-averaged molecules confined in crystal lattices.

Aquaporins fold into a monomeric 'hourglass' or 'dumbbell-like' shaped structure that has been retained in all aquaporins. Individual monomers associate in vivo into functional tetramers, whereby this vertically symmetric structure provides foundation for residence within a lipid bilayer. Two plant aquaporin structures are available in structural databases (as of May 2016), which is that of (i) a predominantly water-permeable plasma membrane intrinsic protein (PIP) aquaporin in open and closed conformational states (PDB IDs: 1Z98, 2B5F and 4IA4) from *Spinacia oleracea* (Tornroth-Horsefield et al. 2006; Frick et al. 2013a, b) and (ii) an open state of a water- and ammonia-permeable tonoplast intrinsic protein (TIP) aquaammoniaporin from *Arabidopsis thaliana* (PDB ID: 5i32) (Kirscht et al. 2016). Detailed structural information on other plant subfamily members is now needed from economically important food plants such as wheat, barley, maize and rice, to provide strong foundations for future smart decisions directed to food production and sustainability.

Surprisingly, limited information is available on the solute permeation specificity determinants of plant aquaporins, although these data in conjunction with structural information are vital strategic tools for modifying their molecular function. Based on predominantly structural studies, it has been suggested that properties and steric occlusions of residues within the specific structural and functional elements are one of the most fundamental characteristics that underlie differences in transport selectivities of aquaporins. These main characteristics include (i) pore dimension parameters including their diameters and overall morphology; (ii) identities and flexibilities of residues lining solute-conducting pores; (iii) chemical configurations of pore constrictions in solute-conducting pores; (iv) properties of pore vestibules and a central pore, also dictated by the residues alongside the fourfold symmetry axis of tetramers; and (v) gating of aquaporins controlled by pH, cation binding, post-translational modifications such as phosphorylation and the dispositions of interacting loops.

We conclude that although structural aquaporin research has significantly progressed in recent years, many questions remain open. For example, are individual protomers within tetramers identical in function, what is the structural basis of permeation of non-electrolytes and ionic species, and the thermodynamic origin of transporting function of solutes, and how exactly have aquaporin proteins evolved during millions years of evolution into their current forms?

2 Aquaporins in Living Systems Including Plants

Plants acquire water from soil through aquaporins or use them as vehicles to dispose of excess of toxic substances (Schnurbusch et al. 2010; Hayes et al. 2013; Xu et al. 2015). Aquaporin molecules, amongst other pathways, are responsible for hydraulic conductance of plants that underlies water uptake together with dissolved mineral nutrients (see also chapter "Plant Aquaporins and Metalloids"). Aquaporins facilitate rapid and selective bidirectional water and uncharged low-molecular-mass solute transport, in response to osmotic and concentration gradients, respectively. The latter does not necessarily rely on an osmotic gradient. This transport, occurring through polytopic aquaporins that span cell membranes, is independent of a supply of external energy (e.g. ATP). Thus, aquaporins are known to be passive transporters, although fundamental to their function are structural flexibility and gating, which may be dependent on the redox state of a cellular environment, on the activity of phosphorylation machinery (controlling the levels of, e.g. ATP) and on membrane and subcellular dynamics.

2.1 *Aquaporins Occur in All Kingdoms of Life*

Aquaporins are spread across all kingdoms of life including archaea, bacteria, protozoa, yeasts, plants and mammals. In archaea and bacteria, typically one aquaporin type is retained, while in eukaryotes gene duplications and horizontal gene transfer events have resulted in occurrence of subfamilies of aquaporins with diversified transport functions, although the canonical hourglass or dumbbell-like shaped architecture has been retained in all aquaporins. The typical examples of duplication and function diversification include aquaporins in fish, mammals and higher plants, in which neo-functionalisation has led to evolution of paralogous proteins with various solute selectivities, gating mechanisms or time and space differential expression (Fotiadis et al. 2001; Zardoya et al. 2002; Abascal et al. 2014). For example, 35, 35 and 39 aquaporins have been described in maize, *Arabidopsis* and rice, respectively (Chaumont et al. 2001; Johanson et al. 2001; Sakurai et al. 2005). These numbers are even higher in non-plant species such as in fish and some land vertebrates, due to several rounds of entire genome duplication during early stages of their evolution (Abascal et al. 2014), although most mammals only require the presence of limited numbers of aquaporins to properly function. Diverse aquaporin isoforms are directed to various subcellular locations and compartments and represent fundamental components for membrane evolution, diversity and differential gene expression. Through these specific membrane aquaporin-containing partitions, plants drive hydrostatic and osmotic forces that help them to maintain water homeostasis, together with hydraulic conductance in roots, stems and other organs (Fricke et al. 1997; Tyerman et al. 1999; Maurel et al. 2008; Chaumont and Tyerman 2014).

2.2 *Plant Aquaporin Sequences and Their Genome-Wide Identification*

Since the first member of the major intrinsic protein (MIP) family was described and its cDNA cloned (Gorin et al. 1984), the first plant MIP from soybean (nodulin 26) was identified (Sandal and Marcker 1988), along with the tonoplast intrinsic protein (TIP) from bean seeds (Johnson et al. 1990), and α-TIP (Höfte et al. 1992) and γ-TIP (Maurel et al. 1993) from *Arabidopsis*, and other plants. Some of these proteins were described as water stress-induced proteins (Höfte et al. 1992) and only later functionally characterised as water channels. These discoveries were followed by a series of informative reviews on physiological function of aquaporins (e.g. Tyerman et al. 1999; Verkman and Mitra 2000; Gomes et al. 2009; Maurel et al. 2008; Chaumont and Tyerman 2014; Li et al. 2014; Mukhopadhyay et al. 2014). These physiological functions include photosynthesis, seed germination, cell elongation, stomata movement, reproduction (Reddy et al. 2015) and responses to a variety of abiotic stresses, such as anoxia (Choi and Roberts 2007), hydrogen peroxide toxicity (Dynowski et al. 2008; Wudick et al. 2015), mineral soil toxicity (boron and arsenic) (Isayenkov and Maathuis 2008; Ma et al. 2008; Kamiya et al. 2009; Schnurbusch et al. 2010; Li et al. 2011; Hayes et al. 2013; Xu et al. 2015), high salt (Zhang et al. 2008; Gao et al. 2010; Hu et al. 2012; Xu et al. 2013) and a low water potential drought (Xu et al. 2014; Li et al. 2015).

Genome-wide identification studies of aquaporin genes are now available from at least fourteen plant species, including *Arabidopsis* (Johanson et al. 2001), maize (Chaumont et al. 2001), rice (Sakurai et al. 2005), poplar (Gupta and Sankararamakrishnan 2009), grapevine (Shelden et al. 2009), cotton (Park et al. 2010), barley (Besse et al. 2011; Tombuloglu et al. 2015), soybean (Zhang et al. 2013), tomato (Reuscher et al. 2013) and bread wheat (Pandey et al. 2013). As a result of recent proliferations of genome sequencing initiatives, several new genome-wide identification studies were conducted in cabbage (Diehn et al. 2015), common bean (Ariani and Geps 2015), sorghum (Reddy et al. 2015) and wheat (Hove et al. 2015). These analyses have provided a rich source of sequence data information for molecular studies that have been conducted through structural bioinformatics (Wang et al. 2005; Deshmukh et al. 2015) and 3D structural (homology or comparative) modelling (Wallace and Roberts 2004; Schnurbusch et al. 2010; Gupta et al. 2012; Verma et al. 2015).

2.3 Classification of Aquaporins

Plant aquaporins are classified within the ancient superfamily of Major Intrinsic Proteins (MIPs) (Saier et al. 2016). Based on sequence homology and subcellular localisation, MIPs constitute five subfamilies, namely, plasma membrane intrinsic proteins (PIPs), tonoplast intrinsic proteins (TIPs), nodulin-26 intrinsic proteins (NIPs), small basic intrinsic proteins (SIPs) and X-intrinsic proteins (XIPs). In recent years, several studies have specifically focussed on molecular evolution and functional divergence of NIP (Liu et al. 2009) and XIP proteins (Bienert et al. 2011; Lopez et al. 2012; Venkatesh et al. 2015). These studies have pointed out that the functional divergence of various classes of aquaporins under selection pressures led to restrictions on the physicochemical properties of key functional amino acid residues, following gene duplication.

3 Three-Dimensional Structures of Aquaporins

3.1 Structural Information on Aquaporins Is Available from All Kingdoms of Life

3D atomic structures of aquaporins are accessible from archaea (Lee et al. 2005), bacteria (Fu et al. 2000; Savage et al. 2003; Savage et al. 2010), protozoa (Newby et al. 2008), yeasts (Fischer et al. 2009; Eriksson et al. 2013), plants (Fotiadis et al. 2000; Törnroth-Horsefield et al. 2006; Frick et al. 2013a, b; Kirscht et al. 2016) and mammals (Sui et al. 2001; Gonen et al. 2004; Harries et al. 2004), including humans (Murata et al. 2000; Viadiu et al. 2007; Horsefield et al. 2008; Ho et al. 2009; Agemark et al. 2012; Frick et al. 2014). For example, a sub-angstrom resolution structure of the *Pichia* water-conducting aquaporin (Eriksson et al. 2013) and a recent high-resolution structure of the *Arabidopsis* aquaammoniaporin (Kirscht et al. 2016) provided an unprecedented view into the landscape of positions of interacting residues and the mode of coordination of water molecules. The water positions that were defined with a high precision in a water-conducting pore (Eriksson et al. 2013), and definitions of tautomeric states of interacting Arg and His residues, provided an abundance of information on water molecule coordination at the entry of the channel. As a result of the availability of high-resolution aquaporin architectures of these structurally similar but functionally distinct MIP and TIP proteins, a plethora of theoretical in silico studies were initiated to investigate molecular dynamics of aquaporins and flow of solutes (Tajkhorshid et al. 2002; Wang et al. 2005; Cordeiro 2015; Han et al. 2015; Verma et al. 2015; Kitchen and Conner 2015). These studies revealed novel information, unavailable to studies of time- and space-averaged molecules confined in crystal lattices, and defined protein dynamics and energy barriers during permeation events of water, ammonia or other solute-conducting aquaporins (Wang et al. 2005; Han et al. 2015; Kirscht et al. 2016).

3.2 An Overall Architecture of Protomers

The 3D structures of aquaporins are highly conserved from archaea to humans. They consist of a circular α-helical bundle with a solute-conducting pore and cytoplasmic (intracellular) and periplasmic (extracellular) conical vestibules. Each monomer is formed by six tilted (crossing angles between 25 and 40°) membrane-spanning α-helices (H1-H3 and H4-H6) and two re-entrant short α-helices (HB and HE) running in two repeats, with five interconnecting loops (LA-LE) that collectively form a right-handed α-helical bundle (Fig. 1). The arrangements of first (H1-H3 and HB) and second (H4-H6 and HE) bipartite segments, α-helices of which are significantly tilted in a membrane, follow a pseudo-twofold axis that runs

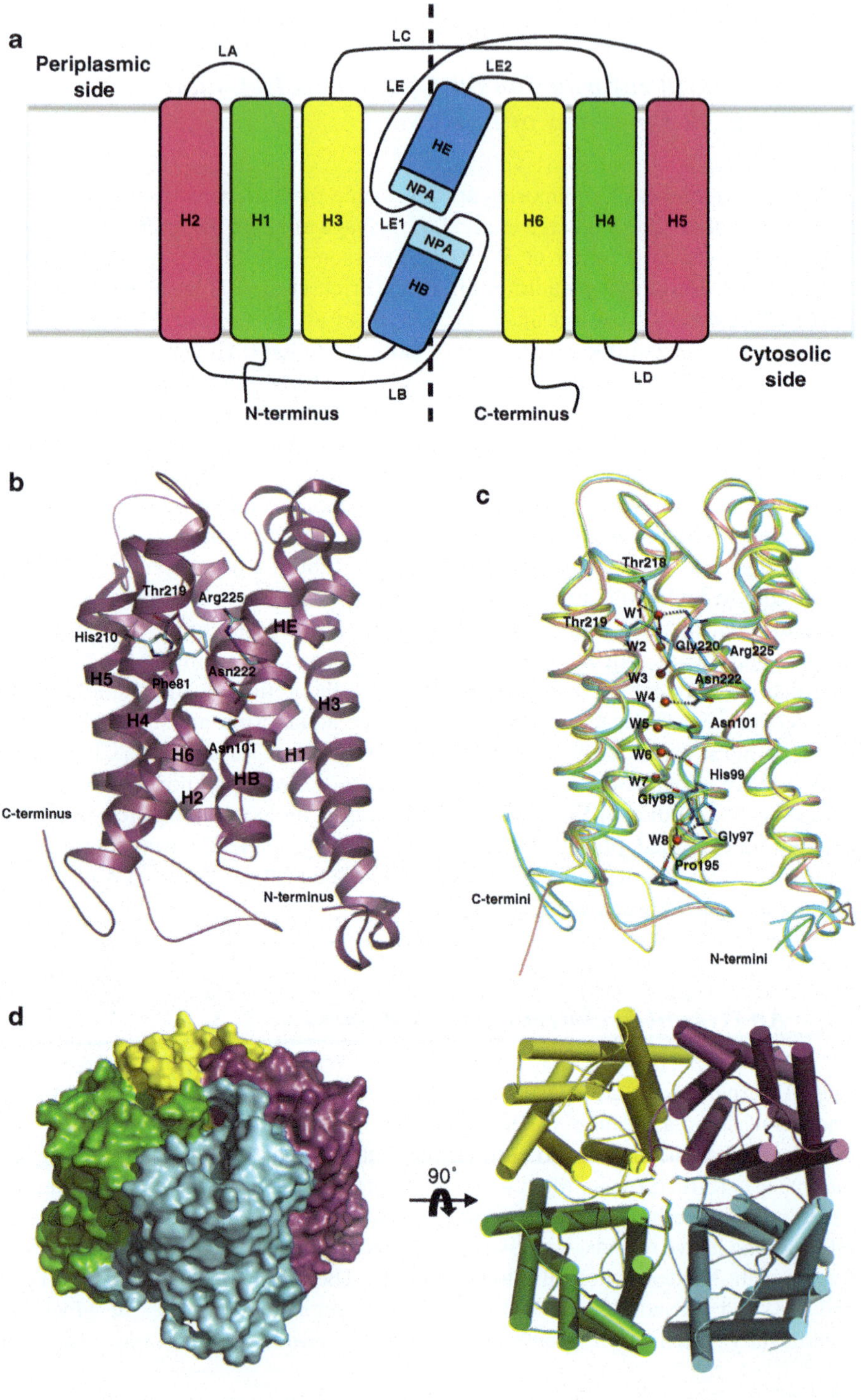
a
Periplasmic side
LA
LC
LE
LE2
HE
NPA
LE1
NPA
HB
H2
H1
H3
H6
H4
H5
Cytosolic side
LD
LB
N-terminus
C-terminus
b
Thr219
Arg225
His210
HE
H5
Phe81
Asn222
H4
H3
H6
Asn101
H1
HB
H2
C-terminus
N-terminus
c
Thr218
Thr219
W1
W2
Gly220
Arg225
W3
Asn222
W4
W5
Asn101
W6
W7
His99
Gly98
W8
Gly97
Pro195
C-termini
N-termini
d
90˚

perpendicularly to a membrane normal plane (Murata et al. 2000) (Fig. 1a; dashed line). In all aquaporins, N- and C-termini are cytoplasmically oriented (Fig. 1a). In some aquaporins, these termini are extended and carry sulfhydryl residues such as Cys (an inhibition site for mercury and other heavy metals) or N-glycosylation, phosphorylation and other post-translation modification sites.

3.3 A Circular Bundle and a Solute-Conducting Pore

The circular bundle of membrane-spanning α-helices encloses a solute-conducting pore (often referred to as a channel or a nanopore), which may be between 20 and 28 Å long and 4 and 6 Å in diameter. For example, in plant aquaporins, the pore narrows down around a selectivity filter region (defined below) (Törnroth-Horsefield et al. 2006; Frick et al. 2013a, b), or remains more uniform throughout the channel (Kirscht et al. 2016), but widens in all aquaporins to conical vestibules at both cytoplasmic and periplasmic sides (Fig. 1b).

A consensus, based on around 20 atomic structures of aquaporins, stipulates that the aromatic selectivity filter represents the narrowest constriction, at least in orthodox (predominantly water conducting) aquaporins (Fu et al. 2000; Sui et al. 2001). The selectivity filter is one of the most important regions underlying aquaporin specificity and represents as a package of four residues positioned near the periplasmic side of the pore. The selectivity filter is about 8–9 Å away from the first Asn-Pro-Ala (NPA) motif and was named the aromatic/Arg (ar/R)/LE1-LE2 constriction region (Fig. 1b). More precisely, this ar/R/LE1-LE2 region consists of one residue each from H2 and H5 α-helices, and two residues positioned on loop LE, located at partitions LE1 and LE2 that flank the NPA motif (Figs. 1a and 2b) (Fu et al. 2000; Sui et al. 2001; Savage et al. 2003). In all solved spinach aquaporins in closed or

Fig. 1 (**a**) A membrane topology diagram of aquaporins. Each protein molecule consists of six transmembrane α-helices (H1-H6) and two re-entrant α-helices (HB and HE), with NPA motifs, shown as cyan boxes. Transmembrane α-helices are connected via five interconnecting loops (LA-LE), whereby partitions LE1 and LE2 flank the second NPA motif, separated by approximately 4–5 Å from the first NPA motif. A *dashed line* indicates bipartite structural repeats of an hourglass aquaporin fold. (**b**) A cartoon representation of the spinach aquaporin SoPIP2;1 in the closed conformation (PDB ID: 1Z98). The selectivity filter residues (Phe81, His210, Thr219 and Arg225) and the two conserved asparagine residues (Asn101 and Asn222) of the NPA motifs are shown as cyan sticks. N- and C-termini are indicated. (**c**) The superposition of SoPIP2;1 structures including closed states at pH 8.0 (PDB ID: 1Z98) and at pH 6.0 (PDB ID: 4IA4), an open state (PDB ID: 2B5F), and the structure with a mercury activation site (PDB ID: 4JC6), shown in *cyan*, *pink*, *yellow* and *green*, respectively. Residues that interact with a single file of water molecules W1-W8, shown as red spheres, are indicated in cyan sticks. Hydrogen bonds between residues and water molecules are shown in *dashed lines*. (**d**) Prediction of a tetrameric assembly of the spinach aquaporin SoPIP2;1 in two orthogonal orientations (*left and right* images are related by 90° rotation to the viewer), whereby cysteine residues (shown in sticks) from each monomer participate in a quaternary assembly

open conformations, protein folds and more specifically pores enclose a single-file chain of water molecules coordinated by surrounding hydrophilic amino acid residues (Fig. 1c, cyan sticks). Recently, the presence of a novel water-filled side pore was defined in the AtTIP2;1 aquaammoniaporin, which is assumed to play a role in

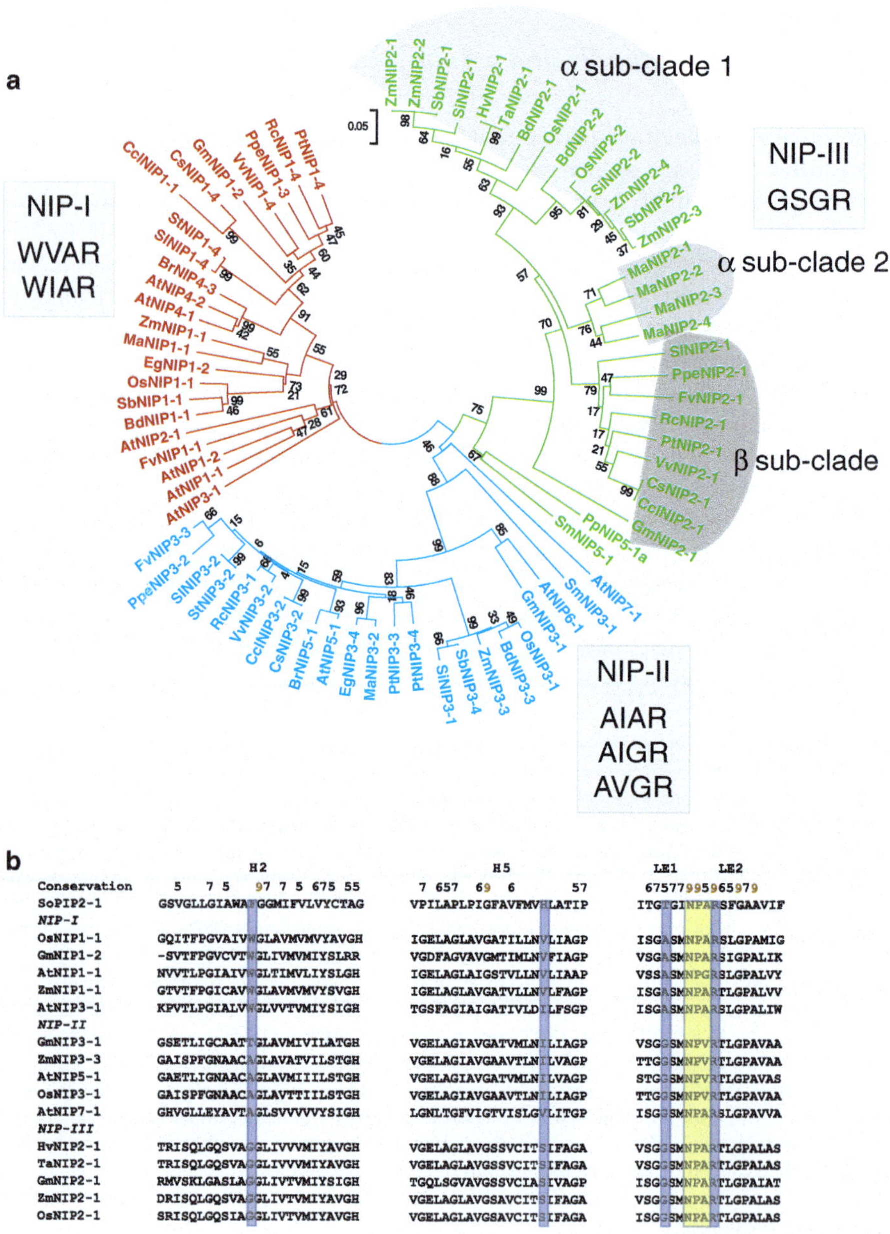

ammonia deprotonation during permeation, as revealed by molecular dynamics simulations (Kirscht et al. 2016). In some aquaporins, a second well-formed constriction is located near to the cytoplasmic vestibule.

3.4 Cytoplasmic and Periplasmic Conical Vestibules

The cylindrical solute-conducting pore is flanked by two shallow, asymmetric vestibules. These are present on each side of the pore that flare into both cytoplasmic and periplasmic spaces and are formed by loop regions at each monomer face and by the N- and C-termini at the cytoplasmic face. The vestibules give a characteristic hourglass shape of aquaporin proteins (Fig. 1b). It was revealed that in nearly every atomic structure, these vestibules contain a contiguous chain of hydrogen-bonded molecules that extend from the surface of vestibules to either an ar/R/LE1-LE2 selectivity filter region of the periplasmic vestibule or to a second constriction near to the cytoplasmic vestibule (Fig. 1b).

3.5 Aquaporins Exist as Functional Tetramers

In native environments, individual monomers form a quaternary tetrameric assembly, in which homo- or hetero-oligomers that act as independent solute-conducting units associate with each other into a tightly fitting extended trapezoid or a

Fig. 2 (**a**) A phylogenetic tree of 75 NIP proteins from *Arabidopsis thaliana* (At), *Brachypodium distachyon* (Bd), *Brassica rapa* (Br), *Cajanus cajan* (Cc), Citrus clementine (Ccl), *Carica papaya* (Cp), *Citrus sinensis* (Cs), *Elaeis guineensis* (Eg), *Fragaria vesca* (Fv), *Glycine max* (Gm), *Hordeum vulgare* (Hv), *Musa acuminate* (Ma), *Oryza sativa* (Os), *Picea abies* (Pa), *Physcomitrella patens* (Pp), *Prunus persica* (Ppe), *Populus trichocarpa* (Pt), *Ricinus communis* (Rc), *Sorghum bicolor* (Sb), *Setaria italica* (Si), *Solanum lycopersicum* (Sl), *Selaginella moellendorffii* (Sm), *Solanum tuberosum* (St), *Triticum aestivum* (Ta), *Vitis vinifera* (Vv) and *Zea mays* (Zm). The tree was constructed by MEGA 6 (Tamura et al. 2013). A bootstrap analysis was performed with 1,000 replicates. Entries (Table 1) are clustered in the three independent clades NIP-I, NIP-II and NIP-III, each with specific selectivity filter signatures. NIP-I clade (in *red*): Trp-Val-Ala-Arg (WVAR) and Trp-Ile-Ala-Arg (WIAR). NIP-II (in *blue*): Ala-Ile-Ala-Arg (AIAR), Ala-Ile-Gly-Arg (AIGR) and Ala-Val-Gly-Arg (AVGR). NIP-III (in *green*): Gly-Ser-Gly-Arg (GSGR). Segregation of α (*lighter grey shades*)- and β (*darkest grey*)-sub-clades consisting of clearly distributed mono- and dicotyledonous sequences, respectively, is indicated. (**b**) A sequence alignment of α-helices H2 and H5 and loop LE of spinach aquaporins (SoPIP2-1) with NIPs from *A. thaliana* (AtNIP), *G. max* (GmNIP), *H. vulgare* (HvNIP2-1), *O. sativa* (OsNIP), *T. aestivum* (TaNIP2-1) and *Z. mays* (ZmNIP). The alignment was performed by ProMals3D (Pei and Grishin 2014). Selectivity filter residues and NPA motifs are shown in *blue* and *yellow*, respectively. Conservation of residues on a scale 5–9 from lower to higher conserved residues is displayed above sequences; 9 in *brown* indicates an absolute conservation

cylindrical wedge (Fig. 1d, tetrameric structures are shown in two orthogonal orientations). Monomers operate in their own right, as demonstrated by studies with mixed active or inactive monomers in *Xenopus laevis* oocytes (Jung et al. 1994). However, by close association, the four monomers form an additional central pore that has been suggested to serve as another route for permeation (Yool et al. 1996; Fu et al. 2000). Individual monomers are related by a fourfold crystallographic axis and interact with each other through neighbouring membrane-spanning α-helices via hydrophobic interactions and hydrogen bonds, such as those of 'hole-to-knob' configurations (Murata et al. 2000; Fu et al. 2000; Sui et al. 2001). Further, interconnecting loops between individual α-helices contribute to mutual inter-monomeric interactions (Fig. 1d, right panel). The tetramers associate with annular or exogenously added lipids, for example, with 1,2-dimyristoyl-sn-glycero-3-phosphocholine (Gonen et al. 2005) or surfactants such as octyl (Fu et al. 2000) and nonyl (Sui et al. 2001) β-D-glucosides, where both lipids and surfactants stabilise supramolecular tetrameric assemblies. These lipid or surfactant interactions have been defined in structures based on 3D (Fu et al. 2000) and two-dimensional (Gonen et al. 2005) crystals and are formed between hydrophobic residues and acyl chains of lipids or between glycosyl moieties of alkyl β-D-glucosides surrounding polar groups and water coordinated molecules. Occasionally, lipid molecules have been found in a central tetrameric pore, formed alongside the fourfold symmetry axis that can be up to 8–10 Å in diameter (Horsefield et al. 2008; Newby et al. 2008).

4 A Structural Basis of Transport by Aquaporins

4.1 Approaches to Measure Solute Transport Selectivity and Kinetic Parameters

Four mainstream approaches have been used to measure selectivity and kinetic parameters of solute permeation of aquaporins: (i) In isolated tissues (e.g. tobacco leaf discs; Uehlein et al. 2003), organelles of living organisms (e.g. endoplasmic reticulum, Noronha et al. 2014) or protoplasts (Ramahaleo et al. 1996; Moshelion et al. 2004; Besserer et al. 2012). (ii) In native vesicles isolated and purified from cells or their membranes (Niemietz and Tyerman 1997; Fang et al. 2002) or in vesicles isolated from membranes of cells with recombinantly expressed aquaporins (Jung et al. 1994; Schnurbusch et al. 2010). (iii) In *X. laevis* oocytes, used for the first time by Preston and co-workers (1992) and subsequently adopted by many researchers (e.g. Dordas et al. 2000). (iv) In liposomes with purified and reconstituted aquaporin proteins (proteo-liposomes) or planar lipid bilayers (Ye and Verkman 1989; Zeidel et al. 1992; Weaver et al. 1994; Verdoucq et al. 2008). Some authors argue that solute permeation measurements using proteo-liposomes are more reliable than those with oocytes (Ho et al. 2009), whereby in the absence of other proteins in a bilayer in the proteo-liposomes, precise kinetic permeation parameters can be derived for wild-type or variant aquaporins and compared. On the other hand, proteins may not be accommodated in the membranes of proteo-liposomes in optimal configurations as these environments are artificial and minimalist in composition. Under ideal circumstances, both non-defined native and fully defined artificial systems should be used when available, for derivation of transport characteristics.

4.2 Solute Selectivity of Aquaporins

Based on permeation function, three major groups of plant aquaporins are recognised: (i) aquaporins that transport water, (ii) aquaglyceroporins that permeate other neutral solutes in addition to water (Borgnia et al. 1999) and (iii) aquaporins that conduct ionic species, based on the evidence of human aquaporins (Yool et al. 1996; Fu et al. 2000; Yu et al. 2006), as discussed below. The group of aquaglyceroporins has been reported to transport a broad range of neutral molecules such as non-electrolyte acetamide (Rivers et al. 1997); long polyols (Tsukaguchi et al. 1999); short polyols including glycerol (1,2,3-propane-triol) (Fu et al. 2000); CO_2 (Uehlein et al. 2003, 2008; Otto et al. 2010; Mori et al. 2014); purines and pyrimidines (Tsukaguchi et al. 1999); non-electrolyte urea (Liu et al. 2003); ammonia and glycerol nitrate (Loqué et al. 2005); silicic acid (Ma et al. 2006; Schnurbush et al. 2010); boric, arsenic and germanic acids (Takano et al. 2006; Kamiya et al. 2009; Schnurbush et al. 2010; Hayes et al. 2013); lactic acid (Choi and Roberts 2007; Bienert et al. 2013); hydrogen peroxide and related oxy-radicals (Dynowski et al. 2008); and selenious acid (Zhao et al. 2010). Although permeation of short and certain long (ribitol, xylitol, D-arabitol and D-sorbitol but not D-mannitol) polyols (Fu et al. 2000) has been detected in numerous studies, permeation of cyclic monosaccharides such as glucose and fructose, or of disaccharides such as sucrose, has never been demonstrated (Tsukaguchi et al. 1998; Fu et al. 2000).

4.3 Rates of Solute Transport and Mechanisms

Aquaporins as water transport facilitators mediate the water flux at rates of approximately $3 \cdot 10^9$ water molecules per second per monomeric unit (Agre and Kozono 2003); these rates are significantly higher than diffusion rates of water molecules through lipid membranes. In nonorthodox aquaporins that permeate other solutes, water transport rates are significantly lower. It has been suggested that steric occlusions of amino acid residues within specific structural and functional elements of aquaporins are one of the most fundamental factors that underlie differences in solute permeation selectivity (Fu et al. 2000; Lee et al. 2005; Kirscht et al. 2016). To this end, in the text below, we will separately discuss three features that collectively contribute to solute transport selectivity: (i) dimensional filtering and roles of periplasmic or cytoplasmic constrictions in permeation of solutes of various volumes;

(ii) chemical filtering of solutes, barriers for ion or proton conductance through pores of monomers and significance of NPA signatures, including roles of dipole moments and electrostatic potentials; and (iii) ion conductance through a central pore of tetramers.

4.4 Dimensional Filtering and the Roles of Constrictions in Permeation of Solutes of Various Volumes

A pathway for solute permeation is shaped by re-entrant α-helices HB and HE that connect to cytoplasmic and periplasmic vestibules, thus generating an hourglass or dumbbell-like shape (Fig. 1c). The solute-conducting channel, which in canonical aquaporins carries a single-file chain of water molecules, is formed by symmetry-related sets of carbonyl groups and hydrophilic side chain residues, both operating as hydrogen bond acceptors, often punctuated by hydrophobic residues alongside the pore. At the pore centre in most aquaporins, the two re-entrant α-helices HB and HE carry NPA motifs, where highly conserved Asn residues, rarely replaced by other residues (Zeuthen et al. 2013; Kirscht et al. 2016), and located at the tip of each re-entrant α-helix, form a part of the surface of the solute-conducting pore (Fig. 1c).

The sequence signatures of aquaporin monomers translated into the structural context underlie the functional properties of aquaporins. Verma et al. (2015) calculated a specific cumulative van der Waals volume (CvV, expressed in Å^3), by adding individual van der Waals volumes of each of the four residues of the ar/R/LE1-LE2 constriction, located close to the periplasmic vestibule. These authors noted large differences in CvV values in several subfamilies of aquaporins. For example, the largest CvV value was calculated for a mammalian aquaporin (ar/R/LE1-LE2 constriction region: Phe-Arg-Tyr-Arg) (572 Å^3), while the lowest CvV values were found for plant SIP (Ser-His-Gly-Ala) (306 Å^3) or protozoan (Ile-Ser-Gly-Ala) (312 Å^3) aquaporins (Verma et al. 2015). However, the ar/R/LE1-LE2 constriction regions Phe-His-Thr-Arg and Trp-Gly-Phe-Arg of the *E. coli* water-selective (AqpZ) and glycerol-selective (AqpF) aquaporins, respectively, exhibit identical CvV values (413 Å^3), so logically it is reasonable to conclude that besides chemical signatures and consequent structural importance of ar/R/LE1-LE2 constriction regions, other structural determinants that are not directly interact with solutes may play essential roles in solute permeation selectivity (Savage et al. 2010). Nevertheless, it might prove advantageous to investigate if additional quantitative parameters correlate with the solute permeation selectivity of aquaporins. The role of a cytoplasmic constriction, located in the proximity of the cytoplasmic vestibule, is less clear based on most structural studies, but this region may operate in a similar manner than that of a periplasmic constriction, regulating solute permeation in an opposite direction.

Further, it has recently been proposed that a specific pattern of residues forming ar/R/LE1-LE2 constriction regions and a precise spacing between NPA motifs control solute-conducting selectivity in plant aquaporins. A bioinformatics analysis of more than 30 aquaporins and experimental measurements of transport rates in *X.*

laevis oocytes (Deshmukh et al. 2015) revealed that permeation of silicic acid was confined to aquaporins with the Gly-Ser-Gly-Arg selectivity filter constriction signature and a precise spacing of 108 residues between NPA motifs. Notably, this Gly-Ser-Gly-Arg signature carried a low CvV value (317 Å^3).

To determine if an observation that a barley NIP-type aquaporin HvNIP2;1 exerts a wide solute selectivity (Schnurbusch et al. 2010) can be linked to its specific sequence and structural features, we conducted the bioinformatics analyses of 75 mono- and dicotyledonous representative sequences of NIP aquaporins (Fig. 2a). These entries (Table 1) formed three independent clades NIP-I, NIP-II and NIP-III

Table 1 The names and GenBank/NCBI accession numbers of 75 nodulin 26-like intrinsic proteins (NIPs) from listed plant species that were used in phylogeny reconstruction (cf. Fig. 2)

Name in the tree	Accession number	Species
AtNIP1-1	CAA16760.2	*Arabidopsis thaliana*
AtNIP1-2	NP_193626.1	*Arabidopsis thaliana*
AtNIP2-1	NP_180986.1	*Arabidopsis thaliana*
AtNIP3-1	NP_174472.2	*Arabidopsis thaliana*
AtNIP4-1	NP_198597.1	*Arabidopsis thaliana*
AtNIP4-2	NP_198598.1	*Arabidopsis thaliana*
AtNIP5-1	NP_192776.1	*Arabidopsis thaliana*
AtNIP6-1	NP_178191.1	*Arabidopsis thaliana*
AtNIP7-1	NP_566271.1	*Arabidopsis thaliana*
BdNIP1-1	XP_003571857.1	*Brachypodium distachyon*
BdNIP2-1	XP_003570658.1	*Brachypodium distachyon*
BdNIP2-2	XP_003564051.1	*Brachypodium distachyon*
BdNIP3-3	XP_003574178.1	*Brachypodium distachyon*
BrNIP4-3	XP_009140163.1	*Brassica rapa*
BrNIP5-1	XP_009134192.1	*Brassica rapa*
CclNIP1-1	XP_006430637.1	*Citrus clementine*
CclNIP2-1	ESR44391.1	*Citrus clementine*
CclNIP3-2	XP_006434369.1	*Citrus clementine*
CsNIP1-4	KDO63097.1	*Citrus sinensis*
CsNIP2-1	XP_006482598.1	*Citrus sinensis*
CsNIP3-2	XP_006472916.1	*Citrus sinensis*
EgNIP1-2	XP_010915460.1	*Elaeis guineensis*
EgNIP3-4	XP_010933763.1	*Elaeis guineensis*
FvNIP1-1	XP_004309621.1	*Fragaria vesca*
FvNIP2-1	XP_004304304.1	*Fragaria vesca*
FvNIP3-3	XP_004309493.1	*Fragaria vesca*
GmNIP1-2	XP_003518381.1	*Glycine max*
GmNIP2-1	XP_003534451.1	*Glycine max*
GmNIP3-1	XP_003547292.1	*Glycine max*
HvNIP2-1	BAH24163	*Hordeum vulgare*
MaNIP1-1	XP_009404528.1	*Musa acuminate*
MaNIP2-1	XP_009381416.1	*Musa acuminate*

(continued)

Table 1 (continued)

Name in the tree	Accession number	Species
MaNIP2-2	XP_009401397.1	*Musa acuminate*
MaNIP2-3	XP_009419139.1	*Musa acuminate*
MaNIP2-4	XP_009403165.1	*Musa acuminate*
MaNIP3-2	XP_009388143.1	*Musa acuminate*
OsNIP1-1	NP_001046375.1	*Oryza sativa*
OsNIP2-1	NP_001048108.1	*Oryza sativa*
OsNIP2-2	BAF19121.1	*Oryza sativa*
OsNIP3-1	Q0IWF3.2	*Oryza sativa*
PpNIP5-1a	XP_001754375.1	*Physcomitrella patens*
PpeNIP1-3	XP_007216120.1	*Prunus persica*
PpeNIP2-1	XP_007216227.1	*Prunus persica*
PpeNIP3-2	XP_007209472.1	*Prunus persica*
PtNIP1-4	XP_006372594.1	*Populus trichocarpa*
PtNIP2-1	XP_002324057.1	*Populus trichocarpa*
PtNIP3-3	XP_002298990.1	*Populus trichocarpa*
PtNIP3-4	XP_002317642.1	*Populus trichocarpa*
RcNIP1-4	XP_002532963.1	*Ricinus communis*
RcNIP2-1	XP_002534417.1	*Ricinus communis*
RcNIP3-1	XP_002518973.1	*Ricinus communis*
SbNIP1-1	XP_002453573.1	*Sorghum bicolor*
SbNIP2-1	XP_002454286	*Sorghum bicolor*
SbNIP2-2	XP_002438105.1	*Sorghum bicolor*
SbNIP3-4	XP_002464380.1	*Sorghum bicolor*
SiNIP2-1	KQL31494.1	*Setaria italic*
SiNIP2-2	KQL10018.1	*Setaria italic*
SiNIP3-1	XP_004982621.1	*Setaria italic*
SlNIP1-4	BAO18645.1	*Solanum lycopersicum*
SlNIP2-1	NP_001274283.1	*Solanum lycopersicum*
SlNIP3-2	NP_001274288.1	*Solanum lycopersicum*
SmNIP3-1	XP_002976312.1	*Selaginella moellendorffii*
SmNIP5-1	XP_002962550.1	*Selaginella moellendorffii*
StNIP1-4	XP_006344325.1	*Solanum tuberosum*
StNIP3-2	NP_001274996.1	*Solanum tuberosum*
TaNIP2-1	ADM47602	*Triticum aestivum*
VvNIP1-4	CBI33542.3	*Vitis vinifera*
VvNIP2-1	XP_002278054.2	*Vitis vinifera*
VvNIP3-2	XP_002276319.1	*Vitis vinifera*
ZmNIP1-1	AFW77428.1	*Zea mays*
ZmNIP2-1	ACF79677.1	*Zea mays*
ZmNIP2-2	ABF67956.1	*Zea mays*
ZmNIP2-3	ACG28405.1	*Zea mays*
ZmNIP2-4	AAK26849.1	*Zea mays*
ZmNIP3-3	NP_001105021.1	*Zea mays*

with different selectivity filter signatures (Mitani et al. 2008; Ma and Yamaji 2015). Members of the NIP-I clade contain Trp-Val-Ala-Arg (WVAR) and Trp-Ile-Ala-Arg (WIAR) motifs, and the NIP-II members have Ala-Ile-Ala-Arg (AIAR), Ala-Ile-Gly-Arg (AIGR) and Ala-Val-Gly-Arg (AVGR) signatures. All NIP-III members, to which the barley NIP-type aquaporin HvNIP2;1 belongs, carry an absolutely conserved Gly-Ser-Gly-Arg (GSGR) signature in their selectivity filters (Fig. 2a in green). In this analysis, we further divided members of NIP-III into two sub-clades, α-sub-clade 1 and α-sub-clade 2 (highlighted in two lighter shades of grey) containing monocotyledonous members, and β-sub-clade (highlighted in darker grey) with dicotyledonous sequences. This suggested that the monocot α-sub-clade has diversified during evolution from the dicot β-sub-clade (Fig. 2a). However, it remains to be established if this clear diversification of selectivity filter motifs can be correlated with a solute permeation specificity of individual aquaporins, classified in specific clades or sub-clades.

4.5 *Chemical Filtering of Solutes, Barriers for Ion or Proton Conductance Through the Pores of Monomers and Significance of NPA Signatures*

It has been suggested, based on crystallographic analyses (Murata et al. 2000; Lee et al. 2005; Ho et al. 2009; Savage et al. 2010) and corroborated by molecular dynamics simulations (Tajkhorshid et al. 2002), that two NPA motifs provide a blocking mechanism against the passage of H^+ and other ions. This mechanism is based on a unique role of Asn residues in the pore, whereby each Asn operates as a hydrogen bond donor that has the ability to polarise the orientation of central water molecules (Savage et al. 2010). In other words, NPA motifs with the macro-dipoles of neighbouring re-entrant α-helices have the ability to flip the dipole moments of water molecules at the centre of conducting pores and to disrupt a single-file chain of water molecules, thus preventing proton conductance through the Grotthuss mechanism (Agmon 1995). Dipole moments and electrostatic potentials of charged ions or protons also ensure that these would experience repulsive forces from many more accessible carbonyl oxygen atoms lining the inner regions of vestibules, selectivity filters and pore regions.

To assure that chemical filtering of solutes is in place, and barriers against ion or proton conductance through monomer pores are operating, a series of hydrogen bond donor carbonyls and other groups pre-align or preselect solute molecules in vestibules that may later be caught in the aquaporin pores. It is assumed that these solutes have already shed their water molecules (Harries et al. 2004; Sui et al. 2001; Ho et al. 2009). A relatively stronger hydrophobicity of vestibules in non-water-conducting aquaporins should improve transport rates of solutes that contain hydrophobic components, and correspondingly the more hydrophilic vestibules of canonical aquaporins would favour the preselection of water molecules (Sui et al. 2001; Savage et al. 2003;

Ho et al. 2009). It has further been proposed that these vestibules are sites for the energetically unfavourable shedding of hydration shells of water molecules from certain solutes during de-solvation, as well as for the increase of effective solute concentrations near the entry into the pore regions (Harries et al. 2004). However, the structural analyses of aquaporin vestibules revealed that central pores operate with a different molecular mechanism. The average distance between water molecules is minimal in the pore (forming a single-file water structure) because of a very high hydrophobicity, while the opposite was found to be true for the vestibule regions of aquaporins, where water adopts a bulk-like state (Han et al. 2015). Based on this premise, it was suggested that the vestibule regions could be effective drug design targets, as these regions are the sites for initial recruitment of solutes and may control their concentrations (Ho et al. 2009; Han et al. 2015). This approach could be tested using aquaporin homology models, based on structural data for closely related experimental structures, to solve the mechanistic problems of aquaporin solute selectivity and for in silico drug design.

4.6 Ion Conductance Through a Central Pore of Tetramers

While conductance of water or other neutral solutes through the central tetrameric pore has been excluded, due to its hydrophobic nature (Fu et al. 2000; Murata et al. 2000), a controversy prevails as to whether a central tetrameric pore conducts ionic species. The reason for this is that the central pore in some aquaporins may be up to 10 Å wide, considerably larger than, for example, the pore in the tetrameric KcsA potassium ion channel (Anderson et al. 1992). It was suggested that a central pore may serve as a potential path for ion permeation (Yool et al. 1996; Fu et al. 2000). To this end, the ion conductivity for a central pore in a human aquaporin has been proposed (especially after cGMP activation), and a proof-of-concept for this hypothesis was supported by molecular dynamics simulations and ion transport measurements in *X. laevis* oocytes (Yu et al. 2006). Notably, through molecular dynamics simulations, cGMP was found to interact with Arg-rich cytoplasmic loop D facilitating its outward movement, which was hypothesised to open a cytoplasmic gate and mediate ion conductance. Further, a homo-tetrameric plasma and inner chloroplast membrane PIP2;1 aquaporin from *Nicotiana tabacum* facilitated CO_2 but did not permeate water (Uehlein et al. 2008). These authors hypothesised that CO_2 could permeate through a central (so-called fifth) pore (Otto et al. 2010). The previous findings were confirmed by Wang et al. (2016), who showed that *Arabidopsis* PIP2;1 permeated CO_2, and served as a key interactor of the carbonic anhydrase βCA4. Importantly, these authors established that extracellular CO_2 signalling was linked to a SLAC1 ion channel regulation upon co-expression of PIP2;1, βCA4, SLAC1 and protein kinases. No molecular dynamics simulation studies have yet been performed on CO_2 transport. In summary, the question of whether the central tetrameric pore conducts ionic species or CO_2 is still highly contentious. This pathway must be more thoroughly investigated for its ion-conducting activity, at least in aquaporins in which the properties of the central pore are predicted to be conducive for this function.

4.7 *Mutational Studies to Alter Transport Selectivity and Rates*

One rapid way to investigate solute selectivity, modify transport rates, is to introduce variations in sequences (Jung et al. 1994; Jahn et al. 2004; Bienert et al. 2013; Hayes et al. 2013; Kirscht et al. 2016) and integrate both transport functional and structural observations. For over more than 20 years of this research, significant information has been gained based on the studies of wild-type and variant plant of aquaporins.

Several point mutations (His180Ala/Arg196Ala and Phe56Ala/His180Ala) in the ar/R/LE1-LE2 selectivity filter of a human water-specific aquaporin 1 allowed conversion of this orthodox water-permeable aquaporin into a more multifunctional aquaporin, permeating other solutes such as urea, glycerol and ammonia. These variations increased the maximal diameter of the constriction of the ar/R/LE1-LE2 selectivity filter by threefold (Beitz et al. 2006). However, surprisingly the Arg196Val substitution (removal of a positive charge from Arg196) allowed proton passage in both directions. Further, Beitz and co-authors (2006) established that protons did not permeate according to the Grotthuss mechanism and concluded in accordance with Zeuthen et al. (2013) that the electrostatic proton barrier in aquaporins depended on both NPA and ar/R/LE1-LE2 constrictions. These findings and those of Hub and de Groot (2008) based on molecular dynamics simulations imply that the ar/R region does not preclude water conductance but affects uncharged solutes conductance, emphasising the importance of the ar/R/LE1-LE2 residues for channel selectivity.

On the other hand, when three selectivity filter signature residues (Phe43Trp/His174Gly/Thr183Phe) of the glycerol-permeating *E. coli* aquaporin (AqpF) were introduced into its water-conducting counterpart (AqpZ), there was no increase in glycerol conductance, although a decrease of water permeability was recorded in both reciprocally mutated aquaporins (Savage et al. 2010). Notable observations were reported by Liu et al. (2005), who in a rat anion-selective aquaporin 6 substituted Asn for Gly in α-helix 2. This mutation resulted in the elimination of anion permeability but also led to elevated water transport when variant proteins were expressed in *X. laevis* oocytes. These observations indicated that each aquaporin is structurally unique and that simple variations of selectivity filter residues may not result in an altered solute selectivity. To proceed forward with designing a desired solute selectivity of aquaporins, one needs to integrate multifaceted knowledge of bioinformatics, molecular modelling and classical molecular dynamics.

Ma and co-workers (2008) investigated the substrate specificity of a rice aquaporin NIP2;1 using *X. laevis* and *Saccharomyces cerevisiae* cells. They isolated two alleles, whereby the allele lsi2-1 had lower accumulation of toxic arsenious acid than the allele lsi2-2 but a higher silicic acid uptake (see also chapter "Plant Aquaporins and Metalloids"). These metalloids differ by 0.62 Å in their atomic radii (Fig. 3), and thus it is conceivable to think that protein variants with different transport rates of essential (silicic acid) and toxic (arsenious acid) metalloids could in principle be engineered. Comparison of sequences indicated that Thr342 could be

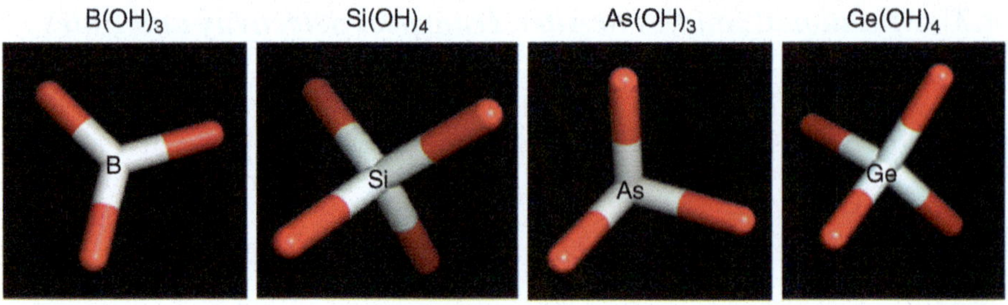

Molecular sizes of metalloids:

$B(OH)_3$	3.43 Å
$As(OH)_3$	3.57 Å
$Si(OH)_4$	4.19 Å
$Ge(OH)_4$	4.48 Å

Fig. 3 Structures and atomic radii of metalloid molecules of boric acid [$B(OH)_3$], silicic acid [$Si(OH)_4$], arsenious acid [$As(OH)_3$] and germanic acids [$Ge(OH)_4$] that are known to be transported by the members of an α-sub-clade of NIP-III aquaporins from monocotyledonous plants. Structures are shown in stick representations. The dimensions of atomic radii are given in Å

mutated to Arg in lsi2-2 that was positioned in the membrane H6 (topology explained in Fig. 1) and not in the pore region of the lsi2-2 protein. This study was extended by Mitani-Ueno et al. (2011), who investigated whether ar/R/LE1-LE2 filter and NPA motifs could be altered to influence the solute transport selectivity of rice NIP2;1 preferring silicic over boric acid, and conversely that of Arabidopsis NIP5;1 with a reversed substrate selectivity. Both proteins also permeate arsenious acid and thus this study also carries biotechnological significance. The individual changes in rice NIP2;1 at the ar/R/LE1 positions did not alter transport of metalloids; however, the H5 mutation led to a loss of transport activity of both metalloids. Conversely, mutations in *Arabidopsis* NIP5;1 did not restore transport of silicic acid, and double mutations in H2 and H5 did not affect transport of arsenious acid. Further, Hayes et al. (2013) performed targeted mutagenesis of the specific residues within the ar/R/LE1-LE2 selectivity filter in barley NIP2;1 to alter its metalloid solute selectivity. Two of the mutations in the H2 position Gly88Ala and Gly88Cys showed a growth restoration in the presence of boric (smallest atomic radius, Fig. 3) and germanic (largest atomic radius) acids; nevertheless, the growth inhibition on arsenious acid (the second smallest atomic radius from the four metalloids) was preserved. These observations suggested that although mutations altered the substrate specificity of barley NIP2;1, metalloid permeation seemed to be controlled by other factors than simply by atomic radii of solutes. Potential controlling factors may entail differences in de-solvation rates within the vestibule regions of aquaporins *prior* to interactions with ar/R/LE1-LE2 selectivity filter residues or differences in overall interaction modes of metalloids with aquaporin molecules. These hypotheses can be tested using molecular dynamics simulation experiments.

The observations outlined above further extend a suggestion that ar/R/LE1-LE2 selectivity filter properties alone do not control solute selectivity of aquaporins and that other structural elements that do not directly interact with solutes may play essential roles in solute permeation specificity (Savage et al. 2010).

Another alternative to identify variations in protein sequences of aquaporins is to search for natural variation in cultivars that have precisely adapted to specific or stress-affected environments. These types of studies are just beginning to appear with aquaporins and other transport systems (e.g. Pallotta et al. 2014). The question then arises as to whether the responses of natural variants for specific stresses, such as drought or mineral toxicity, have already been optimised in crop and other plants through a long history of selection of native variants or are there still opportunities for a significant gain through allelic mining (Langridge et al. 2006). Although the information on natural variation of aquaporins and other transport systems involved in drought or other biotic and abiotic stresses is scarse, a few landmark studies have appeared (Pallotta et al. 2014; Hayes et al. 2015; Nagarajan et al. 2016).

5 Gating Mechanisms of Aquaporins Induced by pH, Cation Binding and Phosphorylation or Lengths of Loops and Mutational Studies

The concept of gating in plant aquaporins was proposed long before (Tyerman et al. 1989; Azaizeh et al. 1992; Tyerman et al. 1999; Yool and Weinstein 2002) both states, i.e. open and closed, of any aquaporin were elucidated at the atomic levels. Later both states of the spinach aquaporin (Fig. 1c) were defined at atomic levels in: (i) closed states at pH 8.0 (PDB ID: 1Z98; Tornroth-Horsefield et al. 2006) and pH 6.0 (PDB ID: 4IA4; Frick et al. 2013a, b) and (ii) an open state (PDB ID: 2B5F; Törnroth-Horsefield et al. 2006). Further, a so-called stochastic model of osmotic water transport was suggested, based on testing of a range of channel sizes and geometries of human aquaporins and their mutants (Zeuthen et al. 2013); this knowledge can directly be linked to the concept of gating.

Two groups of mechanisms that appear to be conserved in plant aquaporins are known to facilitate gating, i.e. the transitions between open and closed states. More precisely, the term gating refers to the opened (activated or conductive) and closed (deactivated or non-conductive) states, whereby these states represent distinct spatial conformations of the same channel. Here, that conformation interchange results in increasing the limiting size of the pore to accommodate solutes. The first group of mechanisms of gating includes pH changes, cation binding and post-translational phosphorylation (Törnroth-Horsefield et al. 2006; Frick et al. 2013a, b). The second group of gating mechanisms is based on loop lengths and their movements (Fischer et al. 2009).

The origin of gating was explored in a spinach aquaporin, for which the atomic structures of both states are available, using single and double Ser115Glu and Ser274Glu phosphorylation mimic variants (Nyblom et al. 2009). Although all mutants crystallised in a closed conformation, the analysis revealed that neither variation mimicked the naturally occurring phosphorylated state of the protein. However, combined functional and structural analyses revealed that in the Ser115Glu variant, the neighbouring Glu31 significantly moved away from its wild-type position, leading to a disruption of the divalent cation (presumed to be Ca^{2+})-binding site that stabilises loop D. These observations highlight the fact that phosphorylation of Ser115 could induce structural rearrangements and thus control opening and closing states of the pore.

The crystal structures of a spinach aquaporin, which have been obtained in several conformational states (a water-closed state at pH 8.0 (Törnroth-Horsefield et al. 2006) and at pH 6.0 (Frick et al. 2013a, b)), revealed a closing mechanism that in the plasma membrane results from a rapid drop of cytosolic pH due to anoxia that occurs during flooding (Tournaire-Roux et al. 2003). The closing mechanism is assumed to involve the interaction of the conserved pH-sensitive His193 residue on cytosolic loop D with the divalent cation (presumed to be Ca^{2+})-binding site. Here, in a protonated state, His adopts an alternative rotameric state and interacts with Asp28 that resides on a short N-terminal α-helix. This closing mechanism is also maintained by dephosphorylation of a closely positioned Ser115 residue on loop B (Frick et al. 2013a, b).

These observations, based on structural analyses of a wild-type and variant spinach aquaporins, indicate that gating mechanisms are linked to movements of loops B and D, post-translational phosphorylating events of Ser residues, protonation states of a His residue and the involvement of a divalent cation-binding site (Törnroth-Horsefield et al. 2006). These studies emphasise the control of gating by several concurrent events to open and close a solute-conducting pore (Nyblom et al. 2009; Frick et al. 2013a, b).

The second group of gating mechanisms is based on loop lengths and their movements alone and was revealed for the first time using the full-length and truncated forms of the yeast aquaporin Aqy1 from *Pichia pastoris*, resolved to 1.15 Å (Fischer et al. 2009). Structural data revealed that the pore of the Aqy1 aquaporin was closed by its own N-terminus. Here, Tyr31 formed a hydrogen bond to a water molecule and the backbone oxygen atoms of nearby Gly residues, located in the vicinity of the pore, consequently obstructing the cytoplasmic entrance to the pore. Additional mutational studies combined with molecular dynamics simulations suggested that water flow through the pore may be regulated by specific arrangements of post-translational regulation sites by phosphorylation and also by mechanosensitive gating. The latter gating could also be related to highly curved membrane environments, where aquaporins may reside. This was confirmed by molecular dynamics simulation, indicating that Aqy1 was regulated by both surface tension and membrane curvature. This type of gating could provide a rapid pressure regulator in response to unexpected cellular shock, aiding adaptation and microbial survival (Fischer et al. 2009), as well as to plants that employ a turgor pressure (Tyerman et al. 1989, 1999).

6 The Structural Knowledge of Aquaporins Has Strategic Significance in Agricultural Biotechnology, Nano-biotechnology and Environmental Sciences

Although transport function is central to plants, limited information is available on a structural basis of the permeation function of plant aquaporins. These investigations have so far been largely driven by genetics and physiology, but the knowledge of molecular function is required if we are to modify the properties of these transport proteins (Schroeder et al. 2013; Chaumont and Tyerman 2014; Nagarajan et al. 2016). Further, modifying the properties of aquaporins depends on a detailed mechanistic knowledge of their behaviour. Even though many aquaporins have been identified, their intrinsic hydrophobic properties made these studies difficult. As of May 2016, from 615 unique membrane proteins (http://blanco.biomol.uci.edu/mpstruc/), only five structures of plant transport proteins are known. These include two aquaporins from *S. oleracea* and *A. thaliana*, a nitrate transporter and a voltage-gated two-pore channel from *A. thaliana* and a SWEET transporter from *Oryza sativa*. Thus, two unique plant aquaporin structures are those of the water-conducting SoPIP2;1 aquaporin from *S. oleracea* (in several conformational states and variant forms) and the AtTIP2;1 aquaammoniaporin from *A. thaliana* (Kirscht et al. 2016).

Surprisingly, limited information is available on solute permeation specificity of plant aquaporins, although these data in conjunction with structural information are vital strategic tools for modifying their molecular function. We therefore need detailed structural data on all subfamilies of plant aquaporins from economically important food plants such as wheat, barley, maize and rice that conduct a variety of solutes, including those of multi-selective NIPs that have importance in food security and safety (Ma et al. 2006; Schnurbusch et al. 2010; Hayes et al. 2013). Targets for this knowledge include, for example, improving the nutritional quality and safety of plant products for humans, such as exclusion of toxic arsenic from food plants (Isayenkov and Maathuis 2008; Kamiya et al. 2009; Li et al. 2011; Hayes et al. 2013; Schnurbusch et al. 2010; Xu et al. 2015). Modifying nutrient fluxes is also important for protection of plants from excessive accumulation of metalloids such as boric acid, which become toxic at high concentrations (Hayes et al. 2015; Nagarajan et al. 2016).

Uncharged ion pairs of mercury, gold, copper and cadmium are also known to perturb plant water status (Belimov et al. 2015) and have been reported to be the potent inhibitors of aquaporins that operate through cysteine-related mechanisms (Niemietz and Tyerman 2002). Heavy metal-induced perturbations of aquaporin function at the plant level have been explained by a decrease of both root and shoot hydraulic conductance, leading to decreasing leaf water potentials and turgor, which may close stomata (Zhu et al. 2005). On the other hand, many aquaporins are mercury insensitive. Remarkably Frick et al. (2013a, b) in a spinach aquaporin observed mercury-increased water permeability, using a non-cysteine-related mechanism, whereby presumably other factors affected the aquaporin; one of them could be the properties of a lipid bilayer.

Finally, an obvious potential application of aquaporins is in nanotechnology by creating stable biomimetic membranes, such as those with embedded robust aquaporin folds that have excellent separation performance and permit rapid water diffusion. Hence, a next important application of aquaporins could be in environmental sciences, more specifically in water desalination, waste-water recovery and fertiliser and soil component retrieval. Application and profitability on an industrial scale would require stable and robust aquaporin structures with highly selective permeation functions and rapid transport rates (Wang et al. 2015).

Acknowledgements This work was supported by the grants from the Australian Research Council (LP120100201 and DP120100900 to M. H.). Jay Rongala and Dr. Julie Hayes (Australian Centre for Plant Functional Genomics, University of Adelaide) are thanked for the assistance with literature and for critically reading the manuscript, respectively. I acknowledge Professor Steve Tyerman and the past members of my laboratory for insightful discussions.

References

Abascal F, Irisarri I, Zardoya R (2014) Diversity and evolution of membrane intrinsic proteins. Biochim Biophys Acta 1840:1468–1481

Agemark M, Kowal J, Kukulski W, Nordén K, Gustavsson N, Johanson U, Engel A, Kjellbom P (2012) Reconstitution of water channel function and 2D-crystallization of human aquaporin 8. Biochim Biophys Acta 1818:839–850

Agmon N (1995) The Grotthuss mechanism. Chem Phys Lett 244:456–462

Agre P, Kozono D (2003) Aquaporin water channels: molecular mechanisms for human diseases. FEBS Lett 555:72–78

Agre P, Sasaki S, Chrispeels J (1993) Aquaporins: a family of water channel proteins. Am J Physiol Ren Physiol 265:F461

Anderson JA, Huprikar SS, Kochian LV, Lucas WJ, Gaber RF (1992) Functional expression of a probable *Arabidopsis thaliana* potassium channel in *Saccharomyces cerevisiae*. Proc Natl Acad Sci U S A 89:3736–3740

Ariani A, Gepts P (2015) Genome-wide identification and characterization of aquaporin gene family in common bean (*Phaseolus vulgaris* L.). Mol Genet 290:1771–1785

Beitz E, Wu B, Holm LM, Schultz JE, Zeuthen T (2006) Point mutations in the aromatic/arginine region in aquaporin 1 allow passage of urea, glycerol, ammonia, and protons. Proc Natl Acad Sci U S A 103:269–274

Belimov AA, Dodd IC, Safronova VI, Malkov NV, Davies WJ, Tikhonovich IA (2015) The cadmium-tolerant pea (*Pisum sativum* L.) mutant SGECdt is more sensitive to mercury: assessing plant water relations. J Exp Bot 66:2359–2369

Besse M, Knipfer T, Miller AJ, Verdeil JL, Jahn TP, Fricke W (2011) Developmental pattern of aquaporin expression in barley (*Hordeum vulgare* L.) leaves. J Exp Bot 62:4127–4142

Besserer A, Burnotte E, Bienert GP, Chevalier AS, Errachid A, Grefen C, Blatt MR, Chaumont F (2012) Selective regulation of maize plasma membrane aquaporin trafficking and activity by the SNARE SYP121. Plant Cell 24:3463–3481

Bienert GP, Bienert MD, Jahn TP, Boutry M, Chaumont F (2011) Solanaceae XIPs are plasma membrane aquaporins that facilitate the transport of many uncharged substrates. Plant J 66:306–317

Bienert GP, Desguin B, Chaumont F, Hols P (2013) Channel-mediated lactic acid transport: a novel function for aquaglyceroporins in bacteria. Biochem J 454:559–570

Borgnia M, Nielsen S, Engel A, Agre P (1999) Cellular and molecular biology of the aquaporin water channels. Annu Rev Biochem 68:425–458

Chaumont F, Tyerman SD (2014) Aquaporins: highly regulated channels controlling plant water relations. Plant Physiol 164:1600–1618

Chaumont F, Barrieu F, Wojcik E, Chrispeels MJ, Jung R (2001) Aquaporins constitute a large and highly divergent protein family in maize. Plant Physiol 125:1206–1215

Choi WG, Roberts DM (2007) *Arabidopsis* NIP2;1, a major intrinsic protein transporter of lactic acid induced by anoxic stress. J Biol Chem 282:24209–24218

Cordeiro RM (2015) Molecular dynamics simulations of the transport of reactive oxygen species by mammalian and plant aquaporins. Biochim Biophys Acta 1850:1786–1794

Deshmukh RK, Vivancos J, Ramakrishnan G, Guérin V, Carpentier G, Sonah H, Labbé C, Isenring P, Belzile FJ, Bélanger RR (2015) A precise spacing between the NPA domains of aquaporins is essential for silicon permeability in plants. Plant J 83:489–500

Diehn TA, Pommerrenig B, Bernhardt N, Hartmann A, Bienert GP (2015) Genome-wide identification of aquaporin encoding genes in *Brassica oleracea* and their phylogenetic sequence comparison to *Brassica* crops and *Arabidopsis*. Front Plant Sci 6:166. doi:10.3389/fpls.2015.00166

Dordas C, Chrispeels MJ, Brown PH (2000) Permeability and channel-mediated transport of boric acid across membrane vesicles isolated from squash roots. Plant Physiol 124:1349–1362

Dynowski M, Schaaf G, Loque D, Moran O, Ludewig U (2008) Plant plasma membrane water channels conduct the signalling molecule H_2O_2. Biochem J 414:53–61

Eriksson UK, Fischer G, Friemann R, Enkavi G, Tajkhorshid E, Neutze R (2013) Subangstrom resolution X-ray structure details aquaporin-water interactions. Science 340:1346–1349

Fang X, Yang B, Matthay MA, Verkman AS (2002) Evidence against aquaporin-1-dependent CO_2 permeability in lung and kidney. J Physiol 542:63–69

Fischer G, Kosinska-Eriksson U, Aponte-Santamaría C, Palmgren M, Geijer C, Hedfalk K, Hohmann S, de Groot BL, Neutze R, Lindkvist-Petersson K (2009) Crystal structure of a yeast aquaporin at 1.15 angstrom reveals a novel gating mechanism. PLoS Biol 7:e1000130

Fotiadis D, Hasler L, Muller DJ, Stahlberg H, Kistler J, Engel A (2000) Surface tongue-and-groove contours on lens MIP facilitate cell-to-cell adherence. J Mol Biol 300:779–789

Fotiadis D, Jenö P, Mini T, Wirtz S, Müller SA, Fraysse L, Kjellbom P, Engel A (2001) Structural characterization of two aquaporins isolated from native spinach leaf plasma membranes. J Biol Chem 276:1707–1714

Frick A, Järvå M, Ekvall M, Uzdavinys P, Nyblom M, Törnroth-Horsefield S (2013a) Mercury increases water permeability of a plant aquaporin through a non-cysteine-related mechanism. Biochem J 454:491–499

Frick A, Järvå M, Törnroth-Horsefield S (2013b) Structural basis for pH gating of plant aquaporins. FEBS Lett 587:989–993

Frick A, Eriksson UK, de Mattia F, Oberg F, Hedfalk K, Neutze R, de Grip WJ, Deen PM, Törnroth-Horsefield S (2014) X-ray structure of human aquaporin 2 and its implications for nephrogenic diabetes insipidus and trafficking. Proc Natl Acad Sci U S A 111:6305–6310

Fricke W, McDonald AJ, Mattson-Djos L (1997) Why do leaves and leaf cells of *N*-limited barley elongate at reduce rate? Planta 202:522–530

Fu D, Libson A, Miercke LJ, Weitzman C, Nollert P, Krucinski J, Stroud RM (2000) Structure of a glycerol-conducting channel and the basis for its selectivity. Science 290:481–486

Gao Z, He X, Zhao B, Zhou C, Liang Y, Ge R, Shen Y, Huang Z (2010) Overexpressing a putative aquaporin gene from wheat, *TaNIP*, enhances salt tolerance in transgenic *Arabidopsis*. Plant Cell Physiol 51:767–775

Gomes D, Agasse A, Thiébaud P, Delrot S, Gerós H, Chaumont F (2009) Aquaporins are multifunctional water and solute transporters highly divergent in living organisms. Biochim Biophys Acta 1788:1213–1228

Gonen T, Sliz P, Kistler J, Cheng Y, Walz T (2004) Aquaporin-0 membrane junctions reveal the structure of a closed water pore. Nature 429:193–197

Gonen T, Cheng Y, Sliz P, Hiroaki Y, Fujiyoshi Y, Harrison SC, Walz T (2005) Lipid-protein interactions in double-layered two-dimensional AQP0 crystals. Nature 438:633–368

Gorin MB, Yancey SB, Cline J, Revel JP, Horwitz J (1984) The major intrinsic protein (MIP) of the bovine lens fiber membrane: characterization and structure based on cDNA cloning. Cell 39:49–59

Gupta AB, Sankararamakrishnan R (2009) Genome-wide analysis of major intrinsic proteins in the tree plant *Populus trichocarpa*: characterization of XIP subfamily of aquaporins from evolutionary perspective. BMC Plant Biol 9:134. doi:10.1186/1471-2229-9-134

Gupta AB, Verma RK, Agarwal V, Vajpai M, Bansal V, Sankararamakrishnan R (2012) MIPModDB: a central resource for the superfamily of major intrinsic proteins. Nucleic Acids Res 40: D362–D369

Han C, Tang D, Kim D (2015) Molecular dynamics simulation on the effect of pore hydrophobicity on water transport through aquaporin-mimic nanopores. Colloids Surf A Physicochem Eng Asp 481:38–42

Harries WE, Akhavan D, Miercke LJ, Khademi S, Stroud RM (2004) The channel architecture of aquaporin 0 at a 2.2-Å resolution. Proc Natl Acad Sci USA 101:14045–14050

Hayes JE, Pallotta M, Baumann U, Berger B, Langridge P, Sutton T (2013) Germanium as a tool to dissect boron toxicity effects in barley and wheat. Funct Plant Biol 40:618–627

Hayes JE, Pallotta M, Garcia M, Öz MT, Rongala J, Sutton T (2015) Diversity in boron toxicity tolerance of Australian barley (*Hordeum vulgare* L.) genotypes. BMC Plant Biol 15:231–234

Ho JD, Yeh R, Sandstrom A, Chorny I, Harries WE, Robbins RA, Miercke LJ, Stroud RM (2009) Crystal structure of human aquaporin 4 at 1.8 Å and its mechanism of conductance. Proc Natl Acad Sci USA 106:7437–7442

Höfte H, Hubbard L, Reizer J, Ludevid D, Herman EM, Chrispeels MJ (1992) Vegetative and seed-specific forms of tonoplast intrinsic protein in the vacuolar membrane of *Arabidopsis thaliana*. Plant Physiol 99:561–570

Horsefield R, Nordén K, Fellert M, Backmark A, Törnroth-Horsefield S, Terwisscha van Scheltinga AC, Kvassman J, Kjellbom P, Johanson U, Neutze R (2008) High-resolution X-ray structure of human aquaporin 5. Proc Natl Acad Sci U S A 105:13327–13332

Hove RM, Ziemann M, Bhave M (2015) Identification and expression analysis of the Barley (*Hordeum vulgare* L.) Aquaporin Gene Family. PLoS ONE 10:e0128025

Hu W, Yuan Q, Wang Y, Cai R, Deng X, Wang J, Zhou S, Chen M, Chen L, Huang C, Ma Z, Yang G, He G (2012) Overexpression of a wheat aquaporin gene, TaAQP8, enhances salt stress tolerance in transgenic tobacco. Plant Cell Physiol 53:2127–2141

Hub JS, de Groot BL (2008) Mechanism of selectivity in aquaporins and aquaglyceroporins. Proc Natl Acad Sci U S A 105:1198–1203

Isayenkov SV, Maathuis FJM (2008) The *Arabidopsis thaliana* aquaglyceroporin AtNIP7;1 is a pathway for arsenite uptake. FEBS Lett 582:1625–1628

Jahn TP, Møller AL, Zeuthen T, Holm LM, Klaerke DA, Mohsin B, Kühlbrandt W, Schjoerring JK (2004) Aquaporin homologues in plants and mammals transport ammonia. FEBS Lett 574: 31–36

Johanson U, Karlsson M, Johansson I, Gustavsson S, Sjövall S, Fraysse L, Weig AR, Kjellbom P (2001) The complete set of genes encoding major intrinsic proteins in *Arabidopsis* provides a framework for a new nomenclature for major intrinsic proteins in plants. Plant Physiol 126:1358–1369

Johnson KD, Höfte H, Chrispeels MJ (1990) An intrinsic tonoplast protein of protein storage vacuoles in seeds is structurally related to a bacterial solute transporter (GlpF). Plant Cell 2:525–532

Jung JS, Preston GM, Smith BL, Guggino WB, Agre P (1994) Molecular structure of the water channel through aquaporin CHIP. The hourglass model. J Biol Chem 269:14648–14654

Kamiya T, Tanaka M, Mitani N, Ma JF, Maeshima M, Fujiwara T (2009) NIP1;1, an aquaporin homolog, determines the arsenite sensitivity of *Arabidopsis thaliana*. J Biol Chem 284:2114–2120

Kirscht A, Kaptan SS, Bienert KP, Chaumont F, Nissen P, de Groot BL, Kjellbom P, Gourdon P, Johanson U (2016) Crystal structure of an ammonia-permeable aquaporin. PLoS Biol 14:e1002411

Kitchen P, Conner AC (2015) Control of the aquaporin-4 channel water permeability by structural dynamics of Aromatic/Arginine selectivity filter residues. Biochemistry (USA) 54:6753–6755

Langridge P, Paltridge N, Fincher G (2006) Functional genomics of abiotic stress tolerance in cereals. Brief Funct Genomic Proteomic 4:343–354

Lee KJ, Kozono D, Remis J, Kitagawa Y, Agre P, Stroud RM (2005) Structural basis for conductance by the archaeal aquaporin AqpM at 1.68Å. Proc Natl Acad Sci USA 102:18932–18937

Li T, Choi WG, Wallace IS, Baudry J, Roberts DM (2011) *Arabidopsis thaliana* NIP7;1: an anther-specific boric acid transporter of the aquaporin superfamily regulated by an unusual tyrosine in helix 2 of the transport pore. Biochemistry 50:6633–6641

Li G, Santoni V, Maurel C (2014) Plant aquaporins: roles in plant physiology. Biochim Biophys Acta 1840:1574–1582

Li J, Ban L, Wen H, Wang Z, Dzyubenko N, Chapurin V, Gao H, Wang X (2015) An aquaporin protein is associated with drought stress tolerance. Biochem Biophys Res Commun 459:208–213

Liu L-H, Ludewig U, Gassert B, Frommer WB, von Wirén N (2003) Urea transport by nitrogen-regulated tonoplast intrinsic proteins in *Arabidopsis*. Plant Physiol 133:1220–1228

Liu K, Kozono D, Kato Y, Agre P, Hazama A, Yasui M (2005) Conversion of aquaporin 6 from an anion channel to a water-selective channel by a single amino acid substitution. Proc Natl Acad Sci U S A 102:2192–2197

Liu Q, Wang H, Zhang Z, Wu J, Feng Y, Zhu Z (2009) Divergence in function and expression of the NOD26-like intrinsic proteins in plants. BMC Genomics 10:313. doi:10.1186/1471-2164-10-313

Lopez D, Bronner G, Brunel N, Auguin D, Bourgerie S, Brignolas F, Carpin S, Tournaire-Roux C, Maurel C, Fumanal B, Martin F, Sakr S, Label P, Julien JL, Gousset-Dupont A, Venisse JS (2012) Insights into Populus XIP aquaporins: evolutionary expansion, protein functionality, and environmental regulation. J Exp Bot 63:2217–2230

Loqué D, Ludewig U, Yuan L, von Wirén N (2005) Tonoplast intrinsic proteins AtTIP2;1 and AtTIP2;3 facilitate NH_3 transport into the vacuole. Plant Physiol 137:671–680

Ma JF, Yamaji N (2015) A cooperative system of silicon transport in plants. Trends Plant Sci 20:435–442

Ma JF, Tamai K, Yamaji N, Mitani N, Konishi S, Katsuhara M, Ishiguro M, Murata Y, Yano M (2006) A silicon transporter in rice. Nature 440:688–691

Ma JF, Yamaji N, Mitani N, Xu XY, Su YH, McGrath SP, Zhao FJ (2008) Transporters of arsenite in rice and their role in arsenic accumulation in rice grain. Proc Natl Acad Sci U S A 105:9931–9935

Maurel C, Reizer J, Schroeder JI, Chrispeels MJ (1993) The vacuolar membrane protein gamma-TIP creates water specific channels in *Xenopus oocytes*. EMBO J 12:2241–2247

Maurel C, Verdoucq L, Luu DT, Santoni V (2008) Plant aquaporins: membrane channels with multiple integrated functions. Annu Rev Plant Biol 59:595–624

Mitani-Ueno N, Yamaji N, Zhao FJ, Ma JF (2011) The aromatic/arginine selectivity filter of NIP aquaporins plays a critical role in substrate selectivity for silicon, boron, and arsenic. J Exp Bot 62:4391–4398

Mori IC, Rhee J, Shibasaka M, Sasano S, Kaneko T, Horie T, Katsuhara M (2014) CO_2 transport by PIP2 aquaporins of barley. Plant Cell Physiol 55:251–257

Moshelion M, Moran N, Chaumont F (2004) Dynamic changes in the osmotic water permeability of protoplast plasma membrane. Plant Physiol 135:2301–2317

Mukhopadhyay R, Bhattacharjee H, Rosen BP (2014) Aquaglyceroporins: generalized metalloid channels. Biochim Biophys Acta 1840:1583–1591

Murata K, Mitsuoka K, Hirai T, Walz T, Agre P, Heymann JB, Engel A, Fujiyoshi Y (2000) Structural determinants of water permeation through aquaporin-1. Nature 407:599–605

Nagarajan Y, Rongala J, Luang S, Shadiac N, Hayes J, Sutton T, Gilliham M, Tyerman SD, McPhee G, Voelcker NH, Mertens HDT, Kirby NM, Sing A, Lee J-G, Yingling YG, Hrmova M (2016) A barley efflux transporter operates in a Na^+-dependent manner, as revealed by a multidisciplinary platform. Plant Cell 28:202–218

Newby ZE, O'Connell J 3rd, Robles-Colmenares Y, Khademi S, Miercke LJ, Stroud RM (2008) Crystal structure of the aquaglyceroporin PfAQP from the malarial parasite Plasmodium falciparum. Nat Struct Mol Biol 15:619–625

Niemietz CM, Tyerman SD (1997) Characterization of water channels in wheat root membrane vesicles. Plant Physiol 115:561–567

Niemietz CM, Tyerman SD (2002) New potent inhibitors of aquaporins: silver and gold compounds inhibit aquaporins of plant and human origin. FEBS Lett 531:443–447

Noronha H, Agasse A, Martins AP, Berny MC, Gomes D, Zarrouk O, Thiebaud P, Delrot S, Soveral G, Chaumont F, Gerós H (2014) The grape aquaporin VvSIP1 transports water across the ER membrane. J Exp Bot 65:981–993

Nyblom M, Frick A, Wang Y, Ekvall M, Hallgren K, Hedfalk K, Neutze R, Tajkhorshid E, Törnroth-Horsefield S (2009) Structural and functional analysis of SoPIP2;1 mutants adds insight into plant aquaporin gating. J Mol Biol 387:653–668

Otto B, Uehlein N, Sdorra S, Fischer M, Ayaz M, Belastegui-Macadam X, Heckwolf M, Lachnit M, Pede N, Priem N, Reinhard A, Siegfart S, Urban M, Kaldenhoff R (2010) Aquaporin tetramer composition modifies the function of tobacco aquaporins. J Biol Chem 285: 31253–31260

Pallotta M, Schnurbusch T, Hayes J, Hay A, Baumann U, Paull J, Langridge P, Sutton T (2014) Molecular basis of adaptation to high soil boron in wheat landraces and elite cultivars. Nature 514:88–91

Pandey B, Sharma P, Pandey DM, Sharma I, Chatrath R (2013) Identification of new aquaporin genes and single nucleotide polymorphism in bread wheat. Evol Bioinformatics Online 9:437–452

Park W, Scheffler BE, Bauer PJ, Campbell BT (2010) Identification of the family of aquaporin genes and their expression in upland cotton (*Gossypium hirsutum* L.). BMC Plant Biol 10:142. doi:10.1186/1471-2229-10-142

Pei J, Grishin NV (2014) PROMALS3D: multiple protein sequence alignment enhanced with evolutionary and three-dimensional structural information. Methods Mol Biol 1079:263–271

Preston GM, Carroll TP, Guggino WB, Agre P (1992) Appearance of water channels in *Xenopus oocytes* expressing red cell CHIP28 protein. Science 256:385–387

Ramahaleo T, Alexandre J, Lassalles JP (1996) Stretch activated channels in plant cells. A new model for osmoelastic coupling. Plant Physiol Biochem 34:327–334

Reddy PS, Rao TSRB, Sharma KK, Vadez V (2015) Genome-wide identification and characterization of the aquaporin gene family in *Sorghum bicolor* (L.). Plant Gene 1:18–28

Reuscher S, Akiyama M, Mori C, Aoki K, Shibata D, Shiratake K (2013) Genome-wide identification and expression analysis of aquaporins in tomato. PLoS ONE 8:e79052

Rivers RL, Dean RM, Chandy G, Hall JE, Roberts DM, Zeidel ML (1997) Functional analysis of nodulin 26, an aquaporin in soybean root nodule symbiosomes. J Biol Chem 272: 16256–16261

Saier MH, Reddy VS, Tsu BV, Ahmed MS, Li C, Moreno-Hagelsieb G (2016) The transporter classification database (TCDB). Nucleic Acids Res 44:D372–D379

Sakurai J, Ishikawa F, Yamaguchi T, Uemura M, Maeshima M (2005) Identification of 33 rice aquaporin genes and analysis of their expression and function. Plant Cell Physiol 46: 1568–1577

Sandal NN, Marcker KA (1988) Soybean nodulin 26 is homologous to the major intrinsic protein of the bovine lens fiber membrane. Nucleic Acids Res 16:9347

Savage DF, Egea PF, Robles-Colmenares Y, O'Connell JD 3rd, Stroud RM (2003) Architecture and selectivity in aquaporins: 2.5Å X-ray structure of aquaporin Z. PLoS Biol 1:E72

Savage DF, O'Connell JD 3rd, Miercke LJ, Finer-Moore J, Stroud RM (2010) Structural context shapes the aquaporin selectivity filter. Proc Natl Acad Sci U S A 107:17164–17169

Schnurbusch T, Hayes J, Hrmova M, Baumann U, Ramesh SA, Tyerman SD, Langridge P, Sutton T (2010) Boron toxicity tolerance in barley through reduced expression of the multifunctional aquaporin HvNIP2;1. Plant Physiol 153:1706–1715

Schroeder JI, Delhaize E, Frommer WB, Guerinot ML, Harrison MJ, Herrera-Estrella L, Horie T, Kochian LV, Munns R, Nishizawa NK, Tsay YF, Sanders D (2013) Using membrane transporters to improve crops for sustainable food production. Nature 497:60–66

Shelden MC, Howitt SM, Kaiser BN, Tyerman SD (2009) Identification and functional characterisation of aquaporins in the grapevine. Funct Plant Biol 36:1065–1078

Sui H, Han BG, Lee JK, Walian P, Jap BK (2001) Structural basis of water-specific transport through the AQP1 water channel. Nature 414:872–878

Tajkhorshid E, Nollert P, Jensen MO, Miercke LJ, O'Connell J, Stroud RM, Schulten K (2002) Control of the selectivity of the aquaporin water channel family by global orientational tuning. Science 296:525–530

Takano J, Wada M, Ludewig U, Schaaf G, von Wirén N, Fujiwara T (2006) The *Arabidopsis* major intrinsic protein NIP5;1 is essential for efficient boron uptake and plant development under boron limitation. Plant Cell 18:1498–1509

Tombuloglu H, Ozean I, Tombuloglu G, Sakcali S, Unver T (2015) Aquaporins in boron-tolerant barley: identification, characterization, and expression analysis. Plant Mol Biol Report. doi:10.1007/s11105-015-0930-6

Törnroth-Horsefield S, Wang Y, Hedfalk K, Johanson U, Karlsson M, Tajkhorshid E, Neutze R, Kjellbom P (2006) Structural mechanism of plant aquaporin gating. Nature 439:688–694

Tournaire-Roux C, Sutka M, Javot H, Gout E, Gerbeau P, Luu DT, Bligny R, Maurel C (2003) Cytosolic pH regulates root water transport during anoxic stress through gating of aquaporins. Nature 425:393–397

Tsukaguchi H, Shayakul C, Berger UV, Mackenzie B, Devidas S, Guggino WB, van Hoek AN, Hediger MA (1998) Molecular characterization of a broad selectivity neutral solute channel. J Biol Chem 273:24737–24743

Tsukaguchi H, Weremowicz S, Morton CC, Hediger MA (1999) Functional and molecular characterization of the human neutral solute channel aquaporin-9. Am J Phys 277:F685–F696

Tyerman SD, Oats P, Gibbs J, Dracup M, Greenway H (1989) Turgor-volume regulation and cellular water relations of *Nicotiana tabacum* roots grown in high salinities. Aust J Plant Physiol 16:517–531

Tyerman SD, Bohnert HJ, Maurel C, Steudle E, Smith JA (1999) Plant aquaporins: their molecular biology, biophysics and significance for plant water relations. J Exp Bot 50:1055–1071

Uehlein N, Lovisolo C, Siefritz F, Kaldenhoff R (2003) The tobacco aquaporin NtAQP1 is a membrane CO_2 pore with physiological functions. Nature 425:734–737

Uehlein N, Otto B, Hanson DT, Fischer M, McDowell N, Kaldenhoff R (2008) Function of *Nicotiana tabacum* aquaporins as chloroplast gas pores challenges the concept of membrane CO2 permeability. Plant Cell 20:648–657

Venkatesh J, Yu JW, Gaston D, Park SW (2015) Molecular evolution and functional divergence of X-intrinsic protein genes in plants. Mol Gen Genomics 290:443–460

Verdoucq L, Grondin A, Maurel C (2008) Structure-function analysis of plant aquaporin AtPIP2;1 gating by divalent cations and protons. Biochem J 415:409–416

Verkman AS, Mitra AK (2000) Structure and function of aquaporin water channels. Am J Physiol Ren Physiol 278:F13–F28

Verma RK, Prabh ND, Sankararamakrishnan R (2015) Intra-helical salt-bridge and helix destabilizing residues within the same helical turn: Role of functionally important loop E half-helix in channel regulation of major intrinsic proteins. Biochim Biophys Acta 1848:1436–1449

Viadiu H, Gonen T, Walz T (2007) Projection map of aquaporin-9 at 7 Å resolution. J Mol Biol 367:80–88

Wallace IS, Roberts DM (2004) Homology modeling of representative subfamilies of Arabidopsis major intrinsic proteins. Classification based on the aromatic/arginine selectivity filter. Plant Physiol 135:1059–1068

Wang Y, Schulten K, Tajkhorshid E (2005) What makes an aquaporin a glycerol channel? A comparative study of AqpZ and GlpF. Structure 13:1107–1118

Wang M, Wang Z, Wang X, Wang S, Ding W, Gao C (2015) Layer-by-layer assembly of aquaporin Z-incorporated biomimetic membranes for water purification. Environ Sci Technol 49:3761–3768

Wang C, Hu H, Qin X, Zeise B, Xu D, Rappel WJ, Boron WF, Schroeder JI (2016) Reconstitution of CO_2 regulation of SLAC1 anion channel and function of CO_2-permeable PIP2;1 aquaporin as CARBONIC ANHYDRASE4 interactor. Plant Cell 28:568–582

Weaver DC, Shomer NH, Louis CF, Roberts DM (1994) Nodulin 26, a nodule-specific symbiosome membrane protein from soybean, is an ion channel. J Biol Chem 269:17858–17862

Wudick MM, Li X, Valentini V, Geldner N, Chory J, Lin J, Maurel C, Luu DT (2015) Subcellular redistribution of root aquaporins induced by hydrogen peroxide. Mol Plant 8:1103–1114

Xu C, Wang M, Zhou L, Quan T, Xia G (2013) Heterologous expression of the wheat aquaporin gene *TaTIP2;2* compromises the abiotic stress tolerance of *Arabidopsis thaliana*. PLoS ONE 8:e79618

Xu Y, Hu W, Liu J, Zhang J, Jia C, Miao H, Xu B, Jin Z (2014) A banana aquaporin gene, *MaPIP1;1*, is involved in tolerance to drought and salt stresses. BMC Plant Biol 14:59

Xu W, Dai W, Yan H, Li S, Shen H, Chen Y, Xu H, Sun Y, He Z, Ma M (2015) *Arabidopsis* NIP3;1 plays an important role in arsenic uptake and root-to-shoot translocation under arsenite stress conditions. Mol Plant 8:722–733

Ye RG, Verkman AS (1989) Simultaneous optical measurement of osmotic and diffusional water permeability in cells and liposomes. Biochemistry 28:824–829

Yool AJ, Weinstein AM (2002) New roles for old holes: ion channel function in aquaporin-1. News Physiol Sci 17:68–72

Yool AJ, Stamer WD, Regan JW (1996) Forskolin stimulation of water and cation permeability in aquaporin 1 water channels. Science 273:1216–1218

Yu J, Yool AJ, Schulten K, Tajkhorshid E (2006) Mechanism of gating and ion conductivity of a possible tetrameric pore in aquaporin-1. Structure 14:1411–1423

Zardoya R, Ding X, Kitagawa Y, Chrispeels MJ (2002) Origin of plant glycerol transporters by horizontal gene transfer and functional recruitment. Proc Natl Acad Sci U S A 99: 14893–14896

Zeidel ML, Ambudkar SV, Smith BL, Agre P (1992) Reconstitution of functional water channels in liposomes containing purified red cell CHIP28 protein. Biochemistry 31:7436–7440

Zeuthen T, Alsterfjord M, Beitz E, MacAulay N (2013) Osmotic water transport in aquaporins: evidence for a stochastic mechanism. J Physiol 591:5017–5029

Zhang da Y, Ali Z, Wang CB, Xu L, Yi JX, Xu ZL, Liu XQ, He XL, Huang YH, Khan IA, Trethowan RM, Ma HX (2013) Genome-wide sequence characterization and expression analysis of major intrinsic proteins in soybean (*Glycine max* L.). PLoS ONE 8:e56312

Zhang J, Deng Z, Cao S, Wang X, Zhang A, Zhang X (2008) Isolation of six novel aquaporin genes from *Triticum aestivum* L. and functional analysis of *TaAQP6* in water redistribution. Plant Mol. Biol Reprod 26:32–45

Zhao XQ, Mitani N, Yamaji N, Shen RF, Ma JF (2010) Involvement of silicon influx transporter OsNIP2;1 in selenite uptake in rice. Plant Physiol 153:1871–1877

Zhu R, Macfie SM, Ding Z (2005) Cadmium-induced plant stress investigated by scanning electrochemical microscopy. J Exp Bot 56:2831–2838

Heteromerization of Plant Aquaporins

Cintia Jozefkowicz, Marie C. Berny, François Chaumont, and Karina Alleva

Abstract The discovery of plasma membrane intrinsic protein (PIP) heterotetramerization has opened a new field of research. This phenomenon was first observed between PIPs belonging to two phylogenetic groups (PIP1 and PIP2) with ubiquitous expression in different plant tissues. These isoforms present few differences in their primary sequence but show major differences in their functionality when expressed in heterologous systems.

Many reports in recent years shed light on the PIP1 and PIP2 interaction as a regulatory mechanism to modulate their trafficking and biological activity. In this regard, PIP heterotetramerization has been proposed as a way of achieving a diversification in the water transport capacity and in the control of net solute transport. Also, acidification conditions were shown to act as a mechanism to control the opening and blockage of these channels in native tissues, and their proton-dependent gating can be affected depending on the presence of PIP2 homotetramers or PIP1-PIP2 heterotetramers in the target membrane.

In the present chapter, we report the state-of-the-art knowledge about PIP heterotetramerization in the context of protein oligomerization. We emphasize the main experiments that help to understand the existence of some relevant structural elements involved in PIP oligomerization and the conditions necessary for these hetero-oligomers to occur in the cell.

C. Jozefkowicz • K. Alleva (✉)
Universidad de Buenos Aires. Consejo Nacional de Investigaciones Científicas y Técnicas. Instituto de Química y Fisicoquímica Biológicas (IQUIFIB). Facultad de Farmacia y Bioquímica, Buenos Aires, Argentina., Buenos Aires, Argentina
e-mail: karina.alleva@gmail.com

M.C. Berny • F. Chaumont
Institut des Sciences de la Vie, Université catholique de Louvain, Croix du Sud 4-L7.07.14, B-1348, Louvain-la-Neuve, Belgium

F. Chaumont, S.D. Tyerman (eds.), *Plant Aquaporins*, Signaling and Communication in Plants, DOI 10.1007/978-3-319-49395-4_2

1 Aquaporin Hetero-oligomerization

Protein-protein interactions play diverse roles in biology, and, in particular, protein multimerization confers many different functional advantages such us diversifying cellular signalling transduction, modulation of biological activity by protein conformational changes, functional diversity and increases in protein stability. Multimerization is common among proteins that generally exist in a crowded environment where many potential binding partners with different surface properties are available. Nooren and Thornton (2003) point out that the specificity for protein-protein interaction derives mainly from the complementarity of shape and chemistry that determine the free energy of binding, but the protein localization also has a role to play. Interestingly, despite the fact that there are many protein-protein interactions between proteins from different families, interactions are also frequent between proteins belonging to the same family. Many proteins are very specific in their choice of partner, assembling as homo- or hetero-oligomers. In the case of homo-oligomers, protein interaction occurs between identical protein chains, while, in the case of hetero-oligomers, the interaction occurs between non-identical protein chains.

Aquaporins are integral membrane proteins that allow the transport of water and non-charged solutes. All aquaporins have six transmembrane helices and N- and C-termini located intracellularly. Despite the fact that each aquaporin monomer forms a channel with its own pore, they have been shown to adopt a quaternary structure organizing mainly as tetrameric homo-oligomers in membranes (see chapter "Structural Basis of the Permeation Function of Plant Aquaporins"). Many mammal aquaporin structures have been resolved, as is the case of AQP1 (Walz et al. 1994; Murata et al. 2000; Sui et al. 2001; Ruiz Carrillo et al. 2014), AQP0 (Gonen et al. 2004; Harries et al. 2004; Palanivelu et al. 2006; Hite et al. 2010), AQP2 (Frick et al. 2014), AQP4 (Ho et al. 2009) and AQP5 (Horsefield et al. 2008). The tetrameric arrangement was first observed for plant AQPs by cryoelectron microscopy of two-dimensional crystals for α-TIP (TIP, tonoplast intrinsic protein) from bean (Daniels et al. 1999) and SoPIP2;1 (PIP, plasma membrane intrinsic protein) from spinach (Kukulski et al. 2005). Later, SoPIP2;1 structure was resolved by X-ray crystallography of 3D crystals in an open conformation to 3.9 Å resolution and in a close conformation to 2.1 Å (Törnroth-Horsefield et al. 2006). More recently, the structure of an ammonia-permeable TIP aquaporin (AtTIP1;2) from *Arabidopsis thaliana* was obtained at 1.8 Å resolution (Kirscht et al. 2016).

All of the resolved structures correspond to homotetrameric assemblies despite hetero-oligomerization having been described for some family members. Among non-plant aquaporins, some cases of hetero-oligomerization have been reported. For example, a mutated AQP2 is able to form a hetero-oligomer with the wild-type form of AQP2 (Sohara et al. 2006). Also, hetero-oligomerization was reported for AQP1, where a non-functional AQP1 mutated in the loop B or E forms mixed oligomers with a truncated AQP1 mutant (D237Z) (Jung et al. 1994). A striking example of hetero-oligomerization among mammal aquaporins is the case of AQP4. This protein exists in two splicing variants, AQP4M1, starting with Met1, and AQP4M23, starting with Met23, which can assemble in the plasma membrane as heterotetramers (Lu et al. 1996; Neely et al. 1999). Interestingly, AQP4M23 homotetramers alone or together with AQP4M23-AQP4M1 heterotetramers can aggregate into supramolecular structures known as orthogonal arrays (OAPs) (Rash et al. 1998; Sorbo et al. 2008; Rossi et al. 2012). It is not clear what is the precise function of this kind of AQP clustering, but the predominant localization of OAPs in cells facing a basal lamina and the type of molecules that interacts with these aggregates of AQP4 suggest that this channel arrangement may be involved in establishing and maintaining cell polarity (Wolburg et al. 2011). Interestingly, the difference in the ability to form OAPs between AQP4M23 and AQP4M1 seems to be due to a specific amino acid sequence located in the native N-terminus (Hiroaki et al. 2006).

In contrast to the reported cases of hetero-oligomerization of mammal aquaporins, where the assembly was obtained by the interaction of different aquaporin splicing variants or between a wild-type AQP and its mutated versions, plant PIP aquaporins have the particularity of forming hetero-oligomers that include different PIP isoforms.

It has been described that proteins that are part of complexes tend to evolve at a relatively slow rate in order to improve the co-evolution with their interacting partners (Mintseris and Weng 2005). Interestingly, the molecular phylogenetic profiling of AQPs from nine genomes of flowering plants has shown that the PIP subfamily has a low evolutionary rate (Soto et al. 2012); this high evolutionary constraint may be due to a functional constraint related to the physical interaction that occurs between different members of the PIP subfamily. In the following sections, the state-of-the-art knowledge of PIP heterotetramerization will be presented and discussed.

2 Plant Plasma Membrane Intrinsic Proteins (PIP): The Paradigmatic Case for Hetero-oligomerization of Aquaporins

The first step in the study of the biological activity of PIP aquaporins consists in water transport assays by expressing them in a heterologous system such as *Xenopus laevis* oocytes (Preston et al. 1992). Intriguingly, while PIP isoforms belonging to the PIP2 group are able to reach the oocyte plasma membrane as functional oligomers to increase the osmotic water permeability coefficient (P_f) of the membrane, most PIP1 isoforms do not (Fetter et al. 2004; Sakurai et al. 2005; Bellati et al. 2010). Only a few cases of functional PIP1 facilitating water diffusion when expressed alone in oocytes have been reported (Tournaire-Roux et al. 2003; Suga and Maeshima 2004; Zhang et al. 2007). The lack of water transport activity when *PIP1* cRNA is injected into the oocytes was first interpreted as PIP1 being inactive or having very low water permeability (Daniels et al. 1994; Yamada et al. 1995; Weig 1997; Johansson et al. 1998; Biela et al. 1999; Chaumont et al. 2000; Marin-Olivier et al. 2000; Moshelion et al. 2002). However later on, it was demonstrated that most PIP1 proteins fail in reaching the oocyte plasma membrane but are retained in intracellular compartments (Fetter et al. 2004; Bienert et al. 2012; Jozefkowicz et al. 2013; Yaneff et al. 2014).

Interestingly, Fetter and co-workers (2004) showed that if maize ZmPIP1;2 is co-expressed with ZmPIP2;1, ZmPIP2;4 or ZmPIP2;5, an increase in P_f, which is dependent on the amount of *ZmPIP1;2* cRNA injected, is observed compared to the P_f of oocytes injected with ZmPIP2;5 cRNA alone (Fetter et al. 2004). Moreover, confocal microscopy analysis of oocytes expressing ZmPIP1;2-GFP alone or ZmPIP1;2-GFP plus ZmPIP2;5 showed that the amount of ZmPIP1;2-GFP present in the plasma membrane is significantly higher in co-expressing cells. A physical interaction was proposed to explain these results. Nickel affinity chromatography purification of ZmPIP2;1 fused to a histidine tag leads to the co-elution of ZmPIP1;2-GFP demonstrating the physical interaction of both channels. Also, immunoprecipitation experiments provided additional evidence for the association of ZmPIP1;2 and ZmPIP2;1 in vivo in maize roots and in suspension cells in the absence of any PIP overexpression. Finally and importantly, when co-expressed in maize protoplasts with ZmPIP2, ZmPIP1 proteins, which are retained in the endoplasmic reticulum (ER) when expressed alone, are re-localized to the plasma membrane (Zelazny et al. 2007).

All of these data demonstrated for the first time that some PIP1 requires a PIP2 partner to reach the plasma membrane. Thus, PIP1-PIP2 interaction is relevant for trafficking and re-localization of PIP1 proteins to the plasma membrane as a consequence of their physical interaction with some PIP2s. A similar conclusion was later drawn in epidermal cells of transgenic *Arabidopsis* roots (Sorieul et al. 2011). So, as a consequence of the physical interaction with PIP2 proteins, PIP1s are correctly targeted to the plasma membrane; otherwise, these PIP1s remain retained in the ER. The observation that PIP1 plasma membrane localization relies on the concomitant presence of PIP2s was extended to many PIP2/PIP1 pairs from several plant species and has been observed in plants, oocytes and yeast expression systems (Zelazny et al. 2007; Mahdieh et al. 2008; Vandeleur et al. 2009; Alleva et al. 2010; Bellati et al. 2010; Otto et al. 2010; Ayadi et al. 2011; Bienert et al. 2012).

Though most PIP1-PIP2 pairs studied can functionally interact (the oocyte P_f increases when they are co-expressed) by physical interaction (contacts between both channels that promote PIP1 re-localization), it is worth mentioning that there are some exceptions. For instance, the co-expression of ZmPIP1;1 with ZmPIP2;5 does not result in a P_f increase greater than the P_f measured after the expression of ZmPIP2;5 alone, indicating that ZmPIP1;1 does not functionally interact with ZmPIP2;5 in *Xenopus* oocytes (Fetter et al. 2004). Also, BvPIP2;1 is not able to functionally interact with BvPIP1;1 (Jozefkowicz et al. 2013). Additionally, in this subset of noninteracting plant aquaporins should be included the pairs OsPIP2;3/OsPIP1;3 (Matsumoto et al. 2009) and PvPIP2;3/PvPIP1;1 (Zhou et al. 2007). While in most cases it is still not clear which are the structural differences between these noninteracting PIP1 and PIP2 pairs and the interacting pairs, in the case of BvPIP2;1, it was shown that the first extracellular loop (named loop A) could be responsible for the lack of interaction with BvPIP1;1 (Jozefkowicz et al. 2013). Furthermore, while the finding of PIP2 and PIP1 forming hetero-oligomeric assemblies has been prevalent, hetero-oligomerization between different PIP2 and different PIP1 isoforms has also been demonstrated, such as ZmPIP2;6 with ZmPIP2;1 (Cavez et al. 2009) and ZmPIP1;1 with ZmPIP1;2 (Fetter et al. 2004).

Interestingly, the amino acid identity between the PIP1 and PIP2 isoforms is quite high (~80 %), with the main differences located at the C- and N-terminal domains and the loop A. However, this high percentage identity between paralogues is not due to the recent emergence of them, since PIP1 and PIP2 subfamilies separated long before the divergence of monocots and dicots occurred; indeed, PIPs representative of the PIP1 and PIP2 groups are found in the moss *Physcomitrella patens* (Danielson and Johanson 2010). The high identity between PIPs can be due to a high evolutionary and functional constraint of this subfamily (Soto et al. 2012).

3 Structure-Function Relationships in PIP Hetero-oligomerization

The precise way in which PIP1 and PIP2 interact is not yet fully elucidated, but strong experimental evidence is emerging and shedding light on this issue. The results obtained by the co-expression of the PIP1 and PIP2 channels, obtained in the first research stages on this topic, were compatible with protein complexes formed either by interactions between different PIP homotetramers or by interactions between different PIP monomers organized within an heterotetramer. Over the years the terms 'hetero-oligomer' and 'heterotetramer' were both used in the research literature, and many times they were even taken as synonymous. Recently, some experimental data indicated that PIP1 and PIP2 co-assemble in heterotetramers (different monomers interacting within the same tetramer). Still, even though the high sequence identity among all the PIP clusters explains the possibility that PIP1 and PIP2 assemble together into the same tetramer, as if they were almost 'identical' molecules, not all of the PIP1-PIP2 pairs follow this behaviour. Indeed, site-directed mutagenesis experiments together with crystallography and molecular modelling revealed the existence of some crucial structural elements involved in PIP oligomerization that can also control hetero-oligomerization.

3.1 Loop A

The prevalent role of the loops in protein function is due to their flexibility and location on the surface of the protein. Besides their role in protein function, intra- and extracellular loops are also important as structural elements mediating protein interactions, especially in membrane proteins. In particular, for PIP aquaporin oligomers, the first extracellular loop A has been reported as mediating the interactions between different monomers within a tetramer. Indeed, a highly conserved cysteine residue and the hinge connecting the first transmembrane domain (TM1) to the loop A have a relevant role in tetrameric organization (Bienert et al. 2012; Jozefkowicz et al. 2013). Early observations by Barone and co-workers (1998) provided experimental evidence that *Beta vulgaris* PIP dimers are connected by a disulphide bridge. In the SoPIP2;1 tetrameric structure, the position of the C-terminus of helix 1 and the N-terminus of helix 2 orients the loop A towards the fourfold centre of the tetramer (Kukulski et al. 2005). In this configuration, the nearness of the four-loop A cysteines suggested the formation of a disulphide bond between two monomers (Kukulski et al. 2005). In accordance with these data, molecular dynamic simulation (MDS) of BvPIP2;1 also showed that the position of loop A is oriented towards the centre of the tetramer, with each of the four loops A being flexible parts of the monomers but having a different solvation pattern and different movement along the MDS (Jozefkowicz et al. 2013).

The demonstration of the existence of a disulphide bond linking the loops A of two monomers was conducted by Bienert et al. (2012) and showed that this disulphide bond is not an artefact of the sample preparation but that they occurred in plant cells. Interestingly, mutation of the loop A cysteine residue does not alter the water channel activity of PIP1 or PIP2 in homo- or heterotetramers, hetero-oligomerization or trafficking to the plasma membrane but does modify its mercury sensitivity (Bienert et al. 2012). Incubation of oocytes expressing ZmPIP2;5 alone with $HgCl_2$, a well-known aquaporin inhibitor, leads to a decrease in the P_f. However, the P_f of oocytes co-expressing ZmPIP1;2Cys85Ser and WT ZmPIP2;5 or ZmPIP2;5Cys75Ser is not inhibited in response to $HgCl_2$ treatment, indicating that when ZmPIP2;5 is co-expressed with ZmPIP1;2Cys85Ser, ZmPIP2;5 becomes Hg^{2+} insensitive. This data suggests that the loop A cysteine of ZmPIP1;2 is involved in the mercury sensitivity of hetero-oligomers and indicates that the conformational arrangement of the PIP2 monomers in the PIP1-PIP2 hetero-oligomers is different from that in the PIP2 homo-oligomers (Bienert et al. 2012). These data point to the potential functional and conformational interaction between PIP1 and PIP2 in hetero-oligomer complexes mediated by the loop A structure (Bienert et al. 2012). The presence of a disulphide bridge could affect the stability of the tetramers in some conditions, but this process needs to be investigated in more detail. The results obtained for BvPIP2;1, a PIP2 unable to interact with BvPIP1;1 and presenting non-conserved amino acid residues in the N-terminus of the loop A, are in accordance with this hypothesis (Jozefkowicz et al. 2013). The mutation of non-conserved loop A residues (NETD) of BvPIP2;1 promoted the recovery of its interaction with BvPIP1;1, resembling the PIP2-PIP1 classical interaction (Jozefkowicz et al. 2013). In agreement, Hayward and Kitaos (2010) stress the importance of the first and last residues of the protein loops, remarking that this constraint might influence the dynamical behaviour of the loop.

Altogether, these data are strong evidence for an important role of the loop A in controlling the interactions between contiguous monomers in PIP homo- or heterotetramers.

3.2 *Loop E*

In all aquaporins, loops B and E containing the NPA motif (asparagine, proline, alanine) form half transmembrane helices that fold into the channel from opposite sides of the membrane. This creates a seventh transmembrane helix, with the NPA motifs located at the centre of the pore (Jung et al. 1994; Murata et al. 2000; Törnroth-Horsefield et al. 2006).

Early studies on AQP1 suggested that loop E together with loop B is not only involved in water permeation through the pore but is also critical for the tetrameric assembly of the channel (Jung et al. 1994; Mathai and Agre 1999). Later, a mutational analysis conducted in plant aquaporins demonstrated the important role of the C-terminal part of loop E in the interaction of PIP monomers (Fetter et al. 2004). To explore the reason for the different behaviour of ZmPIP1;1 and ZmPIP1;2 regarding their functional interaction with ZmPIP2;5 in *Xenopus* oocytes, a ZmPIP1;1 mutant containing the loop E of ZmPIP1;2 (named ZmPIP1;1LE) was constructed (Fetter et al. 2004). The replacement of the loop E of ZmPIP1;1 by that from ZmPIP1;2 modifies the behaviour of the protein when co-expressed with ZmPIP2;5: a positive synergistic effect on the P_f is observed, while it is absent with the WT ZmPIP1;1 (Fetter et al. 2004). The molecular explanation of this functional observation arises from ZmPIP1;1LE MDS, which shows a different position of the loop in ZmPIP1;1LE affecting the structure of the pore. This conformational change in loop E is proposed to affect the structure of the semi-helix E and the TM6, which can in turn impact the oligomerization of this protein (reviewed in Chaumont et al. 2005). However, it was not shown whether loop E from ZmPIP1;2 is required for ZmPIP1;1 interaction with ZmPIP2;5 or only for the activation of ZmPIP1;1 water channel activity. Similarly, exchanging the loop E of GlpF, a bacterial glycerol facilitator of the MIP/aquaporin family, with loop E of the insect AQPcic, alters either oligomer assembly or tetramer stability (Duchesne et al. 2002). Other evidence supporting the involvement of the aquaporin loop E in oligomerization is the crystallographic and molecular modelling data, showing that TM5 participates in the interactions between monomers within the tetramer and can affect the spatial positions of loops B and E in AQP1 and GlpF (Ren et al. 2000; Murata et al. 2000; Fu et al. 2000; Jensen et al. 2001). Moreover, in mammal AQPs, functional analyses of AQP0-AQP2 chimeras have demonstrated that stability of loop E is crucial for the channel activity (Kuwahara et al. 1999; Suga and Maeshima 2004).

3.3 Transmembrane Domains

In addition to the loops, the TM domains from neighbour monomers interact within a tetramer. In AQP1, TM1 and TM2 interact with TM4 and TM5 of an adjacent monomer by coiled-coil interactions in a left-handed fashion (Murata et al. 2000). This type of interaction is common to tightly pack structures, thanks to van der Waals interactions between side chains of close residues. In addition, hydrogen bonds between TM1 and TM5 or between TM2 and TM4 stabilize the structure. The role of specific TM amino acid residue in oligomerization is still unknown, but some TM residues have been shown to turn an inactive PIP1 into an active one without any co-expression in oocytes. It is the case of substitutions in the rice OsPIP1;1 and OsPIP1;3 (A103V and A102V, respectively) that allowed an increase in water channel activity for non-functional PIP1 isoforms (Zhang et al. 2010). In this work, the authors identified by homology modelling and mutagenesis one residue in TM2 of OsPIP1;3 located at the interface between monomers that induced upon mutation a change of conformation within the pore. Indeed, a change of orientation of Ile101 was caused by the substitution of Ala102 into Val and allowed the widening of the constriction region within the pore.

Comparative modelling on the basis of SoPIP2;1 X-ray structure was used to build heterotetramers containing ZmPIP1;2 and ZmPIP2;5 and identify amino acid residues in the TM domains that putatively interact at the interfaces between monomers (Berny et al. 2016). Mutational analysis of these residues showed single residue substitution that either inactivates ZmPIP2;5 (W85A, F92A and F210A) or activates ZmPIP1;2 (Q91L and F220A) without affecting their interaction when express in *Xenopus* oocytes. Interestingly, the activating F220A mutation in TM5 of ZmPIP1;2 inactivates, at the same time, the water channel activity of the interacting ZmPIP2;5 within a heterotetramer (Berny et al. 2016). Altogether, these data highlight the importance of single specific TM amino acid residues in the activity of the channels within a heterotetramer without affecting the interaction between monomers. Multiple mutations might be required to affect the oligomerization state.

4 Co-expression of PIP: A Condition for Oligomerization

The protein localization, concentration and local environment are parameters affecting the interaction and controlling the composition and oligomeric state of protein complexes (Nooren and Thornton 2003). Monomers participating in obligate interactions to form oligomers are supposed to be co-expressed and to co-localize upon synthesis. This time and space synchronization should be valid for homo-oligomers as well as for hetero-oligomers.

The interactions between proteins which allow complex formation are usually driven by the concentration of the components and the free energy of the complex relative to the alternative states (Nooren and Thornton 2003). So, there are different conditions that can control the oligomerization phenomenon, including encountering the interacting surfaces. In this regard, the association of two proteins relies on co-localization in time and space and the adequate concentration of the interacting proteins, where control mechanisms that alter the effective local concentration such as gene expression, protein degradation rates or diffusion rates, among others, are relevant. In the competition between binding partners for protein-protein interactions, different factors can influence oligomerization. As PIP2 and PIP1 can ensemble either as homotetramers or heterotetramers, different elements should be involved in the ruling of the processes that govern each option.

PIPs are expressed in organs and tissues that present large fluxes of water, i.e. vascular tissues, guard cells and suberized endodermis and bundle sheath cells, among others (Gomes et al. 2009; Chaumont and Tyerman 2014). Interestingly, there is not a homogeneous expression pattern for all PIPs, and this has been interpreted as a tuning mechanism by which the plant is able to adapt to different conditions (Gomes et al. 2009). Nonetheless, there are several stimuli that promote a coordinate response in the expression of different PIPs. For example, upon water deprivation, most *Arabidopsis* PIPs expressed in leaves are reported as transcriptionally downregulated, with the exception of AtPIP2;6 which remains constant, and AtPIP1;4 and AtPIP2;5 which are both induced (Alexandersson et al. 2010). A reverse genetic study performed for AtPIPs showed that knock-out of three PIP isoforms belonging to the PIP1 and PIP2 groups contributed individually to the same hydraulic conductivity as the corresponding triple *PIP* mutant (Prado et al. 2013). Microarray studies during grape berry development showed that PIP1 and PIP2 are expressed in the same tissues at the same time, but each one has a particular pattern of expression (Fouquet et al. 2008). All these results indicate synchronization between some *PIP1* and *PIP2* gene expression, but the physical or functional interactions of all the mentioned PIPs have not been studied yet.

Conversely, there are other cases for which information about both tissue expression and protein interaction in heterologous systems were recorded. For instance, *Fragaria x ananassa* PIPs interact to form heterotetramers (Alleva et al. 2010; Yaneff et al. 2014), and both *PIP1* and *PIP2* mRNA are expressed during the whole ripening process in fruits (Alleva et al. 2010). Interestingly, the expression of *FaPIP1* was low in the first ripening stages and later increased; while in the case of *FaPIP2*, the expression was markedly high in the first stages and decreases progressively until the end of ripening or remained approximately constant and low, depending on the cultivar. This dissimilar expression pattern of FaPIP2 and FaPIP1 can be compatible with the reported random stoichiometry suggested for *F. x ananassa* PIP heterotetramers after FaPIP oocyte expression and mathematical analysis of the results (Yaneff et al. 2014). Furthermore, information about the expression pattern of paradigmatic interacting ZmPIPs is also available. Gene and protein expression was studied in maize roots, leaves and stomatal complexes (Hachez et al. 2006, 2008; Heinen et al. 2014). All *ZmPIP* genes, except *ZmPIP2;7*, are expressed in primary roots and leaves, and their expression is dependent on the developmental stage of the organ (Hachez et al. 2006, 2008). In this regard, it was proposed that some specific pairs of PIP1 and PIP2 have a correlation in their expression pattern in accordance with their functional responses (Yaneff et al. 2015).

In addition to transcriptional and translational regulation, PIP localization and stability are also highly regulated (for a review, see Hachez et al. 2013; Chevalier and Chaumont 2015). For instance, after salt exposure to *Arabidopsis* roots, a decrease in PIP1 protein abundance was observed after 30 min, while PIP2 abundance remained constant even after 6 h of exposure to stress conditions. However, PIP1and PIP2 were both reduced after 24 h of salt exposure (Boursiac et al. 2005).

All of these results indicate that, depending on the conditions, PIP1 and PIP2 can be co- or differentially regulated, but PIP physical interaction could represent an additional cooperative way to respond to different physiological processes or even stresses. Indeed, all reports regarding PIP expression in different plants show that both PIP1 and PIP2 proteins are always present together. It is the ratio between protein (or mRNA) amounts of each group that can vary considerably between tissues or cell types, certainly affecting their physiology. For instance, among all the *PIP* transcripts found in maize stomatal complexes, 85 % were PIP1s (Heinen et al.

2014). In this regard, it was empirically proven that *B. vulgaris* PIP1 and PIP2 can assemble in a flexible fashion into tetramers having different stoichiometries depending on the amount of each *PIP1* or *PIP2* cRNA available (Jozefkowicz et al. 2016). Thus, differences in the expression levels of each paralogue condition the existence of different PIP1 and PIP2 homotetramers or heterotetramers.

5 Biological Relevance of PIP Hetero-oligomerization

The hetero-oligomerization of membrane proteins is believed to play a fundamental role in the regulation of cellular function. Oligomerization can increase protein stability, and in particular hetero-oligomerization may allow the diversification of biological activity. In this regard, it was postulated that protein-protein interactions may have evolved to optimize functional efficacy (Nooren and Thornton 2003). However, it is also possible that an interaction with no functional reason evolved, and this interaction survives due to the absence of selective pressure to be rejected from the evolutionary path. Often, the functional rationale of oligomerization is not clear, and a happenstance of oligomerization can be supposed (Nooren and Thornton 2003). However, in the case of PIP1-PIP2 hetero-oligomerization, this kind of supramolecular assembly plays an important role in the subcellular PIP1 localization and modulates the cell membrane hydraulic conductivity (Fig. 1).

→

Fig. 1 PIP heterotetramerization in the plant cell. Scheme of a plant cell where the PIP1 monomers are shown in *orange*, and the PIP2 monomers are shown in *blue*. In the *lower part* of the scheme, the cell is under three different pH units: <6.0, ~ 6.5 and >7.0; in the *upper part* of the scheme, the cell is under pH >7.0. The water capacity of each tetramer is represented as proportional to the length of the *light-blue arrow*, while the solute capacity of each tetramer (i.e. to dioxide carbon) is represented with a *black arrow*. In the *box* on *top* at the *left*, a structural alignment of a representative PIP1 (in *red*) and PIP2 (in *blue*) monomer is shown, pointing out the major differences between these channels (regarding loop A and the N- and C-termini) and some structural elements relevant for heterotetramerization. All reports regarding PIP expression in different tissues or cell types found the simultaneous expression of both PIP1 and PIP2. However, the ratio between mRNA (or protein) amounts of each group can vary considerably between different cell types, conditioning the formation of different PIP1 and PIP2 homo- or heterotetramers. All PIP heterotetramers with different stoichiometries and PIP2 homotetramers are able to reach the plasma membrane, but they show different water transport capacities when heterologous expressed, being the water transport capacity of PIP1-PIP2 heterotetramers higher than the water transport capacity found for PIP2 homotetramers. On the contrary, PIP1 homotetramers are unable to reach the plasma membrane when heterologous expressed. Thus, upon regulation of PIP1-PIP2 expression levels, the plant cell can modulate the localization of different tetrameric species and the osmotic water permeability and substrate permeability of the plasma membrane by the assembly of PIP heterotetramers with different stoichiometries. Since each plant species has a variety of PIP1 and PIP2 isoforms, another point of regulation would be given by the expression of different types of each of these paralogues. Since not all PIP1 PIP2 are capable of interacting, locating PIP1 into the plasma membrane would depend on their capacity to interact with certain PIP2. In this regard, some structural elements involved in PIP1 and PIP2 contacts in PIP heterotetramers – loop A and loop E containing the second NPA motif (show in *yellow*) – and some TM have been shown to be relevant. At the level of the gating of the channel, the intracellular pH changes can differentially regulate the activity of PIP tetramers found in the plasma membrane, either as homo- or heterotetramers.

5.1 *Trafficking*

As mentioned above, immunocytochemistry, immunodetection and the expression of PIP fused to fluorescent proteins showed that most PIP2 aquaporins are localized in the plasma membrane, while most PIP1 aquaporins are found in the ER unless co-expressed with PIP2 (Zelazny et al. 2007; Li et al. 2009; Luu et al. 2012; Besserer et al. 2012; Jozefkowicz et al. 2013). The prevalence of PIP2 homotetramers vs PIP1-PIP2 heterotetramers can be controlled by the regulation of the expression of both paralogues, as described in Sect. 4.

The regulation of PIP1 and PIP2 trafficking to the plasma membrane, as homo- or heterotetramers, is another important mechanism leading to modification of the membrane permeability (Fig. 1) (see also chapter "Plant Aquaporin Trafficking"). Specific trafficking motifs for PIP2 export from the ER have been described. These include a diacidic motif (DxE) located at the N-terminus of some maize and *Arabidopsis* PIP2s (Zelazny et al. 2009; Sorieul et al. 2011) that probably interact with the Sec24 protein of the COPII complex for their recruitment to the moving

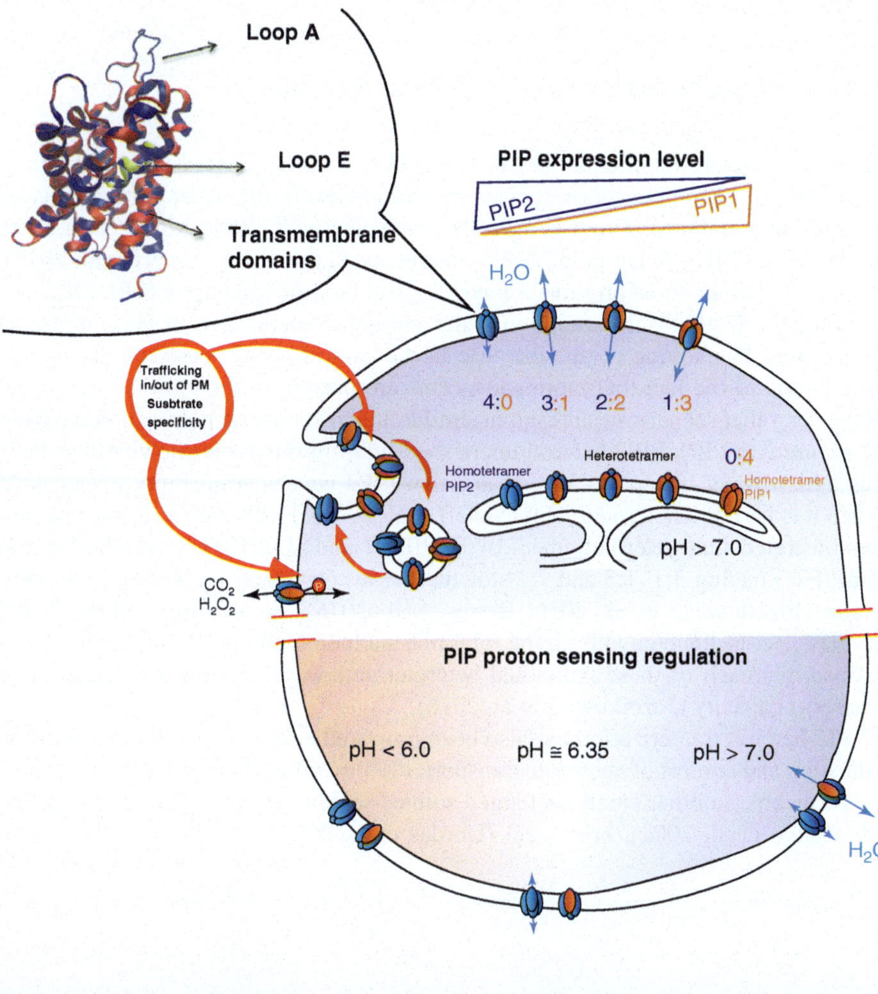

vesicle (reviewed in Hachez et al. 2013; Chevalier and Chaumont 2015). However, ZmPIP2;1 or ZmPIP2;2 can reach the plasma membrane even in the absence of a diacidic motif (Zelazny et al. 2009). Recently, a new LxxxA motif located in the TM3 of ZmPIP2;5 has been shown to be required for the ER export (Chevalier et al. 2014). Interestingly, the addition of ZmPIP2;5 diacidic and/or the LxxxA motifs in ZmPIP1;2 is not enough to induce the ER export of the latter, indicating the presence of specific retention signals in ZmPIP1;2 (Zelazny et al. 2009; Chevalier et al. 2014). The observation that all the hetero-oligomers comprising PIP1-PIP2 tested so far are found in the plasma membrane of plant cells and heterologous systems suggests that (i) the ER retention signal of PIP1 is hidden upon an interaction with PIP2 and (ii) the interaction with PIP2 is enough to ensure the routing of PIP1 to the plasma membrane even in the absence of any export signal in PIP1, meaning that the PIP1-PIP2 interaction does not prevent the PIP2 trafficking motifs from being properly recognized by the trafficking machinery.

From an evolutionary point of view, we can wonder whether most PIP1s have no plasma membrane trafficking motifs due to the fact that these motifs are not required as PIP1s are always co-expressed with PIP2s, which contain the information to direct them together as heterotetramers to the plasma membrane.

5.2 Biological Activity and Substrate Specificity

As it was already mentioned above, when PIP1 is co-expressed with PIP2, their interaction is accompanied in oocytes by an increase in the water permeability of the plasma membrane (Fetter et al. 2004; Vandeleur et al. 2009; Alleva et al. 2010; Bellati et al. 2010; Ayadi et al. 2011; Jozefkowicz et al. 2013; Yaneff et al. 2014), indicating that the most prominent physiological modification upon PIP1-PIP2 hetero-oligomerization is the alteration of the hydraulic conductivity of the cell plasma membrane. The source of this increase in the osmotic water permeability can be simply due to the fact that more aquaporins are present in the plasma membrane. However, other scenarios can result in similar final response. A publication studying *F. x ananassa* PIP1-PIP2 heterotramers shed light on this issue showing that PIP1 has a high water transport capacity and that PIP2 water permeability is enhanced when it is part of a heterotetramer (Yaneff et al. 2014). Furthermore, it was recently demonstrated that heterotetramers of BvPIP2;2 and BvPIP1;1 or ZmPIP1;2 and ZmPIP2;5 having 3:1, 1:3 and 2:2 stoichiometries can coexist at the plasma membrane (Jozefkowicz et al. 2016; Berny et al. 2016). In addition BvPIP2;2 and BvPIP1;1 contribute equally to the total plasma membrane permeability, as it was probed that each of these individual heterotetrameric species present equal water transport capacity (Jozefkowicz et al. 2016).

PIP heterotetramerization has also been proposed as a way of achieving a diversification and control of net solute transport. PIP1 members have been shown to also be permeable to other small uncharged solutes such as glycerol (Biela et al. 1999; Moshelion et al. 2002), boric acid (Dordas et al. 2000) and CO_2 (Uehlein et al.

2003), while PIP2s mainly act as water channels, although some of them can also transport H_2O_2, glycerol (Chaumont et al. 2001; Uehlein et al. 2008; Dynowski et al. 2008; Bienert et al. 2014) and CO_2 (Wang et al. 2016) (Fig. 1). CO_2 transport has been studied in depth in *Nicotiana tabacum* PIP1 and in *Zea mays* PIP1 (Uehlein et al. 2003; Otto et al. 2010; Bienert et al. 2014; Heinen et al. 2014). Different transport profiles have been found for NtPIP2 and NtAQP1 (belonging to the PIP1 group) by analysis of their transport characteristics after co-expression in the yeast *Saccharomyces cerevisiae* (Otto et al. 2010). While the expression of NtPIP2;1 increases the water diffusion through the cell membrane but not the CO_2 diffusion (deduced from CO_2-triggered intracellular acidification of the cells), the expression of NtAQP1 only increases the CO_2 diffusion (Otto et al. 2010). Functional analysis of artificial tetramers with a defined proportion of NtAQP1 and NtPIP2;1 indicates that the presence of a single NtPIP2;1 protein within a heterotetramer allows for an increase in the cell water permeability. On the other hand, an increase in the CO_2 diffusion requires the presence of at least three or four PIP1 in the heterotetramer. These data suggest that the stoichiometry of the heterotetramers can influence the substrate specificity of the complexes.

5.3 *Proton-Sensing Regulation*

Lastly, we consider pH sensing as an additional important modulator of the biological response upon PIP heterotetramerization. PIPs are acid-sensitive water channels both when heterologously expressed and in native tissues (Gerbeau et al. 2002; Tournaire-Roux et al. 2003; Alleva et al. 2006; Verdoucq et al. 2008). On the basis of SoPIP2;1 molecular structure, a gating model was proposed (Törnroth-Horsefield et al. 2006) that shows how conformational changes in the cytosolic loop D promote the opening and closure of the pore and control its water permeability. These conformational modifications are pH dependent since they are mediated by the protonation of a conserved His residue (His193 in SoPIP2;1) (Tournaire-Roux et al. 2003). At acidic pH, the His residue is charged and by interaction with other residues stabilizes the loop D closing the pore (Törnroth-Horsefield et al. 2006; Frick et al. 2013).

A shift in the pH sensing has been reported for PIP1-PIP2 heterotetramers in comparison with the PIP2 homotetramers (Bellati et al. 2010; Jozefkowicz et al. 2013; Yaneff et al. 2014). Water permeability pH dependence presents a low $pH_{0.5}$ (the pH at which the half maximal P_f is found) for PIP2 homotetramers, implying that a high probability of PIP channels in open conformation is found at physiological conditions. On the contrary, for the PIP1-PIP2 heterotetramers, a pH shift of 0.5 towards alkaline values was observed (Fig. 1). Moreover, all PIP1-PIP2 heterotetrameric assemblies of variable stoichiometries present equivalent biological activity in terms of pH gating and cooperative response (Jozefkowicz et al. 2016).

When *B. vulgaris* PIPs were studied, both homotetrameric BvPIP2;2 and BvPIP2;1 presented sigmoidal response curves with a $pH_{0.5}$ of approximately 6.4–6.5, but BvPIP1;1-BvPIP2;2 heterotetramers showed a $pH_{0.5}$ of approximately 6.7–6.8

(Bellati et al. 2010; Jozefkowicz et al. 2013). Interestingly, the plasma membrane of *B. vulgaris* storage roots vesicles shows a pH gating profile with $pH_{0.5}$ values around 6.7 (Alleva et al. 2006). In this way, acidification conditions to control the opening and blockage of water transport in native tissues vary depending on the presence of PIP2 homotetramers or PIP1-PIP2 heterotetramers in the target membrane.

6 Concluding Remarks

The comprehension of the functional relevance of protein oligomerization is not straightforward due to the fact that the proportion of well-studied oligomeric proteins is quite low compared with monomeric proteins. Even harder is the understanding of the biological relevance of hetero-oligomerization. The discovery of the existence of PIP1-PIP2 hetero-oligomers made these proteins of high interest not only for studying the physiological relevance of these complexes at the cell and plant levels in relation to their trafficking, substrate specificity and activity of the channels but also to better understand the biochemical and biophysical mechanisms that drive their physical association to form heterotetramers. They constitute a promising model to elucidate the rules governing protein oligomerization. One of the next challenges will be to obtain high-resolution structural data of heterotetramers.

Acknowledgements This work was supported by grants UBACYT 02042 0159-2013 (to KA) and grants from the Belgian National Fund for Scientific Research (NFRS), the Interuniversity Attraction Poles Programme-Belgian Science Policy (IAP7/29) and the Belgian French community ARC11/16-036 project (to FC). We thank Gerardo Zerbetto De Palma for his contributions with figure design.

References

Alexandersson E, Danielson JAH, Råde J et al (2010) Transcriptional regulation of aquaporins in accessions of Arabidopsis in response to drought stress. Plant J 61:650–660

Alleva K, Marquez M, Villarreal N et al (2010) Cloning, functional characterization, and co-expression studies of a novel aquaporin (FaPIP2;1) of strawberry fruit. J Exp Bot 61:3935–3945

Alleva K, Niemietz CM, Sutka M et al (2006) Plasma membrane of *Beta vulgaris* storage root shows high water channel activity regulated by cytoplasmic pH and a dual range of calcium concentrations. J Exp Bot 57:609–621

Ayadi M, Cavez D, Miled N et al (2011) Identification and characterization of two plasma membrane aquaporins in durum wheat (*Triticum turgidum* L. subsp. durum) and their role in abiotic stress tolerance. Plant Physiol Biochem 49:1029–1039

Barone LM, Mu HH, Shih CJ et al (1998) Distinct biochemical and topological properties of the 31- and 27-kilodalton plasma membrane intrinsic protein subgroups from red beet. Plant Physiol 118:315–322

Bellati J, Alleva K, Soto G et al (2010) Intracellular pH sensing is altered by plasma membrane PIP aquaporin co-expression. Plant Mol Biol 74:105–118

Berny MC, Gilis D, Rooman M, Chaumont F (2016) Single mutations in the transmembrane domains of maize plasma membrane aquaporins affect the activity of monomers within a heterotetramer. Mol Plant. doi:10.1016/j.molp.2016.04.006

Besserer A, Burnotte E, Bienert GP et al (2012) Selective regulation of maize plasma membrane aquaporin trafficking and activity by the SNARE SYP121. Plant Cell 24:3463–3481

Biela A, Grote K, Otto B et al (1999) The *Nicotiana tabacum* plasma membrane aquaporin NtAQP1 is mercury-insensitive and permeable for glycerol. Plant J 18:565–570

Bienert GP, Cavez D, Besserer A et al (2012) A conserved cysteine residue is involved in disulfide bond formation between plant plasma membrane aquaporin monomers. Biochem J 445:101–111

Bienert GP, Heinen RB, Berny MC, Chaumont F (2014) Maize plasma membrane aquaporin ZmPIP2;5, but not ZmPIP1;2, facilitates transmembrane diffusion of hydrogen peroxide. Biochim Biophys Acta 1838:216–222

Boursiac Y, Chen S, Luu D-T et al (2005) Early effects of salinity on water transport in Arabidopsis roots. Molecular and cellular features of aquaporin expression 1. Plant Physiol 139:790–805

Cavez D, Hachez C, Chaumont F (2009) Maize black Mexican sweet suspension cultured cells are a convenient tool for studying aquaporin activity and regulation. Plant Signal Behav 4:890–892

Chaumont F, Tyerman SD (2014) Aquaporins: highly regulated channels controlling plant water relations. Plant Physiol 164:1600–1618

Chaumont F, Barrieu F, Jung R, Chrispeels MJ (2000) Plasma membrane intrinsic proteins from maize cluster in two sequence subgroups with differential aquaporin activity. Plant Physiol 122:1025–1034

Chaumont F, Barrieu F, Wojcik E et al (2001) Aquaporins constitute a large and highly divergent protein family in maize. Plant Physiol 125:1206–1215

Chaumont F, Moshelion M, Daniels MJ (2005) Regulation of plant aquaporin activity. Biol cell under auspices. Eur J Cell Biol 97:749–764

Chevalier AS, Chaumont F (2015) Trafficking of plant plasma membrane aquaporins: multiple regulation levels and complex sorting signals. Plant Cell Physiol 56:819–829

Chevalier AS, Bienert GP, Chaumont F (2014) A new LxxxA motif in the transmembrane Helix3 of maize aquaporins belonging to the plasma membrane intrinsic protein PIP2 group is required for their trafficking to the plasma membrane. Plant Physiol 166:125–138

Daniels MJ, Chrispeels MJ, Yeager M (1999) Projection structure of a plant vacuole membrane aquaporin by electron cryo-crystallography. J Mol Biol 294:1337–1349

Daniels MJ, Mirkov TE, Chrispeels MJ (1994) The plasma membrane of *Arabidopsis thaliana* contains a mercury-insensitive aquaporin that is a homolog of the tonoplast water channel protein TIP. Plant Physiol 106:1325–1333

Danielson JAH, Johanson U (2010) Phylogeny of major intrinsic proteins. Adv Exp Med Biol 679:19–31

Dordas C, Chrispeels MJ, Brown PH (2000) Permeability and channel-mediated transport of boric acid across membrane vesicles isolated from squash roots. Plant Physiol 124:1349–1362

Duchesne L, Pellerin I, Delamarche C et al (2002) Role of C-terminal domain and transmembrane helices 5 and 6 in function and quaternary structure of major intrinsic proteins: analysis of aquaporin/glycerol facilitator chimeric proteins. J Biol Chem 277:20598–20604

Dynowski M, Schaaf G, Loque D et al (2008) Plant plasma membrane water channels conduct the signalling molecule H2O2. Biochem J 414:53–61

Fetter K, Van Wilder V, Moshelion M, Chaumont F (2004) Interactions between plasma membrane aquaporins modulate their water channel activity. Plant Cell 16:215–228

Fouquet R, Léon C, Ollat N, Barrieu F (2008) Identification of grapevine aquaporins and expression analysis in developing berries. Plant Cell Rep 27:1541–1550

Frick A, Eriksson UK, de Mattia F et al (2014) X-ray structure of human aquaporin 2 and its implications for nephrogenic diabetes insipidus and trafficking. Proc Natl Acad Sci U S A 111:6305–6310

Frick A, Järvå M, Törnroth-Horsefield S (2013) Structural basis for pH gating of plant aquaporins. FEBS Lett 587:989–993

Fu D, Libson A, Miercke LJ et al (2000) Structure of a glycerol-conducting channel and the basis for its selectivity. Science 290:481–486

Gerbeau P, Amodeo G, Henzler T et al (2002) The water permeability of Arabidopsis plasma membrane is regulated by divalent cations and pH. Plant J 30:71–81

Gomes D, Agasse A, Thiébaud P et al (2009) Aquaporins are multifunctional water and solute transporters highly divergent in living organisms. Biochim Biophys Acta Biomembr 1788: 1213–1228

Gonen T, Cheng Y, Kistler J, Walz T (2004) Aquaporin-0 membrane junctions form upon proteolytic cleavage. J Mol Biol 342:1337–1345

Hachez C, Besserer A, Chevalier AS, Chaumont F (2013) Insights into plant plasma membrane aquaporin trafficking. Trends Plant Sci 18:1–9

Hachez C, Heinen RB, Draye X, Chaumont F (2008) The expression pattern of plasma membrane aquaporins in maize leaf highlights their role in hydraulic regulation. Plant Mol Biol 68:337–353

Hachez C, Zelazny E, Chaumont F (2006) Modulating the expression of aquaporin genes in planta: a key to understand their physiological functions? Biochim Biophys Acta 1758:1142–1156

Harries WEC, Akhavan D, Miercke LJW et al (2004) The channel architecture of aquaporin 0 at a 2.2-A resolution. Proc Natl Acad Sci U S A 101:14045–14050

Hayward S, Kitao A (2010) The effect of end constraints on protein loop kinematics. Biophys J 98:1976–1985

Heinen RB, Bienert GP, Cohen D et al (2014) Expression and characterization of plasma membrane aquaporins in stomatal complexes of *Zea mays*. Plant Mol Biol 86:335–350

Hiroaki Y, Tani K, Kamegawa A et al (2006) Implications of the aquaporin-4 structure on array formation and cell adhesion. J Mol Biol 355:628–639

Hite RK, Li Z, Walz T (2010) Principles of membrane protein interactions with annular lipids deduced from aquaporin-0 2D crystals. EMBO J 29:1652–1658

Ho JD, Yeh R, Sandstrom A et al (2009) Crystal structure of human aquaporin 4 at 1.8 A and its mechanism of conductance. Proc Natl Acad Sci U S A 106:7437–7442

Horsefield R, Norden K, Fellert M et al (2008) High-resolution x-ray structure of human aquaporin 5. Proc Natl Acad Sci 105:13327–13332

Jensen MO, Tajkhorshid E, Schulten K (2001) The mechanism of glycerol conduction in aquaglyceroporins. Structure 9:1083–1093

Johansson I, Karlsson M, Shukla VK et al (1998) Water transport activity of the plasma membrane aquaporin PM28A is regulated by phosphorylation. Society 10:451–459

Jozefkowicz C, Rosi P, Sigaut L et al (2013) Loop A is critical for the functional interaction of two *Beta vulgaris* PIP aquaporins. PLoS One 8:e57993

Jozefkowicz C, Sigaut L, Scochera F et al (2016) PIP water transport and its pH dependence are regulated by tetramer stoichiometry. Biophys J 110:1312–1321

Jung JS, Preston GM, Smith BL et al (1994) Molecular structure of the water channel through aquaporin CHIP. The hourglass model. J Biol Chem 269:14648–14654

Kirscht A, Kaptan S, Bienert G et al (2016) Crystal structure of an ammonia-permeable aquaporin. PLoS Biol 14:e1002411

Kukulski W, AD S, Johanson U et al (2005) The 5A structure of heterologously expressed plant aquaporin SoPIP2;1. J Mol Biol 350:611–616

Kuwahara M, Shinbo I, Sato K et al (1999) Transmembrane helix 5 is critical for the high water permeability of aquaporin. Biochemistry 38:16340–16346

Li Y, Wu Z, Ma N, Gao J (2009) Regulation of the rose Rh-PIP2;1 promoter by hormones and abiotic stresses in Arabidopsis. Plant Cell Rep 28:185–196

Lu M, Lee MD, Smith BL et al (1996) The human AQP4 gene: definition of the locus encoding two water channel polypeptides in brain. Proc Natl Acad Sci U S A 93:10908–10912

Luu DT, Martiniãre A, Sorieul M et al (2012) Fluorescence recovery after photobleaching reveals high cycling dynamics of plasma membrane aquaporins in Arabidopsis roots under salt stress. Plant J 69:894–905

Mahdieh M, Mostajeran A, Horie T, Katsuhara M (2008) Drought stress alters water relations and expression of PIP-type aquaporin genes in *Nicotiana tabacum* plants. Plant Cell Physiol 49:801–813

Marin-Olivier M, Chevalier T, Fobis-Loisy I et al (2000) Aquaporin PIP genes are not expressed in the stigma papillae in *Brassica oleracea*. Plant J 24:231–240

Mathai JC, Agre P (1999) Hourglass pore-forming domains restrict aquaporin-1 tetramer assembly. Biochemistry 38:923–928. doi:10.1021/bi9823683

Matsumoto T, Lian H-LL, Su W-AA et al (2009) Role of the aquaporin PIP1 subfamily in the chilling tolerance of rice. Plant Cell Physiol 50:216–229

Mintseris J, Weng Z (2005) Structure, function, and evolution of transient and obligate protein-protein interactions. Proc Natl Acad Sci U S A 102:10930–10935

Moshelion M, Becker D, Biela A et al (2002) Plasma membrane aquaporins in the motor cells of Samanea saman: diurnal and circadian regulation. Plant Cell 14:727–739

Murata K, Mitsuoka K, Hirai T et al (2000) Structural determinants of water permeation through aquaporin-1. Nature 407:599–605

Neely JD, Christensen BM, Nielsen S, Agre P (1999) Heterotetrameric composition of aquaporin-4 water channels. Biochemistry 38:11156–11163

Nooren IMA, Thornton JM (2003) Diversity of protein-protein interactions. EMBO J 22: 3486–3492

Otto B, Uehlein N, Sdorra S et al (2010) Aquaporin tetramer composition modifies the function of tobacco aquaporins. J Biol Chem 285:31253–31260

Palanivelu DV, Kozono DE, Engel A et al (2006) Co-axial association of recombinant eye lens aquaporin-0 observed in loosely packed 3D crystals. J Mol Biol 355:605–611

Prado K, Boursiac Y, Tournaire-Roux C et al (2013) Regulation of Arabidopsis leaf hydraulics involves light-dependent phosphorylation of aquaporins in veins. Plant Cell 25:1029–1039

Preston GM, Carroll TP, Guggino WB, Agre P (1992) Appearance of water channels in Xenopus oocytes expressing red cell CHIP28 protein. Science 256:385–387

Rash JE, Yasumura T, Hudson CS et al (1998) Direct immunogold labeling of aquaporin-4 in square arrays of astrocyte and ependymocyte plasma membranes in rat brain and spinal cord. Proc Natl Acad Sci U S A 95:11981–11986

Ren G, Cheng A, Reddy V et al (2000) Three-dimensional fold of the human AQP1 water channel determined at 4 A resolution by electron crystallography of two-dimensional crystals embedded in ice. J Mol Biol 301:369–387

Rossi A, Ratelade J, Papadopoulos MC et al (2012) Neuromyelitis optica IgG does not alter aquaporin-4 water permeability, plasma membrane M1/M23 isoform content, or supramolecular assembly. GLIA 60:2027–2039

Ruiz Carrillo D, To Yiu Ying J, Darwis D et al (2014) Crystallization and preliminary crystallographic analysis of human aquaporin 1 at a resolution of 3.28 Å. Acta Crystallogr Sect F Struct Biol Commun 70:1657–1663

Sakurai J, Ishikawa F, Yamaguchi T et al (2005) Identification of 33 rice aquaporin genes and analysis of their expression and function. Plant Cell Physiol 46:1568–1577

Sohara E, Rai T, Yang S-S et al (2006) Pathogenesis and treatment of autosomal-dominant nephrogenic diabetes insipidus caused by an aquaporin 2 mutation. Proc Natl Acad Sci 103:14217–14222

Sorbo JG, Moe SE, Ottersen OP, Holen T (2008) The molecular composition of square arrays. Biochemistry 47:2631–2637

Sorieul M, Santoni V, Maurel C, Luu D-TT (2011) Mechanisms and effects of retention of over-expressed aquaporin AtPIP2;1 in the endoplasmic reticulum. Traffic 12:473–482

Soto G, Alleva K, Amodeo G et al (2012) New insight into the evolution of aquaporins from flowering plants and vertebrates: orthologous identification and functional transfer is possible. Gene 503:165–176

Suga S, Maeshima M (2004) Water channel activity of radish plasma membrane aquaporins heterologously expressed in yeast and their modification by site-directed mutagenesis. Plant Cell Physiol 45:823–830

Sui H, Han BG, Lee JK et al (2001) Structural basis of water-specific transport through the AQP1 water channel. Nature 414:872–878

Törnroth-Horsefield S, Wang Y, Hedfalk K et al (2006) Structural mechanism of plant aquaporin gating. Nature 439:688–694

Tournaire-Roux C, Sutka M, Javot H et al (2003) Cytosolic pH regulates root water transport during anoxic stress through gating of aquaporins. Nature 425:393–397

Uehlein N, Lovisolo C, Siefritz F, Kaldenhoff R (2003) The tobacco aquaporin NtAQP1 is a membrane CO2 pore with physiological functions. Nature 425:734–737

Uehlein N, Otto B, Hanson DT et al (2008) Function of *Nicotiana tabacum* aquaporins as chloroplast gas pores challenges the concept of membrane CO2 permeability. Plant Cell 20: 648–657

Vandeleur RK, Mayo G, Shelden MC et al (2009) The role of plasma membrane intrinsic protein aquaporins in water transport through roots: diurnal and drought stress responses reveal different strategies between isohydric and anisohydric cultivars of grapevine. Plant Physiol 149:445–460

Verdoucq L, Grondin A, Maurel C (2008) Structure-function analysis of plant aquaporin AtPIP2;1 gating by divalent cations and protons. Biochem J 415:409–416

Walz T, Smith BL, Agre P, Engel A (1994) The three-dimensional structure of human erythrocyte aquaporin CHIP. EMBO J 13:2985–2993

Wang C, Hu H, Qin X et al (2016) Reconstitution of CO2 regulation of SLAC1 anion channel and function of CO2-permeable PIP2;1 aquaporin as carbonic anhydrase 4 interactor. Plant Cell 28:568–582

Weig A (1997) The major intrinsic protein family of Arabidopsis has 23 members that form three distinct groups with functional aquaporins in each group. Plant Physiol 114:1347–1357

Wolburg H, Wolburg-Buchholz K, Fallier-Becker P et al (2011) Structure and functions of aquaporin-4-based orthogonal arrays of particles. Int Rev Cell Mol Biol 287:1–41

Yamada S, Katsuhara M, Kelly WB et al (1995) A family of transcripts encoding water channel proteins: tissue-specific expression in the common ice plant. Plant Cell 7:1129–1142

Yaneff A, Sigaut L, Marquez M et al (2014) Heteromerization of PIP aquaporins affects their intrinsic permeability. Proc Natl Acad Sci 111:231–236

Yaneff A, Vitali V, Amodeo G (2015) PIP1 aquaporins: Intrinsic water channels or PIP2 aquaporin modulators? FEBS Lett 589(23):3508–3515

Zelazny E, Borst JW, Muylaert M et al (2007) FRET imaging in living maize cells reveals that plasma membrane aquaporins interact to regulate their subcellular localization. Proc Natl Acad Sci U S A 104:12359–12364

Zelazny E, Miecielica U, Borst JW et al (2009) An N-terminal diacidic motif is required for the trafficking of maize aquaporins ZmPIP2;4 and ZmPIP2;5 to the plasma membrane. Plant J 57:346–355

Zhang W, Fan L, Wu W-H (2007) Osmo-sensitive and stretch-activated calcium-permeable channels in *Vicia faba* guard cells are regulated by actin dynamics. Plant Physiol 143:1140–1151

Zhang M, Lu S, Li G et al (2010) Identification of a residue in helix 2 of rice plasma membrane intrinsic proteins (PIPs) that influences water permeability. J Biol Chem 285:41982–41992

Zhou Y, Setz N, Niemietz C et al (2007) Aquaporins and unloading of phloem-imported water in coats of developing bean seeds. Plant Cell Environ 30:1566–1577

Plant Aquaporin Trafficking

Junpei Takano, Akira Yoshinari, and Doan-Trung Luu

Abstract Aquaporins transport water and small neutral molecules across different membranes in plant cells and thus play important roles in cellular and whole plant physiology. The high diversity of intracellular localization of aquaporin isoforms is dependent on specific trafficking machineries. ER-to-Golgi trafficking of the plasma membrane intrinsic protein (PIP) isoforms has been shown to be dependent on DxE motifs in N-terminal cytosolic region, LxxxA motif in transmembrane domain 3, phosphorylation in C-terminal cytosolic region, and heteromerization. Stress-induced downregulation of the PIPs in the early secretory pathway was uncovered. Subsets of PIPs and Nodulin 26-like intrinsic proteins (NIPs) showed polar localization in the plasma membrane (PM) in certain cell types for directional transport of water and small neutral molecules such as boric acid and silicic acid. Latest techniques to study the mobility of PIPs revealed immobile nature in the plane of the PM and constitutive cycling between the PM and the endosomes. The roles of clathrin- and microdomain-dependent endocytosis for PIPs were uncovered. When challenged by stress conditions, some PIPs and TIPs showed quick relocalization probably to adjust water status. Vacuolar trafficking of different TIPs was shown to follow multiple routes dependent or independent of Golgi apparatus. These findings greatly advanced our understanding of the trafficking machineries of plant aquaporins, as significant models of plant membrane proteins.

J. Takano (✉) • A. Yoshinari
Graduate School of Life and Environmental Sciences, Osaka Prefecture University,
1-1 Gakuen-cho, Naka-ku, Sakai 599-8531, Japan
e-mail: jtakano@plant.osakafu-u.ac.jp

D.-T. Luu
Biochimie et Physiologie Moléculaire des Plantes, Institut de Biologie Intégrative des Plantes,
UMR 5004 CNRS/UMR 0386 INRA/Montpellier SupAgro/Université Montpellier 2,
34060 Montpellier Cedex 2, France

F. Chaumont, S.D. Tyerman (eds.), *Plant Aquaporins*, Signaling
and Communication in Plants, DOI 10.1007/978-3-319-49395-4_3

1 Introduction

Aquaporins are membrane channel proteins permeating water and/or small neutral molecules. The aquaporins share a basic structure comprising six transmembrane domains linked by five loops with cytosolic N- and C-termini and assembled as tetramers in which each monomer functions as a channel (Törnroth-Horsefield et al. 2006). In plants, aquaporins in different membrane compartments play different roles in various processes at cellular and whole plant levels. Therefore, how aquaporins are transported to their destination is a fundamental question.

Plant aquaporins are classified to seven subfamilies dependent on sequence similarities (see chapter "Structural Basis of the Permeation Function of Plant Aquaporins"). As indicated by the names, the plasma membrane intrinsic proteins (PIPs) and the tonoplast intrinsic proteins (TIPs) function in the plasma membrane (PM) and the vacuolar membrane, respectively. Many of them function in the transport of water, and some of them transport small uncharged molecules besides water. The prototype of Nodulin 26-like intrinsic proteins (NIPs) was identified in the peribacteroid membrane of nitrogen-fixing nodules of legume roots (Fortin et al. 1985), while NIPs were found in the PM and the endoplasmic reticulum (ER) in non-legume plants (see chapter "The Nodulin 26 Instrinsic Protein Subfamily"). Some members of NIPs have efficient transport activity of small uncharged solutes such as boric acid and silicic acid (Ma et al. 2006; Takano et al. 2006; see chapter "Plant Aquaporins and Metalloids"). The small intrinsic proteins (SIPs) were localized to the ER and shown to facilitate water transport (Ishikawa et al. 2005), although their physiological functions remain unknown. The X intrinsic proteins (XIPs) were identified in moss *Physcomitrella patens* and some dicots including poplar and Solanaceae plants tomato and tobacco (Danielson and Johanson 2008; Lopez et al. 2012; Bienert et al. 2011). Solanaceae XIPs were localized in the PM and shown to facilitate transport of various small neutral solutes (Bienert et al. 2011). In moss *P. patens*, additional two subfamilies GlpF-like intrinsic proteins (GIPs) and hybrid intrinsic protein (HIP) were found, although their subcellular localization remains unstudied (Danielson and Johanson 2008; Gustavsson et al. 2005).

As listed above, the members of each subfamily generally share similar patterns of subcellular localization. However, complex patterns of localization have been reported for some isoforms. For example, dual localization was observed for a PIP (*Nt*AQP1) in the PM and the chloroplast inner membrane in leaf mesophyll cells (Uehlein et al. 2008), for AtTIP3;1 and 3;2 in the tonoplast and the PM during seed maturation and germination (Gattolin et al. 2011), and for TIPs in the tonoplast and the symbiosome membrane in developing root nodule of *Medicago truncatula* (Gavrin et al. 2014). Interestingly, some PIPs and NIPs show polar localization in the PM, which is considered to be important for directional transport of water or small uncharged molecules in specific cell types. Some PIPs and TIPs also show dynamic changes of localization in response to environmental conditions. The differential localization between isoforms and the changes of localization are apparently important for plants to adapt to the changing environment. Considering the shared basic structure of aquaporins, the variable localization should be determined by signals embedded in amino acid sequences or specific conformations. The present chapter focuses on the mechanisms and physiological significances of intracellular trafficking of plant aquaporins.

1.1 PM Trafficking of PIPs

1.1.1 ER-to-Golgi Trafficking of PIPs Dependent on Trafficking Signals and Heteromerization

The membrane proteins synthesized by ribosomes are co-translationally inserted into the ER in which they are folded with the help of chaperons. Generally, the membrane protein destined to the PM, tonoplast, or other post-Golgi membrane compartments is transported in the vesicular trafficking network starting from ER to Golgi trafficking. The ER-to-Golgi trafficking is regulated by three complementary mechanisms: retention of immature proteins in the ER, selective packaging of mature proteins into COPII vesicle (ER exit), and retrieval from the Golgi apparatus of immature cargo proteins through COPI vesicles (Geva and Schuldiner 2014). The improperly folded proteins in the ER can be transported into the cytoplasm and degraded by the process known as ER-associated degradation (ERAD; Liu and Li 2014). In the step of ER exit, some cargoes are selectively packaged by direct or indirect binding to the Sec24 subunit of the COPII complex. Therefore, the binding sites of cargo proteins to Sec24 or to cargo receptors, which mediate the interaction to Sec24, function as signals for ER exit. The signals in animal and yeast systems range from diacidic DxE motifs (where x is an undetermined amino acid residue) to conformational epitopes and posttranslational modifications (Venditti et al. 2014; Geva and Schuldiner 2014).

In plant systems, PIP is one of the best-studied cargoes for the ER-to-Golgi trafficking. In ZmPIP2;4 and ZmPIP2;5, a diacidic DIE motif is present at residues 4–6 (Zelazny et al. 2009). In maize mesophyll protoplast, ZmPIP2;4 and PIP2;5 were targeted to the PM when fused to fluorescent proteins, while ZmPIP2;4AIA and ZmPIP2;5AIA were retained in ER (Zelazny et al. 2009). A fluorescence resonance energy transfer (FRET) analysis showed that the ZmPIP2;5AIA mutant still had the ability to form oligomers. These results indicated the importance of the DIE motif not for oligomerization but for ER exit (Fig. 1). Furthermore, in root epidermal cells of *Arabidopsis*, replacement of the DVE motif of AtPIP2;1 to AVE, DVA, AVA, EVE, or DVD resulted in ER retention (Sorieul et al. 2011). This result suggested the requirement of the strict DxE motif rather than just a diacidic motif. The DxE motifs were shown to be a direct binding site to Sec24 in yeast Golgi protein Sys1 (Votsmeier and Gallwitz 2001; Mossessova et al. 2003; Miller et al. 2003). In plant cells, the DxE motifs were found to be important in ER-to-Golgi trafficking of the potassium channel KAT1 (Mikosch et al. 2006) and Golgi-localized proteins GONST1 and CASP (Hanton et al. 2005). Importantly, KAT1 was shown to interact

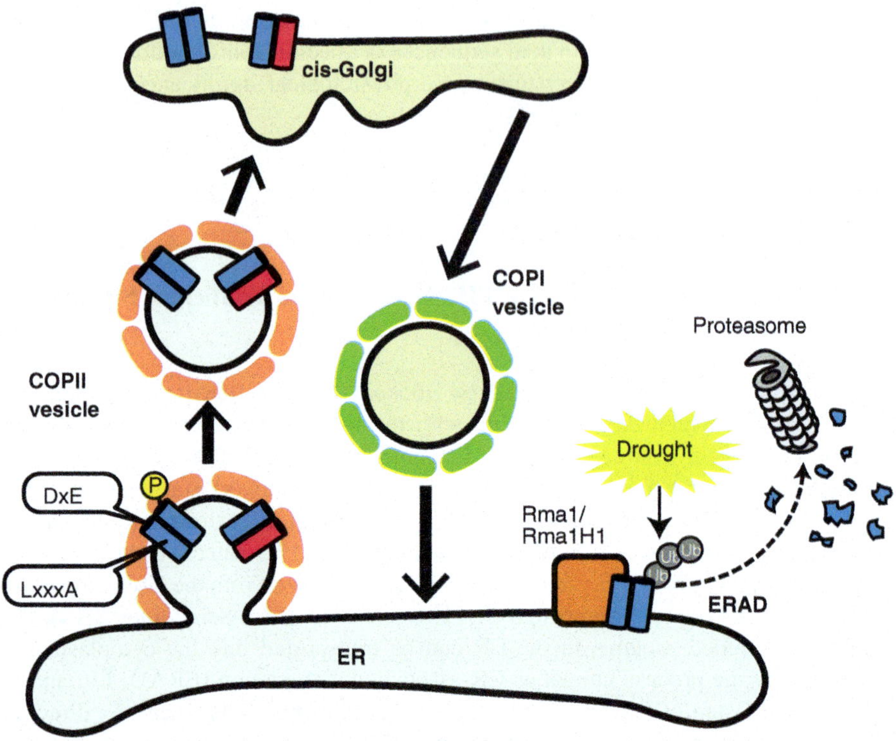

Fig. 1 A model of ER-Golgi trafficking of PIPs. The DxE motif in the N-terminal region, LxxxA motif in the transmembrane domain 3, and phosphorylation of a serine residue in the C-terminal region are implicated for the ER exit of PIP2 isoforms (*magenta cylinders*). PIP1s (*blue cylinders*) alone cannot be targeted to the PM by ER retention/retrieval mechanism, while hetero-oligomers with PIP2 isoforms are targeted to the PM via Golgi apparatus. Under drought stress, E3 ligases Rma1/Rma1H1 in the ER membrane ubiquitinate PIP2s for degradation via the ERAD pathway

with Sec24 by a FRET analysis (Sieben et al. 2008). Therefore, it is likely that some PIP2s are recruited into COPII vesicles via interaction of the DxE motif to Sec24.

The ER-to-Golgi trafficking of PIPs first attracted attention by the finding that the heteromerization of PIP1 and PIP2 isoforms allows ER-to-Golgi trafficking of PIP1s (see chapter "Heteromerization of Plant Aquaporins", Fig. 1). In *Xenopus laevis* oocytes, expression of the maize PIP2s but not PIP1s increased membrane water permeability (P_f) (Chaumont et al. 2000). In this system ZmPIP1;2-GFP was only slightly detected in the PM when expressed alone, while it was significantly increased by co-expression of ZmPIP2;5 (Fetter et al. 2004). ZmPIP1;2 and ZmPIP2;5 were shown to physically interact and synergistically increase the water permeability of the oocytes. The synergistic effect in the oocytes has been reported for PIP1s and PIP2s in various plant species (Chaumont and Tyerman 2014). Subsequently, analysis in maize mesophyll protoplasts showed that ZmPIP1 fusion proteins were retained in the ER when expressed alone, while relocalized to the PM when co-expressed with ZmPIP2s (Zelazny et al. 2007). The physical interactions of ZmPIP1 and ZmPIP2 were demonstrated by a FRET analysis in the protoplasts and co-immunoprecipitation in maize roots and suspension cells. These results suggested that heteromerization with PIP2s is required for the ER-to-Golgi trafficking of PIP1s. This view was confirmed by experiments in *Arabidopsis* roots using AtPIP2;1-GFP and its E6D (DVD) variant which is retained in ER (Sorieul et al. 2011). Co-expression of AtPIP2;1-GFP with AtPIP1;4-mCherry resulted in colocalization of these fusion proteins in the PM, while AtPIP2;1-GFP DVD variant resulted in ER retention of both fusion proteins. Importantly, *Arabidopsis* plants overexpressing the ER-retained variants of AtPIP2;1-GFP significantly reduced root hydraulic conductivity (Lp_r). The ER-retained AtPIP2;1-GFP probably interacted also with other endogenous PIPs and hampered their trafficking to the PM.

An important question underlying the ER-to-Golgi trafficking of PIP1s is whether PIP1s possess ER retention/retrieval signal or do not possess an ER exit signal. It was tested whether the DxE motif found in PIP2s was sufficient to confer ER-to-Golgi trafficking of ZmPIP1;2. The replacement of N-terminal cytosolic region of ZmPIP1;2 to that of ZmPIP2;5 that contains the DxE motif did not confer PM targeting (Zelazny et al. 2009). This result suggests either DxE alone is not sufficient for ER exit or the existence of an ER retention/retrieval signal in other regions of PIP1s.

To further examine the signal for ER-to-Golgi trafficking of PIPs, a protein domain swapping of ZmPIP1;2 and ZmPIP2;5 was conducted and demonstrated the importance of the transmembrane domain 3 (TM3, Chevalier et al. 2014). Furthermore, a LxxxA motif in the TM3 of ZmPIP2;5 was identified to be required for the PM targeting. In this study, chimeric proteins unable to target to the PM by substitution of TM3 (such as mYFP-ZmPIP2;5-$TM3_{PIP1;2}$) were localized in the ER and in punctate structures. The punctate structures overlapped with a Golgi marker and thus raised a question whether the TM3 is involved in the ER-to-Golgi and/or PM trafficking from the Golgi apparatus. The authors suggested that the partial Golgi localization might be a consequence of ER leakage dependent on the presence of the ER exit DxE motif in the N-terminal region of ZmPIP2;5. Indeed, fewer punctate structures were observed for mYFP-ZmPIP2;1-$TM3_{PIP1;2}$, which has no functional DxE motif. Alternatively, the punctate structures overlapping with a Golgi marker might be actually endoplasmic reticulum export sites (ERESs). The ERES and Golgi are often associated and thus are difficult to be distinguished by fluorescent microscopy (da Silva et al. 2004; Takagi et al. 2013). Taken together, it is most likely that the TM3 is required for ER-to-Golgi but not for later secretory trafficking (Fig. 1). The TM-based signal is possibly recognized by a transmembrane cargo receptor that binds to both cargo and Sec24 for packaging the cargo into the COPII vesicles (Dancourt and Barlowe 2010; Cosson et al. 2013; Chevalier and Chaumont 2014). A recent study identified rice cornichon (OsCNIH1), a homolog of cargo receptors Erv14p in yeast and cornichons in *Drosophila* and mammals, as a possible cargo receptor for the Golgi-localized sodium transporter OsHKT1;3 (Rosas-Santiago et al. 2015). In this study, AtPIP2;1 was used as a negative control for interaction of OsCHIH1 and OsHKT1;3 by bimolecular fluorescence complementation (BiFC) assay in tobacco leaves. There might be a specific cargo receptor recognizing the LxxxA motif in the TM3 of PIP2s.

It should be noted that replacement of both the N-terminal region and TM3 of ZmPIP1;2 to those of ZmPIP2;5 did not confer the PM targeting (Chevalier et al. 2014). One possible explanation is that ZmPIP1;2 possesses an ER retention or Golgi-to-ER retrieval signal in a region other than the N-terminus and TM3. The heteromerization of PIP1s with PIP2s might mask the retention/retrieval signal and promote ER-to-Golgi trafficking. For example, arginine-based retrieval signals were identified in cytosolic regions of ATP-sensitive K^+ channels, GABA receptors, and kainate receptors in mammals (Zerangue et al. 1999; Margeta-Mitrovic et al. 2000; Ren et al. 2003). ATP-sensitive K^+ channels have an octameric stoichiometry consisting of four pore-lining inward rectifier α-subunits and four regulatory β-subunits. The arginine-based retrieval signal of each subunit prevents the PM trafficking of unassembled channels, while the octamerization masks the signal and allows the PM trafficking (Zerangue et al. 1999). The arginine-based signal was shown to be recognized by the COPI vesicle coat at the Golgi apparatus for retrieval to the ER (Michelsen et al. 2007). Similar regulations by heteromerization, and with retrieval signals in transmembrane domain, were reported for the yeast iron transporter Ftr1 and Fet3 and mammalian NMDA receptors (Sato et al. 2004; Horak et al. 2008).

1.1.2 ER-to-Golgi Trafficking of *Arabidopsis* PIP2;1 Dependent on Phosphorylation

Another identified factor for ER-to-Golgi trafficking of PIPs is phosphorylation in the C-terminal region of AtPIP2;1 (Prak et al. 2008). A phosphoproteomic analysis identified two phosphorylation sites (S280 and S283) in the C-terminal region of AtPIP2;1. An S280A mutation did not affect the PM localization of AtPIP2;1; however, S283A mutation caused apparent ER retention. Consistently, a phosphomimic mutation S283D allowed PM localization. These results suggest that the phosphorylation of S283 is involved in ER-to-Golgi transport, although the mechanism is as yet unsolved (Fig. 1). The phosphorylation might increase affinity to a COPII subunit for packaging or mask a signal for ER retention/retrieval.

1.1.3 Downregulation of *Arabidopsis* PIP2s in Early Secretary Pathway

Another posttranslational modification, that of ubiquitination, was shown to control the amount of AtPIP2;1 trafficking to the PM (Lee et al. 2009). Rma1H1 from hot pepper and Rma1 from *Arabidopsis* are homologs of a human RING membrane-anchor 1 E3 ubiquitin ligase. Overexpression in *Arabidopsis* protoplasts of Rma1H1 and Rma1, which are both localized in the ER membrane, provoked retention of AtPIP2;1 in the same compartment and reduced its total amount. This reduction effect was inhibited by MG132, a proteasome inhibitor, and also by RNA interference approach targeting *Rma* homologs. Rma1H1 physically interacted with AtPIP2;1, and overexpression of Rma1H1 in transgenic *Arabidopsis* plants resulted in decreased total amount and increased ubiquitination of AtPIP2;1. Therefore, it is likely that Rma1H1 and Rma1 ubiquitinate PIPs to induce degradation by the ERAD process (Fig. 1). Various abiotic stresses, including dehydration, increased the mRNA level of *Rma1H1* in hot pepper plants, and its overexpression in *Arabidopsis* plants greatly enhanced tolerance to drought stress (Lee et al. 2009). Therefore, Rma1H1 and Rma1 are important factors controlling the level of PIPs at the ER in response to various stresses.

In addition, abiotic stress-induced multi-stress regulator was shown to reduce the accumulation of AtPIP2;7 in the PM (Hachez et al. 2014b). The regulator tryptophan-rich sensory protein/translocator (TSPO) is localized in the Golgi apparatus and is downregulated through an autophagic pathway (Vanhee et al. 2011). BiFC analysis showed interaction of TSPO with AtPIP2;7 in the ER and the Golgi apparatus. Their co-expression in transgenic plants or induction of expression of endogenous TSPO by ABA treatment decreased the level of AtPIP2;7 (Hachez et al. 2014b). These data suggest that the PM trafficking of AtPIP2;7 is downregulated by TSPO-mediated degradation via the autophagic pathway in response to abiotic stresses. In *Arabidopsis*, PIP transcripts and protein levels were generally downregulated upon drought stress (Alexandersson et al. 2005). The various regulations at ER and/or Golgi levels in addition to the mRNA level can fine-tune the abundance of PIPs in the PM.

1.1.4 PM Trafficking of PIPs Dependent on SNAREs and a Rab GTPase

Recent studies have uncovered the role of molecules involved in the trafficking to the PM of cargoes such as PIPs. For instance, regulation by SNAREs (soluble N-ethylmaleimide-sensitive factor protein attachment protein receptors) of the syntaxin family has been established (Besserer et al. 2012; Hachez et al. 2014a). The PM-localized SNARE isoform SYP121 has been shown to mediate the trafficking of vesicles between intracellular compartments and the cell surface (Geelen et al. 2002; Tyrrell et al. 2007). Interestingly, the overexpression of a dominant-negative cytosolic fragment of SYP121 (SYP121-Sp2) could impair the targeting to the PM of the K^+ channel KAT1, but not of the H^+-ATPase PMA2 (Sutter et al. 2006; Tyrrell et al. 2007). These results show that such a strategy of dominant-negative expression could be informative and indicate that SYP121 syntaxin exhibits a function for targeting to the PM of specific cargoes through vesicle fusion. Indeed, in maize mesophyll protoplasts, co-expression of an mYFP-ZmPIP2;5 construct and the SYP121-Sp2 impaired the targeting to the PM of the aquaporin isoform (Besserer et al. 2012). As a control, the full-length SYP121 did not exhibit any phenotype. Moreover, expression of Sp2 fragment of either the syntaxins in the PM SYP71 and SYP122 or the syntaxin in the prevacuolar compartment SYP21 had no effect on targeting of the ZmPIP2;5 construct, indicating the specific function of SYP121 on PIP targeting. Functional analysis of the role of SYP121 has been performed by using protoplast swelling assays. Here, protoplasts expressing the ZmPIP2;5 construct alone or co-expressed with the full-length SYP121 exhibited P_f values higher than protoplasts co-expressing ZmPIP2;5 construct with SYP121-Sp2 fragment or expressing only SYP121-Sp2 fragment or mock protoplasts. This series of experiments showed that SYP121 can regulate the function of an aquaporin by controlling the targeting. The authors noticed that although impaired, targeting of the ZmPIP2;5 construct to the PM was not negligible. However, they did not record any increase in P_f consistent with the presence of PIPs in the PM. This observation suggested that SYP121 could have a regulatory function on the intrinsic activity of PIPs. Indeed, the authors used *Xenopus* oocytes swelling assays and observed that when they co-expressed SYP121-Sp2 fragment and ZmPIP2;5 construct, there was reduced P_f values compared to oocytes expressing ZmPIP2;5 construct alone or co-expressed with the full-length SYP121. Importantly, when co-expressed in oocytes with full-length SYP121 or SYP121-Sp2 fragment, the ZmPIP2;5 construct was still localized in the PM, indicating that reduced P_f was related to an inhibition of ZmPIP2;5 intrinsic activity. To go further, affinity chromatography purification, BiFC, and FRET also indicated a direct physical interactions between ZmPIP2;5 and SYP121. This series of experiments shows that SYP121 not only impairs the targeting of PIPs but also physically interacts with them and inhibits their intrinsic activity (Besserer et al. 2012). Such a dual function of SYP121 in trafficking and activity regulation has been also demonstrated on AKT1/KC1 K^+ channel complex (Grefen et al. 2010).

Another SNARE, SYP61, mainly localized in the *trans*-Golgi network/early endosomal compartments (TGN/EE), cycles between this compartment and the PM and has been shown to colocalize with AtPIP2;7 (Hachez et al. 2014a). Physical interactions between SYP61 and AtPIP2;7 have been established. In the *syp61* mutant background, it was observed that a miss targeting of an overexpressed AtPIP2;7 construct to globular or lenticular structures corresponds to ER-derived stacked membrane arrays, suggesting a key function of this SNARE in targeting of PIPs to the PM. Importantly, SYP61 and SYP121 belong to the same complex, and it is believed that they might regulate PIP trafficking (Besserer et al. 2012; Hachez et al. 2014a, b).

As for SNAREs, there is evidence that any molecules with a role in the trafficking between the TGN/EE and the PM might control PIP targeting. For instance, RAS Genes From Rat BrainA1b (RabA1b) is a small GTPase localized in TGN/EE, and the corresponding mutant has been screened for defects in exocytosis (Feraru et al. 2012). Interestingly, the fungal toxin Brefeldin A (BFA) inhibits intracellular trafficking, mainly secretion, and causes accumulation of PM proteins into large aggregates known as BFA compartments or BFA bodies (Geldner et al. 2001). BFA is an inhibitor of a subset of guanine nucleotide exchange factors (GEFs) for the ADP ribosylation factor (ARF) GTPase, which function in recruitment of coat components of vesicles. Washout of BFA allows the gradual disappearance of PM proteins from the BFA bodies and their relocation to the PM. Interestingly, in *bex5*, a mutant of *RabA1b*, washout experiments did not completely suppress the labeling of aggregates by an AtPIP2;1 construct, suggesting a role of this small GTPase in the exocytosis of PIPs (Feraru et al. 2012).

1.2 Dynamic Properties of PIPs

1.2.1 Approaches Developed to Study the Lateral Diffusion of PIPs

Different microscopy approaches such as variable-angle epifluorescence microscopy (VAEM), also named total internal reflection fluorescence microscopy, in combination with fluorescence correlation spectroscopy (FCS) or super-resolution, provided recent quantitative insights into the immobility of AtPIPs in the plane of the PM and their endocytosis ((Hosy et al. 2015; Li et al. 2011), also see (Li et al. 2013) for a review of these techniques in plant cells). However, it has to be mentioned that fluorescence recovery after photobleaching (FRAP) technique was used in pioneer studies to address the mobility of membrane proteins such as AtNIP5;1 and AtPIP2;1 (Takano et al. 2010; Luu et al. 2012; Martinière et al. 2012).

VAEM in combination with single particle tracking (sptVAEM) allowed to track single particles of AtPIP2;1-GFP constructs expressed in epidermal cells of *Arabidopsis* roots at a high spatiotemporal resolution. In a pioneer study using an AtPIP2;1 construct, four types of trajectories and modes of diffusion were examined: Brownian (33.7 ± 3.3 %), directed (27.5 ± 2.4 %), restricted (17.5 ± 2.1 %), or mixed trajectories (21.2 ± 3.1 %) (Li et al. 2011). It was also concluded that the diffusion coefficient of AtPIP2;1-construct particles (2.46×10^{-3} $\mu m^2.s^{-1}$) was ten times lower than that for LTi6a (2.37×10^{-2} $\mu m^2.s^{-1}$), confirming the immobility or the extremely low lateral diffusion for the aquaporins in the PM. FCS, by measuring the intensity of the fluorescence and its variation in a volume, allowed an estimation of the density of the AtPIP2;1-GFP molecules in the PM of 30.3 ± 5.1 μm^{-2} in cells under resting conditions.

VAEM combined with high-density SPT and photoactivated localization microscopy (sptPALM) was used in epidermal cells of *Arabidopsis* roots expressing fusions of mEos2 with PM (AtPIP2;1, LTi6a) or tonoplast (AtTIP1;1) proteins (Hosy et al. 2015). This imaging method provides images with a spatial resolution beyond the diffraction limit, i.e., at the nanometer level, and thus is so-called super-resolution microscopy (Manley et al. 2008). It also provided the diffusion coefficient for each construct: 4.7×10^{-3} μm^2 s^{-1} (AtPIP2;1), 7.7×10^{-2} μm^2 s^{-1} (LTi6a), and 4.8×10^{-1} μm^2 s^{-1} (AtTIP1;1). This analysis revealed that AtPIP2;1 has a mobility seven- to 19-fold lower than that of its mammalian homolog AQP1 (Crane and Verkman 2008) and that LTi6a exhibited a mobility similar to lipids in the PM of plant cells (Dugas et al. 1989). Also, the values obtained for AtPIP2;1 and LTi6a are consistent with previous data by FRAP analysis (Luu et al. 2012) and stressed the immobile nature of the aquaporin. Considering the value of 2.5×10^{-2} μm^2 s^{-1} as a threshold between the immobile and mobile fractions, it was observed that only 22 % of AtPIP2;1 particles were mobile. The authors addressed the molecular mechanisms involved in the high confinement of AtPIP2;1. In an analysis with sptPALM in combination with pharmacological interference, they treated the cells with either oryzalin, latrunculin B, or cytochalasin D, used as an inhibitor of microtubule polymerization, an actin polymerization inhibitor, or an actin depolymerization inducer, respectively. Whereas treatments with oryzalin did not change the confinement of AtPIP2;1 particles, drugs targeting actin filaments provoked a significant increase in their mobility, suggesting that actin is one of the molecules involved in this confinement (Fig. 2a). Gradual plasmolysis of the cells allowing to progressively separate the PM and the cell wall provoked a gradual increase in the mobility of AtPIP2;1 particles. Since plasmolysis may also disrupt actin filaments, it is suggested that these molecules and the cell wall, by its close association with the PM, can immobilize aquaporin AtPIP2;1.

1.2.2 Approaches Developed to Study the Constitutive Cycling

PIPs as other PM cargoes are subjected to constitutive cycling. As canonical proteins of the PM, they have been used as reference markers in several studies which employed pharmacological interference on the constitutive cycling. For instance, the tyrosine analogue, Tyrphostin A23 (A23), is believed to prevent the interaction between the μ2 subunit of the clathrin-binding adaptor protein AP2 complex and cytosolic motifs of cargo PM proteins (Banbury et al. 2003). Indeed, A23 treatments of *Arabidopsis* root cells prevented the labeling of BFA bodies byAtPIP2;1-GFP construct, suggesting an inhibition of its endocytosis by a clathrin-dependent pathway (Dhonukshe et al. 2007). Consistent with this explanation, A23 treatments also increased the density of AtPIP2;1 constructs in the PM (Li et al. 2011). Next, the synthetic auxin analogue, 1-naphthaleneacetic acid (NAA), has also been shown to inhibit the endocytosis of several PM cargoes including AtPIP2;1 (Dhonukshe et al. 2008; Paciorek et al. 2005; Pan et al. 2009). Wortmannin (Wm), an inhibitor of phosphatidylinositol-3-phosphate and phosphatidylinositol-4-phosphate kinases, induces in *Arabidopsis* root cells clustering, fusion, and swelling of TGN vesicles and late endosomes/multivesicular bodies (LE/MVB) (Jaillais et al. 2006; Niemes et al. 2010; Takáč et al. 2012). The labeling of Wm-induced enlarged endosomes by AtPIP2;1 indicates that this aquaporin traffics between TGN and LE/MVB (Jaillais et al. 2006). The polyene antibiotic filipin that has also sterol fluorescence detection properties has been used to show a role for membrane sterols in the endocytosis of AtPIP2;1 (Kleine-Vehn et al. 2006).

As a complement of the pharmacological interference approach using A23, a dominant-negative mutant strategy was employed to prove the role of clathrin-dependent pathway on AtPIP2;1 endocytosis (Dhonukshe et al. 2007). The clathrin hub corresponding to the C-terminal part of clathrin heavy chain was overexpressed in plant cells, provoking the binding to and titering away the clathrin light chains, thus making them unavailable for clathrin cage formation. AtPIP2;1 was unable to label anymore BFA bodies, indicating a disruption of its endocytosis by the clathrin-dependent pathway.

FRAP and photoactivation of a photoactivatable version of GFP (*pa*GFP) were classical techniques for the analysis of the lateral diffusion of membrane proteins. Their use has been extended to study the cycling of PM aquaporins (Luu et al. 2012). The low lateral mobility of AtPIP2;1 in the plane of the PM compared to a variant retained into the ER membranes has been previously shown (Sorieul et al. 2011). FRAP experiments were performed using *Arabidopsis* root epidermal cells stably

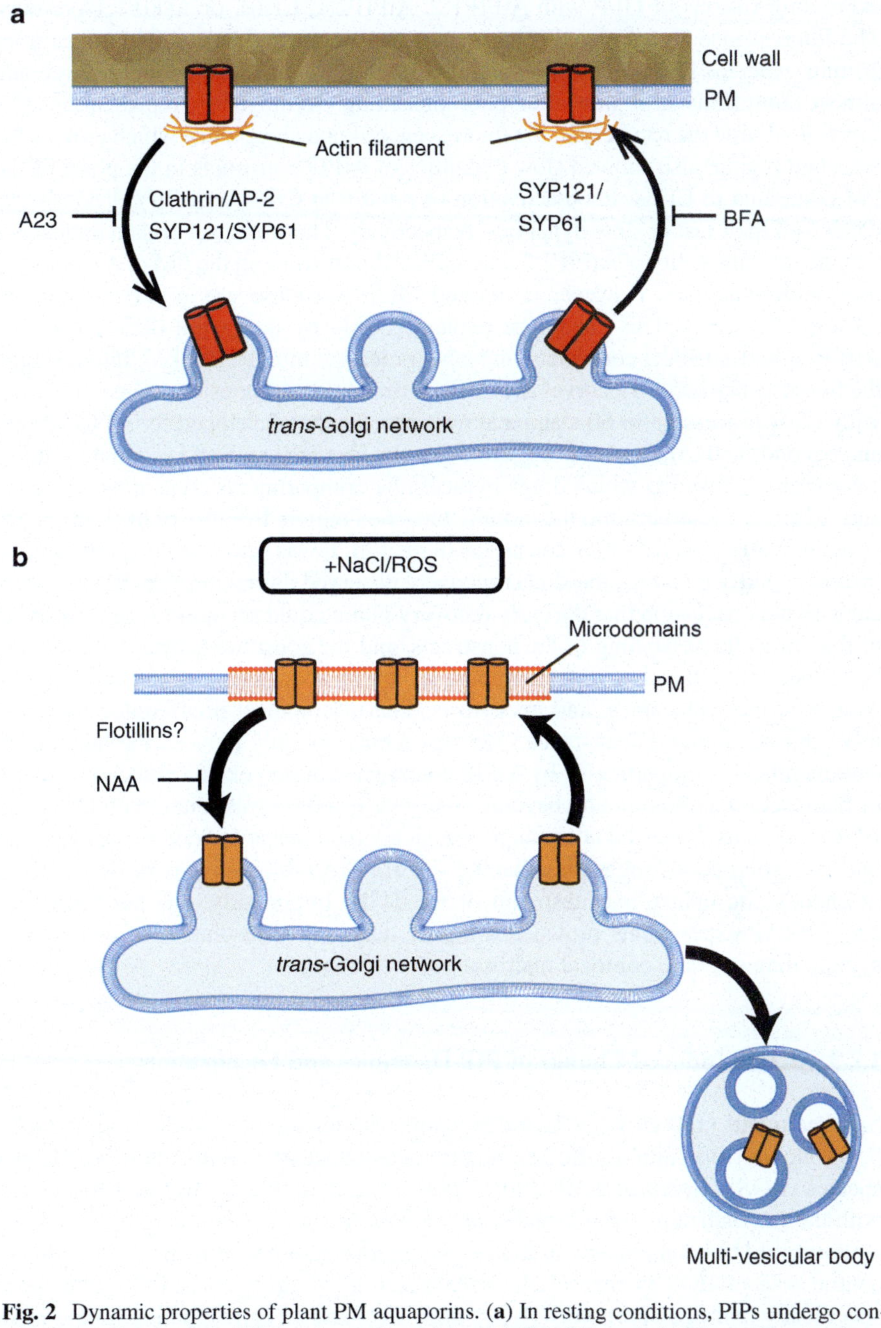

Fig. 2 Dynamic properties of plant PM aquaporins. (**a**) In resting conditions, PIPs undergo constitutive cycling from the TGN to the surface of the cell. SNAREs and clathrin-mediated endocytosis are involved. Their lateral mobility is restricted by a combination of actin filaments and cell wall interactions. (**b**) A marked change in PIP dynamics occurs in stimuli-challenged conditions, where enhanced cycling and higher mobility in the plane of the PM could be observed, concomitant to a trafficking to MVBs. Importantly, a clathrin-independent pathway overcomes the clathrin-mediated pathway, hypothetically involving flotillins. Chemicals used in pharmacological interference approach to block trafficking processes are indicated

expressing fusions of GFP with AtPIP1;2, AtPIP2;1, LTi6a, or AtTIP1;1 proteins. The fluorescence signal of AtPIP constructs had recovered below 60 % even after 30 min, whereas LTi6a construct exhibited a recovery of ~55 % at only 50 sec and almost completely at 7 min after photobleaching. Kymograms, representing as a function of time the recovery of the fluorescence signal along a line which crosses the bleached region, also showed slow dynamics of AtPIP constructs in the plane of the PM compared to LTi6a. Photoactivation of *pa*GFP in LTi6a or AtPIP2;1 constructs confirmed their fast or slow dynamics, respectively. The whole set of data indicates a significant immobility of AtPIP1;2 and AtPIP2;1 constructs in the PM of *Arabidopsis* root epidermal cells. However, a marked FRAP response was observed over the 30 min of records. This cannot be explained only by the lateral diffusion of the AtPIP1;2 and AtPIP2;1 constructs, and other processes must be invoked. Interestingly, the fluorescence recovery curves of AtPIP1;2 and AtPIP2;1 constructs were biphasic, with a fast process up to 60 s and a slower process that developed for up to 30 min and beyond. Thus, the authors suggested that the first process could account for a fast cytoplasmic streaming which drags intracellular compartments containing AtPIP1;2 and AtPIP2;1 constructs into the initially bleached region. Because of the immobility of the AtPIP1;2 and AtPIP2;1 constructs in the PM, it was suggested that constitutive cycling could account for the slower recovery observed during the next 30 min. They came to the conclusion that the early recovery phase could account for an estimation of the intracellular labeling of the constructs, and the constitutive cycling rate could be accounted by the slow recovery phase including sequential steps in endocytosis, sorting in the endosomes, and exocytosis. Pharmacological interference validated this tentative model. For instance, A23 treatments not only reduced the labeling of endomembrane compartments by AtPIP constructs but provoked a reduced recovery of fluorescence at 30 min and beyond. A similar response was found with NAA and BFA treatments. NAA did not change noticeably the intensity of endosomal labeling in these experiments, suggesting that the site of NAA inhibition may be downstream of endocytotic uptake and upstream of the TGN. Importantly, this new approach using FRAP experiments provided a higher quantitative resolution of constitutive cycling than standard confocal microscopy.

1.2.3 Stress-Induced Change of PIP Dynamics and Localization

Salt treatments (150 mM NaCl) or exogenous application of salicylic acid (0.5 mM SA) concomitantly inhibits the Lp_r and provokes an accumulation of reactive oxygen species (ROS) (Boursiac et al. 2008, 2005). These effects are fast and associated with the internalization of the analyzed AtPIP constructs. Importantly, the internalization could be counteracted by ROS-scavenging catalase treatments, indicating the central role of ROS in the cellular relocalization of aquaporins. ROS, and more specifically hydrogen peroxide (H_2O_2), have well-known cell-signaling functions in plants (Apel and Hirt 2004; Foyer and Noctor 2009). An in-depth cell biology analysis of PIP trafficking under ROS stimulus was described by combining several approaches in root cells of *Arabidopsis* (Wudick et al. 2015). Subcellular fractionation experiments revealed that H_2O_2 treatments induced an enhanced localization of AtPIPs in intracellular membranes, as early as 15 min. In parallel, a twofold increase in the lateral mobility of AtPIP2;1 particles and a 40 % decrease of their density in the PM were observed by sptVAEM and FCS, respectively. This indicates not only a change in the mobility of AtPIP2;1 at the cell surface but also a change in their subcellular localization. This change was clearly observed as labeling of intracellular spherical bodies of ~1–2 μm, dot-shaped-like structures, and a more diffuse labeling. Co-expression of AtPIP2;1 constructs with endomembrane markers or staining with the lipophilic styryl dye FM4-64 identified the LE/MVB as the potential compartments of relocalization. This result suggested the possibility of a stress-induced degradation process similar to well-described ones occurring for the borate transporter BOR1, the flagellin receptor FLS2, the iron transporter IRT1, or the auxin efflux facilitator AtPIN2 (Takano et al. 2005; Robatzek 2007; Barberon et al. 2011; Laxmi et al. 2008). Furthermore, a sequestration process could be possible as for the K^+-channel AtKAT1, which is reversibly sequestered upon ABA (Sutter et al. 2007). However, a series of experiments including biochemical and microscopy approaches did not support either the degradation or the reversible sequestration process, suggesting a stress-induced sequestration of PM aquaporins in LE/MVB that is strictly disconnected from vacuolar degradation. The labeling of spherical bodies observed in these stress-induced conditions may correspond to enlarged MVBs described in *Arabidopsis* mutants of SAND-CCZ1 complex involved in the trafficking toward lysosome/vacuole (Ebine et al. 2014; Singh et al. 2014).

Milder concentration of salt (100 mM NaCl) not only significantly reduces Lp_r of *Arabidopsis* plants but affects the distributions of PIPs and TIPs (Boursiac et al. 2005). The effects on PIP dynamics under these mild salt stress conditions were analyzed by using approaches to study the lateral diffusion of PIPs in the plane of the PM and their cycling in root epidermal cells of *Arabidopsis* (Li et al. 2011; Luu et al. 2012). sptVAEM revealed an increase by twofold of the diffusion coefficient of AtPIP2;1 particles and by 60 % proportion of particles with a restricted diffusion mode. Furthermore, the density of AtPIP2;1 construct in the PM was reduced by 46 %. This reduction could be prevented by pretreatments with A23 or methyl-β-cyclodextrin (MβCD), a sterol-disrupting reagent. Thus, these data support the hypothesis that under standard conditions, the endocytosis of PIPs occurs via predominantly the clathrin-dependent pathway, but it is enhanced under salt stress and occurs via both clathrin-mediated and membrane microdomain-mediated pathways (Li et al. 2011). A dissection of the cycling properties of AtPIP1;2 and AtPIP2;1 under salt stress was investigated using a FRAP approach. A twofold increase in the amplitude of fluorescence recovery curves was observed, suggesting an enhanced cycling of PIPs. To validate this hypothesis, the authors observed an earlier labeling of BFA bodies by AtPIP constructs under a concomitant treatment with NaCl and BFA. As a complementary experiment, an earlier decrease in the labeling of the BFA bodies was observed in the washout of BFA with NaCl solutions. A combination of FRAP and pharmacological interference approaches showed that PIP cycling under salt stress was blocked by NAA but had become insensitive to A23. The whole set of data suggests that salt stress enhances both endocytosis and exocytosis of PIPs, and therefore their cycling, and that an endocytotic mechanism independent of clathrin-mediated pathway could occur (Luu et al. 2012). This hypothesis supports the possible involvement of a membrane microdomain-dependent pathway previously uncovered in plant cells (Li et al. 2011).

As previously shown, posttranslational modifications on PIPs such as phosphorylation could alter ER-to-Golgi trafficking in resting conditions. Such a regulatory role has also been uncovered in salt stress conditions (Prak et al. 2008). Salt treatments for 2–4 h caused 30 % decrease in the abundance of phosphorylated S283 form AtPIP2;1. S280A mutation of GFP-AtPIP construct was associated with the labeling of "fuzzy" structures, whereas the S283D mutation allowed a labeling of spherical bodies. An adaptive response of plant root to limit the delivery of PIPs to the PM upon salt stress could be invoked.

1.3 Polar Localization of Aquaporins in Plant Cells

1.3.1 Polar Localization of Mammalian Aquaporins

Polar localization has been observed on several aquaporins in plant cells, although its physiological role in directional transport of substrates has not been shown. Before summarizing the knowledge on plant aquaporins, we introduce the preceding models of polar localization of mammalian aquaporins. In mammalian epithelial cells, three aquaporins are known to show polarity in their localization in the PM. In the renal collecting duct principal cells, AQP2 is targeted to the apical domain of the PM facing the ducts, while AQP3 and AQP4 are localized to basolateral domain of the PM (Edemir et al. 2011; Fig. 3a).

AQP2 is localized to intracellular vesicles, and vasopressin promotes translocation of AQP2 to the apical PM domain for efficient reabsorption of water from the ducts (Nedvetsky et al. 2007). It has been revealed that phosphorylation at S256 and S269 is crucial for the translocation (Hoffert et al. 2008; Van Balkom et al. 2002). Phosphorylation of S256 promotes translocation to the apical PM (Tamma et al. 2011), while phosphorylation at S269, which depends on the prior phosphorylation at S256, represses the clathrin-mediated endocytosis from the apical PM (Moeller

Fig. 3 Models of mechanisms underlying the polar localization of aquaporins in epithelial cells of mammalian renal collecting duct. (**a**) AQP2 is localized in the apical domain of the PM in the epithelial cell and cycled between the PM and Rab11-positive apical recycling endosomes. Rab5- and dynein-dependent transcytosis from basolateral to apical domain is required for the polar localization. AQP3 and AQP4 are distributed to basolateral domain of the PM. Tight junctions restrict the lateral diffusion of the AQPs. (**b**) In the PM, AQP2 is predominantly accumulated in the microdomains. Phosphorylation at S256 and S269 of AQP2 promotes its apical localization. Phosphorylation at S269 inhibits clathrin-mediated endocytosis and retains the localization of AQP2 in the apical PM domain. AQP2-bearing vesicles are transported along actin filaments and fused with the apical PM in VAMP-2 and syntaxin 4-dependent manner. AQP2-bearing vesicles are trafficked by Myosin-Vb motor protein along actin filaments

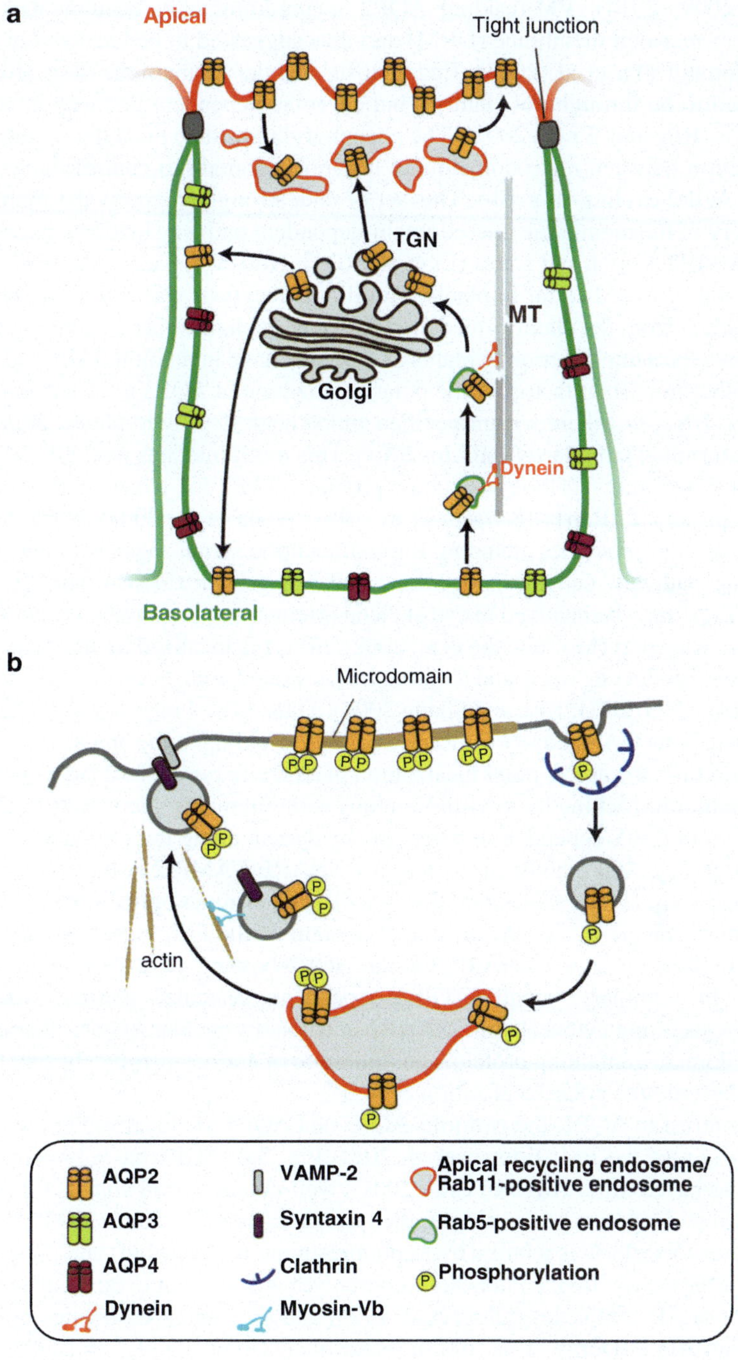
a
Apical
Tight junction
TGN
MT
Golgi
Dynein
Basolateral
b
Microdomain
actin
AQP2
AQP3
AQP4
Dynein
VAMP-2
Syntaxin 4
Clathrin
Myosin-Vb
Apical recycling endosome/
Rab11-positive endosome
Rab5-positive endosome
Phosphorylation

et al. 2009, 2014). PM-residing AQP2 is predominantly accumulated in the detergent-resistant membrane (DRM) and thus suggested to be localized in the PM microdomain (Yu et al. 2008). Endocytosis of microdomain-residing proteins is supposed to be through not clathrin- but caveolae-dependent pathway in mammalian cells (Roy and Wrana 2005). The phosphorylation state of AQP2 might alter its distribution between microdomain and non-microdomain to control the endocytic rate of AQP2 cycling (Fig. 3b). This set of data strongly supports the more recent discovery of the membrane microdomain-dependent pathway involved in endocytosis of AtPIP2;1 upon salt stress (Li et al. 2011). However, a caveolae homolog has not been discovered so far in plants. Flotillin endocytosis was also described to be independent from the clathrin-mediated pathway in plant cells and could hypothetically have the same function in plants as caveolae have in mammals (Li et al. 2012).

Another important factor for the polar localization of AQP2 is a PDZ-interacting motif (x-T/S-x-Φ, where x is unspecified amino acid; Φ is hydrophobic amino acid) in its C-terminal tail (Harris and Lim 2001). This motif interacts with SPA-1, a PDZ-domain containing GTPase-activating protein (GAP) for a small GTPase, Ras-related protein 1 (Rap1) (Kurachi et al. 1997; Noda et al. 2004). Rap1 regulates diverse cellular processes including Ras/extracellular signal-regulated kinase (ERK) signaling pathway, cell morphogenesis, and cell growth (Frische and Zwartkruis 2010). Rap1 may be involved in cytoskeletal orientation (Tsukamoto et al. 1999) and vesicular transport (Moskalenko et al. 2002). SPA-1 is localized to the apical surface of the renal collecting ducts and required for the polar localization of AQP2 as a scaffold (Noda et al. 2004; Kim and Sheng 2004). Thus, SPA-1 regulates AQP2 distribution through its GAP activity for Rap1 and/or scaffolding at the apical surface.

In addition, the apical polar localization of AQP2 requires Rab11-dependent apical vesicular trafficking by myosin Vb along actin filaments (Nedvetsky et al. 2007; Fig. 3b) and Rab5-dependent transcytosis by dynein along microtubules (Yui et al. 2012; Fig. 3a). AQP2 colocalizes with a v-SNARE VAMP-2 at the apical vesicles (Gouraud et al. 2002), suggesting that VAMP-2 determines apical sorting of AQP2-bearing vesicles (Fig. 3b). At the apical domain of the PM, syntaxin 4 was implicated to function as a t-SNARE for the AQP2-bearing vesicles (Gouraud et al. 2002). In summary, small GTPase-dependent vesicular sorting, directional transcytosis along cytoskeletons, SNARE-mediated vesicular fusion, interaction to a PDZ-domain containing protein, and inhibition of endocytosis by phosphorylation are underlying the polar localization of AQP2.

In contrast to AQP2, the isoforms AQP3 and AQP4 are localized to the basolateral domain of the PM (Edemir et al. 2011; Fig. 3a). AQP4 has a tyrosine- and a dileucine-based motif (Madrid et al. 2001), well-characterized signals for clathrin-mediated endocytosis and sorting to the lysosome/vacuole (Traub 2009). The tyrosine motif (YxxΦ; Φ is a bulky hydrophobic amino acid residue) and the dileucine motif ([DE]xxxL[LIM]; [ ] indicates alternative) were shown to be recognized by μ subunits of AP complexes (Ohno et al. 1995) and $\gamma/\sigma1$ and $\alpha/\sigma2$ hemicomplexes of AP-1 and AP-2 (Doray et al. 2007), respectively. Interestingly, an AQP4 variant, whose tyrosine motif was mutated, showed ectopic localization to the apical PM domain and reduced interaction with μ subunit of AP-2 (Madrid et al. 2001). This finding implies that the polar localization of AQP4 to the basolateral PM requires clathrin-mediated endocytosis from the apical side.

AQP3 carries a conserved YRLL motif in its N-terminal region. A variant of AQP3 whose tyrosine is substituted to alanine (ARLL) displayed less polarized localization than the wild-type isoform, while a variant with AAAA substitution showed intracellular localization, suggesting the tyrosine and leucine residues function distinctly (Rai et al. 2006).

Together, polar localization of AQP3 and AQP4 targeted to the basolateral domain requires their endocytic motifs. In contrast to the case of apical localization of AQP2, there is no indication of requirement of phosphorylation.

1.3.2 Polar Localization of Plant Aquaporins

To date, several plant aquaporins have been reported to show polar localization in the PM (Table 1). Immunocytochemistry of maize and rice PIP isoforms identified at least four PIPs polarly localized in the PM, while other PIPs show nonpolar

Table 1 Polar or nonpolar distribution of aquaporin proteins

Protein	Function	Localization	Organism	References
AtPIP1;1	Water transport	Uniform	*Arabidopsis thaliana*	Boursiac et al. 2005
AtPIP2;1	Water transport	Uniform	*Arabidopsis thaliana*	Boursiac et al. 2005
OsPIP2;1	Water transport	Stele side	*Oryza sativa*	Sakurai-Ishikawa et al. 2011
ZmPIP2;1/2;2	Water transport	Uniform/stele side	*Zea mays*	Hachez et al. 2008
ZmPIP2;5	Water transport	Uniform/soil side	*Zea mays*	Hachez et al. 2006
OsPIP2;5	Water transport	Stele side	*Oryza sativa*	Sakurai-Ishikawa et al. 2011
OsLsil	Si transport	Soil side	*Oryza sativa*	Ma et al. 2006
HvLsil	Si transport	Soil side	*Hordeum vulgare*	Chiba et al. 2009
ZmLsil	Si transport	Soil side	*Zea mays*	Mitani et al. 2009
CmLsil	Si transport	Soil side	*Cucurbita moschata*	Mitani et al. 2011
OsLsi6	Si transport	Soil side (root)	*Oryza sativa*	Yamaji et al. 2008; Yamaji and Ma 2009
		Xylem side (xylem parenchyma in leaf)		
		Xylem side (enlarged vascular bundles in node I)		
HvLsi6	Si transport	Soil side (root)	*Hordeum vulgare*	Yamaji et al. 2012
		Xylem side? (xylem parenchyma in leaf)		
		Xylem side (enlarged vascular bundles in node I)		
AtNIP5;1	B transport	Soil side	*Arabidopsis thaliana*	Takano et al. 2010

distribution. In rice root, OsPIP2;1 and OsPIP2;5 were localized in the stele-side domain of the PM (Sakurai-Ishikawa et al. 2011). In maize root, ZmPIP2;5 was localized to the soil-side (distal) domain of the PM in the epidermal cells (Fig. 4a), whereas it was localized in a nonpolar manner in other cell layers (Hachez et al. 2006). In contrast, ZmPIP2;1/2;2 was localized to the inner side (proximal) domain of the PM in leaf epidermal cells (Fig. 4b; Hachez et al. 2006, 2008). These findings suggested that distribution of PIPs in the PM is regulated in cell type-specific manners. Similar to the case of mammalian AQP2, proteomic analyses have identified PIP2 isoforms in DRM in *Arabidopsis thaliana* (Minami et al. 2009; Shahollari et al. 2004) and *Medicago truncatula* (Lefebvre et al. 2007). It is possible that localization to the microdomain is related to the polarized distribution of ZmPIP2;1/2;2/2;5. So far physiological significance and mechanisms of polar localization of aquaporins remain uninvestigated.

Besides PIPs, some members of NIPs were shown to be localized in the PM in a polar manner. In exodermal and endodermal cells of rice roots, OsLsi1 (NIP2;1), a silicic acid channel, and OsLsi2, an anion exporter, are localized in the soil-side and the stele-side domains of the PM, respectively, and cooperatively play a role in translocation of silicon (Si) into the xylem (Ma et al. 2006, 2007; Ma and Yamaji 2015). Si is a beneficial element for plants and highly accumulated in monocotyledonous plants including rice, maize, barley, and wheat (Ma and Yamaji 2006). Si enhances resistance against abiotic and biotic stresses, and Si deficiency causes agricultural problems such as reduced productivity (Ma and Yamaji 2006). Recently, the physiological importance of the polar localization of OsLsi1 and OsLsi2 in rice root was examined in silico by a mathematical model of root cell layers with parameters such as cellular and subcellular distribution of Lsi1 and Lsi2 and location of the Casparian strips (Sakurai et al. 2015). The model suggested that wild-type pattern of polar localization of Lsi1 and Lsi2 and exodermal and endodermal Casparian strips are all required for efficient Si transport in rice (Sakurai et al. 2015).

Lsi6, a homolog of Lsi1, functions as a silicic acid channel in rice, barley, and maize. Interestingly, patterns of expression and subcellular localization of Lsi6 are distinct between tissues. In rice roots, OsLsi6 (OsNIP2;2) is expressed in the epidermis and cortex and localized to the soil-side domain of the PM. However, in leaves and nodes, OsLsi6 is expressed in xylem parenchyma and xylem-transfer cells, respectively, and localized in the polar PM domains toward the side of xylem, for efficient xylem unloading (Yamaji et al. 2008; Yamaji and Ma 2009). Similarly, in the barley node I, HvLsi6 is expressed in the xylem-transfer cells and localized to the PM domain facing xylem of enlarged vascular bundle (X_E, Fig. 5a, b; Yamaji et al. 2012). The other isoform HvLsi2 is expressed in the xylem parenchyma cells adjacent to the xylem-transfer cells and localized to the PM domain at the side of diffuse vascular bundle (DV, Fig. 5c, d). HvLsi6 and HvLsi2 play important roles for reloading Si to the xylem of diffuse vascular bundle from the xylem of enlarged vascular bundle (Yamaji et al. 2012; Ma and Yamaji 2015).

Rice exodermis and endodermis have Casparian strips, which restrict not only apoplastic flow of water and solutes but also lateral diffusion of PM-residing proteins like the case of tight junctions in mammalian epithelial cells (Geldner 2013).

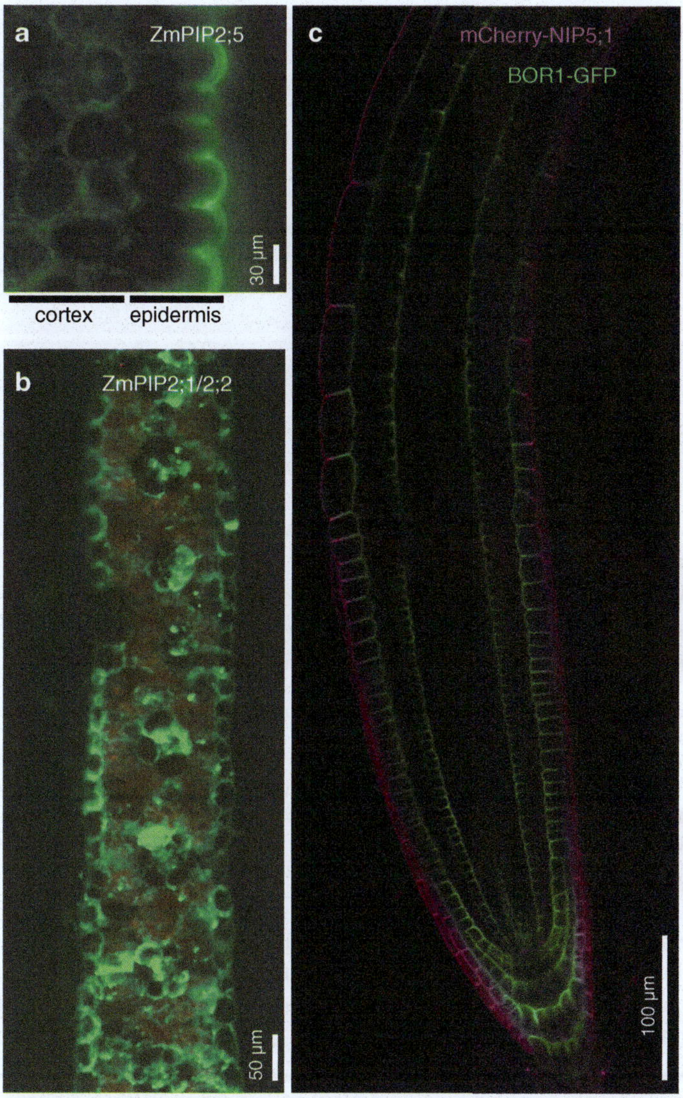

Fig. 4 Polar localization of ZmPIP2;5 in roots, ZmPIP2;1/2;2 in leaves, and AtNIP5;1 in roots. (**a**) Immunofluorescence of ZmPIP2;5 in a maize root. ZmPIP2;5 shows polar localization in the soil-side domain of the PM in the epidermal cells. (**b**) Immunofluorescence of ZmPIP2;1/2;2 in a maize leaf blade. ZmPIP2;1/2;2 shows polar localization in proximal domain of the PM in the epidermal cells. Scale bar = 30 µm (**a**), 50 µm (**b**). Figures are reproduced from Hachez et al. 2006 (**a**) and Hachez et al. 2008 (**b**) with permission. (**c**) A confocal image of a primary root of the transgenic plant carrying *ProAtNIP5;1-mCherry-AtNIP5;1genomic* and *ProAtBOR1,AtBOR1-GFP* grown on low-B media. AtNIP5;1 is expressed which shows polar localization in soil-side domain of the PM in the outermost cell layers, whereas AtBOR1 shows polar localization in stele-side domain in various cells in the root tip (Takano et al. 2010). Scale bar = 100 µm

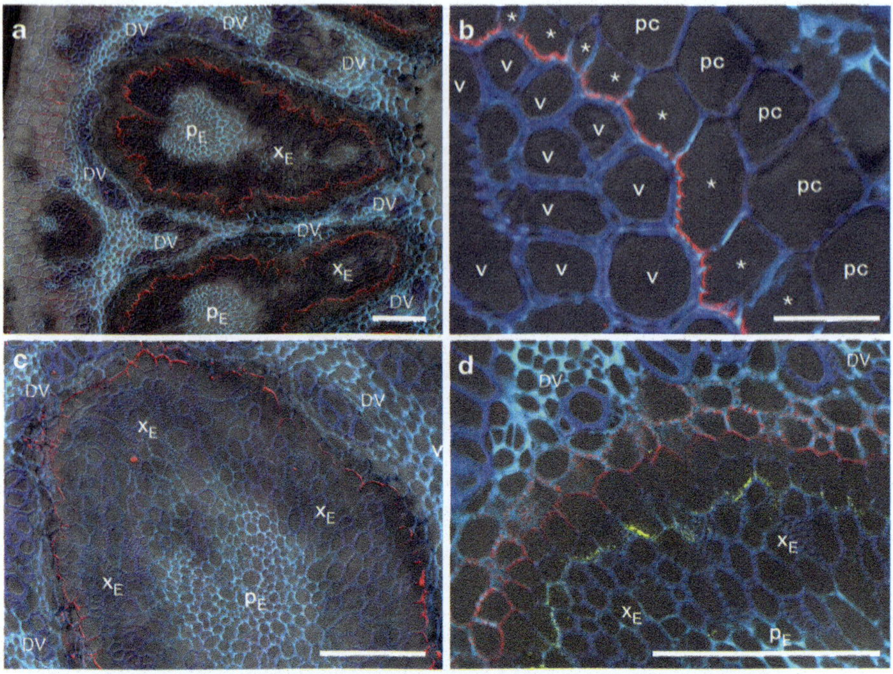

Fig. 5 Localization of HvLsi6 and HvLsi2 in the node I. Immunofluorescence in node I at the flowering stage with anti-HvLsi6 antibody (*red* color in **a**, **b**), anti-HvLsi2 antibody (*red* color in **c**), or double staining with anti-HvLsi6 (*yellow*) and anti-HvLsi2 (*red*) antibodies (**d**). *Blue* and *cyan* colors are UV autofluorescence from the cell wall. x_E, p_E: xylem, and phloem region of enlarged vascular bundle, respectively, *DV* diffuse vascular bundle, *v* xylem vessel, * xylem-transfer cell, *pc* outer parenchyma cell layer next to xylem-transfer cell. Scale bars = 100 μm (**a**, **c**, **d**), 20 μm (**b**) (Reproduced from Yamaji et al. 2012 with permission)

Therefore, one can imagine that the polar localization of OsLsi1 in these cell types is dependent on the presence of the Casparian strip. However, Lsi1 and Lsi6 also showed polar localization in cell types without the Casparian strips (Chiba et al. 2009; Mitani et al. 2009; Yamaji et al. 2012), indicating that the Casparian strip is not necessary for polar localization.

Another example of polar-localized plant aquaporin is NIP5;1, a boric acid channel in *Arabidopsis* (Takano et al. 2006, 2010; see chapter "Plant Aquaporins and Metalloids"). Boron (B) is an essential micronutrient, while excess B is toxic for plants (Marschner 1995). Imaging of GFP fusion revealed that NIP5;1 is localized to the soil-side domain of the PM in outermost cell layers of roots, while a borate efflux transporter BOR1 is localized to the stele-side domain in various cell types, under low-B conditions (Takano et al. 2002, 2010) (Fig. 4c). The opposite polar localization of NIP5;1 and BOR1 is considered to be beneficial for B uptake into root and subsequent translocation to the vascular cylinder under low-B conditions.

Currently, molecular mechanisms underlying polar localization of a subset of PIPs and NIPs are not well understood. As discussed above, a membrane diffusion barrier such as tight junctions in animal epithelial cells is not necessary for maintaining the polar localization in the plant PM. Limited lateral diffusion of the PM-localized proteins (discussed above) is apparently a basis for the maintenance of the polar localization by vesicle trafficking. Although no canonical endocytic motifs have been identified within their cytosolic sequences, other endocytic signals or posttranslational modification such as phosphorylation might regulate the polar localization of plant aquaporins as it does for some mammalian AQP proteins.

1.4 Trafficking of TIPs

1.4.1 Localization of TIPs to the Lytic Vacuole and the Protein Storage Vacuole

TIPs have functions in transport of diverse solutes including water, glycerol, urea, ammonia, or hydrogen peroxide (H_2O_2) across the vacuolar membrane (tonoplast) (Wudick et al. 2009). TIPs are further classified into five groups. Among them, γ-TIP (renamed AtTIP1;1 in *Arabidopsis*) has been used as a marker for the lytic vacuole, and α- and δ-TIP (renamed AtTIP3;1 and AtTIP2;1 in *Arabidopsis*) have been used as markers for the protein storage vacuole (PSV) (Frigerio et al. 2008; Xiang et al. 2013). The lytic vacuole has acidic pH, occurs in mature tissues, and functions similarly to the lysosome in animals. The PSV has less acidic pH and functions as the main protein storage compartment in the storage tissues. There remains a contradiction whether the different TIPs target specifically toward the lytic vacuoles or the PSVs. In *Arabidopsis* mesophyll protoplasts, AtTIP2;1-GFP was colocalized with the isoform *b* of a rice two-pore K^+ channel construct in small PSVs, whereas the construct with the isoform *a* is localized in the tonoplast of the central vacuole (Isayenkov et al. 2011). However, in transgenic *Arabidopsis* plants, different TIPs (AtTIP1;1, AtTIP2;1, and AtTIP3;1) fused to fluorescent proteins coexisted in the same tonoplast of the central vacuole in roots and leaves (Hunter et al. 2007; Gattolin et al. 2009) and in the same tonoplast of PSVs in embryos (Hunter et al. 2007). The differential localization of TIP isoforms to the lytic vacuoles or the PSVs seems to be dependent on tissues and developmental stages.

Interestingly, specifically in early seed maturation and early germination stages, AtTIP3;1 and AtTIP3;2 were shown to localize to the tonoplast of PSVs and also to the PM (Gattolin et al. 2011). This is apparently not an artifact since both N- and C-terminal fluorescent protein fusions of AtTIP3;1 and AtTIP3;2, but not YFP-AtTIP1;1 expressed under the control of the At*TIP3;2* promoter, showed the dual localization. Although the mechanisms of dual targeting are unknown, this finding suggests the distinct trafficking mechanisms for different TIPs and possible importance of the PM localization of TIPs to compensate the absence or low expression of PIPs in these developmental stages.

1.4.2 Trafficking of TIPs in Golgi-Dependent and Golgi-Independent Pathways

The tonoplast targeting of membrane proteins is apparently more complex than the PM targeting via the ER, Golgi apparatus, and TGN. There exist at least two pathways to the tonoplast: Golgi-dependent and Golgi-independent pathways (Pedrazzini et al. 2013; Rojas-Pierce 2013; De Marchis et al. 2013; Viotti 2014). These pathways have been distinguished by sensitivities to the fungal toxin BFA. BFA sensitivity in a cell type is dependent on the sensitivity/insensitivity of ARF-GEF isoforms responsible for each trafficking step (Robinson et al. 2008). For example, in leaf epidermis of tobacco, hypocotyls, and leaves but not roots of *Arabidopsis*, BFA inhibits the retrograde traffic from the Golgi to the ER and induces uncontrolled fusion of Golgi and ER membranes (Brandizzi et al. 2002; Robinson et al. 2008; Rivera-Serrano et al. 2012). In these cells, BFA inhibits the trafficking of soluble cargoes to the vacuole or the apoplast, and some, but not all, tonoplast proteins are investigated (Pedrazzini et al. 2013; Rojas-Pierce 2013; De Marchis et al. 2013). In an early study, using protoplasts from transgenic tobacco leaf cells, the vacuolar traffic of the soluble protein phytohemagglutinin, but not the tonoplast TIP3;1, was inhibited by BFA treatment (Gomez and Chrispeels 1993). In protoplasts of *Arabidopsis*, the Golgi-independent trafficking of AtTIP3;1 was further supported by the finding that Golgi-specific glycosylation was not detected in a modified TIP3;1 variant that contained a potential glycosylation residue (Park et al. 2004). A recent study showed that the tonoplast targeting of AtTIP1;1-YFP but not AtTIP3;1-YFP and GFP-AtTIP2;1 was sensitive to BFA in *Arabidopsis* hypocotyls (Rivera-Serrano et al. 2012). Furthermore, a screen of chemical inhibitors of TIP trafficking in the same study identified a compound, C834, which retained GFP-AtTIP2;1 and AtTIP3;1-YFP in the ER, but did not affect the tonoplast targeting of AtTIP1;1-YFP. Therefore, BFA and C834 treatments are means to differentiate the Golgi-dependent and Golgi-independent pathways as distinct mechanisms (Fig. 6).

1.4.3 Golgi-Dependent Trafficking to the Tonoplast

Although little is known for the Golgi-dependent trafficking of TIPs, studies on various tonoplast proteins revealed multiple pathways to the tonoplast through the Golgi apparatus (Fig. 6). Studies on some tonoplast proteins in plants identified the requirement of dileucine-based motifs and a tyrosine-based motif located in the cytosolic region for tonoplast trafficking (Pedrazzini et al. 2013). As introduced above, dileucine-based motifs (typically [D/E]xxxL[L/I]) and the tyrosine-based signal YxxΦ, where Φ is any bulky hydrophobic residue, are well-established signals for endocytosis and sorting to lysosomes or vacuoles by binding to AP complexes in mammals and yeast (Bonifacino and Traub 2003). Intriguingly, mutations in the dileucine motifs (not limited to the typical ones) of tonoplast proteins, such as a monosaccharide transporter ESL1, a molybdate transporter MoT2, an inositol

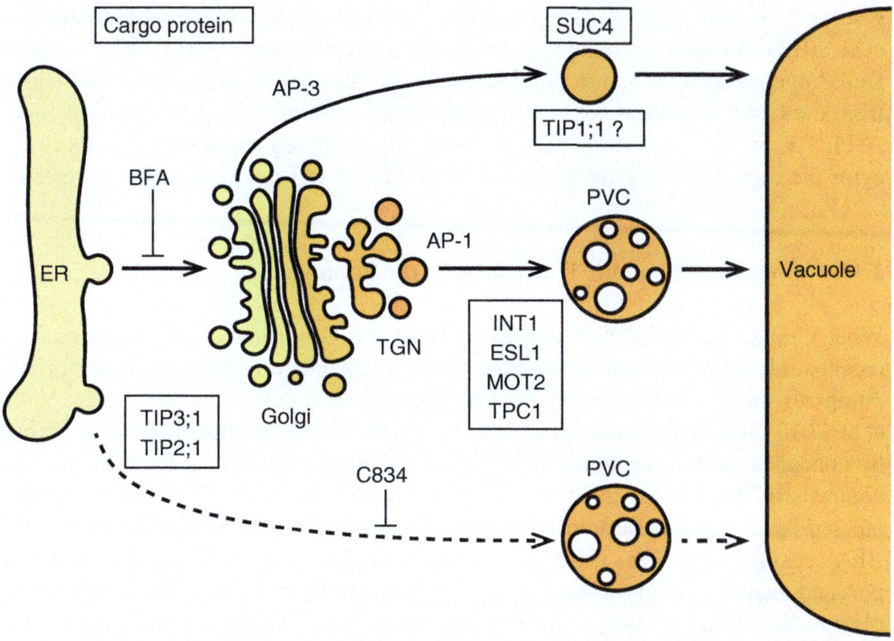

Fig. 6 A model of multiple trafficking pathways to the tonoplast. The trafficking pathway of AtTIP1;1 but not of AtTIP2;1 and AtTIP3;1 is through Golgi. These routes can be differentiated by sensitivities to BFA or C834. The BFA-sensitive, Golgi-dependent pathway is further divided to two pathways, dependent on AP-1 and AP-3 complexes. TIP1;1 does not contain dileucine motifs, which are recognized by the AP-1 complex and found in INT1, PTR2/4/6, ESL1, MOT2, TPC1, and VIT1

transporter INT1, a two-pore channel TPC1, and an iron transporter VIT1, redirected them to the PM (Yamada et al. 2010; Gasber et al. 2011; Wolfenstetter et al. 2012; Larisch et al. 2012; Wang et al. 2014). Furthermore, short amino acid fragments containing the dileucine motif of INT1, PTR2/4/6, or VIT1 were sufficient to redirect PM-targeted proteins to the tonoplast (Wolfenstetter et al. 2012; Komarova et al. 2012; Wang et al. 2014). These results suggest that the tonoplast-targeting mechanism dependent on the dileucine motif is dominant over the PM-targeting mechanism, at least for a subset of membrane proteins. Importantly, the dileucine motif from VIT1 was shown to interact with AP-1 subunits in vitro, and the knock-down of the AP-1 γ-adaptins caused relocation of VIT1 and INT1 to the PM (Wang et al. 2014). Since AP-1 γ-adaptins are mainly localized in the TGN, it was proposed that its complex mediates the targeting of membrane proteins carrying a dileucine motif from the TGN to the tonoplast (Fig. 6).

To add further complexity, there exists another Golgi-dependent vacuolar-sorting pathway, which is dependent on AP-3. The AP-3 complex in *Arabidopsis* is important for biogenesis of the vacuole (Feraru et al. 2010; Zwiewka et al. 2011). The tonoplast trafficking of SUC4 but not of INT1, TPC1, and ESL1 was affected in the AP-3 β-subunit mutant (Larisch et al. 2012; Wolfenstetter et al. 2012). In the AP-3 β-subunit mutant, GFP-SUC4 colocalized with *cis*-Golgi markers (Wolfenstetter et al. 2012). By analogy to the apparent function of yeast AP-3 (Odorizzi et al. 1998; Dell'Angelica 2009), it was proposed that AP-3 is involved in direct trafficking from the Golgi apparatus to the tonoplast (Wolfenstetter et al. 2012; Fig. 6). Since AtTIP1;1, as well as SUC4, does not contain conserved trafficking signals in the cytosolic region, involvement of AP-3 for AtTIP1;1 trafficking needs to be tested.

1.4.4 Golgi-Independent Trafficking to the Tonoplast

What kind of mechanism is conceivable for the Golgi-independent pathway? A recent study carefully investigated the Golgi-independent trafficking of two proton pumps to the lytic vacuoles in the root meristematic cells of *Arabidopsis* (Viotti et al. 2013). The localization of a VHA-a3-mRFP to the tonoplast was not disturbed by concanamycin A treatment, which causes aggregates derived from TGN. This contrasts to the case of INT1 fused to GFP, which was trapped in the aggregates labeled by the endocytic tracer FM4-64. VHA-a3-mRFP was introduced into the BFA-sensitive GNL1/*gnl 1* line (Richter et al. 2007) to block the ER-to-Golgi transport and the post-Golgi transport by BFA treatment. Upon BFA treatment, the tonoplast localization of VHA-a3-mRFA was not affected. These results supported the hypothesis that VHA-a3 trafficking to the tonoplast is Golgi independent. Furthermore, the immunoelectron microscopy analysis of the VHA-a3-GFP and V-PPase AVP1/VHP1 detected both proteins at the tonoplast and double-membrane structures designated as provacuoles, but not in the TGN/EE, Golgi stacks, or ER (Viotti et al. 2013). The formation of the provacuoles was not prevented by complete abolition of Golgi and post-Golgi trafficking in the *gnl* mutant with BFA treatment. Importantly, although at rare incidences, a direct connection between the provacuole and the ER was observed. These results suggested that the biogenesis of the lytic vacuole and trafficking of a subset of tonoplast proteins occur directly from the ER independent of Golgi function. It is tempting to speculate that the tonoplast trafficking of AtTIP3;1 and AtTIP2;1, which is insensitive to BFA, occurs in the similar direct pathway. This hypothesis needs to be tested in the tissue containing PSVs, where native AtTIP3;1 and AtTIP2;1 are expressed. The precursor-accumulating (PAC) vesicles carrying precursors of storage proteins directly from the ER to the PSVs in pumpkin cotyledons (Hara-Nishimura et al. 1998, 2004) or the KDEL vesicles carrying a proform of a KDEL-tailed cysteine protease directly from the ER to the PSVs (Toyooka et al. 2000) might be involved.

1.4.5 Stress-Induced Changes of TIP Localization

The response of aquaporin trafficking to stress conditions was first described for McTIP1;2 in suspension cells and leaves of ice plant (*Mesembryanthemum crystallinum;* Vera-Estrella et al. 2004). Membrane fractionation followed by Western blotting by an anti-McTIP1;2 antibody revealed the changes of localization after mannitol-induced osmotic stress. An immunofluorescence analysis showed localization of McTIP1;2 in the tonoplast under a control condition and its change to unique intracellular spherical structures after the mannitol treatment. The spherical structure labeled by McTIP1;2 was not labeled with tonoplast markers V-PPase, V-ATPase, and AtTIP2;1, PM markers, and LE/MVB markers. The change was accompanied with increased amount of the protein and glycosylation and blocked by the glycosylation inhibitor tunicamycin. These results suggested that osmotic stress induces glycosylation of McTIP1;2 to change the trafficking pathway to spherical structures, possibly to maintain water balance.

Another example of the changes of TIP localization was in *Arabidopsis* root cells after salt treatment (Boursiac et al. 2005). Under control conditions, both isoforms AtTIP1;1-GFP and AtTIP2;1-GFP labeled the tonoplast of the central vacuole, while after the salt treatments, AtTIP1;1-GFP, but not AtTIP2;1-GFP, appeared in intravacuolar spherical structures tentatively identified as tonoplast invaginations. The spherical intravacuolar structures need to be interpreted with caution since this kind of structures can be artificially induced by dimerization of GFP fused with tonoplast proteins when they are abundantly expressed (Segami et al. 2014). However, the spherical structures labeled by AtTIP1;1-GFP were observed only after salt stress (Boursiac et al. 2005), suggesting that at least a change in localization of AtTIP1;1 or vacuole structure occurred. Since AtTIP1 abundance measured by anti-TIP1 antibody was decreased by approximately 20 % within 6 h after salt treatment (Boursiac et al. 2005), this labeling might reflect protein degradation. Alternatively, it might be accompanied by a reshaping of vacuole structure under salt stress.

2 Aquaporins as Markers for Membrane Dynamics

PIPs and TIPs have been used as PM and tonoplast markers for various studies on organelle and membrane dynamics in plant cells. TIPs were also used as protein markers for identification of new small molecule inhibitors for membrane trafficking (Rivera-Serrano et al. 2012). Recently, Arabidopsis AtNIP5;1, a boric acid channel, was utilized as a PM marker for a forward genetics study. Using chemically mutagenized GFP-AtNIP5;1 lines, a fluorescence imaging-based screening isolated mutants in which AtNIP5;1 localized abnormally in root epidermal cells (Uehara et al. 2014). An allelic group of mutants contained intracellular aggregates co-labeled with GFP-AtNIP5;1, endocytic tracer FM4-64, and TGN markers. A positional cloning identified the responsible gene as *UDP-glucose 4-epimerase*, which is important for synthesis and channeling of D-galactose into the Golgi apparatus (Uehara et al. 2014; Seifert et al. 2002). Further analysis of ultrastructure in the *uge4* mutant cells revealed accumulation of high-electron-density vesicles derived from TGN (Wang et al. 2015). It is most likely that accumulation of galactose-depleted xyloglucan caused abnormal function and structure of TGN and caused a "traffic jam" in the membrane traffic system (Kong et al. 2015). The same screen identified an allelic group of mutants in which GFP-AtNIP5;1 is accumulated in the ER in addition to the PM (unpublished). Analysis of the mutant has a potential to identify a novel mechanism of ER exit specific to AtNIP5;1 or related aquaporins.

3 Conclusions and Perspectives

Recent studies have greatly improved knowledge on the localization of plant aquaporins. The localization and trafficking of aquaporins were revealed to be different among isoforms even in the same subfamily or in different cell types. PIP trafficking exhibits similarities and originalities with other PM proteins in plant cells. The PM targeting of PIP1s requires heteromerization with PIP2s. They use the secretory pathway to be targeted to the cell surface, and although fairly immobile in the plane of the PM, they are subjected to the constitutive cycling. Some PIPs and NIPs show polar localization in the PM. Comparison with preceding models in mammal aquaporins will be a key to understand the mechanism of polar trafficking of plant aquaporins. It would be also important to examine the physiological significance of the polar localization by substitution of polar aquaporins with their nonpolar variants. It is still controversial whether different TIP isoforms are differentially targeted to the lytic vacuoles or the PSVs. Systematic studies not only on TIPs but also with other tonoplast and vacuolar proteins using recently identified mutants for membrane trafficking will clarify the mechanisms. The trafficking mechanisms underlying localization of NIPs in the peribacteroid membrane of nitrogen-fixing nodules (Fortin et al. 1987) remain unknown. A change of destination of TIPs from the tonoplast to the symbiosome membrane in nodules of *Medicago truncatula* suggests a dynamic change of the membrane-trafficking system during nodulation (Gavrin et al. 2014). The mechanisms underlying dual localization of a PIP1 homolog (NtAQP1) in the PM and the chloroplast inner membrane in leaf mesophyll cells (Uehlein et al. 2008) await investigation. Aquaporins of the PM and the tonoplast exhibit extraordinary dynamic properties. When challenged to stress stimuli, and in order to adjust water status, plant cells respond by a fast membrane redistribution involving aquaporin relocalization. It would be intriguing to reveal the mechanisms not only on the trafficking but also stress sensing which cause post-translational modifications of aquaporins to change their destinations. Clearly, aquaporin trafficking will continue to be a significant model of plant membrane proteins.

Acknowledgment We thank our colleagues S. Wang for helpful discussions and S. Takada and K. Konishi for generating a transgenic line for Fig. 4c. Work in the group of JT was supported by a Grant-in-Aid for Young Scientists (A, 26712007) from the Ministry of Education, Culture, Sports, Science, and Technology of Japan, the NEXT program from the Japan Society for the Promotion of Science, and the Young Investigators Grant from the Human Frontier Science Program. Work in the group of D-TL was supported by a Marie Curie International Outgoing Fellowship (300150).

References

Alexandersson E, Fraysse L, Sjövall-Larsen S, Gustavsson S, Fellert M, Karlsson M, Johanson U, Kjellbom P (2005) Whole gene family expression and drought stress regulation of aquaporins. Plant Mol Biol 59:469–484

Apel K, Hirt H (2004) Reactive oxygen species: metabolism, oxidative stress, and signal translation. Annu Rev Plant Biol 55:373–399

Banbury DN, Oakley JD, Sessions RB, Banting G (2003) Tyrphostin A23 inhibits internalization of the transferrin receptor by perturbing the interaction between tyrosine motifs and the medium chain subunit of the AP-2 adaptor complex. J Biol Chem 278:12022–12028

Barberon M, Zelazny E, Robert S, Conejero G, Curie C, Friml J, Vert G (2011) Monoubiquitin-dependent endocytosis of the iron-regulated transporter 1 (IRT1) transporter controls iron uptake in plants. Proc Natl Acad Sci U S A 108:450–458

Besserer A, Burnotte E, Bienert GP, Chevalier AS, Errachid A, Grefen C, Blatt MR, Chaumont F (2012) Selective regulation of maize plasma membrane aquaporin trafficking and activity by the SNARE SYP121. Plant Cell 24:3463–3481

Bienert GP, Bienert MD, Jahn TP, Boutry M, Chaumont F (2011) Solanaceae XIPs are plasma membrane aquaporins that facilitate the transport of many uncharged substrates. Plant J 66: 306–317

Bonifacino JS, Traub LM (2003) Signals for sorting of transmembrane proteins to endosomes and lysosomes. Annu Rev Biochem 72:395–447

Boursiac Y, Chen S, Luu D-T, Sorieul M, van den Dries N, Maurel C (2005) Early effects of salinity on water transport in Arabidopsis roots. Molecular and cellular features of aquaporin expression. Plant Physiol 139:790–805

Boursiac Y, Boudet J, Postaire O, Luu D-T, Tournaire-Roux C, Maurel C (2008) Stimulus-induced downregulation of root water transport involves reactive oxygen species-activated cell signalling and plasma membrane intrinsic protein internalization. Plant J 56:207–218

Brandizzi F, Snapp EL, Roberts AG, Lippincott-Schwartz J, Hawes C (2002) Membrane protein transport between the endoplasmic reticulum and the Golgi in tobacco leaves is energy dependent but cytoskeleton independent: evidence from selective photobleaching. Plant Cell 14:1293–1309

Chaumont F, Tyerman SD (2014) Aquaporins: highly regulated channels controlling plant water relations. Plant Physiol 164:1600–1618

Chaumont F, Barrieu F, Jung R, Chrispeels MJ (2000) Plasma membrane intrinsic proteins from maize cluster in two sequence subgroups with differential aquaporin activity. Plant Physiol 122:1025–1034

Chevalier AS, Chaumont F (2014) Trafficking of plant plasma membrane aquaporins: multiple regulation levels and complex sorting signals. Plant Cell Physiol 56:819–829

Chevalier AS, Bienert GP, Chaumont F (2014) A new LxxxA motif in the transmembrane helix3 of maize aquaporins belonging to the plasma membrane intrinsic protein PIP2 group is required for their trafficking to the plasma membrane. Plant Physiol 166:125–138

Chiba Y, Mitani N, Yamaji N, Ma JF (2009) HvLsi1 is a silicon influx transporter in barley. Plant J 57:810–818

Cosson P, Perrin J, Bonifacino JS (2013) Anchors aweigh: protein localization and transport mediated by transmembrane domains. Trends Cell Biol 23:511–517

Crane JM, Verkman AS (2008) Long-range nonanomalous diffusion of quantum dot-labeled aquaporin-1 water channels in the cell plasma membrane. Biophys J 94:702–713

da Silva LLP, Snapp EL, Denecke J, Lippincott-Schwartz J, Hawes C, Brandizzi F (2004) Endoplasmic reticulum export sites and Golgi bodies behave as single mobile secretory units in plant cells. Plant Cell 16:1753–1771

Dancourt J, Barlowe C (2010) Protein sorting receptors in the early secretory pathway. Annu Rev Biochem 79:777–802

Danielson JAH, Johanson U (2008) Unexpected complexity of the aquaporin gene family in the moss *Physcomitrella patens*. BMC Plant Biol 8:45. doi:10.1186/1471-2229-8-45

De Marchis F, Bellucci M, Pompa A (2013) Unconventional pathways of secretory plant proteins from the endoplasmic reticulum to the vacuole bypassing the Golgi complex. Plant Signal Behav 8:1–5

Dell'Angelica EC (2009) AP-3-dependent trafficking and disease: the first decade. Curr Opin Cell Biol 21:552–559

Dhonukshe P, Aniento F, Hwang I, Robinson DG, Mravec J, Stierhof YD, Friml J (2007) Clathrin-mediated constitutive endocytosis of PIN auxin efflux carriers in Arabidopsis. Curr Biol 17:520–527

Dhonukshe P, Grigoriev I, Fischer R, Tominaga M, Robinson DG, Hasek J, Paciorek T, Petrasek J, Seifertova D, Tejos R, Meisel LA, Zazimalova E, Gadella TWJ Jr, Stierhof Y-D, Ueda T, Oiwa K, Akhmanova A, Brock R, Spang A, Friml J (2008) Auxin transport inhibitors impair vesicle motility and actin cytoskeleton dynamics in diverse eukaryotes. Proc Natl Acad Sci 105:4489–4494

Doray B, Lee I, Knisely J, Bu G, Kornfeld S (2007) The γ/σ1 and α/σ2 hemicomplexes of the clathrin adaptors AP-1 and AP-2 harbor the dileucine recognition site. Mol Biol Cell 18:1887–1896

Dugas CM, Li Q, Khan IA, Nothnagel EA (1989) Lateral diffusion in the plasma membrane of maize protoplasts with implications for cell culture. Planta 179:387–396

Ebine K, Inoue T, Ito J, Ito E, Uemura T, Goh T, Abe H, Sato K, Nakano A, Ueda T (2014) Plant vacuolar trafficking occurs through distinctly regulated pathways. Curr Biol 24:1375–1382

Edemir B, Pavenstädt H, Schlatter E, Weide T (2011) Mechanisms of cell polarity and aquaporin sorting in the nephron. Eur J Physiol 461:607–621

Feraru E, Paciorek T, Feraru MI, Zwiewka M, De Groodt R, De Rycke R, Kleine-Vehn J, Friml J (2010) The AP-3 β Adaptin Mediates the Biogenesis and Function of Lytic Vacuoles in Arabidopsis. Plant Cell 22: 2812–2824

Feraru E, Feraru MI, Asaoka R, Paciorek T, De Rycke R, Tanaka H, Nakano A, Friml J (2012) BEX5/RabA1b regulates trans-Golgi network-to-plasma membrane protein trafficking in Arabidopsis. Plant Cell 24:3074–3086

Fetter K, Van Wilder V, Moshelion M, Chaumont F (2004) Interactions between plasma membrane aquaporins modulate their water channel activity. Plant Cell 16:215–228

Fortin MG, Morriso NA, Verma DPS (1985) Specific targeting of membrane nodulins to the bacteroid enclosing compartment in soybean nodules. EMBO J 4:3041–3046

Fortin MG, Morrison NA, Verma DP (1987) Nodulin-26, a peribacteroid membrane nodulin is expressed independently of the development of the peribacteroid compartment. Nuceleic Acid Res 15:813–824

Foyer CH, Noctor G (2009) Redox regulation in photosynthetic organisms: signaling, acclimation, and practical implications. Antioxid Redox Signal 11:861–905

Frigerio L, Hinz G, Robinson DG (2008) Multiple vacuoles in plant cells: rule or exception? Traffic 9:1564–1570

Frische EW, Zwartkruis FJT (2010) Rap1, a mercenary among the Ras-like GTPases. Dev Biol 340:1–9

Gasber A, Klaumann S, Trentmann O, Trampczynska A, Clemens S, Schneider S, Sauer N, Feifer I, Bittner F, Mendel RR, Neuhaus HE (2011) Identification of an Arabidopsis solute carrier critical for intracellular transport and inter-organ allocation of molybdate. Plant Biol 13:710–718

Gattolin S, Sorieul M, Hunter PR, Khonsari RH, Frigerio L (2009) In vivo imaging of the tonoplast intrinsic protein family in Arabidopsis roots. BMC Plant Biol 9:133

Gattolin S, Sorieul M, Frigerio L (2011) Mapping of tonoplast intrinsic proteins in maturing and germinating Arabidopsis seeds reveals dual localization of embryonic TIPs to the tonoplast and plasma membrane. Mol Plant 4:180–189

Gavrin A, Kaiser BN, Geiger D, Tyerman SD, Wen Z, Bisselina T, Fedorova EE (2014) Adjustment of host cells for accommodation of symbiotic bacteria: vacuole defunctionalization, HOPS suppression, and TIP1g retargeting in *Medicago*. Plant Cell 26:3809–3822

Geelen D, Leyman B, Batoko H, Di Sansebastiano G-P, Moore I, Blatt MR (2002) The abscisic acid–reated SNARE homolog NtSyrl contributes to secretion and growth. Plant Cell 14:387–406

Geldner N (2013) The endodermis. Annu Rev Plant Biol 64:531–558

Geldner N, Friml J, Stierhof Y-D, Jurgens G, Palme K (2001) Auxin transport inhibitors block PIN1 cycling and vesicle trafficking. Nature 413:425–428

Geva Y, Schuldiner M (2014) The back and forth of cargo exit from the endoplasmic reticulum. Curr Biol 24:R130–R136. doi:10.1016/j.cub.2013.12.008

Gomez L, Chrispeels M (1993) Tonoplast and soluble vacuolar proteins are targeted by different mechanisms. Plant Cell 5:1113–1124

Gouraud S, Laera A, Calamita G, Carmosino M, Procino G, Rossetto O, Mannucci R, Rosenthal W, Svelto M, Valenti G (2002) Functional involvement of VAMP/synaptobrevin-2 in cAMP-stimulated aquaporin 2 translocation in renal collecting duct cells. J Cell Sci 115:3667–3674

Grefen C, Chen Z, Honsbein A, Donald N, Hills A, Blatt MR (2010) A novel motif essential for SNARE interaction with the K+ channel KC1 and channel gating in Arabidopsis. Plant Cell 22:3076–3092

Gustavsson S, Lebrun A-S, Nordén K, Chaumont F, Johanson U (2005) A novel plant major intrinsic protein in *Physcomitrella patens* most similar to bacterial glycerol channels. Plant Physiol 139:287–295

Hachez C, Moshelion M, Zelazny E, Cavez D, Chaumont F (2006) Localization and quantification of plasma membrane aquaporin expression in maize primary root: a clue to understanding their role as cellular plumbers. Plant Mol Biol 62:305–323

Hachez C, Heinen RB, Draye X, Chaumont F (2008) The expression pattern of plasma membrane aquaporins in maize leaf highlights their role in hydraulic regulation. Plant Mol Biol 68:337–353

Hachez C, Laloux T, Reinhardt H, Cavez D, Degand H, Grefen C, De Rycke R, Inzé D, Blatt MR, Russinova E, Chaumont F (2014a) Arabidopsis SNAREs SYP61 and SYP121 coordinate the trafficking of plasma membrane aquaporin PIP2;7 to modulate the cell membrane water permeability. Plant Cell 26:3132–3147

Hachez C, Veljanovski V, Reinhardt H, Guillaumot D, Vanhee C, Chaumont F, Batoko H (2014b) The Arabidopsis abiotic stress-induced TSPO-related protein reduces cell-surface expression of the aquaporin PIP2;7 through protein-protein interactions and autophagic degradation. Plant Cell 26:4974–4990

Hanton SL, Renna L, Bortolotti LE, Chatre L, Stefano G, Brandizzi F (2005) Diacidic motifs influence the export of transmembrane proteins from the endoplasmic reticulum in plant cells. Plant Cell 17:3081–3093

Hara-Nishimura I, Shimada T, Hatano K, Takeuchi Y, Nishimura M (1998) Transport of storage proteins to protein storage vacuoles is mediated by large precursor-accumulating vesicles. Plant Cell 10:825–836

Hara-Nishimura I, Matsushima R, Shimada T, Nishimura M (2004) Diversity and formation of endoplasmic reticulum-derived compartments in plants. Are these compartments specific to plant cells? Plant Physiol 136:3435–3439

Harris BZ, Lim WA (2001) Mechanism and role of PDZ domains in signaling complex assembly. J Cell Sci 114:3219–3231

Hoffert JD, Fenton RA, Moeller HB, Simons B, Tchapyjnikov D, McDill BW, Yu MJ, Pisitkun T, Chen F, Knepper MA (2008) Vasopressin-stimulated increase in phosphorylation at Ser269 potentiates plasma membrane retention of aquaporin-2. J Biol Chem 283:24617–24627

Horak M, Chang K, Wenthold RJ (2008) Masking of the endoplasmic reticulum retention signals during assembly of the NMDA receptor. J Neurosci 28:3500–3509

Hosy E, Martinière A, Choquet D, Maurel C, Luu D-T (2015) Super-resolved and dynamic imaging of membrane proteins in plant cells reveal contrasting kinetic profiles and multiple confinement mechanisms. Mol Plant 8:339–342

Hunter PR, Craddock CP, Di Benedetto S, Roberts LM, Frigerio L (2007) Fluorescent reporter proteins for the tonoplast and the vacuolar lumen identify a single vacuolar compartment in Arabidopsis cells. Plant Physiol 145:1371–1382

Isayenkov S, Isner J-C, Maathuis FJM (2011) Rice two-pore K+ channels are expressed in different types of vacuoles. Plant Cell 23:756–768

Ishikawa F, Suga S, Uemura T, Sato MH, Maeshima M (2005) Novel type aquaporin SIPs are mainly localized to the ER membrane and show cell-specific expression in *Arabidopsis thaliana*. FEBS Lett 579:5814–5820

Jaillais Y, Fobis-Loisy I, Miege C, Rollin C, Gaude T (2006) AtSNX1 defines an endosome for auxin-carrier trafficking in Arabidopsis. Nature 443:106–109

Kim E, Sheng M (2004) PDZ domain proteins of synapses. Nat Rev Neurosci 5:771–781

Kleine-Vehn J, Dhonukshe P, Swarup R, Bennett M, Friml J (2006) Subcellular trafficking of the Arabidopsis auxin influx carrier AUX1 uses a novel pathway distinct from PIN1. Plant Cell 18:3171–3181

Komarova NY, Meier S, Meier A, Grotemeyer MS, Rentsch D (2012) Determinants for Arabidopsis peptide transporter targeting to the tonoplast or plasma membrane. Traffic 13: 1090–1105

Kong Y, Pena MJ, Renna L, Avic U, Pattathil S, Tuomivaara ST, Li X, Reiter W-D, Brandizzi F, Hahn MG, Darvill AG, York WS, O'Neill M (2015) Galactose-depleted xyloglucan is dysfunctional and leads to dwarfism in Arabidopsis. Plant Physiol 167:1296–1306

Kurachi H, Wada Y, Tsukamoto N, Maeda M, Kubota H, Hattori M, Iwai K, Minato N (1997) Human SPA-1 gene product selectively expressed in lymphoid tissues is a specific GTPase-activating protein for Rap1 and Rap2. J Biol Chem 272:28081–28088

Larisch N, Schulze C, Galione A, Dietrich P (2012) An N-terminal dileucine motif directs two-pore channels to the tonoplast of plant cells. Traffic 13:1012–1022

Laxmi A, Pan J, Morsy M, Chen R (2008) Light plays an essential role in intracellular distribution of auxin efflux carrier PIN2 in *Arabidopsis thaliana*. PLoS ONE 3:e1510

Le Roy C, Wrana JL (2005) Clathrin- and non-clathrin- mediated endocytic regulation of cell signalling. Nat Rev Mol Cell Biol 6:112–126

Lee HK, Cho SK, Son O, Xu Z, Hwang I, Kim WT (2009) Drought stress-induced Rma1H1, a RING membrane-anchor E3 ubiquitin ligase homolog, regulates aquaporin levels via ubiquitination in transgenic Arabidopsis plants. Plant Cell 21:622–641

Lefebvre B, Furt F, Hartmann MA, Michaelson LV, Carde J-P, Sargueil-Boiron F, Rossignol M, Napier JA, Cullimore J, Bessoule JJ, Mongrand S (2007) Characterization of lipid rafts from *Medicago truncatula* root plasma membranes: a proteomic study reveals the presence of a raft-associated redox system. Plant Physiol 144:402–418

Li X, Wang X, Yang Y, Li R, He Q, Fang X, Luu D-T, Maurel C, Lin J (2011) Single-molecule analysis of PIP2;1 dynamics and partitioning reveals multiple modes of Arabidopsis plasma membrane aquaporin regulation. Plant Cell 23:3780–3797

Li R, Liu P, Wan Y, Chen T, Wang Q, Mettbach U, Baluška F, Šamaj J, Fang X, Lucas WJ, Lin J (2012) A membrane microdomain-associated protein, Arabidopsis Flot1, is involved in a clathrin-independent endocytic pathway and is required for seedling development. Plant Cell 24:2105–2122

Li X, Luu D-T, Maurel C, Lin J (2013) Probing plasma membrane dynamics at the single-molecule level. Trends Plant Sci 18:617–624

Liu Y, Li J (2014) Endoplasmic reticulum-mediated protein quality control in Arabidopsis. Front Plant Sci 5:162

Lopez D, Bronner G, Brunel N, Auguin D, Bourgerie S, Brignolas F, Carpin S, Tournaire-Roux C, Maurel C, Fumanal B, Martin F, Sakr S, Label P, Julien JL, A. G-D, JS V (2012) Insights into Populus XIP aquaporins: evolutionary expansion, protein functionality, and environmental regulation. J Exp Bot 63:2217–2230

Luu D-T, Martinière A, Sorieul M, Runions J, Maurel C (2012) Fluorescence recovery after photobleaching reveals high cycling dynamics of plasma membrane aquaporins in Arabidopsis roots under salt stress. Plant J 69:894–905

Ma JF, Tamai K, Yamaji N, Mitani N, Konishi S, Katsuhara M, Ishiguro M, Murata Y, Yano M (2006) A silicon transporter in rice. Nature 440:688–691
Ma JF, Yamaji N (2006) Silicon uptake and accumulation in higher plants. Trends Plant Sci 11: 392–397
Ma JF, Yamaji N (2015) A cooperative system of silicon transport in plants. Trends Plant Sci 20: 435–442
Ma JF, Yamaji N, Mitani N, Tamai K, Konishi S, Fujiwara T, Katsuhara M, Yano M (2007) An efflux transporter of silicon in rice. Nature 448:209–212
Madrid R, Le Maout S, Barrault Â, Janvier K (2001) Polarized trafficking and surface expression of the AQP4 water channel are coordinated by serial and regulated interactions with different clathrin adaptor complexes. EMBO J 20:7008–7021
Manley S, Gillette JM, Patterson GH, Shroff H, Hess HF, Betzig E, Lippincott-Schwartz J (2008) High-density mapping of single-molecule trajectories with photoactivated localization microscopy. Nat Methods 5:155–157
Margeta-Mitrovic M, Jan YN, Jan LY (2000) A trafficking checkpoint controls GABA (B) receptor heterodimerization. Neuron 27:97–106
Marschner H (1995) Mineral nutrition of higher plants, 2nd edn. Academic press, San Diego
Martinière A, Lavagi I, Nageswaran G, Rolfe DJ, Maneta-Peyret L, Luu D-T, Botchway SW, Webb SED, Mongrand S, Maurel C, Martin-Fernandez ML, Kleine-Vehn J, Friml J, Moreau P, Runions J (2012) Cell wall constrains lateral diffusion of plant plasma-membrane proteins. Proc Natl Acad Sci U S A 109:12805–12810
Michelsen K, Schmid V, Metz J, Heusser K, Liebel U, Schwede T, Spang A, Schwappach B (2007) Novel cargo-binding site in the β and δ subunits of coatomer. J Cell Biol 179:209–217
Mikosch M, Hurst AC, Hertel B, Homann U (2006) Diacidic motif is required for efficient transport of the K+ channel KAT1 to the plasma membrane. Plant Physiol 142:923–930
Miller EA, Beilharz TH, Malkus PN, Lee MCS, Hamamoto S, Orci L, Schekman R (2003) Multiple cargo binding sites on the COPII subunit Sec24p ensure capture of diverse membrane proteins into transport vesicles. Cell 114:497–509
Minami A, Fujiwara M, Furuto A, Fukao Y, Yamashita T, Kamo M, Kawamura Y, Uemura M (2009) Alterations in detergent-resistant plasma membrane microdomains in *Arabidopsis thaliana* during cold acclimation. Plant Cell Physiol 50:341–359
Mitani N, Chiba Y, Yamaji N, Ma JF (2009) Identification and characterization of maize and barley Lsi2-like silicon efflux transporters reveals a distinct silicon uptake system from that in rice. Plant Cell 21:2133–2142
Mitani N, Yamaji N, Ago Y, Iwasaki K, Ma JF (2011) Isolation and functional characterization of an influx silicon transporter in two pumpkin cultivars contrasting in silicon accumulation. Plant J 66:231–240
Moeller HB, Knepper MA, Fenton RA (2009) Serine 269 phosphorylated aquaporin-2 is targeted to the apical membrane of collecting duct principal cells. Kidney Int 75:295–303
Moeller HB, Aroankins TS, Slengerik-Hansen J, Pisitkun T, Fenton RA (2014) Phosphorylation and ubiquitylation are opposing processes that regulate endocytosis of the water channel aquaporin-2. J Cell Sci 127:3174–3183
Moskalenko S, Henry DO, Rosse C, Mirey G, Camonis JH, White MA (2002) The exocyst is a Ral effector complex. Nat Cell Biol 4:66–72
Mossessova E, Bickford LC, Goldberg J (2003) SNARE selectivity of the COPII coat. Cell 114:483–495
Nedvetsky PI, Stefan E, Frische S, Santamaria K, Wiesner B, Valenti G, Hammer J a, Nielsen S, Goldenring JR, Rosenthal W, Klussmann E (2007) A role of myosin Vb and Rab11-FIP2 in the aquaporin-2 shuttle. Traffic 8:110–123
Niemes S, Langhans M, Viotti C, Scheuring D, San Wan Yan M, Jiang L, Hillmer S, Robinson DG, Pimpl P (2010) Retromer recycles vacuolar sorting receptors from the trans-Golgi network. Plant J 61:107–121
Noda Y, Horikawa S, Furukawa T, Hirai K, Katayama Y, Asai T, Kuwahara M, Katagiri K, Kinashi T, Hattori M, Minato N, Sasaki S (2004) Aquaporin-2 trafficking is regulated by PDZ-domain containing protein SPA-1. FEBS Lett 568:139–145

Odorizzi G, Cowles CR, Emr SD (1998) The AP-3 complex: a coat of many colours. Trends Cell Biol 8:282–288

Ohno H, Stewart J, Fournier MC, Bosshart H, Rhee I, Miyatake S, Saito T, Gallusser A, Kirchhausen T, Bonifacino JS (1995) Interaction of tyrosine-based sorting signals with clathrin-associated proteins. Science 269:1872–1875

Paciorek T, Zazimalova E, Ruthardt N, Petrasek J, Stierhof YD, Kleine-Vehn J, Morris DA, Emans N, Jurgens G, Geldner N, Friml J (2005) Auxin inhibits endocytosis and promotes its own efflux from cells. Nature 435:1251–1256

Pan J, Fujioka S, Peng J, Chen J, Li G, Chen R (2009) The E3 Ubiquitin Ligase SCFTIR1/AFB and membrane sterols play key roles in auxin regulation of endocytosis, recycling, and plasma membrane accumulation of the auxin efflux transporter PIN2 in *Arabidopsis thaliana*. Plant Cell 21:568–580

Park M, Kim SJ, Vitale A, Hwang I (2004) Identification of the protein storage vacuole and protein targeting to the vacuole in leaf cells of three plant species. Plant Physiol 134:625–639

Pedrazzini E, Komarova NY, Rentsch D, Vitale A (2013) Traffic routes and signals for the tonoplast. Traffic 14:622–628

Prak S, Hem S, Boudet J, Viennois G, Sommerer N, Rossignol M, Maurel C, Santoni V (2008) Multiple phosphorylations in the C-terminal tail of plant plasma membrane aquaporins: role in subcellular trafficking of AtPIP2;1 in response to salt stress. Mol Cell Proteomics 7:1019–1030

Rai T, Sasaki S, Uchida S (2006) Polarized trafficking of the aquaporin-3 water channel is mediated by an NH2-terminal sorting signal. Am J Physiol Cell Physiol 290:C298–C304

Ren Z, Riley NJ, Needleman L a, Sanders JM, Swanson GT, Marshall J (2003) Cell surface expression of GluR5 kainate receptors is regulated by an endoplasmic reticulum retention signal. J Biol Chem 278:52700–52709

Richter S, Geldner N, Schrader J, Wolters H, Stierhof Y-D, Rios G, Koncz C, Robinson DG, Jürgens G (2007) Functional diversification of closely related ARF-GEFs in protein secretion and recycling. Nature 448:488–492

Rivera-Serrano EE, Rodriguez-Welsh MF, Hicks GR, Rojas-Pierce M (2012) A small molecule inhibitor partitions two distinct pathways for trafficking of tonoplast intrinsic proteins in Arabidopsis. PLoS One 7:e44735

Robatzek S (2007) Vesicle trafficking in plant immune responses. Cell Microbiol 9:1–8

Robinson DG, Langhans M, Saint-Jore-Dupas C, Hawes C (2008) BFA effects are tissue and not just plant specific. Trends Plant Sci 13:405–408

Rojas-Pierce M (2013) Targeting of tonoplast proteins to the vacuole. Plant Sci 211:132–136

Rosas-Santiago P, Lagunas-Gomez D, Barkla BJ, Vera-Estrella R, Lalonde S, Jones a, Frommer WB, Zimmermannova O, Sychrova H, Pantoja O (2015) Identification of rice cornichon as a possible cargo receptor for the Golgi-localized sodium transporter OsHKT1;3. J Exp Bot 66:2733–2748

Sakurai G, Satake A, Yamaji N, Mitani-Ueno N, Yokozawa M, FG F, JF M (2015) In silico simulation modeling reveals the importance of the Casparian strip for efficient silicon uptake in rice roots. Plant Cell Physiol. doi:10.1093/pcp/pcv017

Sakurai-Ishikawa J, Murai-Hatano M, Hayashi H, Ahamed A, Fukushi K, Matsumoto T, Kitagawa Y (2011) Transpiration from shoots triggers diurnal changes in root aquaporin expression. Plant Cell Environ 34:1150–1163

Sato M, Sato K, Nakano A (2004) Endoplasmic reticulum quality control of unassembled iron transporter depends on Rer1p-mediated retrieval from the Golgi. Mol Biol Cell 15: 1417–1424

Segami S, Makino S, Miyake A, Asaoka M, Maeshima M (2014) Dynamics of vacuoles and H + -pyrophosphatase visualized by monomeric green fluorescent protein in Arabidopsis: artifactual bulbs and native intravacuolar spherical structures. Plant Cell 26:1–20

Seifert GJ, Barber C, Wells B, Dolan L, Roberts K (2002) Galactose biosynthesis in Arabidopsis: genetic evidence for substrate channeling from UDP-D-galactose into cell wall polymers. Curr Biol 12:1840–1845

Shahollari B, Peskan-Berghöfer T, Oelmüller R (2004) Receptor kinases with leucine-rich repeats are enriched in Triton X-100 insoluble plasma membrane microdomains from plants. Physiol Plant 122:397–403

Sieben C, Mikosch M, Brandizzi F, Homann U (2008) Interaction of the K+–channel KAT1 with the coat protein complex II coat component Sec24 depends on a di-acidic endoplasmic reticulum export motif. Plant J 56:997–1006

Singh MK, Krüger F, Beckmann H, Brumm S, Vermeer Joop EM, Munnik T, Mayer U, Stierhof Y-D, Grefen C, Schumacher K, Jürgens G (2014) Protein delivery to vacuole requires SAND protein-dependent rab GTPase conversion for MVB-vacuole fusion. Curr Biol 24:1383–1389

Sorieul M, Santoni V, Maurel C, Luu D-T (2011) Mechanisms and effects of retention of over-expressed aquaporin AtPIP2;1 in the endoplasmic reticulum. Traffic 12:473–482

Sutter JU, Campanoni P, Tyrrell M, Blatt MR (2006) Selective mobility and sensitivity to SNAREs is exhibited by the Arabidopsis KAT1 K+ channel at the plasma membrane. Plant Cell 18:935–954

Sutter JU, Sieben C, Hartel A, Eisenach C, Thiel G, Blatt MR (2007) Abscisic acid triggers the endocytosis of the Arabidopsis KAT1 K(+) channel and its recycling to the plasma membrane. Curr Biol 17:1396–1402

Takáč T, Pechan T, Šamajová O, Ovečka M, Richter H, Eck C, Niehaus K, Šamaj J (2012) Wortmannin treatment induces changes in Arabidopsis root proteome and post-golgi compartments. J Proteome Res 11:3127–3142

Takagi J, Renna L, Takahashi H, Koumoto Y, Tamura K, Stefano G, Fukao Y, Kondo M, Nishimura M, Shimada T, Brandizzi F, Hara-Nishimura I (2013) MAIGO5 functions in protein export from Golgi-associated endoplasmic reticulum exit sites in Arabidopsis. Plant Cell 25:4658–4675

Takano J, Miwa K, Yuan L, von Wiren N, Fujiwara T (2005) Endocytosis and degradation of BOR1, a boron transporter of *Arabidopsis thaliana*, regulated by boron availability. Proc Natl Acad Sci USA 102:12276 0502060102

Takano J, Noguchi K, Yasumori M, Kobayashi M, Gajdos Z, Miwa K, Hayashi H, Yoneyama T, Fujiwara T (2002) Arabidopsis boron transporter for xylem loading. Nature 21:337–340

Takano J, Wada M, Ludewig U, Schaaf G (2006) The Arabidopsis major intrinsic protein NIP5; 1 is essential for efficient boron uptake and plant development under boron limitation. Plant Cell 18:1498–1509. doi:10.1105/tpc.106.041640.2

Takano J, Tanaka M, Toyoda A, Miwa K, Kasai K, Fuji K, Onouchi H, Naito S, Fujiwara T (2010) Polar localization and degradation of Arabidopsis boron transporters through distinct trafficking pathways. Proc Natl Acad Sci U S A 107:5220–5225

Tamma G, Robben JH, Trimpert C, Boone M, Deen PMT (2011) Regulation of AQP2 localization by S256 and S261 phosphorylation and ubiquitination. Am J Phys Cell Physiol 300:C636–C646

Törnroth-Horsefield S, Wang Y, Hedfalk K, Johanson U, Karlsson M, Tajkhorshid E, Neutze R, Kjellbom P (2006) Structural mechanism of plant aquaporin gating. Nature 439:688–694

Toyooka K, Okamoto T, Minamikawa T (2000) Mass transport of proform of a KDEL- tailed cysteine protease (SH – EP) to protein storage vacuoles by endoplasmic reticulum – derived vesicle is involved in protein mobilization in germinating seeds. J Cell Biol 148:453–463

Traub LM (2009) Tickets to ride: selecting cargo for clathrin-regulated internalization. Nat Rev Mol Cell Biol 10:583–596

Tsukamoto N, Hattori M, Yang H, Bos JL, Minato N (1999) Rap1 GTPase-activating protein SPA-1 negatively regulates cell adhesion. J Biol Chem 274:18463–18469

Tyrrell M, Campanoni P, Sutter JU, Pratelli R, Paneque M, Sokolovski S, Blatt MR (2007) Selective targeting of plasma membrane and tonoplast traffic by inhibitory (dominant-negative) SNARE fragments. Plant J 51:1099–1115

Uehara M, Wang S, Kamiya T, Shigenobu S, Yamaguchi K, Fujiwara T, Naito S, Takano J (2014) Identification and characterization of an Arabidopsis mutant with altered localization of NIP5;1, a plasma membrane boric acid channel, reveals the requirement for D-galactose in endomembrane organization. Plant Cell Physiol 55:704–714

Uehlein N, Otto B, Hanson DT, Fischer M, McDowell N, Kaldenhoff R (2008) Function of *Nicotiana tabacum* aquaporins as chloroplast gas pores challenges the concept of membrane CO2 permeability. Plant Cell 20:648–657

Van Balkom BWM, Savelkoul PJM, Markovich D, Hofman E, Nielsen S, Van Der Sluijs P, Deen PMT (2002) The role of putative phosphorylation sites in the targeting and shuttling of the aquaporin-2 water channel. J Biol Chem 277:41473–41479

Vanhee C, Zapotoczny G, Masquelier D, Ghislain M, Batoko H (2011) The Arabidopsis multistress regulator TSPO is a heme binding membrane protein and a potential scavenger of porphyrins via an autophagy-dependent degradation mechanism. Plant Cell 23:785–805

Venditti R, Wilson C, De Matteis MA (2014) Exiting the ER: what we know and what we don't. Trends Cell Biol 24:9–18

Vera-Estrella R, Barkla BJ, Bohnert HJ, Pantoja O (2004) Novel regulation of aquaporins during osmotic stress. Plant Physiol 135:2318–2329

Viotti C (2014) ER and vacuoles: never been closer. Front Plant Sci 5:20

Viotti C, Krüger F, Krebs M, Neubert C, Fink F, Lupanga U, Scheuring D, Boutté Y, Frescatada-Rosa M, Wolfenstetter S, Sauer N, Hillmer S, Grebe M, Schumacher K (2013) The endoplasmic reticulum is the main membrane source for biogenesis of the lytic vacuole in Arabidopsis. Plant Cell 25:3434–3449

Votsmeier C, Gallwitz D (2001) An acidic sequence of a putative yeast Golgi membrane protein binds COPII and facilitates ER export. EMBO J 20:6742–6750

Wang X, Cai Y, Wang H, Zeng Y, Zhuang X, Li B, Jiang L (2014) Trans-golgi network-located AP1 gamma adaptins mediate dileucine motif-directed vacuolar targeting in Arabidopsis. Plant Cell 26:4102–4118

Wang S, Ito T, Uehara M, Naito S, Takano J (2015) UDP-d-galactose synthesis by UDP-glucose 4-epimerase 4 is required for organization of the trans-Golgi network/early endosome in *Arabidopsis thaliana* root epidermal cells. J Plant Res 128:863–873

Wolfenstetter S, Wirsching P, Dotzauer D, Schneider S, Sauer N (2012) Routes to the tonoplast: the sorting of tonoplast transporters in Arabidopsis mesophyll protoplasts. Plant Cell 24:215–232

Wudick MM, Luu DT, Maurel C (2009) A look inside: localization patterns and functions of intracellular plant aquaporins. New Phytol 184:289–302

Wudick MM, Li X, Valentini V, Geldner N, Chory J, Lin J, Maurel C, Luu D-T (2015) Subcellular redistribution of root aquaporins induced by hydrogen peroxide. Mol Plant 8:1103–1114

Xiang L, Etxeberria E, Van Den Ende W (2013) Vacuolar protein sorting mechanisms in plants. FEBS J 280:979–993

Yamada K, Osakabe Y, Mizoi J, Nakashima K, Fujita Y, Shinozaki K, Yamaguchi-Shinozaki K (2010) Functional analysis of an *Arabidopsis thaliana* abiotic stress-inducible facilitated diffusion transporter for monosaccharides. J Biol Chem 285:1138–1146

Yamaji N, Ma JF (2009) A transporter at the node responsible for intervascular transfer of silicon in rice. Plant Cell 21:2878–2883

Yamaji N, Mitatni N, Ma JF (2008) A transporter regulating silicon distribution in rice shoots. Plant Cell 20:1381–1389

Yamaji N, Chiba Y, Mitani-Ueno N, Feng Ma J (2012) Functional characterization of a silicon transporter gene implicated in silicon distribution in Barley. Plant Physiol 160:1491–1497

Yu M-J, Pisitkun T, Wang G, Aranda JF, Gonzales P A, Tchapyjnikov D, Shen R-F, Alonso M A, Knepper M A (2008) Large-scale quantitative LC-MS/MS analysis of detergent-resistant membrane proteins from rat renal collecting duct. Am J Phys Cell Physiol 295:C661–C678

Yui N, Lu HAJ, Chen Y, Nomura N, Bouley R, Brown D (2012) Basolateral targeting and microtubule dependent transcytosis of the aquaporin-2 water channel. AJP Cell Physiol 02114: 38–48

Zelazny E, Borst JW, Muylaert M, Batoko H, Hemminga MA, Chaumont F (2007) FRET imaging in living maize cells reveals that plasma membrane aquaporins interact to regulate their subcellular localization. Proc Natl Acad Sci U S A 104:12359–12364

Zelazny E, Miecielica U, Borst JW, Hemminga MA, Chaumont F (2009) An N-terminal diacidic motif is required for the trafficking of maize aquaporins ZmPIP2;4 and ZmPIP2;5 to the plasma membrane. Plant J 57:346–355

Zerangue N, Schwappach B, Jan YN, Jan LY (1999) A new ER trafficking signal regulates the subunit stoichiometry of plasma membrane K(ATP) channels. Neuron 22:537–548

Zwiewka M, Feraru E, Möller B, Hwang I, Feraru MI, Kleine-Vehn J, Weijers D, Friml J (2011) The AP-3 adaptor complex is required for vacuolar function in Arabidopsis. Cell Res 21:1711–1722

Plant Aquaporin Posttranslational Regulation

Véronique Santoni

Abstract Posttranslational modifications are mechanisms that modulate and control the functions of proteins. The development of mass spectrometry methodology allows description of the extent of posttranslational modifications affecting plant aquaporins. Hence, more than 70 phosphorylation sites are described in several aquaporin isoforms belonging to PIP, TIP, and NIP groups across different species, and several kinases have been characterized. N-terminal protein modifications also occur on plant aquaporins as well as deamidation, glycosylation, methylation, and ubiquitination. This chapter summarizes the knowledge about aquaporin posttranslational modifications and their implication in aquaporin function.

Abbreviations

ABA	Abscisic acid
AGC kinase	cAMP-dependent protein kinase (PKA), cGMP-dependent protein kinase (PKG), and protein kinase C (PKC) families
Lp_r	Root hydraulic conductivity
MS	Mass spectrometry
NIP	Nodulin26-like intrinsic protein
PIP	Plasma membrane intrinsic protein
PK	Protein kinase
PTM	Posttranslational modification
Ser	Serine residue
Thr	Threonine residue
TIP	Tonoplast intrinsic protein
Ub	Ubiquitin

V. Santoni
Biochimie et Physiologie Moléculaire des Plantes, INRA/CNRS/SupAgro/UM2, UMR 5004, 2 Place Viala, 34060 Montpellier, Cedex1, France
e-mail: veronique.santoni@inra.fr

F. Chaumont, S.D. Tyerman (eds.), *Plant Aquaporins*, Signaling and Communication in Plants, DOI 10.1007/978-3-319-49395-4_4

1 Introduction

Maintaining water balance in plant cells and organs involves control of transmembrane ion and water fluxes and, hence, a tight regulation of ion channels and water channels (aquaporins) in plant cell membranes (Maurel et al. 2015). An excellent mechanism for fine-tuning, the function of channels can be provided by posttranslational modifications (PTMs). PTMs are central to regulate protein structure and function and thereby to modulate and control protein catalytic activity, subcellular localization, stability, and interaction with other partners (Temporini et al. 2008; Jensen 2006). Thus, the determination of the expression of aquaporins and of their PTMs appears essential when studying plant–water relations. During the past 10 years, mass spectrometry (MS)-based proteomics has begun to reveal the true extent of the PTM universe. For many PTMs, including phosphorylation, ubiquitination, glycosylation, and acetylation, tens of thousands of sites can now be confidently identified and localized in the sequence of proteins. The quantification of PTM levels between different cellular states and in specific environmental conditions can be likewise established, with label-free methods showing particular promise. It is also becoming possible to determine the absolute occupancy or stoichiometry of PTMs (Olsen and Mann 2013). The following sections summarize the knowledge regarding plant aquaporin PTMs including phosphorylation, N-terminal modification, deamidation, glycosylation, methylation, and ubiquitination.

2 Phosphorylation of Aquaporins

Protein phosphorylation is a critical regulatory step in signaling networks and is arguably the most widespread protein modification affecting almost all basic cellular processes in various organisms. In recent years, the identification of protein phosphorylation sites has become routine through detection of phosphorylated peptides by MS (Steen et al. 2006). Hence, increased confidence in phosphoproteome analysis came with development of suitable enrichment methods for phosphoproteins and phosphopeptides from complex protein digests and with increased technical performance of mass spectrometers regarding sensitivity, mass accuracy, resolution, and dynamic range (Schulze 2010).

2.1 Phosphorylated Residues in Aquaporins

Recent phosphoproteomic developments allowed the identification of more than 70 phosphorylation sites in several aquaporin isoforms belonging to PIP, TIP, and NIP groups across different species (Table 1, http://phosphat.uni-hohenheim.de/). Serine appears as the major (87 %) phosphorylated residue in plant aquaporins. Eleven and 2 % of phosphorylated sites correspond to threonine and tyrosine residues,

Table 1 Identification of phosphorylation sites in plant aquaporins by mass spectrometry and amino acid sequencing

Protein name	Identifier	Phosphorylated residue	Reference
*At*PIP1;1 and *At*PIP1;2	At3g61430 and At2g45960	Ser24, Ser27	Di Pietro et al. (2013), Vialaret et al. (2014), and Band et al. (2012)
*At*PIP2;1	At3g53420	Ser277, Ser280, Ser283	Sugiyama et al. (2008), Prado et al. (2013), Prak et al. (2008), Di Pietro et al. (2013), Engelsberger and Schulze (2012), Hsu et al. (2009), Kline et al. (2010), Niittylä et al. (2007), Vialaret et al. (2014), Wang et al. (2013a, b), (Wu et al. 2013, 2014), Yang et al. (2013), Qing et al. (2016), Choudhary et al. (2015), Band et al. (2012), Nakagami et al. (2010), Santoni et al. (2003), and Nühse et al. 2004, 2007)
*At*PIP2;2	At2g37170	Thr22, Ser275, Ser278, Ser281	Sugiyama et al. (2008), Prado et al. (2013), Prak et al. (2008), Di Pietro et al. (2013), Engelsberger and Schulze (2012), Hsu et al. (2009), Kline et al. (2010), Niittylä et al. (2007), Vialaret et al. (2014), Wang et al. (2013a, b) Wu et al. (2013, 2014), Yang et al. (2013), Choudhary et al. (2015), Band et al. (2012), Nakagami et al. (2010), Santoni et al. (2003), Nühse et al. (2004, 2007)
*At*PIP2;3	At2g37180	Thr13, Ser275, Ser278, Ser281	Sugiyama et al. (2008), Prado et al. (2013), Prak et al. (2008), Di Pietro et al. (2013), Engelsberger and Schulze (2012), Hsu et al. (2009), Kline et al. (2010), Niittylä et al. (2007), Vialaret et al. (2014), Wang et al. (2013a, b), Wu et al. (2013, 2014), Yang et al. (2013), Choudhary et al. (2015), Band et al. (2012), Nakagami et al. (2010), Santoni et al. (2003), and Nühse et al. 2004, 2007)
*At*PIP2;4	At5g60660	Ser10, Ser280, Ser283, Ser286	Prak et al. (2008), Di Pietro et al. (2013), Engelsberger and Schulze (2012), Hsu et al. (2009), Kline et al. (2010), Stecker et al. (2014), Vialaret et al. (2014), Wang et al. (2013a, b), (Wu et al. 2013, 2014), Choudhary et al. (2015), and Band et al. (2012)
*At*PIP2;5	At3g54820	Ser279, Ser282	Di Pietro et al. (2013), Vialaret et al. (2014), Yang et al. (2013), and Choudhary et al. (2015)

(continued)

Table 1 (continued)

Protein name	Identifier	Phosphorylated residue	Reference
*At*PIP2;6	At2g39010	Thr2, Thr7, Ser11, Ser13, Tyr17, Tyr277, Ser279, Ser282	Niittylä et al. (2007), Wang et al. (2013a, b), Wu et al. (2013), 2014), Xue et al. (2013), Yang et al. (2013), Choudhary et al. (2015), Nakagami et al. (2010), Reiland et al. (2011, 2009, and Whiteman et al. (2008a)
*At*PIP2;7	At4g35100	Ser6, Ser273, Ser276, Thr279	Sugiyama et al. (2008), Prak et al. (2008), Di Pietro et al. (2013), Engelsberger and Schulze (2012), Vialaret et al. (2014), Wang et al. (2013a, b), Wu et al. (2013, 2014), Xue et al. (2013), Choudhary et al. (2015), Nakagami et al. (2010), Nühse et al. (2007), and Reiland et al. (2009)
*At*PIP2;8	At2g16850	Ser271, Ser274, Thr277	Sugiyama et al. (2008), Di Pietro et al. (2013), Kline et al. (2010), Vialaret et al. (2014), Wang et al. (2013a), Nakagami et al. (2010), and Reiland et al. (2011, 2009)
*At*NIP4;1	At5g37810	Ser267	Sugiyama et al. (2008) and Nakagami et al. (2010)
*At*NIP4;2	At5g37820	Ser265, Ser267	Sugiyama et al. (2008), Wang et al. (2013a), and Nakagami et al. (2010)
*At*NIP1;1	At4g19030	Ser45, Ser286	Niittylä et al. 2007), Vialaret et al. (2014), and Nühse et al. (2007)
*At*NIP1;2	At4g18910	Ser283	Niittylä et al. (2007), and Nühse et al. (2007)
*At*NIP2;1	At2g34390	Ser5	Vialaret et al. (2014)
*Bd*PIP1;2-like	I1J0I4	Ser16, Ser28	Lv et al. (2014)
*Bd*PIP2;1-like	I1GUT1	Ser285	Lv et al. (2014)
*Bd*PIP2-5-like	I1IZR7	Ser280, Ser283	Lv et al. (2014)
*Bd*PIP2-7-like	I1ISB9	Ser2	Lv et al. (2014)
*Bd*NIP2-2-like	I1GZJ9	Ser292	Lv et al. (2014)
*Gm*Nodulin-26	P08995	Ser262	Guenther et al. (2003) and Weaver and Roberts (1992)
Hvγ-TIP-like	Q43480	Ser10, Thr248	Endler et al. (2009)
*Os*PIP2;1	Os07g26690	Ser9, Ser288	Nakagami et al. (2010) and Whiteman et al. (2008b)

*Os*PIP2;2	Os02g41860	Ser8, Ser16	Wang et al. (2014)
*Os*PIP2;6	Os04g16450	Ser2, Ser275, Ser278, Ser281	Nakagami et al. (2010), Whiteman et al. (2008b), and Wang et al. (2014)
*Os*PIP2;7	Os09g36930	Ser252 or Thr253	Whiteman et al. (2008b)
*Os*TIP2;2	Os06g22960	Ser2, Ser245	Wang et al. (2014)
*Os*TIP1;1	Os03g05290	Ser10	Wang et al. (2014)
*Pv*TIP3;1	P23958	Ser7	Daniel and Yeager (2005) and Johnson and Chrispeels (1992)
*So*PIP2;1	Q41372	Ser274, Ser277	Johansson et al. (1998)
*Zm*PIP1;1-/1;4	Q41870/Q9XF59/Q9AQU5	Ser16	Van Wilder et al. (2008)
*Zm*PIP2;1/2;2/2;7[a]	Q84RL7/Q9ATM8/Q9ATM4	Ser285, Ser288	Van Wilder et al. (2008) and Hu et al. (2015)
*Zm*PIP2;3/2;4[b]	Q9ATM7/ Q9ATM6	Ser284	Van Wilder et al. (2008)
*Zm*PIP2;6	Q9ATM5	Ser283	Van Wilder et al. (2008)
*Zm*NIP2;2	Q9ATN2	Ser284	Hu et al. (2015)

The table summarizes the phosphorylation sites in plant aquaporins based on literature and PhosPhAt database (http://phosphat.uni-hohenheim.de/; (Durek et al. 2010))

At Arabidopsis thaliana, Bd Brachypodium distachyon, Hv Hordeum vulgare, Os Oryza sativa, Pv Phaseolus vulgaris, Gm Glycine max; So Spinacia oleracea, Zm Zea mays

[a]Because of a strict sequence similarity of a tryptic peptide arising from the C-terminal tail of *Zm*PIP2;1, *Zm*PIP2;2, and *Zm*PIP2;7 (ALGSFRSNA), the phosphorylated residues are Ser285 and Ser288 in *Zm*PIP2;1, Ser287 and Ser290 in *Zm*PIP2;2, and Ser282 and Ser285 in *Zm*PIP2;7

[b]Because of a strict sequence similarity of a tryptic peptides arising from the C-terminal tail of *Zm*PIP2;3 and *Zm*PIP2;4 (LGSYRSNA), the phosphorylated residue is Ser284 in *Zm*PIP2;3 and Ser283 in *Zm*PIP2;4

a

*At*PIP2;1	--	ASGSKSL ------	GSFRSAANV$_{287}$
*At*PIP2;2	--	ASGSKSL ------	GSFRSAANV$_{285}$
*At*PIP2;3	--	ASGSKSL ------	GSFRSAANV$_{285}$
*At*PIP2;4	--	AAAIKALGSFG	SFGSFRSFA$_{291}$
*At*PIP2;5	--	AGAIKAL ------	GSFRSQPHV$_{286}$
*At*PIP2;6	--	AGAMKAY ------	GSVRSQLHELHA$_{289}$
*At*PIP2;7	--	ASAIKAL ------	GSFRSNATN$_{280}$
*At*PIP2;8	--	AAAIKAL ------	ASFRSNPTN$_{278}$
*Bd*PIP2;1	--	AGAIKAL ------	GSFRSNA$_{290}$
*Bd*PIP2;5	--	ASATKL -------	GSSASFGRN$_{287}$
*Os*PIP2;1	--	AGAIKAL ------	GSFRSNA$_{290}$
*Os*PIP2;6	--	AAAIKAL ------	GSFRSNPSN$_{282}$
*Os*PIP2;7	--	GEAAKAL ------	SSFRSTSVTA$_{257}$
*So*PIP2;1	--	AAAIKAL ------	GSFRSNPTN$_{281}$
*Zm*PIP2;1	--	AGAIKAL ------	GSFRSNA$_{290}$
*Zm*PIP2;2	--	AGAIKAL ------	GSFRSNA$_{292}$
*Zm*PIP2;3	--	ASATKL -------	GSYRSNA$_{289}$
*Zm*PIP2;4	--	ASATKL -------	GSYRSNA$_{288}$
*Zm*PIP2;6	--	ASARGY -------	GSFRSNA$_{288}$
*Zm*PIP2;7	--	GSAIKAL ------	GSFRSNA$_{287}$

b

*At*NIP1;1	--	EITK	SGSFLKTVRIGST$_{296}$
*At*NIP1;2	--	EITK	SGSFLKTVRNGSSR$_{294}$
*At*NIP4;1	--	ELTK	SASFLRAVSPSHKGSSSKT$_{283}$
*At*NIP4;2	--	ELTK	SASFLRSVAQKDNASKSDG$_{283}$
*Gm*Nod26	--	ETTK	SASFLKGRAASK$_{271}$

Fig. 1 Alignment of the amino acid sequences of C-termini of PIP2s and NIPs from several species. Multiple amino acid sequence alignment of PIP2s in *Arabidopsis thaliana* (*At*), *Brachypodium distachyon* (*Bd*), *Oryza sativa* (*Os*), *Spinacia oleracea* (*So*), and *Zea mays* (*Zm*) (A) and of NIPs in *Arabidopsis thaliana* and *Glycine max* (*Gm*) (B), the phosphorylation sites of which were experimentally determined (*red* characters; see Table 1 for details). Alignment was performed with Muscle v3.8.31. Highly conserved phosphorylation sites are in bold characters

respectively (Table 1). Such a distribution is similar to the one described for a cell lysate phosphoproteome in *Arabidopsis* (Sugiyama et al. 2008). The N-terminal tail of PIPs belonging to the PIP1 subfamily harbors phosphorylation sites in up to two residues of its N-terminal tail in *Arabidopsis*, *Brachypodium*, and maize. The PIPs belonging to the PIP2 subfamily exhibit N-terminal Ser- and Thr-phosphorylated residues in *Arabidopsis*, *Brachypodium*, and rice. In addition, PIP2s carry several conserved phosphorylation sites in their C-terminal tail (Table 1). Amino acid sequence alignments reveal that two C-terminal phosphorylated Ser residues are highly conserved in PIP2 isoforms across species (Fig. 1a). Interestingly, the C-terminal sequence alignment of NIPs revealed a conserved phosphorylation site at a Ser residue (Fig. 1b). Additional structure–function analyses of TIPs, PIPs, and NIPs in *Xenopus* oocytes (Guenther et al. 2003; Johansson et al. 1998; Maurel et al. 1995) have pointed to the role of several cytosol-exposed phosphorylation sites in controlling water transport. These studies rely on the substitution of a Ser residue by a non-phosphorylable residue. The resulting decrease of water channel activity of the corresponding aquaporin suggests a role for this residue and its phosphorylated form in water transport activity. Using this approach, a critical Ser residue conserved in all plant PIPs was identified in loop *B* (corresponding to Ser128 and Ser121 in *At*PIP1;1 and *At*PIP2;1, respectively) (Johansson et al. 1998, 1996; Amezcua-Romero et al. 2010; Temmei et al. 2005; Van Wilder et al. 2008; Azad et al. 2008) and a phosphorylable residue in loop *D* (Ser203 in *Zm*PIP2;1) (Van Wilder et al. 2008) and in the N-terminus close to the first helix (Ser35) of *Tg*PIP2;2 (Azad et al. 2008). However, such experiments are not sufficient enough to fully conclude on the role of the phosphorylated residue on aquaporin function. For that, additional studies should include the demonstration of a loss of activation by a kinase or an agonist (Johansson et al. 1998; Amezcua-Romero et al. 2010; Van Wilder et al. 2008; Azad et al. 2008) or the expression of a mutated form of the Ser residue in a phospho-mimetic residue (Asp or Glu) (Van Wilder et al. 2008; Grondin et al. 2015; Nyblom et al. 2009; Prado et al. 2013; Prak et al. 2008). Finally, some residues are still hardly detected as phosphorylated residues *in planta*, in particular the Ser residue conserved in plant PIPs loop *B*. A specific antibody raised against a phosphorylated loop *B* peptide from *Zm*PIP2 isoforms was used to detect phosphorylation of loop *B* in plant protein extracts (Aroca et al. 2005, 2007). Now, despite that peptides arising from the PIP loop *B* can be detected by MS, no phosphorylated forms of these peptides have been identified so far by a MS approach. This may be due to an instability of the phosphate group that may be lost during protein sample preparation.

2.2 Role of Aquaporin Phosphorylation

X-ray crystallography of spinach *So*PIP2;1 revealed with a high resolution the structure of an aquaporin in its closed and open states (Nyblom et al. 2009; Tornroth-Horsefield et al. 2006), shedding light onto the original gating properties of PIPs. Closure of *So*PIP2;1 is triggered by Ser115 in the cytosolic loop *B* and Ser274 in the C-terminal region. *So*PIP2;1 atomic structure models explain how phosphorylation of Ser115 in *So*PIP2;1 would destabilize the loop *D*-loop *B* anchor, thereby favoring the open-pore conformation. Interestingly, phosphorylation of Ser274 on *So*PIP2;1 C-terminal tail would act similarly but through a transactivation process, whereby interaction between the C-terminal tail and the loop *D* of an adjacent monomer would be destabilized after Ser274 phosphorylation. Yet, functional reconstitution in proteoliposomes of *So*PIP2;1 forms mutated in loop *B* failed to confirm this model (Nyblom et al. 2009). By contrast, heterologous expression in *Xenopus* oocytes of *So*PIP2;1 or *Tg*PIP2;2 indicated that phosphorylation of Ser115 or Ser274 increases channel activity (Johansson et al. 1998; Azad et al. 2008). This discrepancy may arise from different phospholipidic environments, which were shown to be critical for the water permeability of mammalian AQP4 and AQP0 (Tong et al. 2012, 2013). In addition to gating, PIP phosphorylation can modulate their subcellular localization (reviewed in (Luu and Maurel 2013); see chapter "Plant Aquaporin Trafficking"). In *At*PIP2;1, phosphorylation of Ser283, but not of Ser280, is required for targeting of newly synthesized proteins to the plasma membrane (Prak et al. 2008). This mark also provides an intracellular sorting signal for directing internalized *At*PIP2;1 to specific spherical bodies under salt stress (Prak et al. 2008).

2.3 *Phosphorylation of Aquaporin* In Planta

The role of phosphorylation *in planta* has been initially addressed through pharmacological approaches and more recently making use of quantitative proteomics. Thus, the phosphorylation mediated by staurosporin-sensitive kinases positively regulates the root hydraulic conductivity (Lp_r) in barley roots (Horie et al. 2011). Staurosporin treatments were also found to reduce significantly the Lp_r of barley plants exposed to 100 mM NaCl for 4 h (Horie et al. 2011). In addition, fluoride, a phosphatase inhibitor, was shown to lower the cytosolic calcium concentration and to act as a strong inhibitor of protein dephosphorylation (Gerbeau et al. 2002). In earlier studies, either increase (Gerbeau et al. 2002) or decrease (Kamaluddin and Zwiazek 2003; Calvo-Polanco et al. 2009; Lee et al. 2010) of root and cell hydraulic conductivity by fluoride treatment was reported. Okadaic acid, another phosphatase inhibitor, significantly inhibited the salinity-induced Lp_r repression, suggesting that dephosphorylation is a key event to downregulate Lp_r, which correlates with the role of phosphorylation in modulating Lp_r of barley (Horie et al. 2011; Kaneko et al. 2015). A few studies made use of antibodies against phosphorylated forms of aquaporins to correlate the phosphorylation level of the aquaporin with a hydraulic phenotype (Aroca et al. 2005, 2007; Sánchez-Romera et al. 2015). In particular, the use of anti-phospho-aquaporins showed that jasmonate addition modified the response of root conductance to arbuscular mycorrhizal symbiosis and drought, by regulating in part the phosphorylation state of Ser280 of *Pv*PIP2s (Sánchez-Romera et al. 2015). Since a few years, relationships between changes in phosphorylation state of aquaporins and the regulation of plant tissue hydraulics have been studied using quantitative phosphoproteomic strategies (Table 2). The phosphorylation of the C-terminal tail of *At*PIP2s appears to be highly responsive to numerous sets of constraints including abiotic, osmotic, nutritional, oxidative, as well as hormonal treatments (Prado et al. 2013; Prak et al. 2008; Di Pietro et al. 2013; Engelsberger and Schulze 2012; Hsu et al. 2009; Kline et al. 2010; Niittylä et al. 2007; Stecker et al. 2014; Vialaret et al. 2014; Wang et al. 2013a; Wu et al. 2013, 2014; Xue et al. 2013; Yang et al. 2013) (Table 2). A large-scale proteomics analysis of aquaporins in *Arabidopsis* roots under a broad set of constraints revealed that the overall phosphorylation status of PIPs was positively correlated with root hydraulic conductivity across the whole set of studied treatments (Di Pietro et al. 2013). This holds true in other species such as maize in which the doubly C-terminal phosphorylated form of *Zm*PIP2;7 was decreased upon heat stress and combined heat and drought stress

Table 2 Modulation of aquaporin phosphorylation according to treatments and mutant genetic backgrounds unraveled by quantitative phosphoproteomics in *Arabidopsis*

Treatment and mutant genetic background	Aquaporin	Phospho-residue	Quantitative behavior	Reference
NaCl/P_{starv}/P_{resup}	*At*PIP1;1/1;2	Ser24/Ser27	↓	Di Pietro et al. (2013)
Mannitol/H_2O_2	*At*PIP1;1/1;2	Ser24/Ser27	↑	Di Pietro et al. (2013)
ABA/NaCl/N_{resup}/mannitol/N_{starv}/*ctr1*/	*At*PIP2;1/2;2 /2;3	Ser280 + Ser283	↓	Prak et al. (2008), Di Pietro et al. (2013), Engelsberger and Schulze (2012), Hsu et al. (2009), Kline et al. (2010), Vialaret et al. (2014), Xue et al. (2013), Yang et al. (2013), and Wang et al. (2013b)
Darkness/P_{resup}/*bsk8*	*At*PIP2;1/2;2 /2;3	Ser280 + Ser283	↑	Prado et al. (2013), Di Pietro et al. (2013), and Wu et al. (2014)
Sucrose/NaCl/mannitol/H_2O_2/NO/darkness/suc_{resup}/N_{starv}/	*At*PIP2;1/2;2 /2;3	Ser280	↑	Di Pietro et al. (2013) and Niittylä et al. (2007)
P_{starv}	*At*PIP2;1/2;2 /2;3	Ser280	↓	Di Pietro et al. (2013)
sirk1	*At*PIP2;1/2;2 /2;3	Ser283	↓	Wu et al. (2013)
osmotic	*At*PIP2;4	Ser280 + Ser286	↑	Stecker et al. (2014) and Wang et al. (2013b)
ABA, *sirk1*	*At*PIP2;4	Ser286	↓	Kline et al. (2010) and Wu et al. (2013)
Sucrose/N_{resup}/*sirk1*	*At*PIP2;6	Ser282	↓	Engelsberger and Schulze (2012), Niittylä et al. (2007), and Wu et al. (2013)
Mannitol/H_2O_2/darkness/N_{starv}	*At*PIP2;7	Ser273 + Ser276	↑	Di Pietro et al. (2013)
ABA/P_{starv}	*At*PIP2;7	Ser273 + Ser276	↓	Di Pietro et al. (2013), Kline et al. (2010), and Xue et al. (2013)

(continued)

Table 2 (continued)

Treatment and mutant genetic background	Aquaporin	Phospho-residue	Quantitative behavior	Reference
NO, *bsk8*, *sirk1*	*At*PIP2;7	Ser273	↓	Di Pietro et al. (2013), Wu et al. (2013), and Wu et al. (2014)
Mannitol/darkness/suc_{resup}/P_{starv}/N_{starv}	*At*PIP2;7	Ser273	↑	Di Pietro et al. (2013)
ABA	*At*PIP2;8	Ser271 + Ser274	↓	Kline et al. (2010)

The table summarizes the treatments and genetic backgrounds inducing quantitative variations in the abundance of phosphorylation sites in PIPs. Precise descriptions of treatments, plant tissues, proteomic quantitative methodologies, and the amplitude in the variations of phosphorylation can be found in the references cited in the last column. Oscillations in the abundance of C-terminal diphosphorylated form of *At*PIP2;1-*At*PIP2;4 (i.e., corresponding to Ser280 and Ser283 in *At*PIP2;1) were also observed according to a circadian rhythm (Choudhary et al. 2015)

"/": aquaporin isoforms (column 2) and phospho-residues (column 3) are equivalent and not distinguishable

↑: increased level of phosphorylation, ↓: decreased level of phosphorylation, P_{starv} phosphate starvation, N_{starv} nitrate starvation, N_{resup} nitrate resupply after nitrate starvation, *NO* nitric oxide, suc_{resup} sucrose supply after darkness treatment, *osmotic* osmotic treatment, *bsk8*: At5g41260, *ctr1*: At5g03730, *sirk1*: At5g10020

(Hu et al. 2015). Aquaporin phosphorylation can also be regulated by hormones in particular abscisic acid (ABA) and ethylene (Table 2). The ethylene signaling cascade starts from the membrane-associated ethylene receptors (ETR1, ETR2), ethylene insensitive 4 (EIN4), and ethylene response sensors (ERS1, ERS2) (Hua and Meyerowitz 1998). The physical interaction of ethylene with each receptor complex triggers a suppression of the kinase activity of constitutive triple response 1 (CTR1) (Kieber et al. 1993) that is a putative Raf-like mitogen-activated PK which induces EIN2 dephosphorylation. Consequently, the C-terminus of EIN2 is cleaved off from its N-terminal domain and migrates into the nucleus to stabilize ethylene response transcription factors, ethylene insensitive 3 (EIN3) and ethylene insensitive 3-like 1 (EIL1) (Qiao et al. 2012). The quantitative phosphoproteomics analysis of *ctr1* showed that the presence of the PK CTR1 in *Arabidopsis* cells suppresses the C-terminal double phosphorylation of *At*PIP2;1 under ethylene treatment suggesting that ethylene may partially inactivate the function of CTR1 (Yang et al. 2013). Another recent quantitative phosphoproteomics study that combined protoplast swelling/shrinking experiments and leaf water loss assays on the transgenic plants expressing both the wild-type and Ser280A/Ser283A-mutated PIP2;1 in both Col-0 and *ein3eil1* genetic backgrounds suggests that ethylene increases water transport rate in *Arabidopsis* cells by enhancing Ser280/Ser283 phosphorylation at the C-terminus of PIP2;1 (Qing et al. 2016).

Upon treatment of young plantlets with 50 μM ABA for 5–30 min, the C-terminal double phosphorylation of five PIP isoforms (*At*PIP2;1, *At*PIP2;2, *At*PIP2;3, *At*PIP2;4, *At*PIP2;8) decreased (Kline et al. 2010). Thus, together with the decreased phosphorylation of the C-terminal tail of *At*PIP2s in response to water stress treatments (Prak et al. 2008; Di Pietro et al. 2013; Hsu et al. 2009; Vialaret et al. 2014; Wang et al. 2013a; Xue et al. 2013), these observations are consistent with a role of PIP dephosphorylation in decreasing water flux in response to drought and in preventing rehydration during ABA-regulated physiological processes. However, a transient increased in the phosphorylation of one of the two C-terminal phosphorylated sites (i.e., Ser280 in *At*PIP2;1) could be observed upon osmotic treatments (Di Pietro et al. 2013; Niittylä et al. 2007) suggesting an unanticipated complex regulation of the C-terminal phosphorylation of *At*PIP2;1 in a short time frame upon such treatment exposure. Then, in a more recent study aiming at characterizing posttranslational events tied to the circadian system, a survey of circadian-regulated protein phosphorylation events was carried out in *Arabidopsis* seedlings and revealed circadian oscillations in the phosphorylation state of two conserved C-terminal Ser residues of *At*PIP2;1-*At*PIP2;7 (Choudhary et al. 2015). These results suggest that circadian oscillations in the phosphorylation state of key regulatory Ser residues in aquaporins may contribute to the circadian regulation of water uptake and drive hypocotyl and leaf movement rhythms (Choudhary et al. 2015).

Most importantly, the putative regulations *in planta* require to be functionally validated by expression in transgenic plants of phospho-mimetic and phospho-deficient forms of the aquaporin isoform involved. Thus, in support of quantitative phosphoproteomics, expression studies of phospho-mimetic and phosphorylation-deficient forms of *At*PIP2;1 in transgenic *Arabidopsis* were used (Prado et al. 2013).

In this study, phosphorylation at Ser280 and Ser283 of *At*PIP2;1 was shown to be necessary for mediating leaf hydraulic conductivity enhancement under darkness (Prado et al. 2013). Quantitative proteomic analyses showed that light-dependent regulation of rosette hydraulic conductivity is linked to the double C-terminal phosphorylation of *At*PIP2;1. Expression in *pip2;1* plants of phospho-mimetic and phosphorylation-deficient forms of *At*PIP2;1;1 demonstrated that phosphorylation at these two sites is necessary for rosette hydraulic conductivity enhancement under darkness. These findings establish how regulation of a single aquaporin isoform by phosphorylation in leaf veins critically determines leaf hydraulics (Prado et al. 2013). In a recent work, aquaporins were shown to contribute to ABA-triggered stomatal closure through OST1-mediated phosphorylation (Grondin et al. 2015). Hence, the expression in *pip2;1* plants, of a phospho-mimetic form (mutation of Ser121 in Asp121) but not a phospho-deficient form of PIP2;1 (mutation of Ser121 in Ala121), was shown to constitutively enhance the osmotic water permeability of guard cell protoplasts while suppressing its ABA-dependent activation and was able to restore ABA-dependent stomatal closure in *pip2;1* (see also chapters "Aquaporins and Leaf Water Relations" and "Roles of Aquaporins in Stomata"). Thus, this work demonstrated that ABA-triggered stomatal closure requires an increase in guard cell permeability to water and possibly to hydrogen peroxide, through OST1-dependent phosphorylation of *At*PIP2;1 at Ser121 (Grondin et al. 2015).

2.4 Protein Kinase and Phosphatase Acting on Aquaporins

Characterization of specific protein kinases (PK) and phosphatases involved in printing and removing phosphorylation marks on aquaporins is of importance to understand aquaporin regulation networks. However, the knowledge of such modifying enzymes is still scarce. TIP isoforms from the protein storage vacuole membrane of *Lens culinaris* seeds were shown to be phosphorylated by a 52 kDa Mg^{2+}-dependent PK (Harvengt et al. 2000). Two PKs acting on the phosphorylation sites Ser115 and Ser274 in *So*PIP2;1 were characterized (Sjovall-Larsen et al. 2006). The PK acting on Ser115 was shown to be cytosolic and Mg^{2+} dependent, and the one acting on Ser274 was shown to be a plasma membrane-bound Ca^{2+}-dependent PK (Sjovall-Larsen et al. 2006). A similar Ca^{2+}-dependent PK was described in maize (Van Wilder et al. 2008). A recent work revealed that Open stomata 1 (OST1)/Snf1-related PK 2.6 (SnRK2.6), a PK involved in guard cell ABA signaling, was able to phosphorylate a cytosolic *At*PIP2;1 peptide at Ser121 (Grondin et al. 2015). This work strongly suggested that ABA-triggered stomatal closure requires an increase in guard cell permeability to water and possibly to hydrogen peroxide, through SnrK2.6-dependent phosphorylation of *At*PIP2;1 at Ser121 (Grondin et al. 2015) (see also chapters "Aquaporins and Leaf Water Relations" and "Roles of Aquaporins in Stomata").

A large-scale analysis of sucrose-induced protein phosphorylation recently led to the identification of SIRK1 and BSK8, a leucine-rich repeat receptor-like kinase

(LRR-RLK) and a receptor-like cytoplasmic kinase (RLCK), respectively (Niittylä et al. 2007; Wu et al. 2013). The phosphorylation of five aquaporins (*At*PIP2;1–2;4, *At*PIP2;7) in response to sucrose stimulation was reduced in the mutant *sirk1* upon sucrose stimulation, and *At*PIP2;4 was confirmed to be modulated directly by the receptor kinase SIRK1 (Wu et al. 2013). SIRK1 is thus supposed to be involved in the regulation of sucrose-specific osmotic responses through direct interaction with and activation of an aquaporin via phosphorylation, and it is presumed that the duration of this response is controlled by phosphorylation-dependent receptor internalization (Wu et al. 2013). In addition, interestingly, among the proteins with reduced phosphorylation in *bsk8* plants was the aquaporin *At*PIP2;7 (Wu et al. 2014). Thus, responses of aquaporins to sucrose-induced osmotic changes may involve different PKs. However, an enhancement of the diphosphorylated form of *At*PIP2;1 at Ser280 and Ser283 in *bsk8* plants was noticed, suggesting that BSK8 kinase may also indirectly suppress the phosphorylation of *At*PIP2;1 (Wu et al. 2014). A few other studies suggest that cAMP-dependent PK, cGMP-dependent PK, and PKC families, also called AGC kinases, could be involved in aquaporin phosphorylation. Thus, in *Mc*PIP2;1, both water permeation and phosphorylation status of *Mc*PIP2;1 were shown to be markedly decreased by the inhibitory peptides PKI 14–22 and PKC 20–28, inhibitors of PKA and PKC, respectively (Amezcua-Romero et al. 2010). Activation of endogenous PKA increased the osmotic water permeability coefficient of *Pv*TIP3;1- and *Zm*PIP2;1-expressing oocytes (Maurel et al. 1995; Van Wilder et al. 2008). The targets of PKA were described as both in loop *B* (Ser126) and loop *D* (Ser203) of *Zm*PIP2;1 (Van Wilder et al. 2008). The use of predictive bioinformatic tools for PK motifs also suggested that *At*PIP2s could be substrates of PKA and PKC (Vialaret et al. 2014). However, such PKs remain to be characterized. Overall, these studies reveal that different PKs act on aquaporins and that a single serine residue may be the target of several PKs. Since a large number of environmental conditions modulate the phosphorylation status of aquaporins, additional PKs acting on aquaporins surely remain to be discovered as well as phosphatases. For instance, hypothetical phosphorylation cascades triggered by ethylene and negatively regulating aquaporin phosphorylation have recently been modeled (Yang et al. 2013), but we still ignore the signaling components involved.

3 Other Posttranslational Modifications of Aquaporins

3.1 N-Terminal Modification

Modifications affecting the protein N-terminus, collectively known as *N*-terminal protein modifications, are the earliest modifications which a protein undergoes. They involve the N-terminal methionine excision process, N-α-acetylation, and N-myristoylation. In *Arabidopsis*, the initiating methionine was shown to be co-translationally processed in all PIP isoforms studied: whereas it was Nα-acetylated in members of the *At*PIP1 subclass, methionine was cleaved in *At*PIP2 isoforms

(Santoni et al. 2006). These observations were also described for PIPs in broccoli (Casado-Vela et al. 2010). Excision of the N-terminal methionine has been extensively studied in soluble proteins (Giglione et al. 2004; Moerschell et al. 1990) and to a much lesser extent in membrane proteins. This process was shown to depend on the nature of penultimate residues and occurs in the presence of residues with side chains of reduced steric hindrance (Giglione et al. 2004; Moerschell et al. 1990). In agreement with this rule, *At*PIP2;1, *At*PIP2;2, and *At*PIP2;4, which exhibit an alanine residue at position 2, all had their methionine excised, whereas PIP1s, which have a perfectly conserved glutamic acid residue, did not. N-terminal acetylation, which has been thoroughly studied in soluble mammalian and yeast proteins, depends on N-terminal sequences (Polevoda and Sherman 2003). Recent studies provide evidence that N-terminal methionine excision may play a role in the regulation of protein stability, reviewed in (Giglione et al. 2015). Similarly, recent studies revealed that biological functions of N-terminal α-acetylation are to create specific degrons inducing degradation of the protein by the proteasome, to influence membrane targeting, protein–protein interactions, and changes in proteostasis (see (Giglione et al. 2015) for a recent review). However, additional studies are required to unravel the role of N-terminal modifications in PIP aquaporins.

3.2 Deamidation

Deamidation is the irreversible conversion of the amino acids glutamine and asparagine to glutamic acid and aspartic acid, respectively. This irreversible amino acid conversion results in an increase of approximately 1 Da in the mass of the target protein, an increase in its negative charge, and the release of ammonia. Specific enzymatic deamidation can regulate cellular functions (Chao et al. 2006), and non-enzymatic process may also spontaneously occur during sample preparation (Hao et al. 2011; Li et al. 2008). In several *At*PIPs, deamidation was shown to occur at Asn and Gln residues located in the second extracellular loop (loop *C*) and in the N- and C-terminal tails (Di Pietro et al. 2013). Interestingly, the deamidation of several *At*PIPs showed quantitative variations according to abiotic treatments of plants that modulate aquaporin activity, suggesting that deamidation may occur *in planta* (Di Pietro et al. 2013). In addition, deamidation was observed to be closely associated with phosphorylation in C-terminal *At*PIP2 peptides (Di Pietro et al. 2013). This observation may provide a clue to explain how an apparent increase in *At*PIP2 C-terminal phosphorylation could be associated with a decrease in Lp_r. A typical case is *At*PIP2;7, for which the abundance of the mono-phosphorylated form and that of the same form with an additional deamidation showed opposite behaviors (Di Pietro et al. 2013). However, it remains unknown whether or how neighboring deamidation and phosphorylation could interfere to alter protein function.

Data from the literature suggest that deamidation possibly interferes with protein stability, protein activation, and protein–protein interactions. In mammalian AQP0, several sites of N-terminal truncation have been identified as sites of Asn deamidation

(Ball et al. 2004; Schey et al. 2000). This age-related PTM, together with other PTMs, including phosphorylation, is supposed to serve as a molecular clock (Robinson and Robinson 2001) to regulate AQP0 function. In several *At*PIPs, deamidation was identified at a conserved Asn residue in loop *C* (Di Pietro et al. 2013). In the case of *At*PIP2;7, molecular modeling according to the previously described *So*PIP2;1 structure (Tornroth-Horsefield et al. 2006) indicated that the deamidated Asn152 residue is located at the entrance of the pore, close to Arg224 (Di Pietro et al. 2013). The latter residue contributes to one of the main pore constrictions involved in substrate specificity. By introducing a negative charge at physiological pH, deamidation may interfere with aquaporin action. However, such speculative effects of deamidation on aquaporin function clearly deserve further experimental work.

3.3 Glycosylation

Glycosylation is a complex topic as many glycosylation-branched configurations with different sugar units exist. However, a consensus sequence motif Asn-Xaa-Ser/Thr (Xaa is any amino acid except Pro) has been defined as a prerequisite for N-glycosylation in eucaryotes (Blom et al. 2004). N-glycosylation was predicted for some MIPs, among them *At*PIP2;1 and *Os*NIP2;1, at sites located in the N- and C-termini, loops *A*, *B*, and *E*, and in the second transmembrane domain that remain to be experimentally described and functionally investigated (Hove and Bhave 2011). In addition, N-glycosylation has been experimentally observed in *Gm*NOD26 (Miao et al. 1992) and in *Mc*TIP2;1 (Vera-Estrella et al. 2004). *At*NIPs and *At*TIP4;1 were predicted to have putative O-glycosylation sites, mostly in loop *B* (Hove and Bhave 2011). A possible consensus for O-glycosylation, G-H-I-S-G-A/G-H-X, unique to NIPs only, was also identified (Hove and Bhave 2011). The glycosylation of animal aquaporins has been reported to be involved in the proper routing and membrane insertion of mammalian AQP1 and AQP2 (Baumgarten et al. 1998) and exit of AQP2 from Golgi complex and sorting to plasma membrane (Hendricks et al. 2004). The role of glycosylation in plant aquaporins is much less understood. It was shown that upon a mannitol treatment, *Mc*TIP1;2 may reside in several membrane compartments and that such a redistribution depends on posttranslational modifications including glycosylation (Vera-Estrella et al. 2004).

3.4 Methylation

Protein methylation can occur as either N-methylation of residues such as lysine, arginine, histidine, alanine, proline, glutamine, phenylalanine, asparagine, and methionine or carboxymethylation, i.e., O-methylesterification of glutamic acid and aspartic acid residues. The N-terminal tail of several PIP1 and PIP2 isoforms in *Arabidopsis* and in *Brassica oleracea* was described to be methylated (Santoni et al. 2006; Casado-Vela et al. 2010). In particular, *At*PIP2;1 and *At*PIP2;2 isoforms carried methylated Lys and Glu residues on the cytosolic N-terminal tail. Thus, Lys3 and Glu6 of *At*PIP2;1, although they existed in their native form, occasionally carried dimethyl (Lys3) or monomethyl (Glu6) moieties (Santoni et al. 2006). In *At*PIP2;1, a diacidic motif located on the N-terminal tail likely interacts with the COPII sorting machinery and critically determines their export from the endoplasmic reticulum (Matheson et al. 2006; Sorieul et al. 2011) (see chapter "Plant Aquaporin Trafficking"). The N-terminal methylated residues overlap with this diacidic motif suggesting a role for *At*PIP2;1 methylation in protein subcellular trafficking. Yet, any mutation at these sites had dominating effects on aquaporin trafficking, thereby preventing proper structure–function analyses (Santoni et al. 2006). Two *Arabidopsis* methyltransferases were identified to methylate the N-terminal tail of *At*PIP2;1, specifically at either one of the two methylation sites Lys3 and Glu6 (Sahr et al. 2010). SDG7 (At2g44150), an extranuclear methyltransferase, and OMTF3 (At3g61990), a glutamate methyltransferase, were shown to dimethylate Lys3 and monomethylate Glu6, respectively, in *At*PIP2;1(Sahr et al. 2010). The localization of OMTF3 and SDG7 proteins in the endoplasmic reticulum but not in the plasma membrane was described to be compatible with aquaporin expression and function and pointed to a role for methylation during aquaporin biogenesis or targeting to the PM. However, although SDG7 and OMTF3 meet all enzymatic criteria for methylating *At*PIP2;1 in terms of substrate affinity or preference, the methylation of this aquaporin by the two enzymes had not been proven in vivo, and the true physiological role of the two enzymes awaits elucidation.

3.5 Ubiquitination

Although aquaporin abundance can markedly vary in response to environmental conditions, information on the modes of plant aquaporin degradation has remained scarce. Ubiquitination is the posttranslational attachment of ubiquitin (Ub), a highly conserved 8-kD protein, to a wide range of target proteins (Pickart and Eddins 2004; Mukhopadhyay and Riezman 2007). In the ubiquitination pathway, Ub is attached to substrate proteins in three consecutive steps catalyzed by E1, E2, and E3 enzymes (Kraft et al. 2005; Stone et al. 2005). The ubiquitin chain topology includes the formation of Lys48-linked poly-ubiquitin chains which are able to direct the degradation of tagged proteins by the 26S proteasome. However, recent results in the literature revealed that atypical linkages through lysine residues at positions 6, 11, 27, 29, 33, and 63 perform a range of diverse functions (Walsh and Sadanandom 2014). Using stringent two-step affinity methods to purify Ub-protein conjugates followed by high-sensitivity MS, about 950 ubiquitylation substrates were identified in whole *Arabidopsis thaliana* seedlings, including *At*PIP1;1–1;4, *At*PIP2;1, *At*PIP2;4,

*At*PIP2;7, *At*PIP2;8, *At*TIP1;2, and *At*TIP2;1 (Kim et al. 2013). However, Ub attachment sites were not resolved. The ubiquitylation level of *At*TIP1;2, *At*TIP2;1, *At*PIP1;1, *At*PIP1;4, *At*PIP2;1, and *At*PIP2;7 was shown to increase after treating seedlings with the proteasome inhibitor MG132, strongly suggesting that Ub addition commits these proteins to degradation by the 26S proteasome (Kim et al. 2013). In another study, a RING membrane-anchor E3 ubiquitin ligase from pepper, and its *Arabidopsis* homologs, was shown to localize in the endoplasmic reticulum and to ubiquitinate *At*PIP2;1 leading to its retention in this compartment (Lee et al. 2009) (see chapter "Plant Aquaporin Trafficking"). This process was enhanced under osmotic stress, leading to aquaporin degradation through the proteasome pathway. In addition to a degradation process driven by a putative aquaporin ubiquitination and subsequent proteasomal targeting, a recent study established a novel link between aquaporin degradation and intracellular trafficking in *Arabidopsis* (Hachez et al. 2014) independent of ubiquitination pathway. Thus, *At*PIP2;7 was shown to interact in the endoplasmic reticulum and Golgi, with a membrane protein, named TSPO for tryptophan-rich sensory protein/translocator and serving as multistress regulator (see chapter "Plant Aquaporin Trafficking"). The complex is then directed toward vacuolar degradation, using the autophagosome pathway, this process being stimulated by ABA (Hachez et al. 2014). Thus, plant cells under water stress can use various pathways, including ubiquitination and autophagy pathways, for aquaporin degradation and long-term downregulation of plasma membrane water permeability.

4 Conclusion and Perspective

The advent of highly sensitive MS-based strategies improved our knowledge about PTMs in aquaporins and revealed that aquaporins are highly modified membrane proteins (Fig. 2). In addition, functional studies revealed that PTMs may interfere with aquaporin gating and subcellular localization and degradation (Fig. 2). However, some residues still remain to be identified *in planta* in their modified form, in particular a critical serine residue in loop *B* that strongly interferes with aquaporin intrinsic activity. In addition, the ubiquitinated sites still remain to be identified as well as their role as part of proteasome-mediated degradation and proteasome-independent roles, including the endocytosis of membrane proteins as recently described for a few plant membrane proteins (Leitner et al. 2012; Martins et al. 2015; Barberon et al. 2011). Quantitative proteomics has contributed to our understanding of the functional significance of PTMs, in particular of phosphorylation. However, additional studies are still necessary to fully interpret the role of PTMs. In addition, due to putative interplay between PTMs (Venne et al. 2014), a challenge will be to describe these cross talks and interpret their role as part of regulatory mechanisms of aquaporin function.

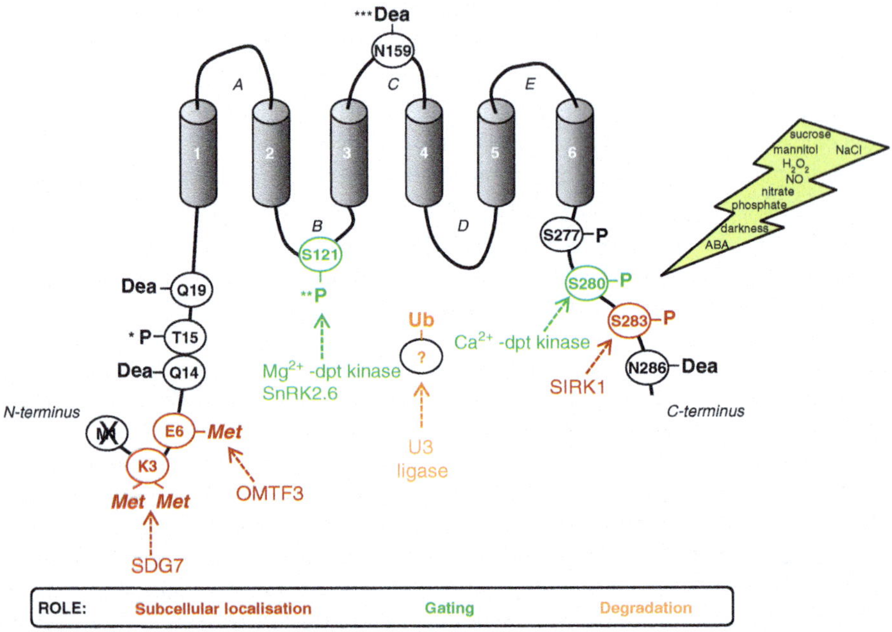

Fig. 2 Multiple covalent posttranslational mechanisms acting on PIP2 in *Arabidopsis*. The figure, adapted from Maurel et al. (2015), shows a schematic representation of *At*PIP2;1 with its six transmembrane domains, five connecting loops (*A–E*), and N- and C-terminal tails bathing in the cytosol. Unless stated, the indicated modifications were experimentally determined and are detailed in Table 1. They include cleavage of the initiating methionine (cross); dimethylation (*Met Met*) of a Lys residue (*K3*); monomethylation (*Met*) of a Glu residue (*E6*) (Santoni et al. 2006); deamidation (*Dea*) of Gln and Asn residues (Q14, Q19, N159, N286) (Di Pietro et al. 2013); phosphorylation of Ser121 (S121) (Tornroth-Horsefield et al. 2006); phosphorylation of C-terminal Ser residues S277, S280, and S283 (see Table 1 for references); and ubiquitination (*Ub*), the ubiquitinated residue(s) being unknown (Lee et al. 2009). *The phosphorylation at Thr15 (*T15*) in *At*PIP2;1 was deduced from amino acid sequence alignment with *At*PIP2;3 for which Thr13 was experimentally shown to be phosphorylated (Wu et al. 2013). **The phosphorylation of loop *B* at Ser121 was inferred from studies on spinach *So*PIP2;1 (Tornroth-Horsefield et al. 2006). ***The deamidation at Asn159 (*N159*) in *At*PIP2;1 was deduced from amino acid sequence alignment with *At*PIP2;7 for which Asn152 was experimentally determined to be deamidated (Di Pietro et al. 2013). Enzymes involved in covalent modification of PIP2 aquaporins in different plant species are indicated: two methyltransferases SDG7 and OMTF3 (Sahr et al. 2010), a U3 ligase (Lee et al. 2009), Mg2+- and Ca2+-dependent protein kinases (Sjovall-Larsen et al. 2006), Snrk2.6 (Grondin et al. 2015), BSK8 (Wu et al. 2014), and SIRK1 (Wu et al. 2013). The figure is color coded to show the main molecular and cellular function [subcellular localization (*red*), gating (*green*), and degradation of the aquaporin (*orange*)] associated to the posttranslational modification. Treatments that modulate the level of phosphorylation of S280 and S283 are indicated (see Table 2 for details and references)

References

Amezcua-Romero JC, Pantoja O, Vera-Estrella R (2010) Ser123 is essential for the water channel activity of McPIP2;1 from *Mesembryanthemum crystallinum*. J Biol Chem 285:16739–16747

Aroca R, Amodeo G, Fernandez-Illescas S, Herman EM, Chaumont F, Chrispeels MJ (2005) The role of aquaporins and membrane damage in chilling and hydrogen peroxide induced changes in the hydraulic conductance of maize roots. Plant Physiol 137:341–353

Aroca R, Porcel R, Ruiz-Lozano JM (2007) How does arbuscular mycorrhizal symbiosis regulate root hydraulic properties and plasma membrane aquaporins in *Phaseolus vulgaris* under drought, cold, or salinity stresses. New Phytol 173:808–819

Azad AK, Katsuhara M, Sawa Y, Ishikawa T, Shibata H (2008) Characterization of four plasma membrane aquaporins in tulip petals: a putative homolog is regulated by phosphorylation. Plant Cell Physiol 49:1196–1208

Ball LE, Garland DL, Crouch RK, Schey KL (2004) Post-translational modifications of aquaporin-0 (AQP0) in the normal human lens: spatial and temporal occurrence. Biochemistry 43:9856–9865

Band LR, Wells DM, Larrieu A, Sun JY, Middleton AM, French AP, Brunoud G, Sato EM, Wilson MH, Peret B et al (2012) Root gravitropism is regulated by a transient lateral auxin gradient controlled by a tipping-point mechanism. Proc Natl Acad Sci U S A 109:4668–4673

Barberon M, Zelazny E, Robert S, Conejero G, Curie C, Friml J, Vert G (2011) Monoubiquitin-dependent endocytosis of the iron-regulated transporter 1 (IRT1) transporter controls iron uptake in plants. Proc Natl Acad Sci U S A 108:E450–E458

Baumgarten R, Van de Pol MHJ, Wetzels JFM, Van Os CH, Deen PMT (1998) Glycosylation is not essential for vasopressin-dependent routing of aquaporin-2 in transfected Madin-Darby, canine kidney cells. J Am Soc Nephrol 9:1553–1559

Blom N, Sicheritz-Ponten T, Gupta R, Gammeltoft S, Brunak S (2004) Prediction of post-translational glycosylation and phosphorylation of proteins from the amino acid sequence. Proteomics 4:1633–1649

Calvo-Polanco M, Zwiazek JJ, Jones MD, MacKinnon MD (2009) Effects of NaCl on responses of ectomycorrhizal black spruce (*Picea mariana*), white spruce (*Picea glauca*) and jack pine (*Pinus banksiana*) to fluoride. Physiol Plant 135:51–61

Casado-Vela J, Muries B, Carvajal M, Iloro I, Elortza F, Martinez-Ballesta MC (2010) Analysis of root plasma membrane aquaporins from *Brassica oleracea*: post-translational modifications, de novo sequencing and detection of isoforms by high resolution mass spectrometry. J Proteome Res 9:3479–3494

Chao XJ, Muff TJ, Park SY, Zhang S, Pollard AM, Ordal GW, Bilwes AM, Crane BR (2006) A receptor-modifying deamidase in complex with a signaling phosphatase reveals reciprocal regulation. Cell 124:561–571

Choudhary MK, Nomura Y, Wang L, Nakagami H, Somers DE (2015) Quantitative circadian phosphoproteomic analysis of *Arabidopsis* reveals extensive clock control of key components in physiological, metabolic and signaling pathways. Mol Cell Proteomics 14:2243–2260

Daniel JA, Yeager M (2005) Phosphorylation of aquaporin PvTIP3;1 defined by mass spectrometry and molecular modeling. Biochemistry 44:14443–14454

Di Pietro M, Vialaret J, Li G-W, Hem S, Prado K, Rossignol M, Maurel C, Santoni V (2013) Coordinated post-translational responses of aquaporins to abiotic and nutritional stimuli in *Arabidopsis* roots. Mol Cell Proteomics 12:3886–3897

Durek P, Schmidt R, Heazlewood JL, Jones A, MacLean D, Nagel A, Kersten B, Schulze WX (2010) PhosPhAt: the *Arabidopsis thaliana* phosphorylation site database. An update. Nucleic Acids Res 38:D828–D834

Endler A, Reiland S, Gerrits B, Schmidt UG, Baginsky S, Martinoia E (2009) In vivo phosphorylation sites of barley tonoplast proteins identified by a phosphoproteomic approach. Proteomics 9:310–321

Engelsberger WR, Schulze WX (2012) Nitrate and ammonium lead to distinct global dynamic phosphorylation patterns when resupplied to nitrogen-starved *Arabidopsis* seedlings. Plant J 69:978–995
Gerbeau P, Amodeo G, Henzler T, Santoni V, Ripoche P, Maurel C (2002) The water permeability of *Arabidopsis* plasma membrane is regulated by divalent cations and pH. Plant J 30:71–81
Giglione C, Boularot A, Meinnel T (2004) Protein N-terminal methionine excision. Cell Mol Life Sci 61:1455–1474
Giglione C, Fieulaine S, Meinnel T (2015) N-terminal protein modifications: bringing back into play the ribosome. Biochimie 114:134–146
Grondin A, Rodrigues O, Verdoucq L, Merlot S, Leonhardt N, Maurel C (2015) Aquaporins contribute to ABA-triggered stomatal closure through OST1-mediated phosphorylation. Plant Cell 27:1–11
Guenther JF, Chanmanivone N, Galetovic MP, Wallace IS, Cobb JA, Roberts DM (2003) Phosphorylation of soybean nodulin 26 on serine 262 enhances water permeability and is regulated developmentally and by osmotic signals. Plant Cell 15:981–991
Hachez C, Veljanovski V, Reinhardt H, Guillaumot D, Vanhee C, Chaumont F, Batoko H (2014) The Arabidopsis abiotic stress-induced TSPO-related protein reduces cell-surface expression of the aquaporin PIP2;7 through protein-protein interactions and autophagic degradation. Plant Cell 26:4974–4990
Hao PL, Ren Y, Alpert AJ, Sze SK (2011) Detection, evaluation and minimization of nonenzymatic deamidation in proteomic sample preparation. Mol Cell Proteomics 10:0111.009381
Harvengt P, Vlerick A, Fuks B, Wattiez R, Ruysschaert J-M, Homblé F (2000) Lentil seed aquaporins form a hetero-oligomer which is phosphorylated by a Mg^{2+}-dependent and Ca^{2+}-regulated kinase. Biochem J 352:183–190
Hendricks G, Koudijs M, van Balkom BWM, Oorschot V, Klumperman J, Deen PMT, van der Sluijs P (2004) Glycosylation is important for cell surface expression of the water channel aquaporin-2 but is not essential for tetramerization in the endoplasmic reticulum. J Biol Chem 279:2975–2983
Horie T, Kaneko T, Sugimoto G, Sasano S, Panda SK, Shibasaka M, Katsuhara M (2011) Mechanisms of water transport mediated by PIP aquaporins and their regulation via phosphorylation events under salinity stress in barley roots. Plant Cell Physiol 52:663–675
Hove RM, Bhave M (2011) Plant aquaporins with non-aqua functions: deciphering the signature sequences. Plant Mol Biol 75:413–430
Hsu JL, Wang LY, Wang SY, Lin CH, Ho KC, Shi FK, Chang IF (2009) Functional phosphoproteomic profiling of phosphorylation sites in membrane fractions of salt-stressed *Arabidopsis thaliana*. Proteome Sci 7:42
Hu X, Wu L, Zhao F, Zhang D, Li N, Zhu G, Li C, Wang W (2015) Phosphoproteomic analysis of the response of maize leaves to drought, heat and their combination stress. Front Plant Sci 5:298
Hua J, Meyerowitz EM (1998) Ethylene responses are negatively regulated by a receptor gene family in *Arabidopsis thaliana*. Cell 94:261–271
Jensen ON (2006) Interpreting the protein language using proteomics. Mol Cell Biol 7:391–403
Johansson I, Karlsson M, Shukla VK, Chrispeels MJ, Larsson C, Kjellbom P (1998) Water transport activity of the plasma membrane aquaporin PM28A is regulated by phosphorylation. Plant Cell 10:451–459
Johansson I, Larsson C, Ek B, Kjellbom P (1996) The major integral proteins of spinach leaf plasma membranes are putative aquaporins and are phosphorylated in response to Ca^{2+} and apoplastic water potential. Plant Cell 8:1181–1191
Johnson KD, Chrispeels MJ (1992) Tonoplast-bound protein kinase phosphorylates tonoplast intrinsic protein. Plant Physiol 100:1787–1795
Kamaluddin M, Zwiazek JJ (2003) Fluoride inhibits root water transport and affects leaf expansion and gas exchange in aspen (*Populus tremuloides*) seedlings. Physiol Plant 117:368–375
Kaneko T, Horie T, Nakahara Y, Tsuji N, Shibasaka M, Katsuhara M (2015) Dynamic regulation of the root hydraulic conductivity of barley plants in response to salinity/osmotic stress. Plant Cell Physiol 56:875–882

Kieber JJ, Rothenberg M, Roman G, Feldmann KA, Ecker JR (1993) CTR1, a negative regulator of the ethylene response pathway in *Arabidopsis*, encodes a member of the RAF family of protein-kinases. Cell 72:427–441

Kim DY, Scalf M, Smith LM, Vierstra RD (2013) Advanced proteomic analyses yield a deep catalog of ubiquitylation targets in *Arabidopsis*. Plant Cell 25:1523–1540

Kline KG, Barrett-Wilt GA, Sussman MR (2010) In planta changes in protein phosphorylation induced by the plant hormone abscisic acid. Proc Natl Acad Sci U S A 36:15986–15991

Kraft E, Stone SL, Ma LG, Su N, Gao Y, Lau OS, Deng XW, Callis J (2005) Genome analysis and functional characterization of the E2 and RING-type E3 ligase ubiquitination enzymes of Arabidopsis. Plant Physiol 139:1597–1611

Lee SH, Calvo-Polanco M, Chung GC, Zwiazek JJ (2010) Role of aquaporins in root water transport of ectomycorrhizal jack pine (*Pinus banksiana*) seedlings exposed to NaCl and fluoride. Plant Cell Environ 33:769–780

Lee HK, Cho SK, Son O, Xu Z, Hwang I, Kim WT (2009) Drought stress-induced Rma1H1, a RING membrane-anchor E3 ubiquitin ligase homolog, regulates aquaporin levels via ubiquitination in transgenic *Arabidopsis* plants. Plant Cell 21:622–641

Leitner J, Petrasek J, Tomanov K, Retzer K, Parezova M, Korbei B, Bachmair A, Zazimalova E, Luschnig C (2012) Lysine(63)-linked ubiquitylation of PIN2 auxin carrier protein governs hormonally controlled adaptation of *Arabidopsis* root growth. Proc Natl Acad Sci U S A 109:8322–8327

Li XJ, Cournoyer JJ, Lin C, O'Cormora PB (2008) Use of O-18 labels to monitor deamidation during protein and peptide sample processing. J Am Soc Mass Spectrom 19:855–864

Luu DT, Maurel C (2013) Aquaporin trafficking in plant cells: an emerging membrane-protein model. Traffic 14:629–635

Lv DW, Subburaj S, Cao M, Yan X, Li XH, Appels R, Sun DF, Ma WJ, Yan YM (2014) Proteome and phosphoproteome characterization reveals new response and defense mechanisms of *Brachypodium distachyon* leaves under salt stress. Mol Cell Proteomics 13:632–652

Martins S, Dohmann EMN, Cayrel A, Johnson A, Fischer W, Pojer F, Satiat-Jeunemaitre B, Jaillais Y, Chory J, Geldner N et al (2015) Internalization and vacuolar targeting of the brassinosteroid hormone receptor BRI1 are regulated by ubiquitination. Nat Commun 6:6151

Matheson LA, Hanton SL, Brandizzi F (2006) Traffic between the plant endoplasmic reticulum and Golgi apparatus: to the Golgi and beyond. Curr Opin Plant Biol 9:601–609

Maurel C, Boursiac Y, Luu D-T, Santoni V, Shahzad Z, Verdoucq L (2015) Aquaporins in plants. Physiol Rev 95:1321–1358

Maurel C, Kado RT, Guern J, Chrispeels MJ (1995) Phosphorylation regulates the water channel activity of the seed-specific aquaporin alpha-TIP. EMBO J 14:3028–3035

Miao G-H, Hong Z, Verma DPS (1992) Topology and phosphorylation of soybean nodulin-26, an intrinsic protein of the peribacteroid membrane. J Cell Biol 118:481–490

Moerschell RP, Hosokawa Y, Tsunasawa S, Sherman F (1990) The specificities of yeast methionine aminopeptidase and acetylation of amino-terminal methionine in vivo. J Biol Chem 265:19638–19643

Mukhopadhyay D, Riezman H (2007) Proteasome-independent functions of ubiquitin in endocytosis and signaling. Science 315:201–205

Nakagami H, Sugiyama N, Mochida K, Daudi A, Yoshida Y, Toyoda T, Tomita M, Ishihama Y, Shirasu K (2010) Large-scale comparative phosphoproteomics identifies conserved phosphorylation sites in plants. Plant Physiol 153:1161–1174

Niittylä T, Fuglsang AT, Palmgren MG, Frommer WB, Schulze WX (2007) Temporal analysis of sucrose-induced phosphorylation changes in plasma membrane proteins of *Arabidopsis*. Mol Cell Proteomics 6:1711–1726

Nühse TS, Bottrill AR, Jones AM, Peck SC (2007) Quantitative phosphoproteomic analysis of plasma membrane proteins reveals regulatory mechanisms of plant innate immune responses. Plant J 51:931–940

Nühse TS, Stensballe A, Jensen ON, Peck SC (2004) Phosphoproteomics of the *Arabidopsis* plasma membrane and a new phosphorylation site database. Plant Cell 16:2394–2405

Nyblom M, Frick A, Wang Y, Ekvall M, Hallgren K, Hedfalk K, Neutze R, Tajkhorshid E, Törnroth-Horsefield S (2009) Structural and functional analysis of SoPIP2;1 mutants adds insight into plant aquaporin gating. J Mol Biol 387:653–668

Olsen JV, Mann M (2013) Status of large-scale analysis of post-translational modifications by mass spectrometry. Mol Cell Proteomics 12:3444–3452

Pickart CM, Eddins MJ (2004) Ubiquitin: structures, functions, mechanisms. Biochim Biophys Acta-Mol Cell Res 1695:55–72

Polevoda B, Sherman F (2003) Composition and function of the eukaryotic N-terminal acetyltransferase subunits. Biochem Biophys Res Commun 308:1–11

Prado K, Boursiac Y, Tournaire-Roux C, Monneuse JM, Postaire O, Da Ines O, Schäffner AR, Hem S, Santoni V, Maurel C (2013) Regulation of *Arabidopsis* leaf hydraulics involves light-dependent phosphorylation of aquaporins in veins. Plant Cell 25:1029–1039

Prak S, Hem S, Boudet J, Viennois J, Sommerer N, Rossignol R, Maurel C, Santoni V (2008) Multiple phosphorylations in the C-terminal tail of plant plasma membrane aquaporins. Role in sub-cellular trafficking of *At*PIP2;1 in response to salt stress. Mol Cell Proteomics 7:1019–1030

Qiao H, Shen ZX, Huang SSC, Schmitz RJ, Urich MA, Briggs SP, Ecker JR (2012) Processing and subcellular trafficking of ER-tethered EIN2 control response to ethylene gas. Science 338:390–393

Qing DJ, Yang Z, Li MZ, Wong WS, Guo GY, Liu SC, Guo HW, Li N (2016) Quantitative and functional phosphoproteomic analysis reveals that ethylene regulates water transport via the C-terminal phosphorylation of aquaporin PIP2; 1 in *Arabidopsis*. Mol Plant 9:158–174

Reiland S, Finazzi G, Endler A, Willig A, Baerenfaller K, Grossmann J, Gerrits B, Rutishauser D, Gruissem W, Rochaix J-D et al (2011) Comparative phosphoproteome profiling reveals a function of the STN8 kinase in fine-tuning of cyclic electron flow (CEF). Proc Natl Acad Sci U S A 108:12955–12960

Reiland S, Messerli G, Baerenfaller K, Gerrits B, Endler A, Grossmann J, Gruissem W, Baginsky S (2009) Large-scale *Arabidopsis* phosphoproteome profiling reveals novel chloroplast kinase substrates and phosphorylation networks. Plant Physiol 150:889–903

Robinson NE, Robinson AB (2001) Deamidation of human proteins. Proc Natl Acad Sci U S A 98:12409–12413

Sahr T, Adam T, Fizames C, Maurel C, Santoni V (2010) O-Carboxyl- and N-methyltransferases active on plant aquaporins. Plant Cell Physiol 51:2092–2104

Sánchez-Romera B, Ruiz-Lozano JM, Zamarreño ÁM, García-Mina JM, Aroca R (2015) Arbuscular mycorrhizal symbiosis and methyl jasmonate avoid the inhibition of root hydraulic conductivity caused by drought. Mycorrhiza 26:111–122

Santoni V, Verdoucq L, Sommerer N, Vinh J, Pflieger D, Maurel C (2006) Methylation of aquaporins in plant plasma membrane. Biochem J 400:189–197

Santoni V, Vinh J, Pflieger D, Sommerer N, Maurel C (2003) A proteomic study reveals novel insights into the diversity of aquaporin forms expressed in the plasma membrane of plant roots. Biochem J 373:289–296

Schey KL, Little M, Fowler JG, Crouch RK (2000) Characterization of human lens major intrinsic protein structure. Invest Ophthalmol Vis Sci 41:175–182

Schulze WX (2010) Proteomics approaches to understand protein phosphorylation in pathway modulation. Curr Opin Plant Biol 13:280–287

Sjovall-Larsen S, Alexandersson E, Johansson I, Karlsson M, Johanson U, Kjellbom P (2006) Purification and characterization of two protein kinases acting on the aquaporin *So*PIP2;1. Biochim Biophys Acta 1758:1157–1164

Sorieul M, Santoni V, Maurel C, Luu DT (2011) Mechanisms and effects of retention of over-expressed aquaporin *At*PIP2;1 in the endoplasmic reticulum. Traffic 12:473–482

Stecker KE, Minkoff BB, Sussman MR (2014) Phosphoproteomic analyses reveal early signaling events in the osmotic stress response. Plant Physiol 165:1171–1187

Steen H, Jebanathirajah JA, Rush J, Morrice N, Kirschner MW (2006) Phosphorylation analysis by mass spectrometry – myths, facts, and the consequences for qualitative and quantitative measurements. Mol Cell Proteomics 5:172–181

Stone SL, Hauksdottir H, Troy A, Herschleb J, Kraft E, Callis J (2005) Functional analysis of the RING-type ubiquitin ligase family of *Arabidopsis*. Plant Physiol 137:13–30

Sugiyama N, Nakagami H, Mochida K, Daudi A, Tomita M, Shirasu K, Ishihama Y (2008) Large-scale phosphorylation mapping reveals the extent of tyrosine phosphorylation in *Arabidopsis*. Mol Syst Biol 4:1–7

Temmei Y, Uchida S, Hoshino D, Kanzawa N, Kuwahara M, Sasaki S, Tsuchiya T (2005) Water channel activities of *Mimosa pudica* plasma membrane intrinsic proteins are regulated by direct interaction and phosphorylation. FEBS Lett 579:4417–4422

Temporini C, Callerli E, Massolini G, Caccialanza G (2008) Integrated analytical strategies for the study of phosphorylation and glycosylation in proteins. Mass Spectrom Rev 27:207–236

Tong JH, Briggs MM, McIntosh TJ (2012) Water permeability of aquaporin-4 channel depends on bilayer composition, thickness, and elasticity. Biophys J 103:1899–1908

Tong JH, Canty JT, Briggs MM, McIntosh TJ (2013) The water permeability of lens aquaporin-0 depends on its lipid bilayer environment. Exp Eye Res 113:32–40

Tornroth-Horsefield S, Wang Y, Hedfalk K, Johanson U, Karlsson M, Tajkhorshid E, Neutze R, Kjellbom P (2006) Structural mechanism of plant aquaporin gating. Nature 439:688–694

Van Wilder V, Miecielica U, Degand H, Derua R, Waelkens E, Chaumont F (2008) Maize plasma membrane aquaporins belonging to the PIP1 and PIP2 subgroups are in vivo phosphorylated. Plant Cell Physiol 49:1364–1377

Venne AS, Kollipara L, Zahedi RP (2014) The next level of complexity: crosstalk of posttranslational modifications. Proteomics 14:513–524

Vera-Estrella R, Barkla BJ, Bonhert HJ, Pantoja O (2004) Novel regulation of aquaporins during osmotic stress. Plant Physiol 135:2318–2329

Vialaret J, Di Pietro M, Hem S, Maurel C, Rossignol M, Santoni V (2014) Phosphorylation dynamics of membrane proteins from *Arabidopsis* roots submitted to salt stress. Proteomics 14:1058–1070

Walsh CK, Sadanandom A (2014) Ubiquitin chain topology in plant cell signaling: a new facet to an evergreen story. Front Plant Sci 5:122

Wang X, Bian Y, Cheng K, Gu LF, Ye M, Zou H, Sun SS, He JX (2013a) A large-scale protein phosphorylation analysis reveals novel phosphorylation motifs and phosphoregulatory networks in *Arabidopsis*. J Proteome 78:486–498

Wang P, Xue L, Batelli G, Lee S, Hou Y-J, Van Oosten MJ, Zhang H, Tao WA, Zhu J-K (2013b) Quantitative phosphoproteomics identifies SnRK2 protein kinase substrates and reveals the effectors of abscisic acid action. Proc Natl Acad Sci U S A 110:11205–11210

Wang K, Zhao Y, Li M, Gao F, Yang MK, Wang X, Li SQ, Yang PF (2014) Analysis of phosphoproteome in rice pistil. Proteomics 14:2319–2334

Weaver CD, Roberts DM (1992) Determination of the site of phosphorylation of nodulin 26 by the calcium-dependent protein kinase from soybean nodules. Biochemistry 31:8954–8959

Whiteman SA, Serazetdinova L, Jones AM, Sanders D, Rathjen J, Peck SC, Maathuis FJ (2008a) Identification of novel proteins and phosphorylation sites in a tonoplast enriched membrane fraction of *Arabidopsis thaliana*. Proteomics 8:3536–3547

Whiteman SA, Nühse TS, Ashford DA, Sanders D, Maathuis FJ (2008b) A proteomic and phosphoproteomic analysis of *Oryza sativa* plasma membrane and vacuolar membrane. Plant J 56:146–156

Wu XN, Sanchez-Rodriguez C, Pertl-Obermeyer H, Obermeyer G, Schulze WX (2013) Sucrose-induced receptor kinase SIRK1 regulates plasma membrane aquaporins in *Arabidopsis*. Mol Cell Proteomics 12:2856–2873

Wu XN, Sklodowski K, Encke B, Schulze WX (2014) A kinase-phosphatase signaling module with BSK8 and BSL2 involved in regulation of sucrose-phosphate synthase. J Proteome Res 13:3397–3409

Xue LWP, Wang L, Renzi E, Radivojac P, Tang H, Arnold R, Zhu JK, Tao WA (2013) Quantitative measurement of phosphoproteome response to osmotic stress in *Arabidopsis* based on Library-Assisted eXtracted Ion Chromatogram (LAXIC). Mol Cell Proteomics 12:2354–2369

Yang Z, Guo G, Zhang M, Liu CY, Hu Q, Lam H, Cheng H, Xue Y, Li J, Li N (2013) Stable isotope metabolic labeling-based quantitative phosphoproteomic analysis of *Arabidopsis* mutants reveals ethylene-regulated time-dependent phosphoproteins and putative substrates of constitutive triple response 1 kinase. Mol Cell Proteomics 12:3559–3582

Plant Aquaporins and Cell Elongation

Wieland Fricke and Thorsten Knipfer

Abstract There exists an increasing number of reports which show that the gene transcript, and in some cases also protein level, of particular aquaporin (AQP) isoforms is higher in growing than in nongrowing plant tissues. This suggests that AQPs play a role in the process of cell expansion. The most likely role of AQPs is that of facilitating water inflow into cells as they expand to a multiple of their original volume. The question is whether this is the major role which AQPs play in expanding cells and whether expanding cells actually need AQPs given the rate at which they expand and the hydraulic conductivity (Lp) of their membranes. These questions are addressed in this chapter by using a combination of molecular (AQP), biophysical (Lp, driving forces and water potential difference), anatomical (apoplastic barriers) and physiological (cell dimensions and relative growth rates) data for growing plant tissues and cells. The focus of analyses is on growing root and leaf tissues and on plasma membrane intrinsic (PIPs) and tonoplast intrinsic proteins (TIPs). It is concluded that a high expression of AQPs and a high Lp in growing plant cells are required more for facilitating water transport at significant (and high) rates *through* cells and tissues rather than for facilitating water transport *into* cells to sustain the (comparatively smaller) water uptake rates required for the volume expansion of these cells.

W. Fricke (✉)
School of Biology and Environmental Sciences, University College Dublin (UCD), Belfield, Dublin 4, Ireland
e-mail: Wieland.fricke@ucd.ie

T. Knipfer
Department of Viticulture and Enology, University of California, Davis, CA 95616-5270, USA
e-mail: knipfer.thorsten@yahoo.de

F. Chaumont, S.D. Tyerman (eds.), *Plant Aquaporins*, Signaling and Communication in Plants, DOI 10.1007/978-3-319-49395-4_5

Abbreviations

AQP	Aquaporin
Lp	Hydraulic conductivity
NIP	Nodule-26 like intrinsic protein
PIP	Plasma membrane intrinsic protein
SIP	Small basic intrinsic protein.
TIP	Tonoplast intrinsic protein
$\Delta\Psi$	Water potential difference
Ψ	Water potential

1 Introduction

1.1 Plants and Animals: The Little Difference

Terrestrial higher plants are sessile organisms. In contrast to most animals, which can move physically and often keep the next generation in close proximity, plants reflect the opposite evolutionary strategy – if there exists anything like an evolutionary 'strategy' – in that they literally form roots and try to spread through the next generation (seeds). Being able (animals) or not being able (plants) to move as vegetative organism has many implications for the design of such an organism. An almost endless list of such implications and differences in design between sessile and mobile multicellular organisms could be listed here, yet possibly the 'single'-most basic differences which impact organ growth and development are the absence (animals) and presence (plants) of a cell wall and a large central vacuole in cells.

Not being able to move as a vegetative organism from A to B means that self-sufficiency in nutrition and temporal and spatial variation in nutrient supply has to be optimised; this also applies to the access of water. It also implies that some cells and tissues of the organism must be exposed directly, and in a highly conductive manner, to an environment in which the water availability and osmotic strength can change quickly. Furthermore, dealing with waste becomes a major issue as waste cannot simply be discharged next to the organism as this would lead in the longer-term to the build-up of toxic concentration of waste products. The solution to these challenges in plants is that all mature living cells are surrounded by a wall and contain a large central vacuole. The wall prevents cells from bursting when cells, such as root surface cells, are exposed to a hypo-osmotic environment. The large central vacuole enables plants to optimise various processes: (i) excess nutrients can be stored transiently; (ii) waste products or toxic compounds can be stored indefinite (for as long as the cell is alive); (iii) water is stored; and (iv), maybe of most relevance to the process of cell expansion, the bulk of protoplast volume is occupied by a low-protein, resource-efficient aqueous compartment – the large central vacuole or, at very early stages of cell development, the sum of many smaller vacuoles ('vacuon'). The presence of a significant vacuolar compartment renders cell expansion 'cheap' in terms of protein and nitrogen use, and as nitrogen often limits plant production in a natural environment, it potentially provides an evolutionary advantage. To complete this comparison of evolutionary strategies between plants and animals, the latter substitute the lack of a cell wall through a tight control of the osmolarity of interstitial and extracellular fluid, and take in food and discharge waste products in either liquid or solid form through specialised body openings. Cells in animals play a minor role in waste product storage, with few exceptions such as liver tissue.

1.2 Meristems

Having cells with a mechanically tough wall has some potential disadvantages. Because plants must be able to respond in their growth and tissue/organ repair/replacement to changes in their environment, this 'plasticity' requires the growth of new tissue which in turn requires the production of new cells. The presence of a cell wall in plants precludes the option that cells can be produced at one part of the body, such as through stem cells in the human body, and migrate to the site of tissue damage and growth. Rather, the ability to produce new cells must persist throughout the lifetime of plant and throughout the plant body. Secondary meristems such as the root lateral meristem, the vascular cambium in stem and the leaf axillary meristem fulfil such a function.

In contrast to secondary meristems, primary meristems facilitate the growth of the main axis of the two primary organs of plant, root and shoot, in the form of apical meristems. These meristems form already early during embryo development. Most studies that are concerned with the function of AQPs in cell growth have focused on primary meristems. Once cells have been produced in the apical meristem, or 'cell division zone', they start to expand to a multiple of their original volume in the so-called cell expansion/elongation zone before they differentiate in the 'differentiation zone' to reach their final form and function. All three zones together will be referred to as 'growth zone' in the following. The spatial and sequential arrangement of meristem, cell expansion and cell differentiation zone makes it easy to study the development-dependent expression of genes, particularly in roots and in grass leaves. One aspect of cell and tissue differentiation which becomes particularly important when trying to understand how growing tissues take up water is the circumstance that neither the water conduction transport paths (xylem, phloem) along the main axis of organ (from root to shoot or from shoot to root) are fully developed along the entire growth zone, nor are possible apoplastic barriers within the radial transport paths (roots, between soil and root xylem; leaf, between leaf xylem and epidermis) (e.g. Hukin et al. 2002; Fricke 2002; Enstone et al. 2003; Hachez et al. 2006; Knipfer and Fricke 2011). This means that accepted views of water flows and hydraulic barriers in mature tissues cannot necessarily be adopted to explain flows and barriers in growing tissue. In fact, as will be shown, our knowledge about the hydraulic architecture of the meristematic and proximal cell expansion zone in roots and leaves is far from complete.

In the following, we will first have a closer look at the hydraulics of growing plant tissues, their architecture and water potential gradients, and what they tell us about the possible limitation of growth through cell and tissue hydraulic properties – and therefore also AQPs. We will then move on to data on the expression of AQPs in growth zones and on the hydraulic conductivity of the plasma membrane of growing leaf cells. Finally, we will use this information together with data on the rate of cell volume expansion to ask the question whether growing plant cells actually need AQPs to take in water at sufficiently high rates to support their volume expansion rates.

2 Water Potential Differences in Growing Plant Tissues

2.1 *Quantitative Relationships*

From the biophysical point of view, the process of cell expansion and factors limiting cell expansion are best described by the Lockhart equation (Lockhart 1965; e.g. Touati et al. 2015). This equation relates the relative growth rate of a cell to (i) the mechanical and hydraulic forces which drive volume expansion and (ii) the mechanical and hydraulic conducting properties which affect the gain in cell volume per unit driving force. The mechanical properties are those of the wall of growing cells, and the hydraulic properties are those of the plasma membrane and, for tissues, of the water conducting path. We are not so much interested here into addressing the question whether growth is limited more through hydraulic or wall properties of cells and tissues (e.g. Boyer et al. 1985; Cosgrove 1993; Boyer 2001; Boyer and Silk 2004; Touati et al. 2015), but rather in the specific hydraulic properties of growing tissues and the contribution that AQPs make to these properties.

The volume flux density of water (J; volume [m^3] of water per unit surface (m^2) and time (s^{-1}); unit: $m^3\ m^{-2}\ s^{-1}$, or $m\ s^{-1}$) from location A to B equals the product of hydraulic conductivity (Lp; unit: $m\ s^{-1}\ MPa^{-1}$) and the driving force (water potential difference between A and B, $\Delta\Psi$; unit: MPa); so $J = Lp \times \Delta\Psi$. Locations A and B could be the apoplast (A) and cytosol (B), respectively, of a cell; they could also be the root environment (A) and xylem (B), respectively, of a root system. Note that $\Delta\Psi$ includes two components, an osmotic component and a hydrostatic pressure component; the osmotic component can only drive water flow between A and B in the presence of a semipermeable structure (containing AQPs; see below) that separates both compartments and largely inhibits the free diffusion of osmotically active solutes (i.e. allows the build-up of an osmotic gradient between A and B). The flux of water which is of relevance when studying cell expansion is that flux which is required to let the growing tissue expand at a certain relative rate. Maximum relative growth rates of roots are typically higher than those of leaf tissues, and values range from about 20 to 40 % h^{-1} (for a review, see Pritchard 1994). It can be seen from the relation '$J = Lp \times \Delta\Psi$' that the smaller Lp is for a given J, the larger $\Delta\Psi$ needs to be. The smaller the Lp is, the less hydraulically conductive is the path between A and B; in other words, a small Lp and a significant $\Delta\Psi$ are indicative of some hydraulic limitation of growth. If so, one could predict that an increase in Lp, such as through increased expression and activity of AQPs – provided that water has to cross at least one membrane – should lead to an increase in J and cell expansion rate, provided $\Delta\Psi$ can be maintained at its original level, for example, by maintaining an osmotic gradient across a semipermeable membrane barrier through active solute transport (Fricke and Flowers 1998). It follows from the above that an analysis of $\Delta\Psi$, for a given J, provides an indication whether growth is potentially limited through Lp and hydraulic properties of cells or not.

2.2 *Water Potential Difference* $\Delta\Psi$

There exists evidence in support and not in support of a significant (ca 0.05 MPa and larger) $\Delta\Psi$, between the plant-internal water source (e.g. leaf xylem) and expanding cell (e.g. leaf epidermis) in growing plant tissues (for reviews, see Cosgrove 1993; Fricke 2002; Boyer and Silk 2004). A significant $\Delta\Psi$ has been reported in particular for the elongation zone of grass leaves and growing hypocotyls and epicotyls (stem sections) (e.g. Boyer et al. 1985; Fricke et al. 1997; Fricke and Flowers 1998; Martre et al. 1999; Tang and Boyer 2002; Touati et al. 2015). The apparent discrepancy between studies may be related to differences in the species analysed and in the growth and experimental conditions used. Significant differences in Ψ have also been reported between the root medium and growing regions of roots (e.g. Miyamoto et al. 2002; Hukin et al. 2002).

If we accept that there can exist a significant $\Delta\Psi$ in some growing plant tissues, does this necessarily mean that a change in plasma membrane or tonoplast AQP activity alters the growth rate of cells and tissues? No, it does not. Figure 1 shows two possible scenarios for a setup where six elongating cells are located, in sequence, along a radial water path. Water is provided just outside cell 1 and moves radially along cells 2–5 to cell 6. In both cases, water moves from cell to cell by crossing membranes, and if water only crosses the plasma membrane (and not also tonoplast), it has to cross 11 plasma membranes, in a polar manner (entering one side and exiting the opposite side of a cell) until it has entered cell 6. The hydraulic resistance of the path calculates as the additive resistance of the 11 plasma membranes and the six sections of wall that have to be crossed, by analogy to Ohm's law for an electric circuit (Van den Honert 1948). In scenario 'a', the hydraulic resistance of the wall sections is negligible compared with that of the plasma membranes, and the overall Lp of flow path is dominated by the Lp of plasma membranes. In scenario 'b', water moves the same path from cell 1 to cell 6, but before water enters cell 1, it has to pass through an apoplast which has a very high hydraulic

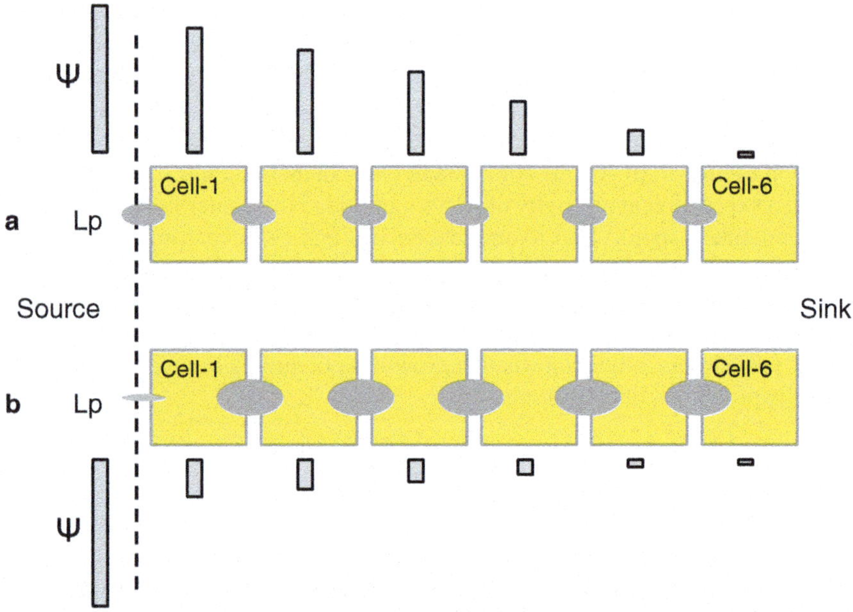

Fig. 1 The significance of a water potential difference across tissues for a hydraulic limitation of growth through cell hydraulic properties. The scheme shows a cross-sectional view of six cells that are arranged next to each other along the radial flow path of water from source (e.g. root medium) to sink (e.g. cell 6 or xylem). Each cell has a hydraulic conductivity (*Lp*) at its membrane, and the size of Lp symbol reflects the size of Lp; the same applies to the symbols used for water potential (*Ψ*), aquaporin (AQP) activity and expression and for plasmodesmatal (PD, symplastic) connection between cells. In (**a**), the Lp of all six cells contributes equally to the difference in Ψ (ΔΨ) across the tissue, and any change in Lp will affect ΔΨ and a hydraulic limitation of growth. In (**b**), the by far smallest Lp (highest resistance to water movement) is found at the entry point of water into cell 1; this could be due to, e.g. suberinisation of the apoplast. The Lp of cells contributes little to ΔΨ, and changes in cell Lp will have little effect on ΔΨ. The value of ΔΨ is similar in (**a**, **b**) and points to a hydraulic limitation of growth, yet only in (**a**) could such a limitation be significantly affected by changes in cell Lp, for example, through changes in AQP activity

resistance (low Lp, as represented by the small size Lp symbols in Fig. 1), e.g. due to intense suberisation (hydrophobic hydraulic barrier). The resistance of the suberised wall section dominates the overall resistance of the flow path, and any alteration of the Lp of plasma membranes will have little effect on the overall Lp. Both scenarios lead to a significant ΔΨ. In ‘a’, the significant ΔΨ is due to an inherently low Lp of the plasma membrane of cells, and upregulation of PIPs may increase Lp and growth. In contrast, in ‘b’, upregulation of PIP activity will have little effect on the overall Lp and growth rate. Thus, the existence of a significant ΔΨ in growing tissues is a pre-requisite but not proof per se that any alteration in plasma membrane AQP activity leads to an altered growth rate of cells. The possible existence of local and significant apoplast barriers to water movement needs to be considered too.

2.3 *Root Growth Zones Are Special*

There exist one fundamental difference between the hydraulic arrangement of growing tissues in roots on the one hand and stem and leaf on the other. Growth zones of stems, such as hypocotyls, and of leaves of monocot crop plants (e.g. wheat, barley and maize), which are exposed to the atmosphere (as opposed to, e.g. leaves of seagrass that are submerged under water), are located along the axial transport route of water from roots to shoot (Fricke 2002; for water flow in submerged plants such as seagrass, *Posidonia australis*, see Tyerman et al. 1984). In contrast, growth zones of roots are located at the very starting point, or even prior to that starting point, where water is transported axially from root to shoot (Fig. 2). Which implications does this have for a hydraulic limitation of growth and a role that AQPs could play?

In roots, transpiration water in xylem moves away from the growth zone and may not reach any of the proximal (closest to the tip) regions. In the latter regions, phloem-delivered water may constitute the main source of water to expanding cells, in addition to water entering from the soil/root interface. In addition, water may move along a symplastic path, through plasmodesmata (Hukin et al. 2002). This is supported by a study on maize roots, which showed that AQP inhibitors have little or no effect on the half-time of water exchange (and by implication plasma membrane Lp) in the growing tip region (Hukin et al. 2002). By linking the expansion of cells in roots to the supply of water, and resources such as carbon and energy through the phloem from shoots, shoot productivity and transpirational surface can be fine-tuned with growth of water and mineral nutrient absorbing root surface. In contrast, the distal portion of root growth zones is more directly linked to xylem water flow. Radial movement of water can occur along the transmembrane path, from cell-to-cell crossing membranes, and AQP activity and AQP inhibitors impact on this water movement (e.g. Hukin et al. 2002; Frensch and Steudle 1989; Knipfer et al. 2011).

3 AQP Expression in Growing Root and Leaf Tissues

A comprehensive literature review of data on AQP expression in growing plant tissues, including root, leaf, stem, petals and hairs, is provided by Obroucheva and Sin'kevich (2010). We will focus here on roots and leaves, with some information also on fibre elongation.

3.1 *Roots*

There exist a few studies in which the expression of the majority of AQP isoforms of a particular plant species has been compared between root and shoot tissue. These studies show two trends: (i) when there are AQP isoforms within a species

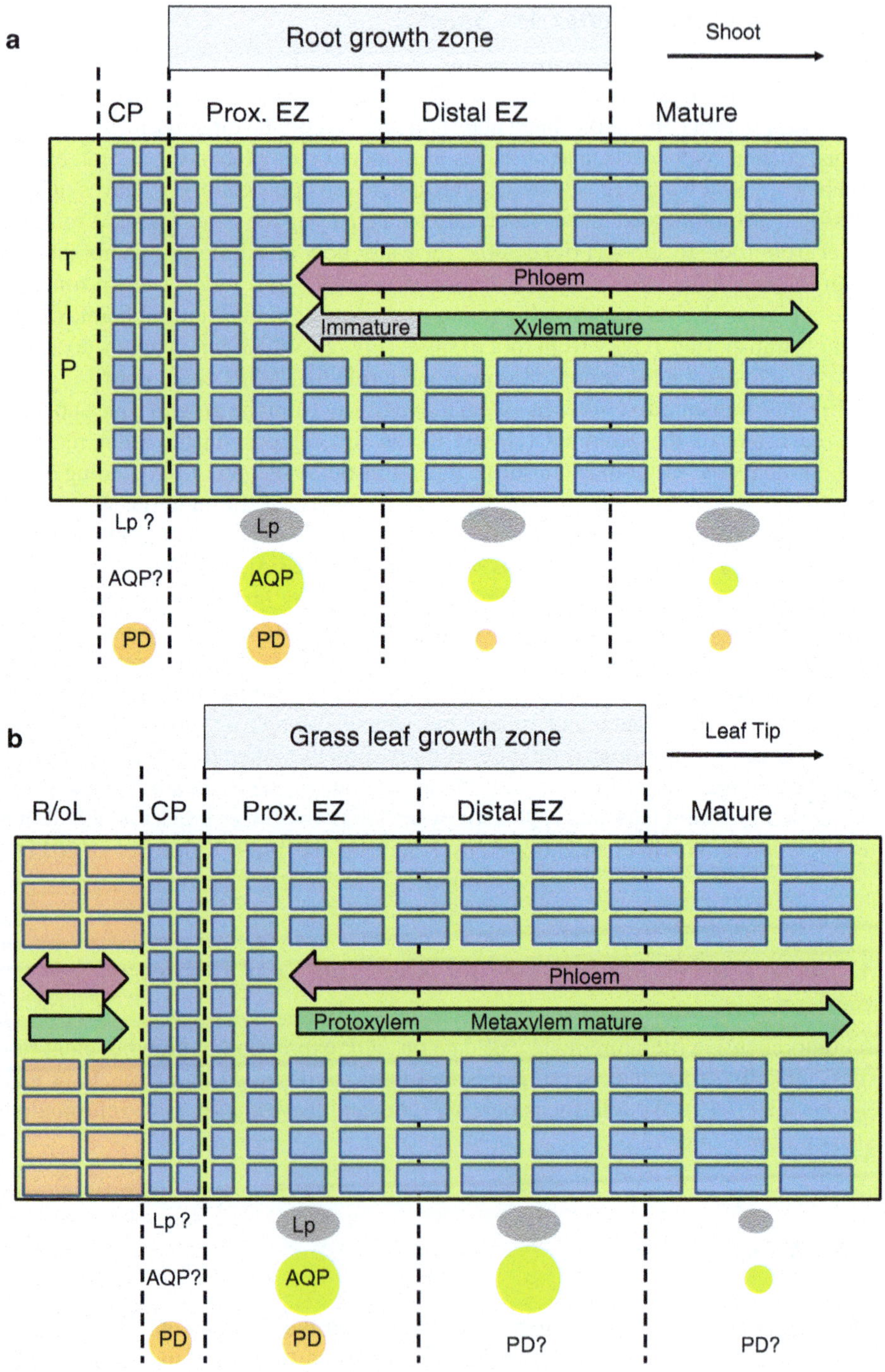
a
Root growth zone
Shoot
CP
Prox. EZ
Distal EZ
Mature
T
I
P
Phloem
Immature
Xylem mature
Lp ?
Lp
AQP?
AQP
PD
PD
b
Grass leaf growth zone
Leaf Tip
R/oL
CP
Prox. EZ
Distal EZ
Mature
Phloem
Protoxylem
Metaxylem mature
Lp ?
Lp
AQP?
AQP
PD
PD
PD?
PD?

that are almost exclusively expressed in one vegetative plant organ, then the exclusive expression occurs mostly in roots and not in shoots, and (ii) roots show a higher total expression of AQPs compared with shoots. For example, relatively (compared with shoot tissue) high or exclusive expression of AQPs in roots was observed for *Oryza sativa* (OsPIP2;3, OsPIP2;5, OsTIP2;1, Sakurai et al. 2005), *Zea mays* (ZmPIP2;1, ZmPIP2;5, ZmPIP1;5, Hachez et al. 2006; Heinen et al. 2009; Lopes et al. 2003), *Arabidopsis thaliana* (AtPIP2;2, AtPIP1;1, AtPIP1;2, AtPIP2;1, AtTIP1;1, AtTIP1;2, Javot et al. 2003; Alexanderson et al. 2005), *Vitis vinifera* (VvPIP1;1, VvPIP2;2, Vandeleur et al. 2009), *Hordeum vulgare* (HvPIP1;2, HvPIP2;1, HvPIP2;2, HvPIP2;5, Katsuhara et al. 2002; Katsuhara 2007; Besse et al. 2011; Knipfer et al. 2011) and *Pisum sativum* (PsPIP2;1, Beaudette et al. 2007). As some of these AQPs were also expressed in growing root tissue, the data show that there is not a single set of AQPs in a particular plant species that facilitates growth in growing root *and* leaf tissue, but that this role can be organ-specific and involve isoform-specific AQPs.

Using *in situ* hybridisation and also immunocytochemical approaches, AQP isoforms have been localised in certain root tissue types (e.g. cortex, epidermis) and in dependence on root developmental stage (Hachez et al. 2006; Sakurai et al. 2005, 2008; Vandeleur et al. 2009; Knipfer et al. 2011). The most complete studies so far exist for rice (Sakurai et al. 2005, 2008), vine (*Vitis vinifera*, Vandeleur et al. 2009; Gambetta et al. 2013), maize (Hachez et al. 2006, 2012), *Arabidopsis* (Brady et al. 2007) and barley (Knipfer et al. 2011). For example, it was shown for maize that ZmPIP2;1/2;2 proteins are most abundant in the stelar cells at the root tip, encompassing the growth zone, whereas in more mature root regions, these two AQP isoforms were preferentially localised to the cortex and epidermis; ZmPIP2;5 protein was in particular found in cortex tissue at the root tip and in the endodermis in mature root tissue (Hachez et al. 2006). In rice, Sakurai et al. (2008) showed that OsPIP1s and OsPIP2;1, OsPIP2;3 and OsPIP2;5 proteins were localised preferentially in the endodermis at the root tip. In more mature root regions, OsPIP2s proteins were distributed rather evenly between tissues, whereas OsTIP2;1 and OsTIP2;2 proteins were localised preferentially in stelar cells. In grapevine, Vandeleur et al. (2009) observed that VvPIP1s and VvPIP2s gene expression and protein level were localised evenly in cortex tissue and vascular tissue at the root tip, but showed lower signals in the cortex of mature root regions. Gambetta et al. (2013) compared the root zone-dependent gene expression of isoforms in grapevine with

---

Fig. 2 Hydraulic architecture of the growth zone of (**a**) roots and (**b**) grass leaves. The scheme gives a longitudinal view of the arrangement of expanding cells and the relative location of tissues (xylem, phloem) through which water can be supplied internally to growing tissues. The size of Lp symbol reflects the size of Lp; the same applies to the symbols used for aquaporin (*AQP*) activity and expression and for plasmodesmatal (*PD*, symplastic) connection between cells. A '*?*' indicates that we lack detailed information here, and *arrows* for phloem and xylem point to possible main directions of flow; for details, see text. *CP* cell production zone (meristem), *prox. EZ* proximal elongation zone, *distal EZ* distal elongation zone, *mature* mature root or leaf region where cells have attained their full size, *TIP* root tip region including the root cap

their homologues in *Arabidopsis* (using data from Brady et al. 2007) and concluded that most AQP isoforms distributed similar in the two species. There was a much higher gene expression of PIPs in the root tip compared with more mature regions along the main root axis. In barley, tissue localisation of the gene expression of six AQP isoforms was tested (*HvPIP2;2*; *HvPIP2;5*, *HvPIP2;7*, *HvPIP1;2*, *HvTIP1;1*, *HvTIP2;3*) (Knipfer et al. 2011). There was generally high expression of AQP genes in the epidermis and protoxylem. Expression in cortex tissue was evident in all root developmental zones and in both seminal and adventitious roots. Expression in the endodermis and stele was observed particularly in less mature adventitious roots, highlighting a potential role in regulating radial water transport. Of all barley AQPs tested, *HvTIP1;1* was the most ubiquitously expressed gene, while *HvPIP2;5* was expressed particularly in cortex tissue.

In barley, adventitious roots had a threefold higher cortex cell hydraulic conductivity and total expression of PIP2s and TIPs compared with seminal roots (Knipfer et al. 2011). This difference in AQP expression was due to higher expression of three aquaporins, *HvPIP2;*2, *HvPIP2;5* and *HvTIP1;1*, all of which display water channel activity (Besse et al. 2011). These aquaporins were expressed in the epidermis, cortex, endodermis and stele of the transition zone of adventitious roots, where cells have completed some of their elongation yet are not fully mature as judged from endodermis development. *HvPIP2;5* and *HvTIP1;1* were the highest-expressed AQPs tested.

In seminal roots of barley, *HvTIP1;1* was expressed lowest in the mature zone, and this coincided with the lowest cortex cell hydraulic conductivity in this root region compared with the immature and transition zone, which, together encompassed the entire growth zone (Knipfer et al. 2011). The expression of three HvPIPs (HvPIP1;2, HvPIP2;2, HvPIP2;5) was compared between growing (immature + transition zone) and nongrowing (mature zone) root regions in seminal roots. None of the three genes tested was expressed higher in growing tissue, despite cell hydraulic conductivity being almost four times higher in growing compared with nongrowing tissue. Other PIPs, which were not tested in the study by Knipfer et al. (2011), may be responsible for the higher cell Lp in growing barley root tissue. What the study shows, though, is that AQPs such as HvPIP2;5 may facilitate water flow in growing root tissue in one type (adventitious roots) but not another type (seminal roots) of roots, even within one species.

Sequence comparison between barley and maize PIPs shows that ZmPIP2;1 and ZmPIP2;2 share highest sequence identity with HvPIP2;5. ZmPIP2;1 is among the highest-expressed PIPs in maize roots, The tissue localisation of ZmPIP2;1 protein changes during maize root development from a predominant location in the stele and endodermis to a location in the cortex and epidermis (Hachez et al. 2006). Such a change in tissue localisation was not observed for *HvPIP2;5* in barley (Knipfer et al. 2011), where gene expression was analysed. *HvPIP2;5* was expressed in cortex tissue in both transition and mature zones. Sakurai et al. (2008), using immunocytochemistry, observed for rice roots that candidate AQPs occurred predominantly in the endodermis and stele, with some protein in the rhizodermis and very little in the cortex. The difference in results between the study on barley (Knipfer et al.

2011) and the studies on maize (Hachez et al. 2006) and rice (Sakurai et al. 2008) may reflect differences between species or the circumstance that AQP gene and protein abundance do not correlate in time and space.

TIP1;1 isoforms are generally the most abundantly expressed members of the TIP family of AQPs (e.g. Alexandersson et al. 2005; Sakurai et al. 2005) and share a high sequence identity among the plant species tested. The ubiquitous and abundant expression of *HvTIP1;1* in barley roots (Knipfer et al. 2011) suggests that this AQP is a 'housekeeping' type of AQP, which provides a 'baseline' level tonoplast hydraulic conductance to guarantee rapid osmotic equilibration between vacuole and cytosol (Maurel et al. 1993). This does not preclude a role of HvTIP1;1 in growth-facilitated water uptake. A complete loss of (water channel) function of HvTIP1;1 is not expected to cause a phenotype in barley (see also Schüssler et al. (2008) for *Arabidopsis*), as another TIP (HvTIP2;3), which shows water channel activity (Besse et al. 2011), is expressed in roots, though with a different tissue (e.g. the cortex, stele, epidermis) pattern (Knipfer et al. 2011). It remains to be shown why multiple TIP isoforms which show water channel activity are co-expressed abundantly in root cells and whether any of these TIP isoforms carries a growth-specific function.

3.2 *Leaves*

Besse et al. (2011) conducted a detailed study on the development-dependent expression of AQPs in growing barley leaves. At the time of study, the entire set of almost 40 barley AQPs (Hove et al. 2015) was not known, and 23 AQPs were studied. Five AQP genes, including one PIP1 (*HvPIP1;2*), were expressed at such low levels in leaf regions that it was difficult to conclude on their pattern of expression. A sixth gene (*HvSIP2;1*) was expressed so uniformly between leaf developmental zones that it turned out to be a suitable reference gene of expression. The 17 remaining genes analysed, which included most known barley PIPs, were expressed particularly in either growing (seven genes) or in emerged, mature leaf tissue (ten genes). It can be concluded from these data that differential expression during leaf development is the rule rather than exception for barley MIPs and that all MIPs that facilitate diffusion of water across the plasma membrane are under developmental or environmental control. There is no obvious reason why control of water channel activity of one particular AQP through post-translational regulation and trafficking (Johansson et al. 1998; Tournaire-Roux et al. 2003; Maurel 2007; Zelazny et al. 2007; Boursiac et al. 2008; Hachez et al. 2014) should not provide sufficient means to meet requirements specific to growing and mature leaf tissue. Therefore, the observation that many different barley AQP isoforms show development-specific expression points to these AQPs fulfilling tissue-specific functions. Individual AQPs with localisation to specific tissues may play important roles during tissue and cell expansive growth in barley, but their relative abundance in whole tissue extracts is lower because of their specific localisation.

None of the AQPs tested in the study by Besse et al. (2011) were expressed highest or lowest in the non-elongation zone – that zone during grass leaf cell development, where cells have ceased to elongate, yet still show some residual expansion in width before being displaced through growth of more basal regions from subtending sheaths into the ambient atmosphere. Instead, the start (elongation zone) and end point (emerged, mature blade) of a cell's ontogeny were accompanied by maximum or minimum expression of a particular AQP isoform (Besse et al. 2011). This contrasts with the only other comparable study, on maize (Hachez et al. 2008), where expression of AQPs was analysed in detail and at high spatial resolution in different developmental zones of a leaf. In the study on maize, expression of PIPs in developing leaves was highest in the zone near the emergence point from the sheath of older leaves, with subsequent decrease in expression in the mature part of blade. A continuous increase in expression during leaf development was only observed for *ZmPIP1;5*. As the studies on maize and barley are the only ones of their kind, it cannot be said which study presents the rule and which the exception with respect to AQP isoform expression during leaf development. The most notable difference between barley and maize is that barley is a C_3 and maize a C_4 plant. How, and whether, this could explain differences in the expression pattern of AQP in leaves of the two species remains to be shown.

It is not known what regulates the differential expression of barley (or maize or any other grass) MIPs during leaf development. The most pronounced difference in microenvironment between the elongation zone and the more mature leaf regions is the intensity and quality of light that reaches cells; also relative humidity in the air next to the elongation zone enclosed by subtending sheaths of older leaves should be considerably higher than ambient. The elongation zone of leaf three is enclosed by sheaths of leaf one and two, whereas the non-elongation zone is only enclosed by the sheath of leaf two. The sheaths are green and photosynthetic. As a result, the light that reaches the non-elongation and, particularly, elongation zone will have a higher ratio of far-red to red light than that striking the emerged and mature blade. This could enable regulation of AQP expression through the phytochrome system in a development- and therefore also growth-dependent manner.

3.3 Roles of Particular AQP Isoforms During Leaf Growth

In barley, *HvPIP2;5*, *HvTIP1;1* and *HvTIP2;3* were expressed abundantly and highest in growing tissue of roots and leaves (Besse et al. 2011). The same was observed for *HvPIP1;1* (identical to the barley AQP annotated as *HvPIP1;6*) in a previous study on barley (Wei et al. 2007). These four MIPs, all of which show water channel activity (Wei et al. 2007; Besse et al. 2011), seem to have a role that is specific to cell growth in barley, irrespective of the organ. In contrast, the water channel HvPIP2;2 was expressed particularly in growing tissue of leaves (Besse et al. 2011) and seems to have a growth-related function that is more leaf-specific.

High expression of TIP1;1 isoforms in meristematic and elongating shoot tissue has been reported, e.g. maize (Chaumont et al. 1998), tulip (*Tulipa gesneriana*, Balk and de Boer 1999), cauliflower (*Brassica oleracea*, Barrieu et al. 1998) and oilseed rape (*Brassica napus*, Frangne et al. 2001), and appears to be a common characteristic associated with growth. *Arabidopsis* plants lacking AtTIP1;1 (and AtTIP1;2) protein do not show any phenotype or change in growth rate under normal growth conditions (Schüssler et al. 2008). This does not preclude a role of AtTIP1;1 or the barley homologue HvTIP1;1 in facilitating water uptake and vacuole enlargement during leaf cell expansion, nor in playing a role in any cell expansion-related event such as lateral root formation (see recent work on role of TIPs in lateral root growth using triple TIP mutants of *Arabidopsis*; Reinhardt et al. 2016). For example, the high expression of the water channel HvTIP2;3 in the growth zone of barley leaves points to some redundancy in function among TIPs (Besse et al. 2011).

In the leaf elongation zone of barley, *HvPIP1;1* and *HvPIP2;5* accounted for 90 % or more of the expression of PIP1s and PIP2s, respectively (Besse et al. 2011). Their closest maize homologues (based on sequence identity), *ZmPIP1;1* and *ZmPIP2;1* together with *ZmPIP2;2*, also accounted for the bulk of expression of PIP1s and PIP2s in the elongation zone of maize (Hachez et al. 2008). It appears from these two studies on grasses that dominant PIP isoforms are conserved in elongating leaf tissue. As these isoforms include members of the PIP1 and PIP2 subfamily, they may regulate cell Lp and growth through formation of PIP1/PIP2 heteromers (for a review, see Chaumont and Tyerman 2014) (see chapter "Heteromerization of Plant Aquaporins").

In barley, the water channel *HvPIP2;5* was expressed abundantly in the leaf elongation zone, and this included the mesophyll in this leaf region (Besse et al. 2011). Mesophyll constitutes most of tissue volume of leaves, and this could explain why *HvPIP2;5* accounted for more than 90 % of PIP2 expression in the elongation zone (Besse et al. 2011). This renders HvPIP2;5, a prime candidate to mediate plasma membrane water flow in growing mesophyll cells. It would also explain the higher cell hydraulic conductivity in growing compared with nongrowing mesophyll tissue in barley, as concluded from swelling assays of the osmotic water permeability of mesophyll protoplasts (Volkov et al. 2007). In growing leaf epidermal cells of barley, the function of HvPIP2;5 appears to be carried out by HvPIP1;1 (HvPIP1;6) and HvPIP2;2, both of which are expressed highest in the epidermis (Wei et al. 2007; Besse et al. 2011). Trans-tonoplast movement of water in growing leaf tissues seems to be facilitated by the abundantly expressed HvTIP1;1 and HvTIP2;3 (Besse et al. 2011).

3.4 The Role of Vascular Bundles in the Hydraulics of Leaf Growth

It is not known whether water reaches epidermal cells in the elongation zone of grass leaves directly through mesophyll or through bundle sheath extensions, from where it diffuses radial within the epidermis. In the latter case, many membranes and hydraulic resistances have to be overcome. A potential hydraulic limitation of cell expansion growth could be avoided by high expression of AQPs such as *HvPIP1;1/1;6* in the epidermis, leading to a higher cell hydraulic conductivity in the epidermis of elongation compared to mature leaf tissue (Volkov et al. 2007). The comparatively low water transport activity of HvPIP1;1/1;6 (Wei et al. 2007) may be partially compensated for by high expression levels of, and heteromerisation (Fetter et al. 2004; Zelazny et al. 2007) (see chapter "Heteromerization of Plant Aquaporins") with, the concurrently expressed *HvPIP2;2* and *HvPIP2;5*.

The mestome sheath of grass leaves can be suberised (O'Brien and Carr 1970; O'Brien and Kuo 1975; for a review, see Lersten 1997; Fricke 2002). The study by Besse et al. (2011) on barley showed that Casparian-band like structures increased during leaf development. The mestome sheath may fulfil a role in leaves that is comparable to that of the endodermis in roots (Fricke 2002; Enstone et al. 2003; Wu et al. 2005; Heinen et al. 2009). This view receives increasing experimental support through studies which emphasise the role of the bundle sheath and bundle sheath AQPs as potential hydraulic bottlenecks, through which the radial movement of water from xylem to substomatal cavity is controlled in transpiring leaves of monocot and dicot plant species (e.g. Shatil-Cohen et al. 2011; Sade et al. 2014).

Increase in expression of HvPIP2;7 during barley leaf development, with smaller expression in elongating and higher expression in mature leaf tissue, in vascular bundles (Besse et al. 2011) could compensate for the formation of any apoplastic barriers by facilitating radial movement of transpiration water through membranes along a cell-to-cell pathway. Such a pathway has been supported by a study on *Tradescantia* (Ye et al. 2008). In rice, *OsPIP2;7* is expressed in leaves predominantly in mesophyll, and not in vascular bundles as in barley. Overexpression of *OsPIP2;7* in rice results in increased transpirational water loss (Li et al. 2008). The data on PIP2;7 expression in barley and rice support a role of this PIP in facilitating radial movement of transpiration water in both species, yet the tissue site where this facilitation occurs differs between barley (vascular bundle) and rice (mesophyll). The considerable expression of *HvPIP2;7* in the non-transpiring non-elongation zone in barley (Besse et al. 2011) might be in preparation of the displacement of cells into the open atmosphere (past the point of emergence from the sheath of leaf two). This displacement can occur in as little as 10 h (Richardson et al. 2005).

Between 98 and 99 of every 100 water molecules that enter the leaf elongation zone of barley along the xylem are lost through the emerged blade; only one to two molecules are used to support cell expansive growth (Fricke 2002). Therefore, it is surprising that water channels such as HvPIP2;5 and HvTIP1;1 are expressed at so much lower levels in mature, transpiring compared with growing, non-transpiring leaf tissue (Besse et al. 2011). Could it be that their water channel activity is an experimental disguise of their true function *in planta* (e.g. Hill et al. 2004) or that the need to rapidly osmotically equilibrate water across membranes is much higher in growing than in mature plant tissue? We do not know.

3.5 Examples of Other Experimental Systems

Fibres, which grow on/around seeds, such as cotton (*Gossypium hirsutum*) seed fibres offer a great experimental system to study the molecular processes accompanying cell elongation. The fibres and fibre cells are large; they are arranged in series, are easy accessible, are easy to observe and are comparatively easy to analyse; and in addition, they are also commercially very important. In *Gossypium hirsutum*, four PIP2s (GhPIP2;3, GhPIP2;4, GhPIP2;5 and GhPIP2;6; Li et al. 2013) and one PIP1 and TIP (GhPIP1;2, GhγTIP; Yang and Cui 2009) have been shown to be expressed particularly in fibres and to peak in expression during a period when fibres and fibre cells elongate at their highest rate. Knockdown of expression of GhPIP2 genes in cotton significantly decreased the rate of fibre elongation (Li et al. 2013). Furthermore, the authors also observed that most of the PIP2s that were expressed particularly in cotton fibre cells were able to form heterotetramers and, through this, increased water channel activity further.

In the milkweed *Calotropis procera*, which produces long seed trichomes, CpPIP2 AQPs were also implicated in the process of fibre cell elongation (Aslam et al. 2013). This conclusion was based on the observation that the expression of CpPIP2s in fibre cells was highest during the period of highest rates of fibre cell elongation. In addition, transgenic tobacco plants that expressed CpPIP2s had a larger number of trichomes on leaves and stem.

In rose (*Rosa hybrid* 'Samantha'), the rose PIP2 RhPIP2;1 was shown to be involved in the ethylene-regulated expansion of petals (Ma et al. 2008). RhPIP2;1 was localised primarily in the abaxial subepidermal cells of petals, and its expression during petal development was highly correlated with petal expansion rate. Furthermore, treatments that reduced the rate of petal expansion, such as ethylene and silencing RhPIP2;1 in transgenic plants, also reduced the gene expression level of RhPIP2;1.

In deepwater rice (*Oryza sativa* L. ssp. indica), shoot internodes must elongate quickly in response to flooding to minimise anoxia stress to leaf tissues. Muto et al. (2011) observed that the gene transcript levels of several OsAQPs (OsTIP1;1, OsTIP2;2, OsPIP1;1, OsPIP2;1, OsPIP2;2) increased significantly in stems in response to submersion. This occurred in parallel to the stimulation of other processes that facilitate elongation growth of cells, such as expression of vacuolar proton pumps (OsVHP1;3). The authors also observed that AQPs (OsNIP2;2 and OsNIP3;1), which are not primarily involved in water transport, but in the transport of substances (silicic acid and boric acid) that can interfere with wall expansion, were reduced in expression in response to flooding. The latter provides a good example for a role of AQPs in cell elongation not related to their water transport function.

4 Membrane Hydraulic Conductivity in Growing Plant Cells

The above section was concerned with identifying AQP isoforms in a range of species, which facilitate water uptake into growing leaf or root tissue. Now we have a closer look at the hydraulic conductivity (Lp) of growing cells. First, we want to know whether growing cells have a higher hydraulic conductivity, at their plasma membrane, compared with mature cells. Then we ask whether such differences in plasma membrane Lp are indicative of growing cells taking in more water per unit time and driving force to support volume expansion.

4.1 Roots

Most of the data on the Lp of higher plant cells and tissues exist for roots. This is because roots provide an easier experimental system compared with leaves: (i) roots can be grown hydroponically and enable an easier application of experimental treatments; (ii) roots are cylinders in shape, which facilitates modelling of water flows; and (iii) the discovery of AQPs in plants has opened up the possibility that plant water flow can be regulated also in the short term through roots rather than through stomata in shoots. There exist surprisingly few data on the plasma membrane Lp of growing root cells mainly due to the experimental challenges involved, in particular for cells that are located very close to the root tip in the proximal half of root elongation zone. The majority of cell Lp data on roots has been obtained on mature tissues, often with the aim to link cell Lp to root tissue Lp and to address the role which AQPs and particular root developmental regions play in root water uptake (e.g. Hachez et al. 2006; Bramley et al. 2009; Ehlert et al. 2009; Knipfer et al. 2011; Gambetta et al. 2013). This has made it possible to conclude on the main path of radial water movement (apoplast versus cell to cell) across the root cylinder and to test the composite model of water transport across roots (Frensch and Steudle 1989; Steudle 2000; Steudle and Peterson 1998). The implicit assumption of these analyses is that a higher expression of AQPs in conjunction with a higher Lp supports the idea that AQPs contribute to root water uptake in a particular root region and that a significant portion of water moves along the transmembrane component of cell-to-cell path. This assumption may apply to mature root regions. However, one has to be careful when applying this assumption to growing and meristematic root regions. This is demonstrated nicely by the studies of Hukin et al. (2002) on maize, Gambetta et al. (2013) on grapevine and Miyamoto et al. (2002) on pea roots.

Hukin et al. (2002) studied the expression of two AQPs, and cell Lp in the presence and absence of the AQP inhibitor Hg and the symplastic connectivity in the tip region of maize roots. The authors analysed locations in the proximal half of the growth zone, half-way along the growth zone (where relative elemental growth rates were highest) and at the distal end of the growth zone and the mature zone just beyond that. The main observations were that (i) Lp of cells averaged about $1.5–2 \times 10^{-7}$ m s^{-1} MPa^{-1} throughout the growth zone and increased two- to three-fold in mature tissue; (ii) Hg reduced cell Lp half-way along and in the distal portion of the growth zone and in the mature zone but not closer to the root tip; and (iii) symplastic continuity was observed only closer to the root tip, in the proximal portion of growth zone. The conclusions that can be drawn from these observations are that maize root AQPs contribute to cell Lp and radial root water flow across the root cylinder in the distal half of the growth zone and mature root regions. However, cells closer to the root tip, where xylem is not fully developed and functional, have a high symplastic connectivity to enable the import of water through the phloem (compared with xylem earlier developed) and transport of water through tissue. The two AQPs studied were expressed at much lower levels in the proximal compared with distal half of the growth zone, yet cell Lp was similar in the two root regions. It is possible that other AQP isoforms that were not studied showed a different pattern of expression, with expression being higher in the proximal portion of the growth zone. The disparity between AQP expression and cell Lp could also be due to the technique used to determine cell Lp: the cell pressure probe. This instrument can be used to induce water flow across the plasma membrane of a cell and follow the subsequent turgor pressure relaxation, which in turn is used to determine cell Lp. What this technique measures is water flow across the plasma membrane (and through plasmodesamata, as water flow through these two structures cannot be measured separated with currently available techniques) of a cell, yet it cannot distinguish between the path of water flow, which could be through simple diffusion through the lipid bilayer, facilitated diffusion through AQPs or through movement of water through plasmodesmata. Thus, the above data could also be interpreted in such a way that the Lp in cells close to the root tip reflects plasmodesmatal water transport, whereas the Lp of cells in more distal regions of the growth zone and in the mature region reflects water transport through AQPs. If this is the case, we cannot simply ask the question whether AQPs contribute to water uptake into expanding root cells, but we must distinguish between earlier and later stages of cell expansion. Also, at earlier stages of cell expansion, water supply appears to be through a root internal source (phloem), whereas at later stages of cell expansion, water supply occurs through an external source (root medium). This has implications for the direction of water movement (early, axial, tip-wards, radial from inner to outer tissues; late, axial, towards shoot; radial from outer to inner tissues) during cell elongation in roots and the driving forces that facilitate water movement through tissues.

Miyamoto et al. (2002) studied the gravitropic bending response of pea roots. Using the cell pressure probe, the authors observed that cell Lp in the part of growth zone where xylem was not fully developed averaged about 1.7×10^{-6} m s^{-1} MPa^{-1}. This value was higher than cell Lp in more mature root regions and about 100 times higher than tissue Lp in the growth zone. The authors explained the latter observation such that the growing cells with a high Lp obtained their water not from the root medium but through xylem vessels that were up to 50 cells (and 100 plasma membranes) located away from those growing cells. This supports some of the above data by Hukin et al. (2002) on maize.

Gambetta et al. (2013) carried out a detailed analysis of the expression of AQPs and tissue Lp in different developmental regions of the woody plant grapevine. The authors also distinguished between the very proximal portion of growth zone, which was rather meristematic, and the more distal portion of growth zone, where most cell elongation occurred. The authors observed much higher expression of most AQPs in the proximal portion of growth zone compared with the other root regions. This coincided with a higher tissue Lp in the proximal part, and Lp could be inhibited through Hg. The authors concluded that even though mature root regions have lower AQP gene expression and tissue Lp compared with the other root regions, they contribute significantly to root water uptake of plants. The authors questioned the significance of the very tip region of roots for root water uptake, despite having the highest tissue Lp and AQP expression level, and raised the possibility that such high values may have little to do with root water uptake but with the volume expansion of cells. The above cited studies of Hukin et al. (2002) and Miyamoto et al. (2002) would support such an interpretation of data. Gambetta et al. (2013) observed highest expression of many AQP isoforms in the proximal and meristematic portion of growth zone, whereas Hukin et al. (2002) observed a much lower expression of the two AQP isoforms studied in proximal compared with distal regions. This difference may be due to the number of AQP isoforms and species studied. Our own data on the expression of AQPs in the proximal portion of growth zone of barley roots (Knipfer, Besse and Fricke, unpublished results) supports the data by Gambetta et al. (2013).

4.2 *Relevance of Cell Lp for Root Growth*

The above analyses highlight four major aspects, which must be considered when studying the role of AQPs in root cell expansion. Firstly, a high cell Lp as determined with the cell pressure probe does not necessarily reflect water flow through AQPs across the plasma membrane; it may actually reflect water flow through plasmodesmata (see Zhang and Tyerman 1991). Secondly, proximal and distal portions of the growth zone must be distinguished during analyses. Thirdly, the direction of water flow and the source of water supply to expanding root cells may differ between proximal and distal portions of the growth zone. Fourthly, and in relation to the previous aspect, the Lp of a growing root cell should be seen more in context of facilitating water flow through a tissue rather than facilitating water uptake at a sufficiently high rate into an individual cell to support that cell's expansion growth. This can be demonstrated through calculations where data on cell Lp, growth rates and cell dimensions are used.

Published values of the Lp of growing root cells are in the region 10^{-6} to 10^{-7} m s^{-1} MPa^{-1} (Miyamoto et al. 2002; Hukin et al. 2002) and generally not that different from values for mature root cells (e.g. Ehlert et al. 2009; Azaizeh et al. 1992; Lee et al. 2012), though differences in Lp between root regions have been reported (e.g. barley, Knipfer et al. 2011). Maximum relative growth rates of root cells are generally higher than those of leaf cells, with peak values half-way along the elongation zone of about 20–40 % h^{-1} (e.g. Miyamoto et al. 2002; Hukin et al. 2002). The volume of a typical root cortex cell half-way along the growth zone is in the range of 30–50 pl ($30–50 \times 10^{-15}$ m^3). The surface area of such a cortex cell, if we assume that it is shaped like a cube with 35 μm of each side, is about 40×10^{-9} m^2 (compare also, e.g. Miyamoto et al. 2002; Hukin et al. 2002). If we take here a relative growth rate of 30 % h^{-1}, a cell volume of 40×10^{-15} m^3 (40 pl) and a surface area of

40×10^{-9} m^2, we obtain a net rate of water uptake into growing cells of (0.3 h$^{-1} \times 40 \times 10^{-15}$ m^3) 12×10^{-15} m^3 h^{-1}, or 3.3×10^{-18} m^3 s^{-1}; the water flow rate per unit cell surface area calculates to (3.3×10^{-18} m^3 s^{-1} divided by 40×10^{-9} m^2) 8.3×10^{-11} m s^{-1}. If we take an Lp of about 5×10^{-7} m s^{-1} MPa^{-1}, we see that a water potential difference of (8.3×10^{-11} m s^{-1} divided by 5×10^{-17} m s^{-1} MPa^{-1}) 1.7×10^{-4} MPa across the plasma membrane of the growing cortex cell is required to sustain such a net water uptake rate for growth. Thus, we calculate here theoretical water potential differences as small as 0.1–0.2 kPa. We do not know how large the actual water potential difference is across the plasma membrane of a growing root cortex cell. On the one hand, cells are thought to be in local water potential equilibrium with their immediately surrounding apoplast, and the water potential difference may be as small as, e.g. 1.7×10^{-4} MPa (0.17 kPa); on the other hand, water potential differences across the root cylinder of larger than 0.1 MPa (100 kPa) have been measured in growing root tissues (e.g. Miyamoto et al. 2002; Hukin et al. 2002). Even if 20 cells were located along the radial path, and provided that the hydraulic resistances of cells are comparable, the resulting water potential step across each cell would be by factor 100 to 1,000 larger than that required to sustain the volume expansion of these cells.

The most likely conclusion from these data is that cell Lp and plasma membrane localised AQPs do not limit the growth of individual cells through restricting their capacity to take in water. Rather, cell Lp and AQPs can limit the growth of root cells through limiting the rate of water supply to these cells through tissues. This conclusion is supported through the observation that the expression of AQPs in growing regions of roots is generally higher in stelar compared with more peripherally located tissues (e.g. Gambettta et al. 2013; Knipfer et al. 2011). The cells in the root stele are mostly much smaller than the cells in the cortex. If it was only for their own water demand during cell expansion, the smaller stelar cells would not need a higher expression of AQPs (and by implication plasma membrane water flow). However, the cells in the stele encounter much high radial water flow densities per unit cell surface compared with, e.g. cortex cells (Bramley et al. 2009), and this is independent of the direction of flow (from the epidermis towards the xylem or from the xylem towards the epidermis). The higher water flow densities are the most likely reason for the higher expression of AQPs in stelar tissue. In other words: a high expression of AQPs in growing stelar tissue relates more to the position of cells than to particular volume expansion rates.

4.3 *Relevance of Cell Lp for Leaf Growth*

There exist some detailed studies on the hydraulics of cell expansion in leaves, in particular leaves of grasses. Most studies have been carried out on maize, tall fescue and barley (e.g. Ehlert et al. 2009; Martre et al. 1999; Fricke et al. 1997; Fricke and Peters 2002; Bouchabké et al. 2006; Parent et al. 2009; Touati et al. 2015). These studies showed that there exist growth-induced water potential differences between growing tissue and leaf internal water source (xylem). The implication is that growth is co-limited by hydraulic properties in addition to a mechanical limitation through wall-yielding properties. For example, the Lp of growing leaf epidermal cells in barley ranged from 0.4 (Volkov et al. 2007) to 2×10^{-6} m s^{-1} MPa^{-1} (Touati et al. 2015) and was slightly, though significantly, larger than the Lp of mature epidermal cells (Volkov et al. 2007). The same applies to the osmotic water permeability of mesophyll protoplasts when comparing growing with mature leaf regions (Volkov et al. 2007). One can, similar to the above calculations for roots, estimate the water potential difference across the plasma membrane of a growing leaf cell required to support water uptake for growth, given the dimension, growth rate and Lp of the cell. The conclusion is the same as for roots (Touati et al. 2015), in that the measured water potential difference across a tissue exceeds the one required to sustain growth by a factor of 1,000 or more. These data suggest that cell Lp and any growth-dependent expression of AQPs in leaves match more the need for trans-tissue transport of water than for sustaining volume expansion rates of an individual cell. The Lp values in the leaf epidermis are slightly at odds with this conclusion: as the epidermis is at the end of the water transport path from leaf xylem to leaf periphery, any high cell Lp in the epidermis contributes little to speeding up the overall transport of water from xylem to epidermis. So, why is cell Lp in growing epidermal cells high and also higher than in the mature leaf region, and why does this coincide with the tissue-specific expression of AQPs, as shown for barley (HvPIP1;1/1;6)? One possible explanation is that the cell Lp reflects extensive symplastic movement of water between epidermal cells, though previous studies on barley make this explanation unlikely (Fricke 2000). Another explanation is that water reaches the epidermis along bundle sheath extensions and then has to pass through 10, 20 or more lateral-located epidermal cells in succession to reach the end of its transport route (Fricke 2000); similarly, water may move over significant distances axially along a file of epidermal cells, from cell to cell, rather than being supplied to epidermal cells throughout along a radial transport route from xylem, via mesophyll/bundle sheath extension to epidermis. A third explanation could be that close to 100 % of the difference in water potential difference between epidermis and leaf xylem (ca 0.2–0.3 MPa) is generated upstream of the transport route of water, for example, at the parenchymatous and mestome sheath of vascular bundles (Fricke 2002; Heinen et al. 2009). In this case, the difference in water potential between two adjacent cells in the epidermis, or between adjacent mesophyll and epidermal cells, may actually be so small that it requires the high cell Lp observed for growing cells.

As for roots, we could ask for leaves whether the high cell Lp in the growth zone has also the potential function to speed up water transport in the proximal portion of growth zone, which is close to the meristem. This question is of particular interest when studying grasses, as these have an intercalary 'apical' meristem located at the base of leaves during the early vegetative growth stage of plants. This means that the cell production zone and portion of growth zone containing undifferentiated cells including xylem are located between the water supply route from the root (xylem) and more distally located growing and mature leaf region (Fig. 2). There exist some anatomical and biophysical studies on the hydraulics of this proximal meristematic region, yet we do not know for sure how water gets through the very base of growing grass leaves (for a review, see Fricke 2002); at least the author is not aware of any study that answers this question conclusively. Therefore, it is possible that similar to roots, growing cells close to the leaf meristem require a high cell Lp to facilitate the movement of water through tissues. Detailed analyses of the Lp of cells in this leaf region are required.

5 AQPs: Facilitators of Water Uptake into Growing Plant Cells or…?

The above considerations allow several conclusions as to any involvement of AQPs in cell expansion in plants.

(i) There exist AQP isoforms that are expressed particularly in growing compared with nongrowing tissues, in roots and leaves and in all species examined so far.
(ii) Higher expression of AQPs in growing compared with nongrowing tissues is often associated with a higher cell Lp, and sometimes also tissue Lp, in growing tissue.
(iii) The portion of growth zone closest to the meristem – the 'proximal' growth zone – seems to be hydraulically isolated from the major internal source (xylem) of water supply, particularly in roots. This has implications for the interpretation of data on AQPs and cell Lp and renders the (very) proximal growth zone different from the more distal regions of the growth zone, which are located further away from the meristem.
(iv) Growing cells have a cell Lp that should easily satisfy their need for water uptake associated with volume expansion. Rather, the high cell Lp is required to facilitate water transport at a sufficient rate through the tissue. Exceptions from this rule could be cells that are located at the very periphery (epidermis) of growing organs.
(v) It is concluded from the above that any specific expression of AQPs and high values of cell Lp in growing tissues reflect less any growth-specific requirements of individual cells but are a consequence of the overall developmental state and water flow pattern in these tissues. The primary role of AQPs in growing plant tissues is not so much the net transport of water *into* cells but *through* cells.
(vi) To test the above conclusion requires the isolation of growing cells from their usual tissue environment and a comparison of the AQP expression pattern and cell Lp between such cells and cells that are retained within tissues. Also, detailed data are needed on the hydraulic architecture of the very proximal portion of growth zone in roots and leaves. In addition, analyses of cell/tissue growth in multiple AQP knockout/knockdown lines would be interesting to address this question (e.g. compare Reinhardt et al. 2016), especially if this can be combined with a downregulation of AQP expression in a tissue-/cell-specific manner.

References

Alexandersson E, Fraysse L, Sjovall-Larsen S, Gustavsson S, Fellert M, Karlsson M, Johanson U, Kjelbom P (2005) Whole gene family expression and drought stress regulation of aquaporins. Plant Mol Biol 59:469–484

Aslam U, Bashir A, Khatoon A, Cheema HMN (2013) Identification and characterization of plasma membrane aquaporins from *Calotropis procera*. J Zhejiang Univ-SC B 14(7):586–595

Azaizeh H, Gunse B, Steudle E (1992) Effects of NaCl and $CaCl_2$ on water transport across root cells of maize (*Zea mays* L.) seedlings. Plant Physiol 99:886–894

Balk PA, de Boer AD (1999) Rapid stalk elongation in tulip (*Tulipa gesneriana* L. cv. Apeldoorn) and the combined action of cold-induced invertase and the water-channel protein γTIP. Planta 209:346–354

Barrieu F, Thomas D, Marty-Mazars D, Charbonnier M, Marty F (1998) Tonoplast intrinsic proteins from cauliflower (*Brassica oleracea* L. var. *botrytis*): immunological analysis, cDNA cloning and evidence for expression in meristematic tissues. Planta 204:335–344

Beaudette PC, Chlup M, Yee J, Emery RJ (2007) Relationships of root conductivity and aquaporin gene expression in *Pisum sativum*: diurnal patterns and the response to $HgCl_2$ and ABA. J Exp Bot 58:1291–1300

Besse M, Knipfer T, Miller AJ, Verdeil J-L, Jahn TP, Fricke W (2011) Developmental pattern of aquaporin expression in barley (*Hordeum vulgare* L.) leaves. J Exp Bot 62:4127–4142

Bouchabké O, Tardieu F, Simonneau T (2006) Leaf growth and turgor in growing cells of maize (*Zea mays* L.) respond to evaporative demand under moderate irrigation but not in water-saturated soil. Plant Cell Environ 29:1138–1148

Boursiac Y, Boudet J, Postaire O, Luu DT, Tournaire-Roux C, Maurel C (2008) Stimulus-induced downregulation of root water transport involves reactive oxygen species-activated cell signalling and plasma membrane intrinsic protein internalization. Plant J 56:207–218

Boyer JS (2001) Growth-induced water potentials originate from wall yielding during growth. J Exp Bot 52:1483–1488

Boyer JS, Silk WK (2004) Hydraulics of plant growth. Funct Plant Biol 31:761–773

Boyer JS, Cavalieri AJ, Schulze ED (1985) Control of the rate of cell enlargement: excision, wall relaxation, and growth-induced water potentials. Planta 163:527–543

Brady SM, Orlando DA, Lee JY, Wang JY, Koch J, Dinneny JR, Mace D, Ohler U, Benfey PN (2007) A high-resolution root spatiotemporal map reveals dominant expression patterns. Science 318:801–806

Bramley H, Turner NC, Turner DW, Tyerman SD (2009) Roles of morphology, anatomy, and aquaporins in determining contrasting hydraulic behavior of roots. Plant Physiol 150:348–364

Chaumont F, Tyerman SD (2014) Aquaporins: highly regulated channels controlling plant water relations. Plant Physiol 4:1600–1618

Chaumont F, Barrieu F, Herman EM, Chrispeels MJ (1998) Characterization of a maize tonoplast aquaporin expressed in zones of cell division and elongation. Plant Physiol 117:1143–1152

Cosgrove DJ (1993) Water uptake by growing cells: an assessment of the controlling roles of wall relaxation, solute uptake, and hydraulic conductance. Int J Plant Sci 154:10–21

Ehlert C, Maurel C, Tardieu F, Simonneau T (2009) Aquaporin-mediated reduction in maize root hydraulic conductivity impacts cell turgor and leaf elongation even without changing transpiration. Plant Physiol 150:1093–1104

Enstone DE, Peterson CA, Ma F (2003) Root endodermis and exodermis: structure, function, and responses to the environment. J Plant Growth Regul 21:335–351

Fetter K, Van Wilder V, Moshelion M, Chaumont F (2004) Interactions between plasma membrane aquaporins modulate their water channel activity. Plant Cell 16:215–228

Frangne N, Maeshima M, Schäffner AR, Mandel T, Martinoia E, Bonnemain JL (2001) Expression and distribution of a vacuolar aquaporin in young and mature leaf tissues of *Brassica napus* in relation to water fluxes. Planta 212:270–278

Frensch J, Steudle E (1989) Axial and radial hydraulic resistance to roots of maize (*Zea mays* L.). Plant Physiol 91:719–726
Fricke W (2000) Water movement between epidermal cells of barley leaves – a symplastic connection? Plant Cell Environ 23:991–997
Fricke W (2002) Botanical briefing review: biophysical limitation of cell elongation in cereal leaves. Ann Bot 90:1–11
Fricke W, Flowers TJ (1998) Control of leaf cell elongation in barley. Generation rates of osmotic pressure and turgor, and growth-associated water potential gradients. Planta 206:53–65
Fricke W, Peters WS (2002) The biophysics of leaf growth in salt-stressed barley, a study at the cell level. Plant Physiol 129:1–15
Fricke W, McDonald AJS, Mattson-Djos L (1997) Why do leaves and leaf cells of N-limited barley elongate at reduced rates? Planta 202:522–530
Gambetta GA, Fei J, Rost TL, Knipfer T, Matthews MA, Shackel KA, Walker MA, McElrone AJ (2013) Water uptake along the length of grapevine fine roots: developmental anatomy, tissue-specific aquaporin expression, and pathways of water transport. Plant Physiol 163:1254–1265
Hachez C, Moshelion M, Zelazny E, Cavez D, Chaumont F (2006) Localization and quantification of plasma membrane aquaporin expression in maize primary root: a clue to understanding their role as cellular plumbers. Plant Mol Biol 62:305–323
Hachez C, Heinen RB, Draye X, Chaumont F (2008) The expression pattern of plasma membrane aquaporins in maize leaf highlights their role in hydraulic regulation. Plant Mol Biol 68:337–353
Hachez C, Veselov D, Ye Q, Reinhardt H, Knipfer T, Fricke W, Chaumont F (2012) Short-term control of maize cell and root water permeability through plasma membrane aquaporin isoforms. Plant Cell Environ 35:185–198
Hachez C, Veljanovski V, Reinhardt H, Guillaumont, Vanhee C, Chaumont F, Batako H (2014) The *Arabidopsis* abiotic stress-induced TSPO-related protein reduces cell-surface expression of the aquaporin PIP2;7 through protein-protein interactions and autophagic degradation. Plant Cell 26:4974–4990
Heinen RB, Ye Q, Chaumont F (2009) Role of aquaporins in leaf physiology. J Exp Bot 60: 2971–2985
Hill AE, Shachar-Hill B, Shachar-Hill Y (2004) What are aquaporins for? J Membr Biol 197: 1–32
Hove RM, Ziemann M, Bhave M (2015) Identification and expression analysis of the *barley* (*Hordeum vulgare* L.) *aquaporin* gene family. PLoS ONE 10(6):e0128025. doi:10.1371/journal.pone.0128025
Hukin D, Doering-Saad C, Thomas CR, Pritchard J (2002) Sensitivity of cell hydraulic conductivity to mercury is coincident with symplasmic isolation and expression of plasmalemma aquaporin genes in growing maize roots. Planta 215:1047–1056
Javot H, Lauvergeat V, Santoni V, Martin-Laurent F, Güçlü J, Vinh J, Heyes J, Franck KI, Schäffner AR, Bouchez D, Maurel C (2003) Role of a single aquaporin isoform in root water uptake. Plant Cell 15:509–522
Johansson I, Karlsson M, Shukla VK, Chrispeels MJ, Larsson C, Kjellbom P (1998) Water transport activity of the plasma membrane aquaporin PM28A is regulated by phosphorylation. Plant Cell 10:451–460
Katsuhara M, Akiyama Y, Koshio K, Shibasaka M, Kasamo K (2002) Functional analysis of water channels in barley roots. Plant Cell Physiol 43:885–893
Katsuhara M (2007) Molecular mechanisms of water uptake and transport in plant roots: research progress with water channel aquaporins. Plant Root 1:22–26
Knipfer T, Fricke W (2011) Water uptake by seminal and adventitious roots in relation to whole-plant water flow in barley (*Hordeum vulgare* L.). J Exp Bot 62:717–733
Knipfer T, Besse M, Verdeil J-L, Fricke W (2011) Aquaporin-facilitated water uptake in barley (*Hordeum vulgare* L.) roots. J Exp Bot 62:4115–4126

Lee SH, Chung GC, Jang JY, Ahn SJ, Zwiazek JJ (2012) Overexpression of PIP2;5 aquaporin alleviates effects of low root temperature on cell hydraulic conductivity and growth in Arabidopsis. Plant Physiol 159:479–488

Lersten NR (1997) Occurrence of endodermis with a casparian strip in stem and leaf. Bot Rev 63:265–272

Li GW, Zhang MH, Cai WM, Sun WN, Su WA (2008) Characterization of OsPIP2;7, a water channel protein in rice. Plant Cell Physiol 49:1851–1858

Li DD, Ruan XM, Zhang J, Wu YJ, Wang XL, Li XB (2013) Cotton plasma membrane intrinsic protein 2s (PIP2s) selectively interact to regulate their water channel activities and are required for fibre development. New Phytol 199(3):695–707

Lockhart JA (1965) An analysis of irreversible plant cell growth. J Theor Biol 8:264–275

Lopez F, Bousser A, Sissoeff I, Gaspar M, Lachaise B, Hoarau J, Mahe A (2003) Diurnal regulation of water transport and aquaporin gene expression in maize roots: contribution of PIP2 proteins. Plant Cell Physiol 44:1384–1395

Ma N, Xue JQ, Li YH, Liu XJ, Dai FW, Jia WS, Luo YB, Gao JP (2008) Rh-PIP2;1, a rose aquaporin gene, is involved in ethylene-regulated petal expansion. Plant Physiol 148:894–907

Martre P, Bogeat-Triboulot MB, Durand JL (1999) Measurement of a growth-induced water potential gradient in tall fescue leaves. New Phytol 142:435–439

Maurel C (2007) Plant aquaporins: novel functions and regulation properties. FEBS Lett 581: 2227–2236

Maurel C, Reizer J, Schroeder JI, Chrispeels MJ (1993) The vacuolar membrane protein gamma-TIP creates water specific channels in *Xenopus* oocytes. EMBO J 12:2241–2247

Miyamoto N, Ookawa T, Takahashi H, Hirasawa T (2002) Water uptake and hydraulic properties of elongation cells in hydrotropically bending roots of *Pisum sativum* L. Plant Cell Physiol 43:393–401

Muto Y, Segami S, Hayashi H, Sakurai J, Murai-Hatano M, Hattori Y, Ashikari M, Maeshima M (2011) Vacuolar proton pumps and aquaporins involved in rapid internode elongation of deepwater rice. Biosci Biotechnol Biochem 75:114–122

O'Brien TP, Carr DJ (1970) A suberized layer in the cell walls of the bundle sheath of grasses. Aust J Biol Sci 23:275–287

O'Brien TP, Kuo J (1975) Development of the suberized lamella in the mestome sheath of wheat leaves. Aust J Bot 23:783–794

Obroucheva NV, Sin'kevich IA (2010) Aquaporins and cell growth. Russ J Plant Physiol 57:153–165

Parent B, Hachez C, Redondo E, Simonneau T, Chaumont F, Tardieu F (2009) Drought and ABA effects on aquaporin content translate into changes in hydraulic conductivity and leaf growth rate: a trans-scale approach. Plant Physiol 149:2000–2012

Pritchard J (1994) The control of cell expansion in roots. New Phytol 127:3–26

Reinhardt H, Hachez C, Bienert MD, Beebo A, Swarup K, Voß U, Bouhidel K, Frigerio L, Schjoerring LK, Bennett MJ, Chaumont F (2016) Tonoplast aquaporins facilitate lateral root emergence. Plant Physiol 170:1640–1654

Richardson A, Franke R, Kerstiens G, Jarvis M, Schreiber L, Fricke W (2005) Cuticular wax deposition in barley leaves commences in relation to the point of emergence from sheaths of older leaves. Planta 222:472–483

Sade N, Shatil-Cohen A, Attia Z, Maurel C, Boursiac Y, Kelly G et al (2014) The role of plasma membrane aquaporins in regulating the bundle sheath-mesophyll continuum and leaf hydraulics. Plant Physiol 166:1609–1620

Sakurai J, Ishikawa F, Yamaguchi T, Uemura M, Maeshima M (2005) Identification of 33 aquaporin genes and analysis of their expression and function. Plant Cell Physiol 46:1568–1577

Sakurai J, Ahamed A, Murai M, Maeshima M, Uemura M (2008) Tissue and cell-specific localization of rice aquaporins and their water transport activities. Plant Cell Physiol 49:30–39

Schüssler MD, Alexandersson E, Bienert GP, Kichey T, Laursen KH, Johanson U, Kjellbom P, Schjoerring JK, Jahn TP (2008) The effects of the loss of TIP1;1 and TIP1;2 aquaporins in *Arabidopsis thaliana*. Plant J 56:756–767

Shatil-Cohen A, Attia Z, Moshelion M (2011) Bundle-sheath cell regulation of xylem-mesophyll water transport via aquaporins under drought stress: a target of xylem-borne ABA? Plant J 67: 72–80

Steudle E (2000) Water uptake by plant roots:an integration of views. Plant Soil 226:46–56

Steudle E, Peterson CA (1998) How does water get through roots? J Exp Bot 49:775–788

Tang AC, Boyer JS (2002) Growth-induced water potentials and the growth of maize leaves. J Exp Bot 53:489–503

Touati M, Knipfer T, Visnovitz T, Kameli A, Fricke W (2015) Limitation of cell elongation in barley (*Hordeum vulgare* L.) leaves through mechanical and tissue-hydraulic properties. Plant Cell Physiol 56:1364–1373

Tournaire-Roux C, Sutka M, Javot H, Gout E, Gerbeau P, Luu DT, Bligny R, Maurel C (2003) Cytosolic pH regulates root water transport during anoxic stress through gating of aquaporins. Nature 425:393–397

Tyerman SD, Hatcher AI, West RJ, Larkum AWD (1984) *Posidonia australis* growing in altered salinities – leaf growth, regulation of turgor and the development of osmotic gradients. Aust J Plant Physiol 11:35–47

Van den Honert TH (1948) Water transport in plants as a catenary process. Disc Faraday Soc 3:146–153

Vandeleur RK, Mayo G, Shelden MC, Gilliham M, Kaiser BN, Tyerman SD (2009) The role of plasma membrane intrinsic protein aquaporins in water transport through roots: diurnal and drought stress responses reveal different strategies between isohydric and anisohydric cultivars of grapevine. Plant Physiol 149:445–460

Volkov V, Hachez C, Moshelion M, Draye X, Chaumont F, Fricke W (2007) Water permeability differs between growing and non-growing barley leaf tissues. J Exp Bot 58:377–390

Wei W, Alexandersson E, Golldack D, Miller AJ, Kjellbom PO, Fricke W (2007) HvPIP1;6, a barley (*Hordeum vulgare* L.) plasma membrane water channel particularly expressed in growing compared with non-growing leaf tissues. Plant Cell Physiol 48:1132–1147

Wu X, Lin J, Lin Q, Wang J, Schreiber L (2005) Casparian strips in needles are more solute-permeable than endodermal transport barriers in roots of *Pinus bungeana*. Plant Cell Physiol 46:1799–1808

Yang S, Cui L (2009) The action of aquaporins in cell elongation, salt stress and photosynthesis. Chin J Biotechnol 25:321–327

Ye Q, Holbrook NM, Zwieniecki MA (2008) Cell-to-cell pathway dominates xylem-epidermis hydraulic connection in *Tradescantia fluminensis* (Vell. Conc.) leaves. Planta 227:1311–1317

Zelazny E, Borst JW, Muylaert M, Batoko H, Hemminga MA, Chaumont F (2007) FRET imaging in living maize cells reveals that plasma membrane aquaporins interact to regulate their subcellular localization. Proc Natl Acad Sci U S A 104:12359–12364

Zhang WH, Tyerman SD (1991) Effect of low O_2 concentration and azide on hydraulic conductivity and osmotic volume of the cortical-cells of wheat roots. Aust J Plant Physiol 18:603–613

Aquaporins and Root Water Uptake

Gregory A. Gambetta, Thorsten Knipfer, Wieland Fricke, and Andrew J. McElrone

Abstract Water is one of the most critical resources limiting plant growth and crop productivity, and root water uptake is an important aspect of plant physiology governing plant water use and stress tolerance. Pathways of root water uptake are complex and are affected by root structure and physiological responses of the tissue. Water travels from the soil to the root xylem through the apoplast (i.e., cell wall space) and/or cell-to-cell, but hydraulic barriers in the apoplast (e.g., suberized structures in the endodermis) can force water to traverse cell membranes at some points along this path. Anytime water crosses a cell membrane, its transport can be affected by the activity of membrane-intrinsic water channel proteins (aquaporins). We review how aquaporins can play an important role in affecting root water transport properties (hydraulic conductivity, Lp), and thus alter water uptake, plant water status, nutrient acquisition, growth, and transpiration. Plants have the capacity to regulate aquaporin activity through a variety of mechanisms (e.g., pH, phosphorylation, internalization, oxidative gating), which may provide a rapid and reversible means of regulating root Lp. Changes in root Lp via the modulation of aquaporin activity is thought to contribute to root responses to a broad range of stresses including drought, salt, nutrient deficiency, and cold. Given their role in contributing to stress tolerance, aquaporins may serve as future targets for improving crop performance in stressful environments.

G.A. Gambetta (✉)
Bordeaux Sciences Agro, Institut des Sciences de la Vigne et du Vin, Ecophysiologie et Génomique Fonctionnelle de la Vigne, UMR 1287, F- 33140 Villenave d'Ornon, France
e-mail: gregory.gambetta@agro-bordeaux.fr

T. Knipfer
Department of Viticulture and Enology, University of California, Davis, CA 95616, USA

W. Fricke
School of Biology and Environmental Science, Science Centre West, University College Dublin, Belfield, Dublin 4, Ireland

A.J. McElrone
Department of Viticulture and Enology, University of California, Davis, CA 95616, USA

United States Department of Agriculture-Agricultural Research Service, Crops Pathology and Genetics Research Unit, Davis, CA 95616, USA

F. Chaumont, S.D. Tyerman (eds.), *Plant Aquaporins*, Signaling and Communication in Plants, DOI 10.1007/978-3-319-49395-4_6

1 Introduction

Water is one of the most limiting factors to plant growth and productivity, and a primary determinant of vegetation distributions worldwide. Plants growing in natural ecosystems must cope with periodic droughts due to limited precipitation in mesic regions or chronically dry conditions in arid habitats. This coping often involves the ability to efficiently adjust water flow at the whole plant and cellular level (for review, Hsiao 1973). In agroecosystems, water scarcity (i.e., when demands exceed supplies) threatens crop production in dry growing regions across the globe due to growing competition between municipal, industrial, conservation, and agricultural entities for limited high-quality water supplies. Changing climatic conditions could exacerbate this situation; projections for the twenty-first century predict that more frequent and more severe drought events will be accompanied by warmer summers resulting in record soil moisture deficits for some regions (Cayan et al. 2010). Conservation irrigation techniques (e.g., deficit irrigation or dry farming) are often used under conditions of limited water availability, but these strategies can lead to the emergence of other abiotic stressors. When combined with impaired irrigation water quality, limited rainfall, and high evapotranspiration, deficit irrigation strategies can cause salt to accumulate in crop root zones (UN-FAO 2002). In order to sustain crop production under these conditions, growers require plant material that tolerates a variety of abiotic stressors. A major challenge in developing stress-resistant plant material is elucidating how roots can better meet increased evapotranspiration demands of the canopy with lower soil water availability and increased soil salinity.

For cultivated crops, drought resistance is defined as the ability of a plant to grow satisfactorily and maintain or enhance yield and fruit quality when exposed to periods of water deficit (Blum 2011). One fundamental function of a root system is to match canopy water demands needed to sustain transpirational water loss associated with carbon assimilation and plant growth (Kramer and Boyer 1995). Root systems can contribute to drought resistance by maintaining an expansive and active/absorptive surface area that maintains water uptake capacity as soils dry. Commercial rootstocks are thought to impart drought resistance to scion by partitioning roots deeper in the soil profile (e.g., Padgett-Johnson et al. 2003). While expansive growth is important, the functionality of young non-woody fine roots is likely paramount because they contribute disproportionately to root system water uptake (Queen 1967; Gambetta et al. 2013). However, this portion of the root system may also be the weak link when subjected to drying soil (North and Nobel 1991). The importance of young root proliferation associated with a higher Lp has been suggested for a drought-resistant rootstock (e.g., Alsina et al. 2011) and likely involves the regulation of aquaporins (membrane-intrinsic water channel proteins that facilitate transmembrane water transport) to maintain water permeability and growth of the root tips. A greater understanding of how root hydraulic traits at both the cellular and organ level affect plant performance under these water deficit and other abiotic stress conditions can improve the breeding of stress-tolerant crops, improve irrigation management, and enhance water conservation (Passioura 2002; Comas et al. 2013; Barrios-Masias and Jackson 2014; Seversike et al. 2014).

A root system consists of a complex network of individual roots that can differ in maturity and vary in developmental status along their length. As opposed to older woody roots, the newest portions of non-woody fine roots are considered the most permeable portion of a root system to water and are thought to have the greatest ability to absorb water (McCully 1999). Water absorbed by young fine roots moves radially, crossing several living and metabolically active cell layers before entering the xylem long-distance transport pathway to the leaves. Transport efficiency across these cell layers can be affected by the activity, density, and location of aquaporins (Maurel et al. 2015; Gambetta et al. 2013; Hachez et al. 2012).

2 Root Water Uptake and Pathways of Water Transport

2.1 Root Water Uptake: Composite Transport Model

Upon absorption by the root, water first crosses the epidermis and then moves toward the center of the root crossing the cortex and endodermis before arriving at the xylem. Based on the composite model of water transport across roots (Steudle and Peterson 1998; Steudle 2000), water can flow along two pathways in parallel between the soil and the root, i.e., along the apoplast and/or from cell-to-cell encountering a major hydraulic barrier at the endodermis (Fig. 1). The apoplastic path involves a substantial movement of water through the cell wall up to the endodermis, where water is either forced through endodermal cell membranes (if Casparian bands containing hydrophobic suberin depositions are present) or continues to move along the radial walls of endodermal cells. The cell-to-cell path involves mainly a flow of water across membranes (transcellular), which can occur by simple diffusion through the lipid bilayer and by facilitated diffusion through aquaporins and/or transport across plasmodesmata (symplastic). The actual pathway of water transport across the root is extremely complex, and the relative contribution of each pathway likely varies with species, plant development, and growth conditions. It should be noted that aquaporins can only make a significant contribution to root water uptake when the dominating flow path through the root cylinder is from cell-to-cell (Knipfer and Fricke 2010, 2011).

The rate of water flow (Q) through the root depends on the effective surface area for water uptake (A) and the biophysical force that drives the flow (i.e., water potential difference, $\Delta\Psi$). By definition, a hydraulic conductance L (=$Q/\Delta\Psi$) is dependent on root surface area (Fig. 2). To obtain an estimate of the intrinsic root hydraulic property, involving changes in aquaporin activity, L needs to be divided by A which gives the root hydraulic conductivity Lp [=$Q/(\Delta\Psi *A)$]. The Lp depends on the hydraulic properties of the axial flow path along mature xylem vessels and the radial flow paths across the root cylinder. It is generally accepted that the radial path of

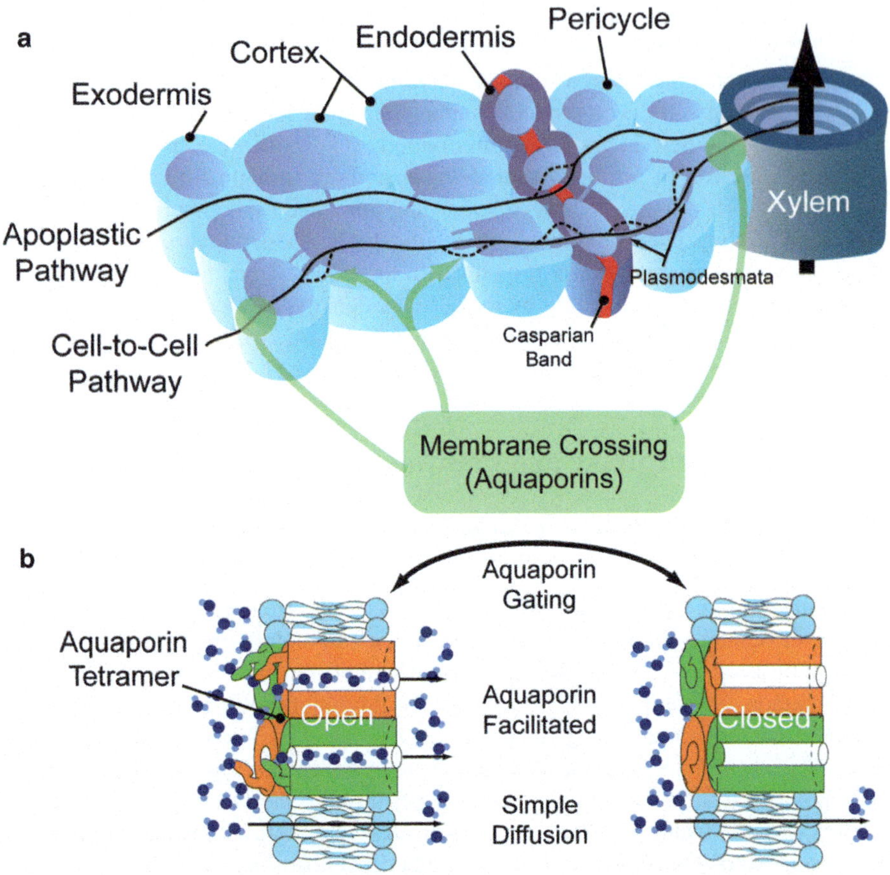

Fig. 1 (**a**) The composite model of water transport proposes that water can flow from the soil to the root xylem along two parallel pathways. One pathway is the apoplastic pathway (through the cell walls) and the other is the cell-to-cell pathway. Water flow through the apoplast can encounter a hydraulic barrier at the endodermis (Casparian band in *red*) which can force water flow across endodermal cell membranes if suberized (*dotted line*). The cell-to-cell pathway involves the flow of water through plasmodesmata and/or across membranes (*dotted lines*; i.e., the transcellular pathway). (**b**) Water crosses cell membranes by simple diffusion and/or by facilitated diffusion through aquaporins. Aquaporins function as homo- and heterotetramers, and their contribution to water transport can be regulated by their abundance, location, and interaction between different aquaporin isoforms. In addition, aquaporin proteins undergo gating, where various stimuli (e.g., pH, reactive oxygen species) can modulate the proteins between open and closed configurations, thus rapidly controlling their activity

apoplastic and cell-to-cell water movements constitutes the hydraulic bottleneck within roots (Frensch and Steudle 1989; Knipfer and Fricke 2011). For this reason, the radial flow path is particularly suited to adjust root water uptake.

Radial water flow through the root cylinder along the apoplast and cell-to-cell pathway is driven by a difference in Ψ between soil and root xylem. This water

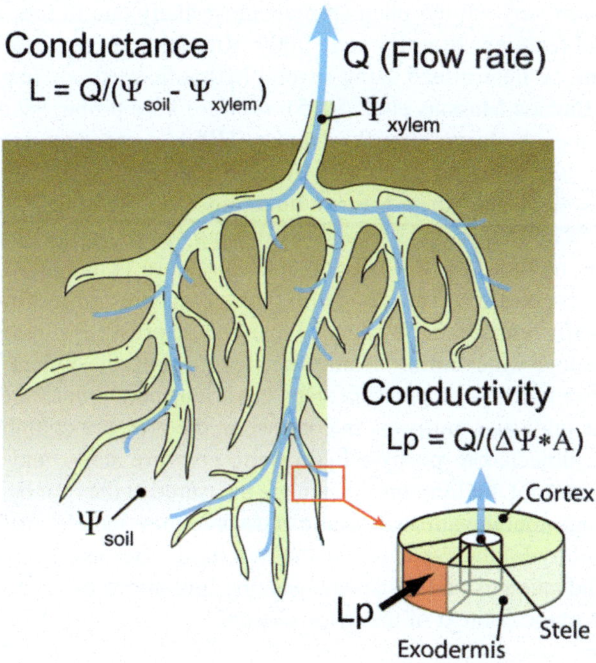

Fig. 2 The hydraulic property of a root system can be described by hydraulic conductance (L) or conductivity (Lp). Hydraulic conductance (L) is the relationship between the rate of water flow (Q, $m^3 s^{-1}$) relative to the biophysical force that drives the flow (i.e., water potential gradient, $\Delta\Psi$, MPa) and is expressed by the equation $L=Q/\Delta\Psi$; $m^3 s^{-1} MPa^{-1}$. L is dependent on root surface area (m^2). L is divided by effective area for water uptake (A) to give the root hydraulic conductivity Lp [$=Q/(\Delta\Psi * A)$, $m s^{-1} MPa^{-1}$] which represents the intrinsic root hydraulic property per unit surface area. Lp is a surface-independent size. As such, Lp can be used to assess the contribution of aquaporins to water uptake in root systems of various sizes

potential difference ($\Delta\Psi = \Delta P + \Delta\pi$) is composed of a hydrostatic (ΔP) and osmotic ($\Delta\pi$) pressure component. The osmotic component drives water uptake only in the presence of a semipermeable barrier (i.e., osmotic Lp), and since such barriers are predominantly provided by cellular membranes but not by the relatively more permeable apoplast, osmotic gradients drive water uptake across cellular membranes at some point along the pathway. In contrast, ΔP drives water flow along both the apoplastic pathway by bulk flow and cell-to-cell pathway following a gradient in water potential (i.e., hydrostatic Lp). The contribution of these two pathways to the flow will depend on their resistances relative to each other. The relative sizes of osmotic and hydrostatic Lp indicate which path contributes most to radial water uptake. If osmotic and hydrostatic root Lp values are of similar magnitude, it is assumed that most water flows along the cell-to-cell path; if hydrostatic Lp exceeds osmotic Lp several fold, most water flows along the apoplast path (Steudle and Peterson 1998). The contributions of the apoplast and cell-to-cell pathways to root water uptake can vary among plant species, most likely due to root morphological and anatomical features (Bramley et al. 2009; Knipfer and Fricke 2010, 2011).

Root Lp can be determined using several hydraulic methods (e.g., Knipfer and Fricke 2011; Barrios-Massias et al. 2015): root pressure probe, vacuum perfusion, pressure chamber, high-pressure flow meter (HPFM), and root exudation. All of these methods allow measurements of hydraulic properties on excised individual roots or root systems, but they differ in direction and magnitude of induced water flow and nature and magnitude of applied driving force. For example, in the root pressure probe method, pulses of hydrostatic pressure (< 0.1 MPa) are typically applied to the excised root base, which induces transient in-/outflows of water. In the vacuum perfusion method, a negative hydrostatic pressure step is applied to the root base inducing water inflow from root medium into the xylem. In the HFPM method, positive hydrostatic pressure steps of various magnitudes are applied to the root base inducing water outflows. In contrast, in the pressure chamber method, the entire root is subjected to positive hydrostatic pressure steps, and corresponding water flows from root medium into xylem are measured at the root base. Application of physiological nonrelevant hydrostatic pressures may induce artifacts related to the existence of hydraulic capacitors in the root tissue. Because each of these methods has their advantages and disadvantages, the limitations of each method should be considered in the context of the research goal.

2.2 *Aquaporins: The Contribution to Root Lp*

Much work over the last two decades has demonstrated the role of aquaporins in affecting root hydraulic properties. The magnitude of aquaporin contribution to root hydraulic conductivity is highly variable across species, ranging from 0 to 90 %, with higher values typically reported for herbaceous species (Table 1; reviewed in Javot and Maurel 2002). Across studies there is great variability in the experimental approach utilized both with respect to the type of inhibitor and the methodology of measuring Lp (Table 1). Studies have also demonstrated the importance of aquaporins to plant vigor and water relations of herbaceous species. For transgenic tobacco growing under favorable conditions, constitutive overexpression of *Arabidopsis* AtPIP1b increased transpiration rates and plant vigor (Aharon et al. 2003). More recently Sade et al. (2010) demonstrated that constitutively overexpressing NtAQP1 in *Arabidopsis* and tomato plants can enhance transpiration, photosynthesis, and shoot growth rates under favorable growing conditions. Lovisolo et al. (2007) assessed links between aquaporin gene expression and rootstock effects in perennial woody plants and found a positive correlation between aquaporin expression and root-specific hydraulic conductance (i.e., per gram of root dry weight), which was actually higher in olive dwarfing rootstocks. However, whole root system hydraulic conductance was greater in high vigor plants due to greater root biomass (Lovisolo et al. 2007). This same research group found differential aquaporin activity, measured by mercurial inhibition, under drought conditions for grapevine

Table 1 Summary of selected studies of aquaporin inhibition in roots utilizing different inhibitors and methodologies of measuring Lp

Species	Inhibition (%)	Inhibition methodology	Lp methodology	Reference
Allium cepa	57–84	$HgCl_2$	Hydrostatic (transpiration)	Barrowclough et al. (2000)
Arabidopsis thaliana	87	Azide	Hydrostatic (pressure)	Tournaire-Roux et al. (2003)
Arabidopsis thaliana	0	Transgenic (AtPIP2;2 knockout)	Hydrostatic (pressure)	Javot et al. (2003)
Arabidopsis thaliana	12	Transgenic (AtPIP2;2 knockout)	Osmotic (free exudation)	Javot et al. (2003)
Arabidopsis thaliana	20–30	Transgenic (AtPIP1;2 knockout)	Hydrostatic (pressure)	Postaire et al. (2010)
Beta vulgaris	80	$HgCl_2$	Osmotic (solutions)	Amodeo et al. (1999)
Bumelia lanuginosa	10–44	H_2O_2	Hydrostatic (pressure)	McElrone et al. (2007)
Capsicum annuum	66	$HgCl_2$	Osmotic (free exudation)	Carvajal et al. (1999)
Helianthus annuus	55–73	$HgCl_2$	Hydrostatic (pressure)	Adiredjo et al. (2014)
Hordeum vulgare	90	$HgCl_2$	Osmotic (solutions)	Tazawa et al. (1997)
Hordeum vulgare	40–74	$HgCl_2$	Osmotic (free exudation)	Knipfer et al. (2011)
Lycopersicon esculentum	57	$HgCl_2$	Hydrostatic (pressure)	Maggio and Joly (1995)
Nicotiana tabacum	42	Transgenic (NtAQP1 antisense)	Hydrostatic (HPFM)	Siefritz et al. (2002)
Cylindropuntia acanthocarpa	32 (distal)	$HgCl_2$	Hydrostatic (vacuum)	Martre et al. (2001b)
Cylindropuntia acanthocarpa	0 (midroot)	$HgCl_2$	Hydrostatic (vacuum)	Martre et al. (2001b)
Populus tremuloides	47	$HgCl_2$	Hydrostatic (pressure)	Wan and Zwiazek (1999)
Quercus fusiformis	13–50	H_2O_2	Hydrostatic (pressure)	McElrone et al. (2007)
Triticum aestivum	66	$HgCl_2$	Osmotic (free exudation)	Carvajal et al. (1996)
Vitis vinifera (root system)	4–41	H_2O_2	Hydrostatic (pressure)	Gambetta et al. (2012)
Vitis vinifera (individual roots)	8–19	H_2O_2	Hydrostatic (pressure)	Gambetta et al. (2012)
Vitis vinifera (individual roots)	9–23	H_2O_2	Osmotic (solutions)	Gambetta et al. (2012)
Vitis vinifera (individual roots)	0–5	H_2O_2	Hydrostatic (pressure)	Gambetta et al. (2013)
Vitis vinifera (individual roots)	5–45	H_2O_2	Osmotic (solutions)	Gambetta et al. (2013)
Zea mays	51–61	H_2O_2, acid load	Hydrostatic (vacuum)	Ehlert et al. (2009)

rootstocks derived from varied *Vitis* species parentage (Lovisolo et al. 2008). Previous work has documented greater inherent aquaporin expression and activity and maintenance of higher root hydraulic conductivity under low soil moisture availability for drought-resistant rootstocks compared to sensitive counterparts (Lovisolo et al. 2008; Gambetta et al. 2012; Barrios-Masias et al. 2015).

Root aquaporins can contribute substantially to water uptake if the dominant pathway through the root cylinder is from cell-to-cell. This should be tested first when addressing the role of root aquaporins in water uptake (Knipfer and Fricke 2010, 2011). In a plant, where most water is transported along the cell-to-cell path, any significant increase in aquaporin activity leads to an increase in root Lp. Such an increase in Lp through changes in aquaporin activity could be achieved more rapidly and with lower metabolic inputs than an increase in water-conducting surface area (A) (Suku et al. 2014). However, literature suggests that during plant development, water uptake increases mainly through an increase in A while changes in Lp (potentially through aquaporins) were only observed early in development (bean, Fiscus and Markhart 1979; barley, Suku et al. 2014).

Moreover, the analysis of aquaporin activity in roots may be further complicated by artificially induced changes in aquaporin activity related to sample preparation for hydraulic measurements. Vandeleur et al. (2014) found that shoot topping rapidly reduced root hydraulic conductance of grapevine and soybean by 50–60 %, which was associated with a significant reduction in the expression of several aquaporins. In soybean, there was a five to tenfold reduction in GmPIP1;6 expression over 0.5–1 h which was sustained over the period of reduced conductance. Meng and Fricke (unpublished data) also observed a 50 % decrease in PIP aquaporin expression in rice within 5 min for shoot topping. Given these observations, it can be speculated that measurements of Lp on excised individual roots or root systems may systematically underestimate the real contribution of aquaporins to root water transport in the intact plant because the measurements would be biased toward a larger contribution of the apoplastic pathway.

On a cellular level, aquaporins are water-conducting channels that are gated (Tornroth-Horsefield et al. 2006; Maurel et al. 2008) (Fig. 1). An observed increase/decrease in water flow through aquaporins can have multiple causes: (i) the number of aquaporin proteins in the membrane can change through, e.g., increased synthesis or altered trafficking (see Chapter "Plant Aquaporin Trafficking"); (ii) aquaporins can be gated between an open/closed state (see Chapter "Structural Basis of Permeation Function of Plant Aquaporins"); and/or (iii) their water-conducting capability can change depending on the interaction of different aquaporin isoforms (see Chapter "Heteromerization of Plant Aquaporins"). Aquaporin activity can be regulated through a range of mechanisms and in response to external factors. For example, water channel activity (open state) can be increased through phosphorylation and decreased through dephosphorylation (Johansson et al. 1996, 1998; Guenther et al. 2003; Wei et al. 2007) (see Chapter "Plant Aquaporin Post-translational Regulation"). Also, aquaporin activity can be regulated through intracellular pH and acidification (Gerbeau et al. 2002). For example, pH-dependent regulation provides an important means for roots to adjust water uptake in response to anoxia (Tournaire-Roux et al. 2003). This mechanism has been exploited to inhibit aquaporin activity experimentally through intracellular acidosis using propionic acid or azide (Tournaire-Roux et al. 2003). Hydrogen peroxide (H_2O_2) can affect aquaporin activity through an oxidative gating mechanism (Henzler et al. 2004; Ye and Steudle 2006; Fenton reaction: $Fe^{2+} + H_2O_2 = Fe^{3+} + OH + {*}OH$), or H_2O_2 can cause an internalization of aquaporins from root cell membranes (Boursiac et al. 2008). In response to H_2O_2, root and cell Lp decreases (Martinez-Ballesta et al. 2006; Boursiac et al. 2008; Ye and Steudle 2006). Aquaporins can also be regulated by other proteins. For example, recent work in *Arabidopsis* demonstrated that the stress-induced TSPO protein directly interacts with AtPIP2;7 and limits its abundance in the plasma membrane (Hachez et al. 2014). Experimentally, the most commonly used aquaporin inhibitor is mercuric chloride ($HgCl_2$) (Preston et al. 1993; Henzler and Steudle 1995; Tazawa et al. 1997; Katsuhara et al. 2002; Hukin et al. 2002, Knipfer et al. 2011). $HgCl_2$ is an efficient inhibitor for aquaporins, and there are only a few reports on aquaporin isoforms which are not inhibited (Daniels et al. 1994) although it is highly toxic. The applicability of aquaporin inhibitors such as H_2O_2, phloretin, silver (Ag^+ as $AgNO_3$), and gold (as $HAuCl_4$) appears to be more limited (Javot and Maurel 2002; Niemietz and Tyerman 2002; Maurel et al. 2008; Dordas et al. 2000; Moshelion et al. 2002; Volkov et al. 2007).

Inhibition studies involving nonspecific inhibitors (e.g., $HgCl_2$, H_2O_2, azide, etc.) should also be interpreted cautiously. The use of these inhibitors can only prove that the process of water transport across a root can be altered in its activity. Any cause of this inhibition, whether it involves aquaporin-specific effects, reduced trafficking of plasma membrane vesicles containing various membrane proteins including aquaporins, or other unknown mechanisms cannot be differentiated using these methods. In theory, inhibition experiments should involve "control" experiments where one tests other protein-mediated transport activities like potassium channels or H^+-ATPase activity in parallel, but in practice this is not possible for the vast majority of researchers. It is also interesting to note that few studies have shown that the inhibition of Lp by $HgCl_2$ or H_2O_2 can be fully, or close to fully, reversed through sulfhydryl reagents. In contrast, inhibition with azide has been shown to be almost entirely reversible (Tournaire-Roux et al. 2003; Postaire et al. 2010; Grondin et al. 2016).

2.3 *Root Tissue-Specific Expression/Activity of Aquaporins*

Relatively high or exclusive expression of aquaporins in roots as compared to shoot tissue was observed in several plant species (e.g., Sakurai et al. 2005; Hachez et al. 2006, Javot et al. 2003; Alexandersson et al. 2005; Heinen et al. 2009; Vandeleur et al. 2009; Besse et al. 2011; Beaudette et al. 2007). Also, aquaporin expression can vary along the main axis of roots but is commonly highest in the root tip (Knipfer et al. 2011; Gambetta et al. 2013). These findings infer that root-specific functions are associated with a higher or exclusive expression of aquaporin isoforms in roots.

Tissue localization of aquaporins has been studied using *in situ* hybridization, immunocytochemical approaches, and tissue laser microdissection (Hachez et al. 2006; Sakurai et al. 2005, 2008; Vandeleur et al. 2009; Besse et al. 2011; Gambetta et al. 2013). For example, it was shown for maize that ZmPIP2;1 and ZmPIP1;2 protein were most abundant in stellar cells; in more mature root regions, ZmPIP2;1 and ZmPIP1;2 were preferentially localized to the cortex and epidermis. ZmPIP2;5 was in particular found in cortex tissue at the root tip and in the endodermis in mature root tissue (Hachez et al. 2006). In barley, expression in the endodermis and stele was observed, highlighting a potential role in regulating radial water transport (Knipfer et al. 2011). In grapevines, high aquaporin activity and root Lp were associated with peak *VviPIP* expression in the root tip, and similar radial patterns of expression (Gambetta et al. 2013).

3 Root Aquaporins and Stress

Plants need to adapt to changes in external water availability in the short term (e.g., seconds to minutes to hours) and longer term (days to weeks). Prior to the discovery of aquaporins, it was widely accepted that the only means through which a short-term control of plant water balance could be achieved was through stomatal control. Roots were considered to allow regulation of water balance only in the longer term, for example, through changes in root surface area, morphology, or suberization of root internal and surface tissues. However, the discovery of aquaporins has revealed that similar to stomatal pores in leaves, roots have pores through which water flow can be regulated, even though these pores are molecular and not microscopic in dimension compared with stomata. This means that water flow through the plant can be regulated at either end, root, shoot, or both, and over both the short and/or longer term. Root aquaporins will be involved in both types of responses, and one may speculate whether aquaporins are more important in the short-term response due to their rapid and dynamic gating mechanism. The distinction between shorter- and longer-term responses becomes particularly important when studying aquaporin function in stress experiments which are carried out in a laboratory environment. For example, application of salt stress can lead lead to a more rapid osmotic stress on plants, and application of drought to pot-grown plants will cause a faster change in soil water potential, compared with field environments. In the short term, a closure of aquaporins could be beneficial as it can minimize loss of water to a root environment in which the osmotic strength (e.g., through salt addition) has increased; in the longer term, the same decrease in aquaporin activity can limit root water uptake, leaf gas exchange, and yield. This is just one example, yet it shows that the interpretation of data in literature on stress responses of root aquaporins needs to be viewed in the context of the experimental design (see also Chapter "Aquaporins and Abiotic Stress").

Stress induces changes in root Lp, anatomically and on a cellular scale, in response to a variety of different stresses. Drought, salt, cold, and nutrient stress often results in decreased root Lp (Clarkson 2000; Aroca et al. 2011). This decrease in Lp could serve several functions, which include influencing daily stomatal closure via the production of a hydraulic signal (Christmann et al. 2007) and limiting nighttime water loss back to the soil. The extent to which root system Lp contributes to whole plant hydraulic conductance varies, but can be greater than 50 % (Boyer 1971; Neumann et al. 1974; Martre et al. 2001a); thus decreases in root Lp under stress can potentially play a significant role in limiting water movement through the transpiring plant.

3.1 *Aquaporins and Drought Stress*

Over the long term, root Lp decreases under drought stress (reviewed in Aroca et al. 2011) (Fig. 3). In some cases increases in Lp have been reported but only over short (i.e., hours) time frames (Lian 2004; Siemens and Zwiazek 2004). The short-term increase is consistent with increases in abscisic acid (ABA) under drought, positive effects of ABA on Lp, and the upregulation of aquaporin gene expression by ABA (Hose et al. 2000; Aroca et al. 2006; Parent et al. 2009; Mahdieh and Mostajeran 2009). However, little is known regarding the integration of ABA, aquaporins, and root Lp across different time scales. ABA action is nuanced and can sometimes result in decreases in Lp and aquaporin gene expression in a developmental and concentration-specific manner (Beaudette et al. 2007).

The decrease in root Lp in response to drought likely results from a combination of decreases in the activity of aquaporins and structural changes, namely, through increased or accelerated deposition of suberin (Martre et al. 2001b; North et al. 2004; Hachez et al. 2012; Barrios-Masias et al. 2015). The contribution of decreased aquaporin activity and/or structural changes to decreases in Lp may be specific to particular root portions. North et al. (2004) reported decreases in root Lp of desert agave subjected to drought that were proportional to the contribution of aquaporins under well-watered conditions; however, other root portions exhibited decreases in Lp without any contribution of aquaporins under either control or drought conditions. This suggests that in some root portions, the decrease in Lp results from a purely structural mechanism.

Numerous studies have examined changes in aquaporin gene expression under drought stress, but drawing firm conclusions from these studies has proven difficult. The observed changes in gene expression are variable between studies and among the different aquaporin isogenes (reviewed in Aroca et al. 2011). Interpretation is complicated by the fact that aquaporins form heterotetramers resulting in variable effects on cell membrane hydraulic conductivity (Fetter et al. 2004; Vandeleur et al. 2009) (see Chapter "Heteromerization of Plant Aquaporins") and that aquaporins have tissue-specific expression patterns (discussed above). One technical consideration is that drought stress studies utilize a variety of approaches to induce the stress that include treatment with non-metabolizable solutes, decreased soil water content with daily watering, and complete water withholding among others. These differences may result in differential responses with respect to changes in aquaporin expression, activity, and root Lp.

The manipulation of specific aquaporin genes in transgenic plants has resulted in equally complex outcomes (Siefritz et al. 2002; Aharon et al. 2003; Lian 2004; Jang et al. 2007a, b; Peng et al. 2007; Sade et al. 2009; Wang et al. 2011). Some works

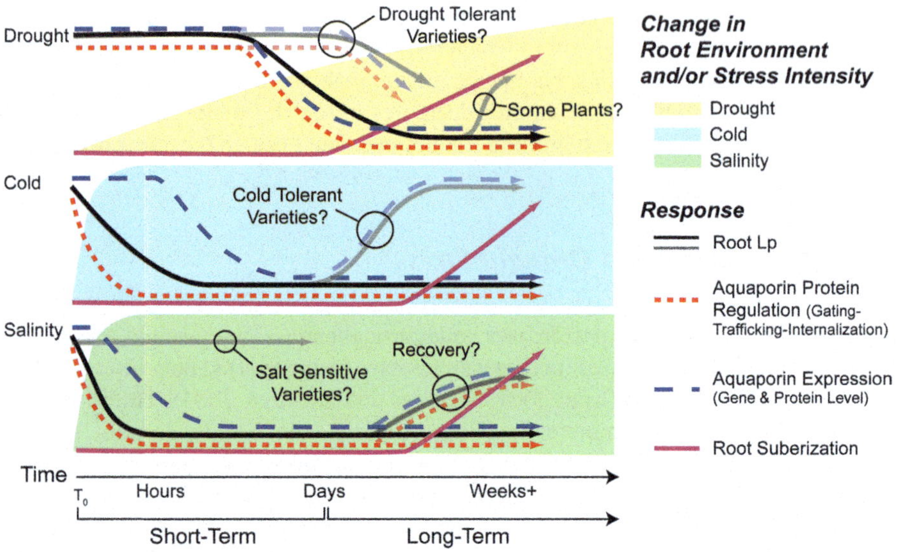

Fig. 3 Hypothetical framework for the integration of short- and long-term responses of root water uptake to various stresses and underlying mechanisms involving the regulation of root Lp and aquaporins. The x-axis represents time. The speed of the change in the root environment and intensity of stress is not the same between stresses, especially in experimentation carried out in the laboratory environment. Shaded curves represent the rate of the stress; where cold and salinity stresses are nearly instantaneous upon application of the stress, drought takes time to develop and it intensifies over time. Changes in Lp, aquaporin protein regulation, aquaporin expression, and suberization are shown over the two time frames

have found that changing the expression of specific aquaporin genes results in increased tolerance to drought or dehydration stress (Sade et al. 2009), while others have found either no, mixed, or negative effects (Aharon et al. 2003; Jang et al. 2007a, b; Peng et al. 2007; Wang et al. 2011). These contrasting results may arise from the complexities of aquaporin function introduced above (i.e., heterotetramerization) which make the effect of manipulating a single isogene, or expressing a heterologous isogene in a different species, difficult to predict. Furthermore, manipulating the expression of a single aquaporin isogene has been shown to change the expression of other endogenous aquaporin isogenes (Jang et al. 2007a, b).

3.2 *Aquaporins and Cold Stress*

Cold stress (typically referring to temperatures less than 15°C but not freezing) rapidly decreases root Lp, but in particular in plant varieties that are cold tolerant Lp recovers over time (Lee et al. 2004; Aroca et al. 2005; Yu et al. 2006; Murai-Hatano et al. 2008) (Fig. 3). The initial decrease in Lp is associated with a rapid downregulation of the expression of most aquaporin isogenes in a variety of species

(Jang et al. 2004; Aroca et al. 2005; Sakurai et al. 2005; Yu et al. 2006) although some studies demonstrate that protein abundance does not decrease as a result (Aroca et al. 2005; Murai-Hatano et al. 2008). This difference suggests that aquaporin contribution to the initial decrease in Lp is due to decreased aquaporin activity via post-translational modification.

The decreases in root Lp in response to cold stress appear to result from both the inhibition of aquaporin activity as well as other factors. Studies have suggested a role for membrane damage and reactive oxygen species, specifically hydrogen peroxide, in facilitating the decreases in Lp (Lee et al. 2004; Aroca et al. 2005). Hydrogen peroxide could potentially act directly on aquaporin activity and/or mediate changes in pH via the plasma membrane proton ATPases as proposed by Lee et al. (2004). Over longer time frames, cold stress can lead to increased suberin accumulation in roots, analogous to drought, likely contributing to decreased Lp (Lee et al. 2005).

In specific cold-tolerant plants, Lp recovers over time and sometimes to values even greater than those prior to the stress event (Aroca et al. 2005; Lee et al. 2005). The recovery of Lp has been associated with increases in aquaporin gene expression (Yu et al. 2006), protein level (Aroca et al. 2005), and contribution of aquaporins to the recovered Lp (Lee et al. 2005). Taken together this suggests that aquaporins play an important role in the recovery of Lp to cold stress. However, similar increases in aquaporin protein abundance during recovery have been observed in cold-intolerant varieties that do not recover Lp suggesting that increased aquaporin activity alone may not be sufficient for Lp recovery (Aroca et al. 2005).

3.3 Aquaporins and Salt Stress

Similar to drought and cold, salt stress (or salinity stress) initially leads to decreased root Lp (reviewed in Aroca et al. 2011) and these decreases have been associated with the downregulation of aquaporin gene expression (Kawasaki et al. 2001; Boursiac et al. 2005; Zhu et al. 2005) (Fig. 3). In contrast, some specific aquaporin isogenes have been shown to be salt inducible in some species (Suga et al. 2002; Bae et al. 2011). Early studies demonstrated that salt stress decreased root Lp and that the residual Lp, contrary to the Lp of non-stressed plants, was little or not at all reduced by aquaporin inhibitors. This observation not only supports a role of aquaporins in root water uptake of non-stressed plants but also suggests that most of the salt-induced decrease in root Lp is due to a reduction in aquaporin activity (Carvajal et al. 1999; Martínez-Ballesta et al. 2003). In *Arabidopsis*, decreases in Lp are associated with decreases in gene expression, protein abundance of specific isogenes, and relocalization of aquaporin proteins into endosome-like structures (Boursiac et al. 2005). Some studies suggest that Lp recovers to some extent after longer exposure to salt (Martínez-Ballesta et al. 2003; Wan 2010), but the role of aquaporins in facilitating this recovery is unknown.

Recently, Kaneko et al. (2015), studying rapid changes in root Lp and aquaporin function in barley cultivars which differ in salt tolerance, concluded that the more tolerant cultivars showed a rapid (1 h) decrease in root Lp following addition of salt. The Lp recovered transiently (4 h) only to decrease again and remain at a low level during the following day. A more salt-sensitive cultivar did not show a decrease in Lp. The authors explained their finding through a combination of a rapid post-translational regulation (dephosphorylation, membrane internalization, degradation) of aquaporin function, coupled with a delayed decrease in aquaporin gene or protein level. Also, the authors proposed that there exists something like a time-dependent "checkpoint" where root Lp either stays low following an initial decrease and transient recovery or approaches the prestress level. Lee and Zwiazek (2015) observed a rapid decrease in root cell Lp in response to low concentrations (10 mM) of salt in wild-type *Arabidopsis* plants but not in plants overexpressing AtPIP2;5. Based on cell pressure probe and gene expression analyses, combined with pharmacological treatments, the authors concluded that changes in the phosphorylation status of aquaporins are among the initial targets of NaCl stress. These two studies highlight the importance of distinguishing between early and late responses of root Lp and aquaporin function to salinity. Even though salt tolerance of crops is defined through yield and longer-term performance in a saline root environment, tolerance mechanisms involving the regulation of root (cell) Lp and aquaporin function are nearly instantaneous.

Like drought stress, increased salt tolerance has been a target of transgenic manipulation of aquaporin genes. Similarly, these approaches have yielded mixed results because of the complexity of aquaporin action discussed above. Some studies have shown that overexpression of particular aquaporin isogenes increases seed germination and/growth under saline conditions (Peng et al. 2007; Gao et al. 2010) while other studies have shown negative effects on salt tolerance (Wang et al. 2011).

4 Nutrient Uptake and Growth

4.1 Aquaporins and Nutrient Uptake

Despite the strong link between root water and nutrient uptake, studies that combine root hydraulics and nutrient treatments are surprisingly rare (see chapter Root Hydraulic and Aquaporin Responses to N Availability). Changes in root hydraulic conductivity, likely involving aquaporins, appear to be tightly linked with the availability of many different nutrients (reviewed in Wang et al. 2016). Studies in various plant species have demonstrated that nitrogen, phosphorous, and sulfur deficiencies lead to a rapid decrease in root Lp, which is reversible when the nutrient is resupplied (Carvajal et al. 1996; Clarkson 2000; Shangguan et al. 2005; Gorska et al. 2008). Some studies have found an absence of a reducing effect of aquaporin inhibition on root Lp following a decrease in Lp in response to nutrient deficiency, indicating that aquaporins no longer contribute to root water uptake (Carvajal et al. 1996; Clarkson 2000), but more detailed characterizations of the role of aquaporins in mediating these changes in root Lp are needed. In corn (*Zea mays*) roots, the rapid increase in Lp associated with the addition of nitrate occurred in the absence of changes in aquaporin gene expression (Gorska et al. 2008) suggesting that if aquaporins do contribute to these changes in Lp those changes likely take place at the posttranslational level. Finally, some studies have demonstrated an interaction between the regulation of root Lp via aquaporins, nutrient uptake, and drought stress response (Liu et al. 2014; Ding et al. 2015).

4.2 Aquaporins and Plant Growth

Recently, more studies have attempted to draw connections between the role of aquaporins in controlling Lp and plant water status, and the resulting effect on growth. In leaves, greater Lp at the cell level corresponds with greater leaf elongation rates, although the role of aquaporins remains unclear (Volkov et al. 2007). There is evidence that aquaporin-mediated changes in root Lp can also affect leaf growth at a distance through effects on leaf turgor (Parent et al. 2009; Ehlert et al. 2009). Caldeira et al. (2014) demonstrated that diurnal oscillations of leaf growth and water status correlated with oscillations in whole plant hydraulic conductance and the expression of root aquaporins. At the level of the whole plant, constitutive overexpression of a tomato tonoplast aquaporin resulted in increased transpiration, plant biomass, and fruit yield under both drought and normal growing conditions (Sade et al. 2009).

There is also emerging evidence linking aquaporins with growth within the root itself. Differences in root cortical cell Lp are associated with differential root growth controlling gravitropism (Miyamoto et al. 2005). Recently a role for aquaporins in facilitating lateral root emergence has been described demonstrating links between auxin-related growth and tissue hydraulics (Péret et al. 2012). In roots, it is difficult to disentangle hydraulic changes and growth effects since increases in growth likely result in increased hydraulic capacity. Transgenic approaches aimed at assessing effects on stress responses have found tolerant phenotypes that were brought about by enhanced growth (Peng et al. 2007), but more detailed hydraulic analyses are needed to assess to what extent this is a direct result of increased Lp mediated by aquaporins.

4.3 Aquaporins as Tools for Breeding in Plants

Aquaporins have garnered a lot of attention for breeding stress-tolerant plants, as evidenced by the numerous attempts at manipulating their expression in the search of stress-tolerant phenotypes (discussed above). At the same time, the unpredictability of the outcomes emphasizes the complexity of aquaporin function, and it remains unclear to what extent aquaporins will fulfill their promise as effective targets for plant breeding. This may be related to limitations of measuring root Lp under field conditions and the overall complexity of root responses to various abiotic conditions. Measurements of root hydraulic conductance (L) are not suited for assessment of aquaporin activity because L is a root-surface area-dependent size and does not reflect the intrinsic hydraulic property of the root (Fig. 2; Suku et al. 2014). Therefore, a better understanding of the role of aquaporins in controlling root hydraulic function is needed. Leveraging this knowledge in the context of utilizing the natural diversity of hydraulic behaviors within a given plant species may prove fruitful. Indeed this is of great current interest (reviewed in Chaumont and Tyerman 2014; Moshelion et al. 2015).

Currently, assessing root hydraulic properties is technically demanding and laborious making the development of effective high-throughput phenotyping tools difficult. Current high-throughput root phenotyping is aimed almost entirely at assessing root architecture via imaging (Kuijken et al. 2015). The development of high-throughput root hydraulics phenotyping is a huge challenge, but holds the potential to open many new avenues with regard to root hydraulics, aquaporins, plant breeding, and rootstock selection.

References

Adiredjo AL, Navaud O, Grieu P, Lamaze T (2014) Hydraulic conductivity and contribution of aquaporins to water uptake in roots of four sunflower genotypes. Bot Stud 55:75. doi:10.1186/s40529-014-0075-1

Aharon R, Shahak Y, Wininger S et al (2003) Overexpression of a plasma membrane aquaporin in transgenic tobacco improves plant vigor under favorable growth conditions but not under drought or salt stress. Plant Cell 15:439–447

Alexandersson E, Fraysse L, Sjovall-Larsen S et al (2005) Whole gene family expression and drought stress regulation of aquaporins. Plant Mol Biol 59:469–484

Alsina MM, Smart DR, Bauerle T et al (2011) Seasonal changes of whole root system conductance by a drought-tolerant grape root system. J Exp Bot 62:99–109

Amodeo G, Dorr R, Vallejo A et al (1999) Radial and axial water transport in the sugar beet storage root. J Exp Bot 50:509–516. doi:10.1093/jxb/50.333.509

Aroca R, Amodeo G, Fernández-Illescas S et al (2005) The role of aquaporins and membrane damage in chilling and hydrogen peroxide induced changes in the hydraulic conductance of maize roots. Plant Physiol 137:341–353. doi:10.1104/pp.104.051045

Aroca R, Ferrante A, Vernieri P, Chrispeels MJ (2006) Drought, abscisic acid and transpiration rate effects on the regulation of PIP aquaporin gene expression and abundance in *Phaseolus vulgaris* plants. Ann Bot 98:1301–1310. doi:10.1093/aob/mcl219

Aroca R, Porcel R, Ruiz-Lozano JM (2011) Regulation of root water uptake under abiotic stress conditions. J Exp Bot 63:43–57. doi:10.1093/jxb/err266

Bae E-K, Lee H, Lee J-S, Noh E-W (2011) Drought, salt and wounding stress induce the expression of the plasma membrane intrinsic protein 1 gene in poplar (Populus alba×P. tremula var. glandulosa). Gene 483:43–48. doi:10.1016/j.gene.2011.05.015

Barrios-Masias FH, Jackson LE (2014) California processing tomatoes: morphological, physiological and phenological traits associated with crop improvement during the last 80 years. Eur J Agron 53:45–55

Barrios-Masias FH, Knipfer T, McElrone AJ (2015) Differential responses of grapevine rootstocks to water stress are associated with adjustments in fine root hydraulic physiology and suberization. J Exp Bot 66:erv324. doi:10.1093/jxb/erv324

Barrowclough DE, Peterson CA, Steudle E (2000) Radial hydraulic conductivity along developing onion roots. J Exp Bot 51:547–557. doi:10.1093/jexbot/51.344.547

Beaudette PC, Chlup M, Yee J, Emery RJN (2007) Relationships of root conductivity and aquaporin gene expression in *Pisum sativum*: diurnal patterns and the response to HgCl2 and ABA. J Exp Bot 58:1291–1300. doi:10.1093/jxb/erl289

Besse M, Knipfer T, Miller AJ et al (2011) Developmental pattern of aquaporin expression in barley (*Hordeum vulgare* L.) leaves. J Exp Bot 62:4127–4142

Blum A (2011) Plant breeding for water-limited environments. Springer, New York

Boursiac Y, Chen S, Luu D-T et al (2005) Early effects of salinity on water transport in Arabidopsis roots. Molecular and cellular features of aquaporin expression. Plant Physiol 139:790–805. doi:10.1104/pp.105.065029

Boursiac Y, Boudet J, Postaire O et al (2008) Stimulus-induced downregulation of root water transport involves reactive oxygen species-activated cell signalling and plasma membrane intrinsic protein internalization. Plant J 56:207–218

Boyer JS (1971) Resistances to water transport in soybean, bean, and sunflower1. Crop Sci 11:403. doi:10.2135/cropsci1971.0011183X001100030028x

Bramley H, Turner NC, Turner DW, Tyerman SD (2009) Roles of morphology, anatomy, and aquaporins in determining contrasting hydraulic behavior of roots. Plant Physiol 150:348–364

Caldeira CF, Jeanguenin L, Chaumont F, Tardieu F (2014) Circadian rhythms of hydraulic conductance and growth are enhanced by drought and improve plant performance. Nat Commun 5:5365. doi:10.1038/ncomms6365

Carvajal M, Cooke D, Clarkson D (1996) Responses of wheat plants to nutrient deprivation may involve the regulation of water-channel function. Planta. doi:10.1007/BF00195729

Carvajal M, Martinez V, Alcaraz CF (1999) Physiological function of water channels as affected by salinityinrootsofpaprikapepper.PhysiolPlant105:95–101.doi:10.1034/j.1399-3054.1999.105115.x

Cayan DR, Das T, Pierce DW et al (2010) Future dryness in the southwest US and the hydrology of the early 21st century drought. Proc Natl Acad Sci U S A 107:21271–21276

Chaumont F, Tyerman SD (2014) Aquaporins: highly regulated channels controlling plant water relations. Plant Physiol 164:1600–1618. doi:10.1104/pp.113.233791

Christmann A, Weiler EW, Steudle E, Grill E (2007) A hydraulic signal in root-to-shoot signalling of water shortage. Plant J 52:167–174. doi:10.1111/j.1365-313X.2007.03234.x

Clarkson DT (2000) Root hydraulic conductance: diurnal aquaporin expression and the effects of nutrient stress. J Exp Bot 51:61–70. doi:10.1093/jexbot/51.342.61

Comas L, Becker S, Cruz VMV et al (2013) Root traits contributing to plant productivity under drought. Front Plant Sci 4:442

Daniels MJ, Mirkov TE, Chrispeels MJ (1994) The plasma membrane of *Arabidopsis thaliana* contains a mercury-insensitive aquaporin that is a homolog of the tonoplast water channel protein TIP. Plant Physiol 106:1325–1333

Ding L, Gao C, Li Y, et al (2015) The enhanced drought tolerance of rice plants under ammonium is related to aquaporin (AQP). Plant Sci 234:14–21. doi: 10.1016/j.plantsci.2015.01.016

Dordas C, Chrispeels MJ, Brown PH (2000) Permeability and channel-mediated transport of boric acid across membrane vesicles isolated from squash roots. Plant Physiol 124:1349–1361

Ehlert C, Maurel C, Tardieu F, Simonneau T (2009) Aquaporin-mediated reduction in maize root hydraulic conductivity impacts cell turgor and leaf elongation even without changing transpiration. Plant Physiol 150:1093–1104. doi:10.1104/pp.108.131458

Fetter K, Van Wilder V, Moshelion M, Chaumont F (2004) Interactions between plasma membrane aquaporins modulate their water channel activity. Plant Cell 16:215–228. doi:10.1105/tpc.017194

Fiscus EL, Markhart AH (1979) Relationships between root system water transport properties and plant size in Phaseolus. Plant Physiology 64:770–773

Frensch J, Steudle E (1989) Axial and radial hydraulic resistance to roots of maize (*Zea mays* L.). Plant Physiol 91:719–726

Gambetta GA, Manuck CM, Drucker ST et al (2012) The relationship between root hydraulics and scion vigour across Vitis rootstocks: what role do root aquaporins play? J Exp Bot 63:6445–6455

Gambetta GA, Fei J, Rost TL et al (2013) Water uptake along the length of grapevine fine roots: developmental anatomy, tissue-specific aquaporin expression, and pathways of water transport. Plant Physiol 163:1254–1265

Gao Z, He X, Zhao B et al (2010) Overexpressing a putative aquaporin gene from wheat, TaNIP, enhances salt tolerance in transgenic Arabidopsis. Plant Cell Physiol 51:767–775. doi:10.1093/pcp/pcq036

Gerbeau P, Amodeo G, Henzler T et al (2002) The water permeability of Arabidopsis plasma membrane is regulated by divalent cations and pH. Plant J 30:71–81

Gorska A, Zwieniecka A, Holbrook NM, Zwieniecki MA (2008) Nitrate induction of root hydraulic conductivity in maize is not correlated with aquaporin expression. Planta 228:989–998. doi:10.1007/s00425-008-0798-x

Grondin A, Mauleon R, Vadez V, Henry A (2016) Root aquaporins contribute to whole plant water fluxes under drought stress in rice (*Oryza sativa* L.). Plant Cell Environ 39:347–365. doi:10.1111/pce.12616

Guenther JF, Chanmanivone N, Galetovic MP et al (2003) Phosphorylation of soybean nodulin 26 on serine 262 enhances water permeability and is regulated developmentally and by osmotic signals. Plant Cell 15:981–991

Hachez C, Moshelion M, Zelazny E et al (2006) Localization and quantification of plasma membrane aquaporin expression in maize primary root: a clue to understanding their role as cellular plumbers. Plant Mol Biol 62:305–323

Hachez C, Veselov D, Ye Q et al (2012) Short-term control of maize cell and root water permeability through plasma membrane aquaporin isoforms. Plant Cell Environ 35:185–198. doi:10.1111/j.1365-3040.2011.02429.x

Hachez C, Veljanovski V, Reinhardt H et al (2014) The Arabidopsis abiotic stress-induced TSPO-related protein reduces cell-surface expression of the aquaporin PIP2;7 through protein-protein interactions and autophagic degradation. Plant Cell 26:4974–4990. doi:10.1105/tpc.114.134080

Heinen RB, Ye Q, Chaumont F (2009) Role of aquaporins in leaf physiology. J Exp Bot 60:2971–2985. doi:10.1093/jxb/erp171

Henzler T, Steudle E (1995) Reversible closing of water channels in Chara internodes provides evidence for a composite transport model of the plasma membrane. J Exp Bot 46:199–209

Henzler T, Ye Q, Steudle E (2004) Oxidative gating of water channels (aquaporins) in Chara by hydroxyl radicals. Plant Cell Environ 27:1184–1195

Hose E, Steudle E, Hartung W (2000) Abscisic acid and hydraulic conductivity of maize roots: a study using cell- and root-pressure probes. Planta 211:874–882. doi:10.1007/s004250000412

Hsiao T (1973) Plant responses to water stress. Annu Rev Plant Physiol 24:519–570

Hukin D, Doering-Saad C, Thomas CR, Pritchard J (2002) Sensitivity of cell hydraulic conductivity to mercury is coincident with symplasmic isolation and expression of plasmalemma aquaporin genes in growing maize roots. Planta 215:1047–1056

Jang JY, Kim DG, Kim YO et al (2004) An expression analysis of a gene family encoding plasma membrane aquaporins in response to abiotic stresses in *Arabidopsis thaliana*. Plant Mol Biol 54:713–725. doi:10.1023/B:PLAN.0000040900.61345.a6

Jang JY, Lee SH, Rhee JY et al (2007a) Transgenic arabidopsis and tobacco plants overexpressing an aquaporin respond differently to various abiotic stresses. Plant Mol Biol 64:621–632. doi:10.1007/s11103-007-9181-8

Jang JY, Rhee JY, Kim DG et al (2007b) Ectopic expression of a foreign aquaporin disrupts the natural expression patterns of endogenous aquaporin genes and alters plant responses to different stress conditions. Plant Cell Physiol 48:1331–1339. doi:10.1093/pcp/pcm101

Javot H, Maurel C (2002) The role of aquaporins in root water uptake. Ann Bot 90:301–313. doi:10.1093/aob/mcf199

Javot H, Lauvergeat V, Santoni V et al (2003) Role of a single aquaporin isoform in root water uptake. Plant Cell 15:509–522 http://dx.doi.org/10.1105/tpc.008888

Johansson I, Larsson C, Ek B, Kjellbom P (1996) The major integral proteins of spinach leaf plasma membranes are putative aquaporins and are phosphorylated in response to Ca2+ and apoplastic water potential. Plant Cell 8:1181–1191

Johansson I, Karlsson M, Shukla VK, Chrispeels MJ, Larsson C, Kjellbom P (1998) Water transport activity of the plasma membrane aquaporin PM28A is regulated by phosphorylation. Plant Cell 10:451–460

Kaneko T, Horie T, Nakahara Y et al (2015) Dynamic regulation of the root hydraulic conductivity of barley plants in response to salinity/osmotic stress. Plant Cell Physiol 56:875–882. doi:10.1093/pcp/pcv013

Katsuhara M, Akiyama Y, Koshio K et al (2002) Functional analysis of water channels in barley roots. Plant Cell Physiol 43:885–893

Kawasaki S, Borchert C, Deyholos M et al (2001) Gene expression profiles during the initial phase of salt stress in rice. Plant Cell 13:889–905

Knipfer T, Fricke W (2010) Root pressure and a solute reflection coefficient close to unity exclude a purely apoplastic pathway of radial water transport in barley (*Hordeum vulgare* L.). New Phytol 187:159–170

Knipfer T, Fricke W (2011) Water uptake by seminal and adventitious roots in relation to whole-plant water flow in barley (*Hordeum vulgare* L.). J Exp Bot 62:717–733

Knipfer T, Besse M, Verdeil J-L, Fricke W (2011) Aquaporin-facilitated water uptake in barley (*Hordeum vulgare* L.) roots. J Exp Bot 62:4115–4126

Kramer PJ, Boyer JS (1995) Water relations of plants and soils. Academic Press, San Diego

Kuijken RCP, van Eeuwijk FA, Marcelis LFM, Bouwmeester HJ (2015) Root phenotyping: from component trait in the lab to breeding. J Exp Bot 66:5389–5401. doi:10.1093/jxb/erv239

Lee SH, Zwiazek JJ (2015) Regulation of aquaporin-mediated water transport in Arabidopsis roots exposed to NaCl. Plant Cell Physiol 56:750–758. doi:10.1093/pcp/pcv003

Lee SH, Singh AP, Chung GC (2004) Rapid accumulation of hydrogen peroxide in cucumber roots due to exposure to low temperature appears to mediate decreases in water transport. J Exp Bot 55:1733–1741. doi:10.1093/jxb/erh189

Lee SH, Chung GC, Steudle E (2005) Gating of aquaporins by low temperature in roots of chilling-sensitive cucumber and chilling-tolerant figleaf gourd. J Exp Bot 56:985–995. doi:10.1093/jxb/eri092

Lian H-L (2004) The role of aquaporin RWC3 in drought avoidance in rice. Plant Cell Physiol 45:481–489. doi:10.1093/pcp/pch058

Liu P, Yin L, Deng X, et al (2014) Aquaporin-mediated increase in root hydraulic conductance is involved in silicon-induced improved root water uptake under osmotic stress in Sorghum bicolor L. J Exp Bot 65:4747–4756. doi: 10.1093/jxb/eru220

Lovisolo C, Secchi F, Nardini A et al (2007) Expression of PIP1 and PIP2 aquaporins is enhanced in olive dwarf genotypes and is related to root and leaf hydraulic conductance. Physiol Plant 130:543–551

Lovisolo C, Tramontini S, Flexas J, Schubert A (2008) Mercurial inhibition of root hydraulic conductance in *Vitis* spp. rootstocks under water stress. Environ Exp Bot 63:178–182

Maggio A, Joly RJ (1995) Effects of mercuric chloride on the hydraulic conductivity of tomato root systems (evidence for a channel-mediated water pathway). Plant Physiol 109:331–335

Mahdieh M, Mostajeran A (2009) Abscisic acid regulates root hydraulic conductance via aquaporin expression modulation in *Nicotiana tabacum*. J Plant Physiol 166:1993–2003. doi:10.1016/j.jplph.2009.06.001

Martínez-Ballesta MC, Aparicio F, Pallás V et al (2003) Influence of saline stress on root hydraulic conductance and PIP expression in Arabidopsis. J Plant Physiol 160:689–697

Martinez-Ballesta MC, Silva C, Lopez-Berenguer C et al (2006) Plant aquaporins: new perspectives on water and nutrient uptake in saline environment. Plant Biol 8:535–546

Martre P, Cochard H, Durand J-L (2001a) Hydraulic architecture and water flow in growing grass tillers (*Festuca arundinacea* Schreb.). Plant Cell Environ 24:65–76. doi:10.1046/j.1365-3040.2001.00657.x

Martre P, North GB, Nobel PS (2001b) Hydraulic conductance and mercury-sensitive water transport for roots of *Opuntia acanthocarpa* in relation to soil drying and rewetting. Plant Physiol 126:352–362

Maurel C, Verdoucq L, Luu DT, Santoni V (2008) Plant aquaporins: membrane channels with multiple integrated functions. Annu Rev Plant Biol 59:595–624

Maurel C, Boursiac Y, Luu DT et al (2015) Aquaporins in plants. Physiol Rev 95:1321–1358

McCully ME (1999) Roots in soil: unearthing the complexities of roots and their rhizospheres. Annu Rev Plant Physiol Plant Mol Biol 50:695–718

McElrone AJ, Bichler J, Pockman WT et al (2007) Aquaporin-mediated changes in hydraulic conductivity of deep tree roots accessed via caves. Plant Cell Environ 30:1411–1421. doi:10.1111/j.1365-3040.2007.01714.x
Miyamoto N, Katsuhara M, Ookawa T et al (2005) Hydraulic conductivity and aquaporins of cortical cells in gravitropically bending roots of *Pisum sativum* L. Plant Prot Sci 8:515–524. doi:10.1626/pps.8.515
Moshelion M, Becker D, Biela A et al (2002) Plasma membrane aquaporins in the motor cells of *Samanea saman*: diurnal and circadian regulation. Plant Cell 14:727–739
Moshelion M, Halperin O, Wallach R et al (2015) Role of aquaporins in determining transpiration and photosynthesis in water-stressed plants: crop water-use efficiency, growth and yield. Plant Cell Environ 38:1785–1793. doi:10.1111/pce.12410
Murai-Hatano M, Kuwagata T, Sakurai J et al (2008) Effect of low root temperature on hydraulic conductivity of rice plants and the possible role of aquaporins. Plant Cell Physiol 49:1294–1305. doi:10.1093/pcp/pcn104
Neumann HH, Thurtell GW, Stevenson KR (1974) In situ measurements of leaf water potential and resistance to water flow in corn, soybean, and sunflower at several transpiration rates. Can J Plant Sci 54:175–184. doi:10.4141/cjps74-027
Niemietz CM, Tyerman SD (2002) New potent inhibitors of aquaporins: silver and gold compounds inhibit aquaporins of plant and human origin. FEBS Lett 531:443–447
North GB, Nobel PS (1991) Changes in hydraulic conductivity and anatomy caused by drying and rewetting roots of *Agave deserti (agavaceae)*. Am J Bot 78:906–915
North GB, Martre P, Nobel PS (2004) Aquaporins account for variations in hydraulic conductance for metabolically active root regions of *Agave deserti* in wet, dry, and rewetted soil. Plant Cell Environ 27:219–228. doi:10.1111/j.1365-3040.2003.01137.x
Padgett-Johnson M, Williams LE, Walker MA (2003) Vine water relations, gas exchange, and vegetative growth of seventeen Vitis species grown under irrigated and nonirrigated conditions in California. J Am Soc Hortic Sci 128:269–276
Parent B, Hachez C, Redondo E et al (2009) Drought and abscisic acid effects on aquaporin content translate into changes in hydraulic conductivity and leaf growth rate: a trans-scale approach. Plant Physiol 149:2000–2012. doi:10.1104/pp.108.130682
Passioura JB (2002) Environmental biology and crop improvement. Funct Plant Biol 29:537–546
Peng Y, Lin W, Cai W, Arora R (2007) Overexpression of a *Panax ginseng* tonoplast aquaporin alters salt tolerance, drought tolerance and cold acclimation ability in transgenic Arabidopsis plants. Planta 226:729–740. doi:10.1007/s00425-007-0520-4
Péret B, Li G, Zhao J et al (2012) Auxin regulates aquaporin function to facilitate lateral root emergence. Nat Cell Biol 14:991–998. doi:10.1038/ncb2573
Postaire O, Tournaire-Roux C, Grondin A et al (2010) A PIP1 aquaporin contributes to hydrostatic pressure-induced water transport in both the root and rosette of Arabidopsis. Plant Physiol 152:1418–1430
Preston GM, Jung JS, Guggino WB, Agre P (1993) The mercury-sensitive residue at cysteine 189 in the CHIP28 water channel. J Biol Chem 268:17–20
Queen WH (1967) Radial movement of water and 32P through suberized and unsuberized roots of grape. PhD thesis, Duke University, Durham, NC
Sade N, Vinocur BJ, Diber A et al (2009) Improving plant stress tolerance and yield production: is the tonoplast aquaporin SlTIP2;2 a key to isohydric to anisohydric conversion? New Phytol 181:651–661. doi:10.1111/j.1469-8137.2008.02689.x
Sade N, Gebretsadik M, Seligmann R et al (2010) The role of tobacco aquaporin1 in improving water use efficiency, hydraulic conductivity, and yield production under salt stress. Plant Physiol 152:245–254
Sakurai J, Ishikawa F, Yamaguchi T et al (2005) Identification of 33 rice aquaporin genes and analysis of their expression and function. Plant Cell Physiol 46:1568–1577. doi:10.1093/pcp/pci172
Sakurai J, Ahamed A, Murai M et al (2008) Tissue and cell-specific localization of rice aquaporins and their water transport activities. Plant Cell Physiol 49:30–39

Seversike T, Sermons S, Sinclair T et al (2014) Physiological properties of a drought-resistant wild soybean genotype: transpiration control with soil drying and expression of root morphology. Plant Soil 374:359–370

Shangguan Z-P, Lei T-W, Shao M-A, Xue Q-W (2005) Effects of phosphorus nutrient on the hydraulic conductivity of Sorghum (*Sorghum vulgare* Pers.) seedling roots under water deficiency. J Integr Plant Biol 47:421–427. doi:10.1111/j.1744-7909.2005.00069.x

Siefritz F, Tyree MT, Lovisolo C et al (2002) PIP1 plasma membrane aquaporins in tobacco: from cellular effects to function in plants. Plant Cell 14:869–876

Siemens JA, Zwiazek JJ (2004) Changes in root water flow properties of solution culture-grown trembling aspen (*Populus tremuloides*) seedlings under different intensities of water-deficit stress. Physiol Plant 121:44–49. doi:10.1111/j.0031-9317.2004.00291.x

Steudle E (2000) Water uptake by plant roots: an integration of views. Plant Soil 226:46–56

Steudle E, Peterson CA (1998) How does water get through roots? J Exp Bot 49:775–788

Suga S, Komatsu S, Maeshima M (2002) Aquaporin isoforms responsive to salt and water stresses and phytohormones in radish seedlings. Plant Cell Physiol 43:1229–1237

Suku S, Knipfer T, Fricke W (2014) Do root hydraulic properties change during the early vegetative stage of plant development in barley (*Hordeum vulgare*)? Ann Bot 113:385–402

Tazawa M, Ohkuma E, Shibasaka M, Nakashima S (1997) Mercurial-sensitive water transport in barley roots. J Plant Res 110:435–442

Tornroth-Horsefield S, Wang Y, Hedfalk K et al (2006) Structural mechanism of plant aquaporin gating. Nature 439:688–694

Tournaire-Roux C, Sutka M, Javot H et al (2003) Cytosolic pH regulates root water transport during anoxic stress through gating of aquaporins. Nature 425:393–397

UN-FAO. 2002. Crops and drops: making the best use of water for agriculture. Rome: Food and Agriculture Organization of the United Nations. Available at: ftp://ftp.fao.org/docrep/fao/005/y3918e/y3918e00.pdf. Accessed 11 Oct 2011..

Vandeleur RK, Mayo G, Shelden MC et al (2009) The role of plasma membrane intrinsic protein aquaporins in water transport through roots: diurnal and drought stress responses reveal different strategies between isohydric and anisohydric cultivars of grapevine. Plant Physiol 149:445–460. doi:10.1104/pp.108.128645

Vandeleur RK, Sullivan W, Athman A et al (2014) Rapid shoot-to-root signalling regulates root hydraulic conductance via aquaporins. Plant Cell Environ 37:520–538. doi:10.1111/pce.12175

Volkov V, Hachez C, Moshelion M et al (2007) Water permeability differs between growing and non-growing barley leaf tissues. J Exp Bot 58:377–390. doi:10.1093/jxb/erl203

Wan X (2010) Osmotic effects of NaCl on cell hydraulic conductivity of corn roots. Acta Biochim Biophys Sin Shanghai 42:351–357. doi:10.1093/abbs/gmq029

Wan X, Zwiazek JJ (1999) Mercuric chloride effects on root water transport in aspen seedlings. Plant Physiol 121:939–946

Wang X, Li Y, Ji W, et al (2011) A novel *Glycine soja* tonoplast intrinsic protein gene responds to abiotic stress and depresses salt and dehydration tolerance in transgenic *Arabidopsis thaliana*. J Plant Physiol 168:1241–8. doi: 10.1016/j.jplph.2011.01.016

Wang M, Ding L, Gao L, et al (2016) The Interactions of aquaporins and mineral nutrients in higher plants. Int J Mol Sci 17:1229. doi: 10.3390/ijms17081229

Wei W, Alexandersson E, Golldack D et al (2007) HvPIP1;6, a barley (*Hordeum vulgare* L.) plasma membrane water channel particularly expressed in growing compared with non-growing leaf tissues. Plant Cell Physiol 48:1132–1147

Ye Q, Steudle E (2006) Oxidative gating of water channels (aquaporins) in corn roots. Plant Cell Environ 29:459–470

Yu X, Peng YH, Zhang MH et al (2006) Water relations and an expression analysis of plasma membrane intrinsic proteins in sensitive and tolerant rice during chilling and recovery. Cell Res 16:599–608. doi:10.1038/sj.cr.7310077

Zhu C, Schraut D, Hartung W, Schäffner AR (2005) Differential responses of maize MIP genes to salt stress and ABA. J Exp Bot 56:2971–2981. doi:10.1093/jxb/eri294

Aquaporins and Leaf Water Relations

Christophe Maurel and Karine Prado

Abstract Leaf water relations are a key factor of plant growth and productivity. Water is delivered to the leaf through its vasculature. It then exits xylem vessels by crossing living cells prior to vaporization and diffusion through stomatal apertures. The present chapter shows how the leaf aquaporin equipment contributes to the water transport capacity of inner leaf tissues (leaf hydraulic conductance, K_{leaf}) with a major role in the vascular bundles. Aquaporins provide optimal and locally adjusted water supply to the leaf during transpiration and leaf growth and movements. The respective roles of leaf vasculature and aquaporins in leaf hydraulic changes in response to endogenous or environmental stimuli are discussed. It is established that regulation of aquaporins at gene expression and protein phosphorylation levels mediates the effects of light, circadian rhythms and water and salt stress on K_{leaf}. However, the signaling mechanisms acting upstream are as yet unknown.

The water relations of leaves are at the crossroad of key plant physiological functions. The stomata present at the leaf surface mediate most of the gas exchange between the plant and atmosphere. Their regulated aperture allows a crucial tradeoff between carbon dioxide (CO_2) absorption and water loss by transpiration. Under high evaporative demand or low soil water availability, transpiration can markedly challenge the leaf water status, thereby impacting overall plant growth and

C. Maurel (✉)
Biochimie et Physiologie Moléculaire des Plantes, Unité Mixte de Recherche 5004, CNRS/INRA/Montpellier SupAgro/Université de Montpellier, 34060 Montpellier, Cedex 2, France
e-mail: christophe.maurel@cnrs.fr

K. Prado
Biochimie et Physiologie Moléculaire des Plantes, Unité Mixte de Recherche 5004, CNRS/INRA/Montpellier SupAgro/Université de Montpellier, 34060 Montpellier, Cedex 2, France

School of Biological Sciences, University of Edinburgh, Max Born Crescent, Edinburgh EH9 3BF, UK

F. Chaumont, S.D. Tyerman (eds.), *Plant Aquaporins*, Signaling and Communication in Plants, DOI 10.1007/978-3-319-49395-4_7

productivity. In particular, photosynthetic carbon fixation requires a proper leaf hydration and can be dramatically reduced under drought. Expansive growth of leaves, which ultimately determines the ability of the plant to capture light, is also highly sensitive to the leaf water status.

The water transport capacity of inner leaf tissues (leaf hydraulic conductance, K_{leaf}) is a key player of leaf water relations. It allows a proper water import from the stem and optimized redistribution within the lamina. K_{leaf} can vary by up to 65-fold between plant species (Prado and Maurel 2013; Sack and Holbrook 2006). Differences in leaf anatomy and, in particular, in hydraulic design of their vasculature contribute to these large interspecific differences. However, nonvascular, living tissues can also determine key features of leaf hydraulics. For instance, K_{leaf} within a given plant species can show marked differences during development or under contrasting physiological conditions due to the expression and regulation of aquaporins (Prado and Maurel 2013; Sack and Holbrook 2006).

This present chapter discusses how aquaporins function in leaf tissues and allow a dynamic adjustment of leaf hydraulics in response to endogenous or environmental stimuli. Complementary information can be found in recent reviews (Chaumont and Tyerman 2014; Heinen et al. 2009; Maurel et al. 2015; Prado and Maurel 2013). The role of leaf aquaporins in CO_2 transport and carbon fixation will be presented in Chap. 10.

1 Principles of Leaf Hydraulics

1.1 Water Transport Pathways

After uptake by the root and transport through the stem xylem vessels, water (xylem sap) is delivered to the leaf through its vasculature. In brief, petiole xylem leads to the midrib that branches into progressively smaller veins embedded in the leaf mesophyll. Water then exits the vessels to cross the living cells forming the lamina, prior to vaporization in interstitial air spaces or substomatal chambers and diffusion through stomatal apertures. In this representation, liquid water transport is successively mediated through vascular and extravascular pathways, which thereby function in series. The respective hydraulic resistances of the two pathways are usually of the same order of magnitude (Sack and Holbrook 2006). While most intense water flows are observed under transpiring conditions during the day and under high evaporative demand, water is also delivered from vascular tissues to the leaf lamina at night or when expansive growth dominates. Conversely, a fraction of leaf water can be exported through phloem translocation.

Plant leaf vasculature shows a highly organized hierarchy of vein orders and species-specific branching or reticulation patterns. The significance of leaf venation with respect to hydraulics has been discussed previously (Sack and Scoffoni 2013). In all cases, vein distribution and density seem to be optimized for distributing water evenly across the leaf. The extravascular pathway first includes several cell types in the vascular bundle (xylem parenchyma cells, bundle sheath cells). Bundle sheath extensions provide, in addition, a direct delivery of water to the epidermis, whereas mesophyll cells mediate water transport from the veins to the substomatal chambers. Thus, leaf water transport can occur along multiple and composite paths. Nevertheless, bundle sheath cells, which are wrapped around the veins, appear as an obligatory passage for all these paths.

At the subcellular level, the paths used for liquid water transport, from cell to cell (transcellular and symplastic paths) or through the cell walls (apoplastic path), are similar to those operating in other organs and, as in the root, are still disputed. Low cell packing and the presence of air spaces in the mesophyll suggested that the apoplastic path may predominate in this tissue. In contrast, transcellular water transport may be crucial in vascular bundles, due to a tighter organization and differentiation of apoplastic barriers (Ache et al. 2010).

1.2 Leaf Hydraulic Measurements

K_{leaf} links, at the organ level, the flow of liquid water across inner leaf tissues to the driving force, i.e., the difference in water potential between the petiole and substomatal chambers. Three main methods have been developed for measuring K_{leaf}: the evaporative flux method, the high-pressure flow method, and the vacuum pump method. A critical comparison of these techniques has been published elsewhere (Prado and Maurel 2013; Sack et al. 2002). Although each method has specific pitfalls, their careful manipulation can yield very similar K_{leaf} values.

2 The Leaf Aquaporin Equipment

2.1 Expression Patterns

In agreement with the high isoform multiplicity of plant aquaporins, transcriptomic and proteomic studies have revealed the complex aquaporin equipment of plant leaves (Alexandersson et al. 2005; Hachez et al. 2008; Monneuse et al. 2011). In maize leaves, for instance, transcripts for 12 out 13 PIP isoforms were present, with the two most abundant ones (*ZmPIP1;1*, *ZmPIP2;1*) accounting for 60 % of *PIP* transcripts. While some PIPs, such as tobacco *Nt*AQP1, show strong expression in the spongy mesophyll parenchyma (Otto and Kaldenhoff 2000), a preferential expression of aquaporins was observed in the vascular bundles of most plant species investigated (Frangne et al. 2001; Hachez et al. 2008; Kirch et al. 2000; Prado et al. 2013). Aquaporins can also be found in phloem companion cells and epidermal and guard cells (Hachez et al. 2008).

Genome-wide co-expression analyses in developing maize leaves have revealed strong links between aquaporins and nutrient homeostasis and transport (Yue et al. 2012). Thus, the various expression patterns of aquaporins in leaves are indicative of isoform-specific roles of aquaporins in transcellular water transport or cell osmoregulation (Hachez et al. 2008; Heinen et al. 2009). Aquaporins can transport physiologically relevant molecules other than water, such as CO_2, ammonia (NH_3), and hydrogen peroxide (H_2O_2). An upregulation of PIP1;3 expression by H_2O_2 has been observed in *Arabidopsis* leaves (Hooijmaijers et al. 2012).

2.2 Overall Contribution of Aquaporins to K_{leaf}

Whereas axial transport of water in xylem vessels does not involve living cell structures, aquaporins possibly account for a large part of downstream extravascular pathways. Aquaporin gene silencing using microRNA constructs in transgenic *Arabidopsis* has revealed the overall contribution of aquaporins to ~35 % and ~50 % of hydraulic conductivity of whole rosettes (K_{ros}) or individual leaves (K_{leaf}), respectively (Sade et al. 2014). Surprisingly, earlier experiments using antisense inhibition of *PIP1* or *PIP2* genes had failed to reveal a role for these aquaporins in leaf water transport in control conditions, whereas a marked impact of aquaporin inhibition on K_{leaf} was observed under water stress (Martre et al. 2002). Aquaporin research in plants and animals critically lacks specific chemical blockers. Although they are potentially toxic, mercury and azide block plant aquaporins through distinct modes of action. Consistent with the genetic studies above, their use indicated a contribution by 25–50 % of aquaporins to K_{leaf} in sunflower, grapevine, various deciduous trees, or *Arabidopsis* (Aasamaa and Sober 2005; Nardini and Salleo 2005; Postaire et al. 2010; Pou et al. 2013).

2.3 Tissue-Specific Functions of Leaf Aquaporins

Several recent studies have investigated which cell layers are hydraulically limiting during extravascular transport, the underlying idea being that these cells should be a preferential site for aquaporin expression and regulation. One first approach was to search for correlations between K_{leaf} and the water permeability of tissue-specific protoplasts. *Arabidopsis* leaves subjected to exogenous ABA or changes in irradiance revealed parallel changes in K_{leaf} and osmotic water permeability of bundle sheath protoplasts (Prado et al. 2013; Shatil-Cohen et al. 2011). Xylem parenchyma protoplasts also showed a consistent response to irradiance, whereas water permeability in isolated mesophyll protoplasts was not correlated to K_{leaf} under various ABA or light treatments. These data support the idea that vascular bundles, rather than the mesophyll, represent a hydraulically limiting structure in the extravascular pathway. Recent genetic approaches have brought more direct evidence for tissue-specific function of leaf aquaporins. Single PIP knockout mutants of *Arabidopsis* have revealed that, in this species, three aquaporin isoforms (*At*PIP1;2, *At*PIP2;1, *At*PIP2;6) can individually contribute to ~20 % of K_{ros} (Postaire et al. 2010; Prado et al. 2013). Their common expression in the veins was interpreted to mean that the living cells of these tissues can be hydraulically limiting. Another strategy was to express microRNAs in bundle sheath cells using a SCR promoter (Sade et al. 2014). Although a marked decreased in K_{leaf} (−65 %) was induced in the transgenic plants, it was associated with a concomitant inhibition of bundle sheath and mesophyll protoplast water permeability. Thus, it could not be concluded which tissue water permeability was the rate-limiting step. Another strategy was to genetically complement aquaporin-deficient plants by using aquaporin constructs under the control of vein- or bundle sheath-specific promoters. The partial or full recovery of K_{leaf} or K_{ros} in these plants provides supportive evidence for the importance of PIP functions in veins (Prado et al. 2013; Sade et al. 2015).

These findings do not exclude, however, a minor contribution of the mesophyll to whole leaf water transport. Aquaporin may also contribute to osmoregulation of mesophyll cells, in relation to their high metabolic activity (Morillon and Chrispeels 2001; Prado and Maurel 2013).

3 Key Physiological Roles of Leaf Aquaporins

3.1 Transpiration

The function of aquaporins during plant transpiration represents a key question in plant water relations. Overall, K_{leaf} and stomatal conductance show a tight coupling by diverse and most often unknown mechanisms. For instance, the regulation of aquaporins by light and circadian mechanisms may favor a constant adjustment of K_{leaf} to the transpiration demand. Consistent with this, low air humidity enhanced the K_{leaf} of *Arabidopsis*, by up to threefold and in <1 h (Levin et al. 2007). Interestingly, Pou et al. (2013) observed a strong correlation across control and water stress conditions between the expression level of a TIP2 homolog and stomatal conductance in grapevine. It is not clear whether this tonoplast aquaporin truly contributes to transcellular water transport during transpiration or whether it plays an osmoregulatory role in cells challenged by a high rate of water transport through the leaf. A paradoxical observation was made in mesophyll protoplasts of *Arabidopsis*, where water permeability was strongly but negatively correlated to the plant transpiration regime: thus, protoplast water permeability was maximal at reduced transpiration (Morillon and Chrispeels 2001).

A high evaporative demand can also lead to coordinated hydraulic responses throughout the whole plant body. Transpiration was shown to enhance the expression and function of aquaporins in rice and poplar roots, by signaling mechanisms that remain to be discovered (Laur and Hacke 2013; Sakurai-Ishikawa et al. 2011).

3.2 Leaf Growth

Expansive growth is primarily driven by cell turgor and thereby highly sensitive to the cell and tissue water status (Pantin et al. 2011) (see also chapter "Plant Aquaporins and Cell Elongation"). The idea that expansive growth of the leaf can be limited by water transfer from vascular to peripheral tissues was initially raised by the detection of local growth-induced water potential gradients (Tang and Boyer 2002). This idea is now well formalized, using hydraulic modeling of whole plants under varying water availability in the soil or the atmosphere (Caldeira et al. 2014). At the molecular and cellular levels, it is supported by the finding that, in barley and maize, the leaf growth zone shows enhanced expression of specific leaf aquaporins isoforms together with a high cell hydraulic conductivity (Hachez et al. 2008; Volkov et al. 2007). We note that the expression and function of aquaporins in roots can also impact leaf expansive growth (Caldeira et al. 2014; Ehlert et al. 2009), through effects on leaf xylem water potential.

3.3 Leaf Movements

Plants show diurnal leaf movements for optimizing exposure to incident light. In *Samanea saman*, these movements are mediated through circadian regulation of *PIP* expression and osmotic water permeability in leaf motor cells (Moshelion et al. 2002). Tobacco plants expressing an antisense copy of NtAQP1 revealed a role for this aquaporin in the differential elongation of the upper and lower sides of the petiole thereby contributing to leaf unfolding (Siefritz et al. 2004). In *Rhododendron* leaves, extracellular freezing under subfreezing temperature results in leaf water redistribution and thermonasty (curling). This response was shown to be associated to downregulation of *PIP2* genes (Chen et al. 2013).

3.4 Leaf Water Uptake and Secretion

Whereas roots usually account for most if not all of water uptake, leaf water uptake can be crucial in certain plant species for response to extreme environmental conditions. In epiphytes, for instance, leaf trichomes allow capturing air moisture under drought conditions (Ohrui et al. 2007), whereas in conifers, absorption of melting snow favors embolism refilling after winter (Laur and Hacke 2014a). These processes are associated with enhanced expression of aquaporins at the sites of water absorption. In halophytes such as the tropical mangrove tree *Avicennia officinalis*, excess salt can be secreted by specialized salt glands forming in the leaf epidermis. Salt-induced expression of aquaporins in these glands together with mercury inhibition experiments have suggested that these aquaporins may contribute to both secretion and water reabsorption (Tan et al. 2013; Tyerman 2013).

4 Regulation of Leaf Hydraulics

4.1 Light and Circadian Rhythms

The daily variations of light regime represent the most common regulators of K_{leaf} studied across plant species. In most cases, diurnal rhythms of leaf hydraulics were studied on a 24-h time window (Cochard et al. 2007; Lo Gullo et al. 2005), and K_{leaf} was the highest at high irradiance, consistent with a peak in evaporative demand and transpiration at midday. A few studies showed that leaf hydraulic conductivity was also regulated by the circadian clock. In particular, experiments with sunflower plants under constant darkness revealed that, at least in this species, K_{leaf} is under control of the circadian clock and peaks at the subjective midday (Nardini et al. 2005).

The mode of leaf aquaporin regulation by irradiance may vary depending on species. In walnut, for instance, light-dependent variations in K_{leaf} were tightly associated with parallel changes in transcript abundance of several *PIP1* and *PIP2* genes (Baaziz et al. 2012; Cochard et al. 2007). In maize, most *PIP* genes showed diurnal regulation with a peak in expression during the first hours of the light period (Hachez et al. 2008). The aquaporin mRNA levels of *Samanea saman* leaves were regulated diurnally in phase with leaflet movements, and *SsAQP2* transcription was under circadian control (Moshelion et al. 2002). NtAQP1 abundance was also under circadian regulation in tobacco leaves (Siefritz et al. 2004). In contrast, phosphorylation of a single aquaporin, *At*PIP2;1, was shown to mediate light-dependent regulation of K_{ros} in *Arabidopsis* (Prado et al. 2013). Quantitative proteomics revealed that phosphorylation at two adjacent C-terminal sites (Ser280, Ser283) was enhanced, together with K_{ros}, in response to a dark treatment. Transformation of a *pip2;1* mutant with *At*PIP2;1 forms carrying phosphomimetic mutations of these residues was sufficient to restore the responsiveness of K_{ros} to light whereas *At*PIP2;1 forms carrying phosphorylation-deficient mutations did not.

4.2 Water and Salt Stress

Soil water deficit (drought) results in a decrease in K_{leaf} in many species including *Arabidopsis* (Shatil-Cohen et al. 2011) and poplar (*Populus trichocarpa*) (Laur and Hacke 2014b). Under severe conditions, this decrease is mediated through both xylem embolism in leaf veins (see chapter "Role of Aquaporins in the Maintenance of Xylem Hydraulic Capacity") and downregulation of the aquaporin pathway. These two actions may result in a drop in leaf water potential thereby promoting stomatal closure (Pantin et al. 2013). Regulation of K_{leaf} in *Arabidopsis* under drought is mediated by ABA (Pantin et al. 2013; Shatil-Cohen et al. 2011), and a possible role of aquaporin dephosphorylation has been invoked (Kline et al. 2010; Prado et al. 2013).

Genetic and aquaporin expression analyses have shown that leaf aquaporins of both the PIP and TIP subfamilies support the recovery of droughted plants upon rewatering (Laur and Hacke 2014b; Martre et al. 2002). In particular, aquaporins expressed in the xylem parenchyma may contribute to embolism refilling and smooth delivery of water to the whole leaf blade.

4.3 Other Signals

Leaves are exposed to numerous types of biotic aggressions, by herbivories or microbial pathogens. The associated hydraulic responses have been barely investigated and will deserve more attention in the future. For instance, partial defoliation of poplar seedlings resulted in a five to tenfold increase in expression of a specific PIP isoform, which paralleled an increase in transpiration and lamina hydraulic conductivity (Liu et al. 2014).

5 Conclusions

The hydraulics of leaves has been rather difficult to comprehend, both biophysically and physiologically, due to entangled contributions of the vasculature and living structures (aquaporins), the former showing a complex reticulate organization. The vascular system of leaves mediates crucial hydraulic changes during development and, through localized embolisms, in leaf response to extreme drought or low temperatures. Aquaporins play complementary roles and appear to be at work in multiple physiological conditions. In particular, their capacity to mediate diurnal and spatially restricted adjustments of leaf water transport may be a key factor of plant performance and productivity.

Despite recent progress, we are still far from a clear understanding of leaf hydraulics. Mathematical models have addressed the impact of venation pattern (including vein hierarchy and conductivities) on leaf hydraulics in relation to leaf construction costs (McKown et al. 2010). Models that go beyond a classical Ohm's law analogy are also under development (Rockwell et al. 2014). However, we still lack a leaf hydraulic model that would include a comprehensive description of aquaporin function and regulation.

Although much remains to be learned about aquaporin functions in leaves, two directions should deserve a specific attention in future research. One is water transport in tree leaves, due to its high agronomical and ecological importance. Molecular studies are emerging, and the recent development of genetic tools in poplar (Secchi and Zwieniecki 2013) opens interesting perspectives to understand the role of aquaporins in tree leaf response to extreme drought episodes. Focus should also be put onto the signaling mechanisms that mediate the regulation of aquaporins by endogenous or environmental signals. In particular, the coupling of leaf aquaporins to stomatal functions (see chapter "Roles of Aquaporins in Stomata") and their dual role in water and CO_2 transport (see chapter "Plant Aquaporins and CO_2") represent central questions for future studies.

References

Aasamaa K, Sober A (2005) Seasonal courses of maximum hydraulic conductance in shoots of six temperate deciduous tree species. Funct Plant Biol 32:1077–1087

Ache P, Bauer H, Kollist H, Al-Rasheid KA, Lautner S, Hartung W, Hedrich R (2010) Stomatal action directly feeds back on leaf turgor: new insights into the regulation of the plant water status from non-invasive pressure probe measurements. Plant J 62:1072–1082

Alexandersson E, Fraysse L, Sjovall-Larsen S, Gustavsson S, Fellert M, Karlsson M, Johanson U, Kjellbom P (2005) Whole gene family expression and drought stress regulation of aquaporins. Plant Mol Biol 59:469–484

Baaziz KB, Lopez D, Rabot A, Combes D, Gousset A, Bouzid S, Cochard H, Sakr S, Venisse JS (2012) Light-mediated K_{leaf} induction and contribution of both the PIP1s and PIP2s aquaporins in five tree species: walnut (*Juglans regia*) case study. Tree Physiol 32:423–434

Caldeira CF, Bosio M, Parent B, Jeanguenin L, Chaumont F, Tardieu F (2014) A hydraulic model is compatible with rapid changes in leaf elongation under fluctuating evaporative demand and soil water status. Plant Physiol 164:1718–1730

Chaumont F, Tyerman SD (2014) Aquaporins: highly regulated channels controlling plant water relations. Plant Physiol 164:1600–1618

Chen K, Wang X, Fessehaie A, Yin Y, Arora R (2013) Is expression of aquaporins (plasma membrane intrinsic protein 2s, PIP2s) associated with thermonasty (leaf-curling) in Rhododendron? J Plant Physiol 170:1447–1454

Cochard H, Venisse JS, Barigah TS, Brunel N, Herbette S, Guilliot A, Tyree MT, Sakr S (2007) New insights into the understanding of variable hydraulic conductances in leaves. Evidence for a possible implication of plasma membrane aquaporins. Plant Physiol 143:122–133

Ehlert C, Maurel C, Tardieu F, Simonneau T (2009) Aquaporin-mediated reduction in maize root hydraulic conductivity impacts cell turgor and leaf elongation even without changing transpiration. Plant Physiol 150:1093–1104

Frangne N, Maeshima M, Schaffner AR, Mandel T, Martinoia E, Bonnemain JL (2001) Expression and distribution of a vaculoar aquaporin in young and mature leaf tissues of *Brassica napus* in relation to water fluxes. Planta 212:270–278

Hachez C, Heinen RB, Draye X, Chaumont F (2008) The expression pattern of plasma membrane aquaporins in maize leaf highlights their role in hydraulic regulation. Plant Mol Biol 68:337–353

Heinen RB, Ye Q, Chaumont F (2009) Role of aquaporins in leaf physiology. J Exp Bot 60: 2971–2985

Hooijmaijers C, Rhee JY, Kwak KJ, Chung GC, Horie T, Katsuhara M, Kang H (2012) Hydrogen peroxide permeability of plasma membrane aquaporins of *Arabidopsis thaliana*. J Plant Res 125:147–153

Kirch H-H, Vera-Estrella R, Golldack D, Quigley F, Michalowski CB, Barkla BJ, Bohnert HJ (2000) Expression of water channel proteins in *Mesembryanthemum crystallinum*. Plant Physiol 123:111–124

Kline KG, Barrett-Wilt GA, Sussman MR (2010) *In planta* changes in protein phosphorylation induced by the plant hormone abscisic acid. Proc Natl Acad Sci U S A 107:15986–15991

Laur J, Hacke UG (2013) Transpirational demand affects aquaporin expression in poplar roots. J Exp Bot 64:2283–2293

Laur J, Hacke UG (2014a) Exploring *Picea glauca* aquaporins in the context of needle water uptake and xylem refilling. New Phytol 203:388–400

Laur J, Hacke UG (2014b) The role of water channel proteins in facilitating recovery of leaf hydraulic conductance from water stress in *Populus trichocarpa*. PLoS One 9:e111751

Levin M, Lemcoff JH, Cohen S, Kapulnik Y (2007) Low air humidity increases leaf-specific hydraulic conductance of *Arabidopsis thaliana* (L.) Heynh (Brassicaceae). J Exp Bot 58:3711–3718

Liu J, Equiza MA, Navarro-Rodenas A, Lee SH, Zwiazek JJ (2014) Hydraulic adjustments in aspen (*Populus tremuloides*) seedlings following defoliation involve root and leaf aquaporins. Planta 240:553–564

Lo Gullo MA, Nardini A, Trifilo P, Salleo S (2005) Diurnal and seasonal variations in leaf hydraulic conductance in evergreen and deciduous trees. Tree Physiol 25:505–512

Martre P, Morillon R, Barrieu F, North GB, Nobel PS, Chrispeels MJ (2002) Plasma membrane aquaporins play a significant role during recovery from water deficit. Plant Physiol 130: 2101–2110

Maurel C, Boursiac Y, Luu DT, Santoni V, Shahzad Z, Verdoucq L (2015) Aquaporins in plants. Physiol Rev 95:1321–1358

McKown AD, Cochard H, Sack L (2010) Decoding leaf hydraulics with a spatially explicit model: principles of venation architecture and implications for its evolution. Am Nat 175:447–460

Monneuse JM, Sugano M, Becue T, Santoni V, Hem S, Rossignol M (2011) Towards the profiling of the *Arabidopsis thaliana* plasma membrane transportome by targeted proteomics. Proteomics 11:1789–1797

Morillon R, Chrispeels MJ (2001) The role of ABA and the transpiration stream in the regulation of the osmotic water permeability of leaf cells. Proc Natl Acad Sci U S A 98:14138–14143

Moshelion M, Becker D, Biela A, Uehlein N, Hedrich R, Otto B, Levi H, Moran N, Kaldenhoff R (2002) Plasma membrane aquaporins in the motor cells of *Samanea saman*: diurnal and circadian regulation. Plant Cell 14:727–739

Nardini A, Salleo S (2005) Water stress-induced modifications of leaf hydraulic architecture in sunflower: co-ordination with gas exchange. J Exp Bot 56:3093–3101

Nardini A, Salleo S, Andri S (2005) Circadian regulation of leaf hydraulic conductance in sunflower (*Helianthus annuus* L. cv Margot). Plant Cell Environ 28:750–759

Ohrui T, Nobira H, Sakata Y, Taji T, Yamamoto C, Nishida K, Yamakawa T, Sasuga Y, Yaguchi Y, Takenaga H, Tanaka S (2007) Foliar trichome- and aquaporin-aided water uptake in a drought-resistant epiphyte *Tillandsia ionantha* Planchon. Planta 227:47–56

Otto B, Kaldenhoff R (2000) Cell-specific expression of the mercury-insensitive plasma-membrane aquaporin NtAQP1 from *Nicotiana tabacum*. Planta 211:167–172

Pantin F, Simonneau T, Rolland G, Dauzat M, Muller B (2011) Control of leaf expansion: a developmental switch from metabolics to hydraulics. Plant Physiol 156:803–815

Pantin F, Monnet F, Jannaud D, Costa JM, Renaud J, Muller B, Simonneau T, Genty B (2013) The dual effect of abscisic acid on stomata. New Phytol 197:65–72

Postaire O, Tournaire-Roux C, Grondin A, Boursiac Y, Morillon R, Schäffner T, Maurel C (2010) A PIP1 aquaporin contributes to hydrostatic pressure-induced water transport in both the root and rosette of Arabidopsis. Plant Physiol 152:1418–1430

Pou A, Medrano H, Flexas J, Tyerman SD (2013) A putative role for TIP and PIP aquaporins in dynamics of leaf hydraulic and stomatal conductances in grapevine under water stress and re-watering. Plant Cell Environ 36:828–843

Prado K, Maurel C (2013) Regulation of leaf hydraulics: from molecular to whole plant levels. Front Plant Sci 4:255

Prado K, Boursiac Y, Tournaire-Roux C, Monneuse J-M, Postaire O, Ines O, Schäffner AR, Hem S, Santoni V, Maurel C (2013) Regulation of *Arabidopsis* leaf hydraulics involves light-dependent phosphorylation of aquaporins in veins. Plant Cell 25:1029–1039

Rockwell FE, Holbrook NM, Stroock AD (2014) Leaf hydraulics II: vascularized tissues. J Theor Biol 340:267–284

Sack L, Holbrook NM (2006) Leaf hydraulics. Annu Rev Plant Biol 57:361–381

Sack L, Scoffoni C (2013) Leaf venation: structure, function, development, evolution, ecology and applications in the past, present and future. New Phytol 198:983–1000

Sack L, Melcher PJ, Zwieniecki MA, Holbrook NM (2002) The hydraulic conductance of the angiosperm leaf lamina: a comparison of three measurement methods. J Exp Bot 53: 2177–2184

Sade N, Shatil-Cohen A, Attia Z, Maurel C, Boursiac Y, Kelly G, Granot D, Yaaran A, Lerner S, Moshelion M (2014) The role of plasma membrane aquaporins in regulating the bundle sheath-mesophyll continuum and leaf hydraulics. Plant Physiol 166:1609–1620

Sade N, Shatil-Cohen A, Moshelion M (2015) Bundle-sheath aquaporins play a role in controlling Arabidopsis leaf hydraulic conductivity. Plant Signal Behav 10:e1017177

Sakurai-Ishikawa J, Murai-Hatano M, Hayashi H, Ahamed A, Fukushi K, Matsumoto T, Kitagawa Y (2011) Transpiration from shoots triggers diurnal changes in root aquaporin expression. Plant Cell Environ 34:1150–1163

Secchi F, Zwieniecki MA (2013) The physiological response of *Populus tremula* x *alba* leaves to the down-regulation of PIP1 aquaporin gene expression under no water stress. Front Plant Sci 4:507

Shatil-Cohen A, Attia Z, Moshelion M (2011) Bundle-sheath cell regulation of xylem-mesophyll water transport via aquaporins under drought stress: a target of xylem-borne ABA? Plant J 67: 72–80

Siefritz F, Otto B, Bienert GP, Krol A, Kaldenhoff R (2004) The plasma membrane aquaporin NtAQP1 is a key component of the leaf unfolding mechanism in tobacco. Plant J 37:147–155

Tan WK, Lin Q, Lim TM, Kumar P, Loh CS (2013) Dynamic secretion changes in the salt glands of the mangrove tree species *Avicennia officinalis* in response to a changing saline environment. Plant Cell Environ 36:1410–1422

Tang AC, Boyer JS (2002) Growth-induced water potentials and the growth of maize leaves. J Exp Bot 53:489–503

Tyerman SD (2013) The devil in the detail of secretions. Plant Cell Environ 36:1407–1409

Volkov V, Hachez C, Moshelion M, Draye X, Chaumont F, Fricke W (2007) Water permeability differs between growing and non-growing barley leaf tissues. J Exp Bot 58:377–390

Yue X, Zhao X, Fei Y, Zhang X (2012) Correlation of aquaporins and transmembrane solute transporters revealed by genome-wide analysis in developing maize leaf. Comp Funct Genom 2012:546930

Roles of Aquaporins in Stomata

Charles Hachez*, Thomas Milhiet*, Robert B. Heinen, and François Chaumont

Abstract Stomata can be regarded as tightly regulated hydraulically driven valves that control the fluxes of water vapor and carbon dioxide between the plant and the atmosphere. In this chapter, we will focus on the mechanisms and regulation of the movement of fully developed stomata, which requires rapid and controlled fluxes of ions and water. Guard cells are symplastically isolated from their neighboring cells, implying that the regulation of transmembrane water movement is central to the control of their aperture/closure mechanism. Such hydraulic regulation of stomatal movement can be modulated by the activity of aquaporins, acting as water and small uncharged solute facilitators. Despite the existence of a wide range of transcriptomic and proteomic data showing that multiple plasma membrane aquaporins are expressed in these structures, there is currently only a limited number of experimental data supporting a functional involvement of these water channels in stomatal movements. The present review will highlight the main reverse genetics data linking the modulation of aquaporin activity to the control of the aperture of stomata.

1 Introduction

Stomata are microscopic pores in the epidermis of the aerial parts of virtually all extant land plants. They are bordered by two specialized cells, known as guard cells, which control the aperture of the pore (called an ostiole) following endogenous and environmental signals. Understanding the mechanistic aspects of stomatal movements has raised the interest of plant biologists for as early as the eighteenth century. First statements of observations of variable apertures of the *breathing holes* can indeed be traced back to the German botanist Johannes Hedwig at the end of the eighteenth century (Hedwig 1793). In 1812, Moldenhauer noticed three major signals affecting stomatal movement: light, humidity, and time of the day (Moldenhauer 1812). A few decades later, von Mohl put forward the hypothesis that guard cells open the stomatal

*Equal contribution

C. Hachez • T. Milhiet • R.B. Heinen • F. Chaumont (✉)
Institut des Sciences de la Vie, Université catholique de Louvain,
Croix du Sud 4-L7.07.14, 1348 Louvain-la-Neuve, Belgium
e-mail: francois.chaumont@uclouvain.be

F. Chaumont, S.D. Tyerman (eds.), *Plant Aquaporins*, Signaling and Communication in Plants, DOI 10.1007/978-3-319-49395-4_8

pore thanks to an increase in their turgor pressure and close it by collapsing down (Von Mohl 1856). The concept of turgor-dependent aperture mechanism is therefore nearly as old as the discovery of stomata themselves. However, the molecular mechanisms responsible for such turgor regulation were still totally ignored.

The major role of stomata is to act as a physical checking point for gas exchange, balancing the undesirable loss of water vapor via evapotranspiration with the essential CO_2 uptake, indispensable for photosynthesis. These structures are therefore key players affecting plant development and biomass production. In this respect, the impact of stomata at the global scale is considerable on both plant-mediated water and carbon fluxes. Indeed, the closely coupled controls of stomata on CO_2 and water vapor diffusion processes impact the global distribution of these compounds in the atmosphere (Hetherington and Woodward 2003). Plants optimize CO_2/H_2O fluxes to suit prevailing environmental conditions by (i) regulating the number of developing stomata in the epidermis and (ii) fine-tuning the stomatal movement. The process and genes underlying stomatal development have been intensively investigated for the last decade (reviewed in Dong and Bergmann 2010; Bergmann and Sack 2007; Casson and Hetherington 2010; Dow et al. 2014). Stomatal opening occurs via the activation of plasma membrane and tonoplast transporters, resulting in solute accumulation in guard cells and consequently to water movement into the cells, which increases their turgor. A swelling guard cell changes shape due to the physical properties of its cell walls. This process leads to the opening of the pore. In contrast, an osmotically driven turgor decrease restores the initial closed state. Stomatal aperture and closure are affected by both endogenous and exogenous cues like perceived light intensity, leaf CO_2 concentration, vapor pressure deficit, temperature, and water status of the plant (Kim et al. 2010; Daszkowska-Golec and Szarejko 2013; Andrès et al. 2014). Under constant environmental conditions, the circadian clock was shown to regulate stomatal movements and to influence the sensitivity of the guard cells to extracellular signals (for review, see Webb 2003).

In this chapter, we will focus on the mechanisms regulating the movement of fully developed stomatal complexes that are central players in plant water relations and carbon assimilation. Experimental data highlighting the role of aquaporins in mediating this process will be pinpointed.

2 The Stomatal Complex: Morphological Characteristics

Two major morphological types of stomata exist. Stomatal complexes of grasses exhibit typical dumbbell-shaped guard cells surrounded by two subsidiary cells (Fig. 1a), while kidney-shaped guard cells are found in other species (Fig. 1b). The thin and linear dumbbell-shaped stomata of grasses, with or without subsidiary cells, are considered to be more evolutionary advanced compared to their kidney-shaped relatives (Hetherington and Woodward 2003). Their design translates smaller changes in guard cell volume to larger apertures with probably little energy waste (Raschke 1979; Hetherington and Woodward 2003). The efficient and fast

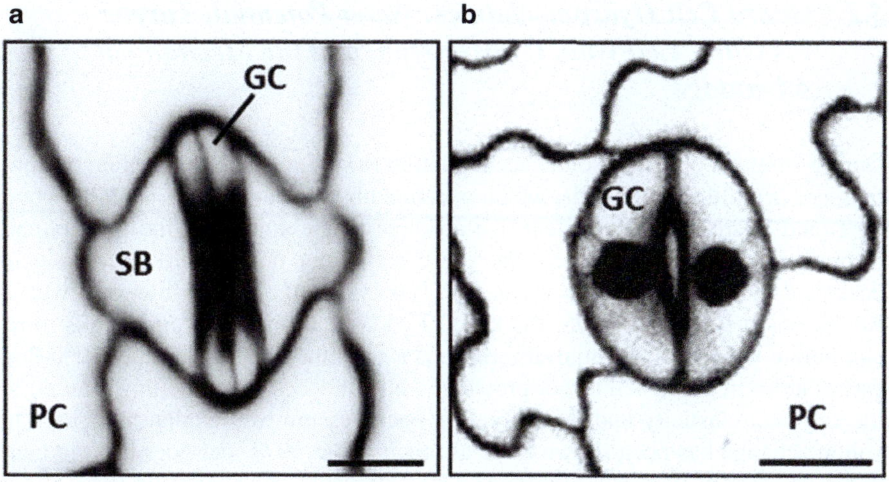

Fig. 1 Morphology of stomata in model mono- or dicotyledonous species as revealed by propidium iodide (*PI*) staining of leaf epidermal cells. (**a**) *Zea mays* stomata complex composed of two dumbbell-shaped guard cells (*GC*) flanked by two subsidiary cells (*SB*). (**b**) *Arabidopsis thaliana* stomata composed of two kidney-shaped guard cells (*GC*). Note that nuclei were stained in GC due to intense PI labeling. In both mono- and dicotyledonous species, stomata are separated from each other in the leaf epidermis by at least one pavement cell (*PC*). Scale bars: 10 μm

stomatal response to fluctuating environmental conditions in grasses is believed to enhance their water use efficiency compared to non-grass species (Franks and Farquhar 2007; Grantz and Assmann 1991; Hetherington and Woodward 2003).

3 The Opening/Closure Mechanism

3.1 The Basic Mechanism

Like most plant movements, stomatal opening and closure mechanisms are based on hydraulic forces (Roelfsema and Hedrich 2005), and such gating mechanism shows a complete reversibility. During stomatal opening, the activation of plasma membrane and tonoplast transporters results in solute accumulation in the guard cell, building up an osmotic gradient. To reestablish the perturbed osmotic equilibrium, water follows the solutes into the guard cells which increases their turgor pressure. Since mature guard cells lack plasmodesmatal connections with their neighboring cells (Erwee et al. 1985; Willmer and Sexton 1979), signaling molecules (including H_2O_2 that permeates aquaporins), ions (K^+, Cl^-), and water most likely reach the guard cell via the surrounding apoplast. As fine-tuned transmembrane water movement is a requisite of such aperture mechanism, water movement might be regulated by channels present in the guard cell membranes (see below).

3.2 *Guard Cell Hydromechanics: Water Potential, Turgor Pressure, Cell Wall, Cytoskeleton, and the Mechanical Advantage*

During stomatal opening, solute accumulation (*n*) increases the prevailing osmotic pressure (π) in the guard cells, which is defined by the equation $\pi = nRT/V$, where R, T, and V are the gas constant, the absolute temperature, and the cell volume, respectively. As a consequence, the guard cell water potential (Ψ) decreases as described by the equation $\Psi = P - \pi$ with *P* being the turgor pressure (reviewed in Roelfsema and Hedrich 2005; Buckley 2005). Water enters the guard cells to re-equilibrate the water potential difference $\Delta\Psi$ with the apoplast. The water inflow causes a rise in guard cell turgor pressure, which induces their swelling, limited by the cell wall elasticity and the increasing backpressure of the adjacent cells. The volume change has been shown to be as much as 40–50 % and goes hand in hand with a cell surface increase (Franks et al. 2001; Raschke and Dickerson 1973; Shope et al. 2003). To accommodate such large changes in surface area, the plasma membrane, whose elasticity is limited to 3–5 % (Wolfe and Steponkus 1983; Morris and Homann 2001), is internalized and remobilized as the cells shrink and swell (Shope et al. 2003; Shope and Mott 2006).

Stomatal aperture (a^{-}_{s}) is positively related to the turgor pressure of guard cells (P_g), but negatively related to the pressure of adjacent subsidiary or epidermal cells (P_e) (reviewed in Buckley 2005; Roelfsema and Hedrich 2005). Raschke noticed the collapsing of subsidiary cells during transient pore opening in *Zea mays* leaves, when dipping the leaf in a mannitol solution (Raschke 1970). The decreasing water potential released the backpressure on the guard cells exerted by the subsidiary cells and facilitated stomatal opening.

3.3 *Aquaporins Are Involved in the Control of Water Flows During Stomatal Movements*

The regulation of stomata aperture is fast (guard cells can adjust their volume by up to 40 % in a few tens of minutes, Franks et al. 2001) and totally reversible. It relies on the controlled shrinking/swelling of guard cells as a consequence of osmotically driven water movement. Such variations in cell volume depend, among other things, on finely tuned efflux/influx of water into guard cells based on a facilitated diffusion mechanism. Given the fact that guard cells are symplastically isolated from their neighboring cells, water movement in or out of the guard cells can only occur via the transmembrane path. By ensuring a higher water permeability to biological membranes, it is tempting to speculate that aquaporins may play a role in rapid changes of turgor/osmotic potentials implicated in stomatal movements. In addition to facilitating water diffusion across the plasma membrane and tonoplast, several plant aquaporins have been involved in the diffusion of small uncharged solutes, including H_2O_2, or CO_2, solutes playing an important role in stomatal regulation (for reviews, see Bienert and Chaumont 2011; Heinen et al. 2009a; Chaumont and Tyerman 2014). This includes aquaporins belonging to the plasma membrane intrinsic proteins (PIPs).

4 Regulation of Stomatal Movement

Guard cells respond within a few minutes to a broad range of signals so that the plant can rapidly adapt to changing environmental conditions or react to threatening stresses (Franks et al. 2001). Multiple endogenous and external factors such as the circadian rhythm, light, CO_2 concentration, stress hormones and secondary messengers (ABA, Ca^{2+}, H_2O_2), drought, and pathogens attacks affect and/or regulate stomatal movement that involves controlled ion and water effluxes. The reader should however keep in mind that the stomatal response is a result of complex cross talks and interactions between different signaling pathways. The detailed mechanisms of stomatal regulation are beyond the scope of this review, but we chose to briefly emphasize the effects of the circadian clock, CO_2 concentration, and abscisic acid (ABA) signaling on stomatal movements.

4.1 Circadian Regulation

Under constant conditions, the circadian clock regulates stomatal movements and increases guard cell sensitivity to endogenous or environmental stimuli (reviewed in Webb 2003). In general, well-watered C_3 and C_4 plants open their stomata during the day and close them at night following the circadian clock. The movements anticipate light–dark transitions and persist under continuous light or darkness. It is worth noting that the expression pattern of plant PIP aquaporins follows the same diurnal pattern, which could help to finely tune the cell membrane water permeability in stomata (Heinen et al. 2014).

4.2 CO_2 Signaling

In the short term, stomata close in response to elevated CO_2 concentrations. Such closure probably involves, among other things, a transient modification of the cell water permeability of guard cells allowing a fast efflux from the cell. The kinase protein HIGH LEAF TEMPERATURE1 (HT1) is the first identified molecular regulator in the CO_2 pathway that negatively regulates CO_2-induced stomatal closing (Hashimoto et al. 2006) through phosphorylation and inhibition of the OST1 protein kinase (Tian et al. 2015), which is required for further CO_2 signal transduction in plants (Merilo et al. 2013; Xue et al. 2011). RHC1, a MATE-type transporter protein, acts as a bicarbonate sensor (Tian et al. 2015) and inhibits HT1 thereby preventing the inhibition of OST1 at elevated CO_2 levels.

4.3 ABA Signaling

ABA is a key hormone regulating plant water status and stomatal movement. The cellular and molecular mechanisms underlying ABA-induced stomatal closure have been extensively studied (Popko et al. 2010; Wilkinson and Davies 1997; Assmann 2003; Hubbard et al. 2010; Lim et al. 2015). Many secondary messengers intervene in the ABA signaling network, such as G proteins, cytosolic pH, intracellular Ca^{2+}, nitric oxide (NO), reactive oxygen species (ROS), and lipid-derived signaling intermediates (phosphoinositides). For an extensive review on the role of ABA in stomatal response to (a)biotic stress, the reader can usefully refer to Lim et al. 2015.

Under stress conditions, the ABA concentration in leaves increases rapidly (Cutler et al. 2010; Hubbard et al. 2010). ABA functions there as a chemical messenger inducing stomatal closure through (in)activation of water and ion channels by protein kinases and phosphatases (Cutler et al. 2010; Hubbard et al. 2010; Gowing et al. 1993; Pei et al. 2000; Schroeder and Hagiwara 1989; Grondin et al. 2015; Lee et al. 2009). The common belief that abscisic acid (ABA) is a xylem-transported hormone that is synthesized in the roots, while acting in the shoot to close stomata in response to a decrease in plant water status, has been challenged by several studies showing that foliage-derived ABA is the predominant source of ABA during drought stress (Christmann et al. 2005, 2007; McAdam et al. 2016). A model in which the leaf senses the root dehydration by the drop in hydraulic pressure and consequently synthesizes ABA on-site was proposed (Christmann et al. 2005; Christmann et al. 2007). ABA might also be involved in the stomatal response to changes in relative air humidity (Hartung et al. 1988). After a drop in air humidity, stomata initially open shortly as a consequence of a general turgor loss and a decreased mechanical advantage of the epidermal pavement cells. This short-term (5–15 min) response, also called the wrong-way response, is followed by stomatal closure under prolonged low air humidity to avoid excessive water loss (reviewed in (Buckley 2005). Altogether, ABA accelerates the plasma membrane depolarization and induces a massive ion and water efflux leading to stomatal closure. Plant mutants affected in ABA synthesis and transduction pathway do not close their stomata in response to drought stress and show a wilty phenotype, which could be rescued by application of exogenous ABA (Xie et al. 2006) (reviewed in Belin et al. 2010).

ABA-triggered stomatal closure was recently shown to require an increase in guard cell permeability to water (Grondin et al. 2015). As will be discussed below, this could occur through phosphorylation-mediated activation of PIP aquaporins in the plasma membrane via a specific kinase protein. Among other phosphorylation events, OST1-dependent phosphorylation of PIP2;1 at Ser-121 indeed activates that water channel. By activating the water channels via such specific phosphorylation events, this model postulates that this posttranslational modification would facilitate greater water efflux from guard cells in response to ABA signaling (Grondin et al. 2015). This could occur via a conformational change impacting the gating of the pore or via an impact on the trafficking or stability of the protein (see chapters "Plant Aquaporin Trafficking" and "Plant Aquaporin Posttranslational Regulation"). Finally, this could also affect the ability of the PIP2 to interact with other PIP isoforms and thereby regulates its activity (see chapter "Heteromerization of Plant Aquaporins").

5 Many Aquaporins Are Expressed in Stomata

Many studies, including guard cell-specific transcriptomic and proteomic approaches, have shown that multiple PIP aquaporins were expressed in guard cells of various plant species (Kaldenhoff et al. 1995; Sarda et al. 1997; Sun et al. 2001; Huang et al. 2002; Leonhardt et al. 2004; Cui et al. 2008; Fraysse et al. 2005; Wei et al. 2007; Hachez et al. 2008; Uehlein et al. 2003, 2008; Flexas et al. 2006; Heinen et al. 2014).

One of the earliest evidence of aquaporin expression in stomata was reported by Kaldenhoff and co-workers, using both GUS fusion and immunodetection to show that AtPIP1;2 was expressed in guard cells of *Arabidopsis thaliana*, although not restricted to this particular cell type (Kaldenhoff et al. 1995). Sarda et al. (1997) reported the mRNA accumulation of two tonoplast aquaporin genes, *SunTIP7* and *SunTIP20,* in sunflower guard cells. Whereas *SunTIP20* expression did not seem related to stomatal movements, the authors observed a *SunTIP7* mRNA accumulation increase at the end of the day suggesting a role in the process of stomatal closure by helping water to exit the vacuole of guard cells. Another *TIP* aquaporin expression was detected by *in situ* hybridization in guard cells of seedlings and mature organs of *Picea abies* (Oliviusson et al. 2001). The expression of the plasma membrane aquaporin SoPIP1;1 was detected in the guard cells of spinach using immunogold labeling (Fraysse et al. 2005). This isoform was also found in phloem and mesophyll cells but showed an interesting localization encircling the guard cells. On the other hand, this study further showed that a PIP isoform could be specifically not expressed in guard cells, like SoPIP1;2. A similar case was observed for AtPIP2;7 whose expression was also specifically null in stomata (Hachez et al. 2014). The expression in guard cells has also been reported for broad bean (*Vicia faba*) VfPIP1 (Cui et al. 2008).

For cereals, HvPIP1;6 was found to be expressed in barley guard cells using *in situ* PCR (Wei et al. 2007). In maize, Hachez et al. (2008) described the expression of ZmPIP1;2 and ZmPIP2;1/2;2 in stomatal complexes using immunocytochemistry. A transcriptomic study of *PIP* gene expression in maize laser-microdissected stomatal complexes (guard cells and subsidiary cells) showed that almost every PIP was expressed, except *ZmPIP2;7* (Heinen et al. 2014). However, the expression of seven of them accounts for more than 98 % of the total *PIP* transcripts. This study was conducted on whole leaf tissue, on peeled epidermis, and on isolated stomata during night and day, thus allowing to detect isoforms specifically expressed in stomatal complexes. For instance, *ZmPIP1;1* was strongly expressed in the leaf mature zone but less in stomata. Most *PIP* genes followed the same expression pattern, with a basal level at night in both whole leaf tissue and isolated stomata. Interestingly, *ZmPIP1;6* was the only isoform found to be specifically expressed in stomata, only during the day. Along with ZmPIP1;5, ZmPIP1;6 was characterized as a water channel when co-expressed with a PIP2 in *Xenopus* oocytes and also as a CO_2 diffusion facilitator across membranes when expressed alone in yeast cells (Heinen et al. 2014). These two isoforms could either play a role in water fluxes in guard cells or help CO_2 assimilation through stomata as suggested for NtAQP1 (Uehlein et al. 2003, 2008; Flexas et al. 2006). They could also be key players in CO_2 sensing in interaction with carbonic anhydrases, as recently demonstrated for the CO_2-permeable aquaporin AtPIP2;1 (see below) (Wang et al. 2016).

6 Reverse Genetics as a Tool to Probe Aquaporin Role in Stomatal Gating Mechanisms

While expression data may give some hints about the importance of aquaporins in stomata regulation, there is only limited evidence using non-transgenic approaches of the functional involvement of aquaporins in stomatal movement, reverse genetics being the best way to prove such a role. When the expression of a single aquaporin isoform is deregulated, either by overexpression, silencing, or knockout, plants usually use compensation mechanisms to adapt their physiology. Such adaptation may involve an altered density of stomata formed in the leaf epidermis (Aharon et al. 2003; Ding et al. 2004; Li et al. 2015) or altered regulation and/or physical properties of the stomatal apparatus (Bi et al. 2015; Heinen et al. 2009; Cui et al. 2008).

Contrasting examples from the literature, mostly about PIP aquaporins, will be discussed thereafter, highlighting the key findings linking the modulation of aquaporin expression through reverse genetics approaches to stomatal density, morphology, and gating mechanism (summarized in Table 1). For instance, tobacco plants overexpressing the *Arabidopsis AtPIP1;2* grew better than the wild-type (WT) plants under favorable conditions (Aharon et al. 2003). A higher transpiration was observed due to a higher stomatal density at both sides of the leaves, indicating that plants took advantage of the heterologous aquaporin expression to use more water, but had to adapt morphologically. This modification of stomatal density did not confer any advantage under drought stress as transgenic plants wilted faster than WT (Aharon et al. 2003). A positive correlation between stomatal conductance (g_s) and *NtAQP1* expression was demonstrated in tobacco using both overexpressing and silenced lines, although no morphological changes were observed (Uehlein et al. 2003, 2008; Flexas et al. 2006). Mesophyll conductance was enhanced by overexpressing the ice plant aquaporin McMIPB in tobacco, leading to a higher photosynthetic rate under well-watered conditions, but g_s did not differ between

Table 1 Effects of deregulation of plant aquaporin expression

Author	Species	Aquaporin subfamily	Transgene	Type of deregulation	Effect
Martre et al. (2002)	*Arabidopsis thaliana*	PIP1 and PIP2	RNAi against PIP1 and PIP2 (antisense)	Silencing	Lower leaf water potential after re-watering, but no difference in stomatal conductance
Cui et al. (2008)	*Arabidopsis thaliana*	PIP1	*VfPIP1* (heterologous)	Overexpression	Decreased transpiration, stomata close faster after dark or ABA treatment
Sade et al. (2014)	*Arabidopsis thaliana*	PIP1	*NtAQP1* (heterologous)	Overexpression	Increased stomatal conductance, mesophyll conductance, and photosynthesis
				Expression restricted to photosynthetic tissue	Increased stomatal conductance, mesophyll conductance, and photosynthesis
				Expression restricted to stomata	Stomatal conductance, mesophyll conductance, and photosynthesis not different from WT
Sade et al. (2014)	*Arabidopsis thaliana*	PIP1 and PIP2	RNAi against PIP1 (amiRNA)	Silencing	Decreased stomatal conductance, mesophyll conductance and photosynthesis for the most affected line
Li et al. (2015)	*Arabidopsis thaliana*	PIP2	*AcPIP2* (heterologous)	Overexpression	Decreased stomata density
Grondin et al. (2015)	*Arabidopsis thaliana*	PIP2	*AtPIP2;1* (homologous)	Knockout	Stomatal closure and guard cell protoplast permeability lack of response to ABA

(continued)

Table 1 (continued)

Author	Species	Aquaporin subfamily	Transgene	Type of deregulation	Effect
Wang et al. (2016)	*Arabidopsis thaliana*	PIP2	*AtPIP2;1* (homologous)	Knockout	No effect on stomatal closure in response to ABA
Aharon et al. (2003)	*Nicotiana tabacum*	PIP1	*AtPIP1;2* (heterologous)	Overexpression	Increased transpiration and photosynthetic efficiency, higher stomatal density
Uehlein et al. (2003)	*Nicotiana tabacum*	PIP1	*NtAQP1* (homologous)	Overexpression	Increased stomatal conductance and net photosynthesis, especially at high CO_2 concentration
			RNAi against *NtAQP1*	Silencing	Decreased stomatal conductance and net photosynthesis, effect disappear at high CO_2 concentration
Ding et al. (2004)	*Nicotiana tabacum*	PIP1	*AqpL1* (heterologous)	Overexpression	Greater stomatal density in young tissues, stomata more open at light
Flexas et al. (2006)	*Nicotiana tabacum*	PIP1	*NtAQP1* (homologous)	Overexpression	Increased stomatal conductance, mesophyll conductance, and photosynthesis
				Silencing	Decreased stomatal conductance, mesophyll conductance, and photosynthesis
Kawase et al. (2013)	*Nicotiana tabacum*	PIP1	*McMIPB* (heterologous)	Overexpression	Increased photosynthesis rate and mesophyll conductance, no change in stomatal conductance

Table 1 (continued)

Author	Species	Aquaporin subfamily	Transgene	Type of deregulation	Effect
Secchi and Zwieniecki (2013)	*Populus tremula x alba*	PIP1	RNAi against PIP1 (antisense)	Silencing	Decreased mesophyll conductance
Bi et al. (2015)	*Populus* × *canescens*	PIP1 and PIP2	RNAi against PIP1 or PIP2 (antisense)	Silencing	Increased stomatal conductance and photosynthesis, increased stomata size
Hanba et al. (2004)	*Oryza sativa*	PIP2	*HvPIP2;1* (heterologous)	Overexpression	Increased stomatal and mesophyll conductance, decreased stomata size and density
Perrone et al. (2012)	*Vitis vinifera*	PIP2	VvPIP2;4 (homologous)	Overexpression	Increased mesophyll conductance

transgenic and WT plants (Kawase et al. 2013). Interestingly, immunolocalization of the McMIPB protein in WT plants revealed its absence in stomata, contrary to overexpressing plants where McMIPB presented a strong signal in chloroplasts in both mesophyll and guard cells. Ding et al. (2004) reported a higher stomatal density in younger leaves of tobacco plants overexpressing a *Lilium PIP1* (*AqpL1*) and that stomata opened slightly faster with light than WT by studying the stomatal aperture on collodion printings of leaf surface. It was also proved that stomata of transgenic *Arabidopsis* isolated epidermis expressing the broad bean *VfPIP1* closed significantly faster than those of control plants when subjected to ABA or dark treatment (Cui et al. 2008). On the other hand, density and size of stomatal apparatus in rice plants overexpressing the barley *HvPIP2;1* were reduced while g_s, CO_2 internal conductance and CO_2 assimilation were higher than in WT plants (Hanba et al. 2004). This could be explained by the fact that HvPIP2;1 was characterized as a CO_2 transporter (Mori et al. 2014). Stomatal density was also reduced in *Arabidopsis* plants overexpressing *AcPIP2* from *Atriplex canescens* while the growth rate was enhanced (Li et al. 2015). The overexpression of *VvPIP2;4* in grapevine induced an increase in g_s under normal conditions (Perrone et al. 2012). As VvPIP2;4 is not present in guard cells in WT plants, the authors proposed a direct effect of this isoform in guard cell turgor pressure. Constitutive overexpression of aquaporins can result in a wide range of phenotypes, depending on the selected isoform and promoter strength. It shows that modulating aquaporin expression clearly impacts the plant physiology, even though endogenous aquaporin expression pattern might also change upon heterologous expression of specific isoform and should be carefully investigated, but does not provide clear evidence of direct implication in one physiological process. For that matter, silencing approaches or milder tissue-specific expression were also used to study aquaporins' implication in stomata-related traits.

Silencing both *PIP1* and *PIP2* aquaporin genes in poplar lines resulted in wider and longer guard cells compared to WT and in a higher proportion of open stomata per leaf area (Bi et al. 2015). Furthermore, in contrast with the usually positive correlation between aquaporin expression and g_s, trees with reduced *PIP* expression had a greater g_s and transpiration rate compared to WT, but had growth defects despite showing a better CO_2 net assimilation. Using a proteomic approach, the authors discovered that *PIP* down-expression induced an upregulation of proteins involved in synthesis and signaling of ABA, trafficking or in cell wall synthesis, indicating a drought stress response that explains the reduced growth. This is however not in accordance with results found by Secchi and Zwieniecki (2013) also using poplar *PIP1* RNAi lines in which no morphological or g_s differences were observed, although mesophyll conductance to CO_2 was greatly reduced. Sade et al. (2014) used a very interesting tissue-specific approach to determine the contribution of *NtAQP1* heterologous expression on several gas-exchange parameters in *Arabidopsis* plants. The authors observed an increase in g_s in normal conditions when targeting the expression of *NtAQP1* to photosynthetic tissues, but surprisingly not in the plants in which the expression was restricted to stomata, that behave like WT in both normal and salt stress conditions. Moreover, no difference in photosynthetic rate or g_s were observed between constitutive and mesophyll-specific expression of *NtAQP1* while the expression was much less important in the latter case but still sufficient to induce a similar increase in mesophyll CO_2 conductance (Sade et al. 2014). According to this study, *NtAQP1* overexpression in photosynthetic tissues had an indirect role on stomata opening by raising mesophyll conductance and photosynthetic rate without the need for higher expression of aquaporins in the guard cells. The use of artificial micro-RNAs to reduce *PIP1* gene expression in *Arabidopsis* led to the generation of lines deregulated in *PIP1* but also *PIP2* expression (Sade et al. 2014). Both lines showed a decrease in g_s, whereas only the most affected line in *PIP1* expression exhibited a significant decrease of mesophyll conductance compared to WT. This was not the case for *Arabidopsis* plants silenced for both *PIP1* and *PIP2* genes which showed no significant difference in g_s in normal conditions, during soil drying or after soil re-watering (Martre et al. 2002).

The first functional evidence of direct implication of aquaporins in stomatal movements was provided by a recent study using *Arabidopsis pip2;1* knockout plants (Grondin et al. 2015). The authors showed that the stomata in isolated epidermal peals of knockout plants responded correctly to light but showed a reduced response to ABA, which should have induced a rapid stomatal closure, as seen for the WT plants. The permeability of guard cell protoplasts was measured and showed that ABA triggered a twofold increase in P_f for WT plants, which was abrogated in the *pip2;1* mutants. The authors concluded that AtPIP2;1, was necessary for ABA-dependent closure and dispensable for CO_2- or light-induced stomatal movements. Based on these observations, a model whereby the stomatal closing response to ABA involves an increase in guard cell water permeability mediated by AtPIP2;1 was proposed (Grondin et al. 2015). However, the authors also highlighted the putative role of AtPIP2;1 in ROS signaling in response to ABA as this isoform is able to facilitate H_2O_2 diffusion across membrane, which is one important messenger in ABA signaling (Dynowski et al. 2008; Bienert et al. 2014).

These results were however contradicted by a recent study (Wang et al. 2016). In this study, genotype-blind stomatal movement imaging analyses of individually mapped stomata of leaf epidermal layers showed that stomata from *pip2;1* single mutant lines (including the mutant line used in the Grondin's study) retained intact responses to ABA and closed to similar levels as the wild type 1 h after ABA treatment. These authors reported that *pip2;1* mutation alone was insufficient to impair the ABA-induced stomatal closing pathway. Such contradictory results could be explained by different experimental growing conditions or measurements methodology (Wang et al. 2016) but raise question regarding the pivotal role of AtPIP2;1 alone in mediating ABA-induced stomatal closing.

A direct interaction between AtPIP2;1 and the carbonic anhydrase βCA4, implicated in stomatal movements in response to CO_2 changes was also described (Wang et al. 2016). This interaction allowed a greater rise in CO_2 permeability of *Xenopus* oocytes compared to each interactor injected separately, proving that AtPIP2;1 is a functional CO_2 channel and could act synergistically with other proteins to increase the CO_2 permeability in plant tissues.

7 Concluding Remarks: Cell-Type-Specific Transgenic Approaches Are Needed to Investigate the Role of Specific Aquaporin Genes in Stomata

Modulating endogenous aquaporin expression levels most often leads to similar observations, although similar phenotypes can have different origins. This is especially true when working with plants as different as *A. thaliana*, *O. sativa*, or *P. trichocarpa* and with aquaporin isoforms facilitating the passage of a wide range of small uncharged molecules. Aquaporins can play a role as water channels or as CO_2 or H_2O_2 facilitators, making the interpretation of physiological observations tricky. As water, H_2O_2, and CO_2 fluxes are strongly impacting leaf physiology, it is difficult to distinguish between an impact on photosynthesis efficiency and an impact on transpiration.

Recent advances in Crispr-CAS9 technology offer the possibility to knock out specific aquaporin genes in virtually any plant species via genome editing and could be an interesting way to probe the role of specific aquaporin isoforms in plant physiology. It is however worth noting that using T-DNA or genome editing approaches will affect gene expression in the whole plant, thereby complicating the interpretation of their role in specific processes. Given the various phenotypes observed when investigating the role of specific aquaporin isoforms in stomata physiology, cell-specific approaches might be required. In the future, it would be interesting to specifically silence/overexpress PIP aquaporins in mature guard cells and to measure the impact of such silencing/overexpression on stomatal gating mechanisms. This will require the identification of guard cell (and/or subsidiary cell)-specific promoters. Such promoters are already available for some plants such as maize, rice, or *Arabidopsis* (Liu et al. 2009; Yang et al. 2008).

Acknowledgments This work was supported by grants from the Belgian National Fund for Scientific Research (FNRS), the Interuniversity Attraction Poles Programme, the Belgian Science Policy (IAP7/29), and the Belgian French community ARC11/16-036 project.

References

Aharon R, Shahak Y, Wininger S, Bendov R, Kapulnik Y, Galili G (2003) Overexpression of a plasma membrane aquaporin in transgenic tobacco improves plant vigor under favorable growth conditions but not under drought or salt stress. Plant Cell 15:439–447

Andrés Z, Pérez-Hormaeche J, Leidi EO, Schlücking K, Steinhorst L, McLachlan DH, Schumacher K, Hetherington AM, Kudla J, Cubero B, Pardo JM (2014) Control of vacuolar dynamics and regulation of stomatal aperture by tonoplast potassium uptake. Proc Natl Acad Sci 111:1806–1814

Assmann SM (2003) OPEN STOMATA1 opens the door to ABA signaling in Arabidopsis guard cells. Trends Plant Sci 8:151–153

Belin C, Thomine S, Schroeder JI (2010) Water balance and the regulation of stomatal movements. In: Pareek A, Sopory SK, Bohnert HJ, Govindjee (eds) Abiotic stress adaptation in plants. Springer, The Netherlands, pp 283–305

Bergmann DC, Sack FD (2007) Stomatal development. Annu Rev Plant Biol 58:163–181

Bi Z, Merl-Pham J, Uehlein N, Zimmer I, Mühlhans S, Aichler M, Walch AK, Kaldenhoff R, Palme K, Schnitzler JP, Block K (2015) RNAi-mediated downregulation of poplar plasma membrane intrinsic proteins (PIPs) changes plasma membrane proteome composition and affects leaf physiology. J Proteomics 128:321–332

Bienert GP, Chaumont F (2011) Plant aquaporins: roles in water homeostasis, nutrition, and signaling processes. In: Geisler M, Venema K (eds) Transporters and pumps in plant signaling, Signaling and Communication in Plants, vol 7. Springer, Berlin/Heidelberg, pp 3–36

Bienert GP, Heinen RB, Berny MC, Chaumont F (2014) Maize plasma membrane aquaporin ZmPIP2;5, but not ZmPIP1;2, facilitates transmembrane diffusion of hydrogen peroxide. Biochim Biophys Acta 1838:216–222

Buckley TN (2005) The control of stomata by water balance. New Phytol 168(2):275–292

Casson SA, Hetherington AM (2010) Environmental regulation of stomatal development. Curr Opin Plant Biol 13:90–95

Chaumont F, Tyerman SD (2014) Aquaporins: highly regulated channels controlling plant water relations. Plant Physiol 164:1600–1618

Christmann A, Hoffmann T, Teplova I, Grill E, Müller A (2005) Generation of active pools of abscisic acid revealed by in vivo imaging of water-stressed arabidopsis. Plant Physiol 137:209–219

Christmann A, Weiler EW, Steudle E, Grill E (2007) A hydraulic signal in root-to-shoot signalling of water shortage. Plant J 52:167–174

Cui XH, Hao FS, Chen H, Chen J, Wang XC (2008) Expression of the *Vicia faba* VfPIP1 gene in Arabidopsis thaliana plants improves their drought resistance. J Plant Res 121:207–214

Cutler SR, Rodriguez PL, Finkelstein RR, Abrams SR (2010) Abscisic acid: emergence of a core signaling network. Annu Rev Plant Biol 61:651–679

Daszkowska-Golec A, Szarejko I (2013) Open or close the gate – stomata action under the control of phytohormones in drought stress conditions. Front Plant Sci 4:138

Ding X, Iwasaki I, Kitagawa Y (2004) Overexpression of a lily PIP1 gene in tobacco increased the osmotic water permeability of leaf cells. Plant Cell Environ 27:177–186

Dong J, Bergmann DC (2010) Stomatal patterning and development. Curr Top Dev Biol 91:267–297
Dow GJ, Bergmann DC, Berry JA (2014) An integrated model of stomatal development and leaf physiology. New Phytol 201:1218–1226
Dynowski M, Schaaf G, Loque D, Moran O, Ludewig U (2008) Plant plasma membrane water channels conduct the signalling molecule H_2O_2. Biochem J 414:53–61
Erwee MG, Goodwin PB, Bel AJE (1985) Cell-cell communication in the leaves of *Commelina cyanea* and other plants. Plant Cell Environ 8:173–178
Flexas J, Ribas-Carbo M, Hanson DT, Bota J, Otto B, Cifre J, McDowell N, Medrano H, Kaldenhoff R (2006) Tobacco aquaporin NtAQP1 is involved in mesophyll conductance to CO_2 in vivo. Plant J 48:427–439
Franks PJ, Farquhar GD (2007) The mechanical diversity of stomata and its significance in gas-exchange control. Plant Physiol 143(1):78–87
Franks PJ, Buckley TN, Shope JC, Mott KA (2001) Guard cell volume and pressure measured concurrently by confocal microscopy and the cell pressure probe. Plant Physiol 125:1577–1584
Fraysse LC, Wells B, Cann MC M, Kjellbom P (2005) Specific plasma membrane aquaporins of the PIP1 subfamily are expressed in sieve elements and guard cells. Biol Cell 97:519–534
Gowing DJG, Davies WJ, Trejo CL, Jones HG (1993) Xylem-transported chemical signals and the regulation of plant growth and physiology. Phil Trans Biol Sci 341:41–47
Grantz DA, Assmann SM (1991) Stomatal response to blue-light – water-use efficiency in sugarcane and soybean. Plant Cell Environ 14:683–690
Grondin A, Rodrigues O, Verdoucq L, Merlot S, Leonhardt N, Maurel C (2015) Aquaporins contribute to ABA-triggered stomatal closure through OST1-mediated phosphorylation. Plant Cell 27:1945–1954
Hachez C, Heinen RB, Draye X, Chaumont F (2008) The expression pattern of plasma membrane aquaporins in maize leaf highlights their role in hydraulic regulation. Plant Mol Biol 6:337–353
Hachez C, Laloux T, Reinhardt H, Cavez D, Degand H, Grefen C, Rycke R, Inze D, Blatt MR, Russinova E, Chaumont F (2014) Arabidopsis SNAREs SYP61 and SYP121 coordinate the trafficking of plasma membrane aquaporin PIP2;7 to modulate the cell membrane water permeability. Plant Cell 26:3132–3147
Hanba YT, Shibasaka M, Hayashi Y, Hayakawa T, Kasamo K, Terashima I, Katsuhara M (2004) Overexpression of the barley aquaporin HvPIP2;1 increases internal CO2 conductance and CO2 assimillation in the leaves of transgenic rice plants. Plant Cell Physiol 45:521–529
Hartung W, Radin JW, Hendrix DL (1988) Abscisic acid movement into the apoplastic solution of water-stressed cotton leaves. Plant Physiol 83:908–913
Hashimoto M, Negi J, Young J, Israelsson M, Schroeder JI, Iba K (2006) Arabidopsis HT1 kinase controls stomatal movements in response to CO_2. Nat Cell Biol 8:391–397
Hedwig DJ (1793) D. Johann Hedwig's Sammlung seiner zerstreuten Abhandlungen und Beobachtungen über botanisch-ökonomische Gegenstände. Erstes Bändchen mit fünf illuminierten Kupfertafeln, Leipzig
Heinen RB, Ye Q, Chaumont F (2009) Role of aquaporins in leaf physiology. J Exp Bot 60:2971–2985
Heinen RB, Bienert GP, Cohen D, Chevalier AS, Uehlein N, Hachez C, Kaldenhoff R, Thiec D, Chaumont F (2014) Expression and characterization of plasma membrane aquaporins in stomatal complexes of *Zea mays*. Plant Mol Biol 86:335–350
Hetherington AM, Woodward FI (2003) The role of stomata in sensing and driving environmental change. Nature 424:901–908
Huang RF, Zhu MJ, Kang Y, Chen J, Wang XC (2002) Identification of plasma membrane aquaporin in guard cells of *Vicia faba* and its role in stomatal movement. Acta Bot Sin 44:42–48
Hubbard KE, Nishimura N, Hitomi K, Getzoff ED, Schroeder JI (2010) Early abscisic acid signal transduction mechanisms: newly discovered components and newly emerging questions. Genes Dev 24:1695–1708
Kaldenhoff R, Kolling A, Meyers J, Karmann U, Ruppel G, Richter G (1995) The blue light-responsive AthH2 gene of Arabidopsis thaliana is primarily expressed in expanding as well

as in differentiating cells and encodes a putative channel protein of the plasmalemma. Plant J 7:87–95

Kawase M, Hanba YT, Katsuhara M (2013) The photosynthetic response of tobacco plants overexpressing ice plant aquaporin McMIPB to a soil water deficit and high vapor pressure deficit. J Plant Res 126:517–527

Kim T-H, Bohmer M, H H, Nishimura N, Schroeder JI (2010) Guard cell signal transduction network: advances in understanding abscisic acid, CO_2, and Ca^{2+} signaling. Annu Rev Plant Biol 61:561–591

Lee SC, Lan W, Buchanan BB, Luan S (2009) A protein kinase-phosphatase pair interacts with an ion channel to regulate ABA signaling in plant guard cells. Proc Natl Acad Sci U S A 106:21419–21424

Leonhardt N, Kwak JM, Robert N, Waner D, Leonhardt G, Schroeder JI (2004) Microarray expression analyses of Arabidopsis guard cells and isolation of a recessive abscisic acid hypersensitive protein phosphatase 2C mutant. Plant Cell 16:596–615

Li J, Yu G, Sun X, Liu Y, Liu J, Zhang X, Jia C, Pan H (2015) AcPIP2, a plasma membrane intrinsic protein from halophyte *Atriplex canescens*, enhances plant growth rate and abiotic stress tolerance when overexpressed in Arabidopsis thaliana. Plant Cell Rep 34:1401–1415

Lim CW, Baek W, Jung J, Kim JH, Lee SC (2015) Function of ABA in stomatal defense against biotic and drought stresses. Int J Mol Sci 16:15251–15270

Liu T, Ohashi-Ito K, Bergmann DC (2009) Orthologs of Arabidopsis thaliana stomatal bHLH genes and regulation of stomatal development in grasses. Development 136:2265–2276

Martre P, Morillon R, Barrieu F, North GB, Nobel PS, Chrispeels MJ (2002) Plasma membrane aquaporins play a significant role during recovery from water deficit. Plant Physiol 130:2101–2110

McAdam SA, Manzi M, Ross JJ, Brodribb TJ, Gomez-Cadenas A (2016) Uprooting an abscisic acid paradigm: shoots are the primary source. Plant Signal Behav 11(6):e1169359

Merilo E, Laanemets K, Hu H, Xue S, Jakobson L, Tulva I, Gonzalez-Guzman M, Rodriguez PL, Schroeder JI, Brosche M, Kollist H (2013) PYR/RCAR receptors contribute to ozone-, reduced air humidity-, darkness-, and CO2-induced stomatal regulation. Plant Physiol 162:1652–1668

Mohl H (1856) Welche Ursachen bewirken die Erweiterung und Verengung der Spaltöffnungen? Bot Ztg 14:697–704

Moldenhauer JJP (1812) Beiträge zur Anatomie der Pflanzen. Königliche Schulbuchdruckerei, Kiel

Mori IC, Rhee J, Shibasaka M, Sasano S, Kaneko T, Horie T, Katsuhara M (2014) CO2 transport by PIP2 aquaporins of barley. Plant Cell Physiol 55:251–257

Morris CE, Homann U (2001) Cell surface area regulation and membrane tension. J Membr Biol 179:79–102

Oliviusson P, Salaj J, Hakman I (2001) Expression pattern of transcripts encoding water channel-like proteins in Norway spruce (*Picea abies*). Plant Mol Biol 46:289–299

Pei Z-M, Murata Y, Benning G, Thomine S, Klusener B, Allen GJ, Grill E, Schroeder JI (2000) Calcium channels activated by hydrogen peroxide mediate abscisic acid signalling in guard cells. Nature 406:731–734

Perrone I, Gambino G, Chitarra W, Vitali M, Pagliarani C, Riccomagno N, Balestrini R, Kaldenhoff R, Uehlein N, Gribaudo I, Schubert A, Lovisolo C (2012) The grapevine root-specific aquaporin VvPIP2;4N controls root hydraulic conductance and leaf gas exchange under well-watered conditions but not under water stress. Plant Physiol 160:965–977

Popko J, Hansch R, Mendel RR, Polle A, Teichmann T (2010) The role of abscisic acid and auxin in the response of poplar to abiotic stress. Plant Biol (Stuttg) 12:242–258

Raschke K (1970) Stomatal response to pressure changes and interruptions in the water supply of detached leaves of *Zea mays* L. Plant Physiol 45:415–423

Raschke K (1979) Movements of stomata. In: Haupt W, Feinleib ME (eds) Encyclopedia of plant physiology, vol 7. Springer, Berlin, pp 383–441

Raschke K, Dickerson M (1973) Changes in shape and volume of guard cells during stomatal movement. J Plant Res 1972:149–153

Roelfsema MRG, Hedrich R (2005) In the light of stomatal opening: new insights into 'the Watergate'. New Phytol 167:665–691
Sade N, Shatil-Cohen A, Attia Z, Maurel C, Boursiac Y, Kelly G, Granot D, Yaaran A, Lerner S, Moshelion M (2014) The role of plasma membrane aquaporins in regulating the bundle sheath-mesophyll continuum and leaf hydraulics. Plant Physiol 166:1609–1620
Sarda X, Tousch D, Ferrare K, Legrand E, Dupuis JM, Casse-Delbart F, Lamaze T (1997) Two TIP-like genes encoding aquaporins are expressed in sunflower guard cells. Plant J 12:1103–1111
Schroeder JI, Hagiwara S (1989) Cytosolic calcium regulates ion channels in the plasma membrane of *Vicia faba* guard cells. Nature 338:427–430
Secchi F, Zwieniecki MA (2013) The physiological response of Populus tremula x alba leaves to the down-regulation of pip1 aquaporin gene expression under no water stress. Front Plant Sci 4:507. doi:10.3389/fpls.2013.00507
Shope JC, Mott KA (2006) Membrane trafficking and osmotically induced volume changes in guard cells. J Exp Bot 57:4123–4131
Shope JC, DeWald DB, Mott KA (2003) Changes in surface area of intact guard cells are correlated with membrane internalization. Plant Physiol 133:1314–1321
Sun MH, Xu W, Zhu YF, Su WA, Tang ZC (2001) A simple method for in situ hybridization to RNA in guard cells of *Vicia faba* L.: the expression of aquaporins in guard cells. Plant Mol Biol Rep 19:129–135
Tian W, Hou C, Ren Z, Pan Y, Jia J, Zhang H, Bai F, Zhang P, Zhu H, He Y, Luo S, Li L, Luan S (2015) A molecular pathway for CO_2 response in Arabidopsis guard cells. Nat Commun 6:6057
Uehlein N, Lovisolo C, Siefritz F, Kaldenhoff R (2003) The tobacco aquaporin NtAQP1 is a membrane CO_2 pore with physiological functions. Nature 425:734–737
Uehlein N, Otto B, Hanson DT, Fischer M, McDowell N, Kaldenhoff R (2008) Function of Nicotiana tabacum aquaporins as chloroplast gas pores challenges the concept of membrane CO_2 permeability. Plant Cell 20:648–657
Wang C, Hu H, Qin X, Zeise B, Xu D, Rappel WJ, Boron WF, Schroeder JI (2016) Reconstitution of CO_2 regulation of SLAC1 anion channel and function of CO_2-permeable PIP2;1 aquaporin as CARBONIC ANHYDRASE4 interactor. Plant Cell 28:568–582
Webb AAR (2003) The physiology of circadian rhythms in plants. New Phytol 160:281–303
Wei W, Alexandersson E, Golldack D, Miller AJ, Kjellbom PO, Fricke W (2007) HvPIP1;6, a barley (*Hordeum vulgare* L.) plasma membrane water channel particularly expressed in growing compared with non-growing leaf tissues. Plant Cell Physiol 48:1132–1147
Wilkinson S, Davies WJ (1997) Xylem sap pH increase: a drought signal received at the apoplastic face of the guard cell that involves the suppression of saturable abscisic acid uptake by the epidermal symplast. Plant Physiol 113:559–573
Willmer CM, Sexton R (1979) Stomata and plasmodesmata. Protoplasma 100:113–124
Wolfe J, Steponkus PL (1983) Mechanical properties of the plasma membrane of isolated plant protoplasts: mechanism of hyperosmotic and extracellular freezing injury. Plant Physiol 71:276–285
Xie X, Wang Y, Williamson L, Holroyd GH, Tagliavia C, Murchie E, Theobald J, Knight MR, Davies WJ, Leyser HMO, Hetherington AM (2006) The identification of genes involved in the stomatal response to reduced atmospheric relative humidity. Curr Biol 16:882–887
Xue S, Hu H, Ries A, Merilo E, Kollist H, Schroeder JI (2011) Central functions of bicarbonate in S-type anion channel activation and OST1 protein kinase in CO2 signal transduction in guard cell. EMBO J 30:1645–1658
Yang Y, Costa A, Leonhardt N, Siegel RS, Schroeder JI (2008) Isolation of a strong Arabidopsis guard cell promoter and its potential as a research tool. Plant Methods 4:6

Plant Aquaporins and Abiotic Stress

Nir Sade and Menachem Moshelion

Abstract The global shortage of fresh water is one of our most severe ecological and agronomical problems, leading to dry and saline lands that reduce plant growth and crop yield and have a major effect on plant physiology and water management. Aquaporins (AQPs) are thought to be the main transporters of water as well as small and uncharged solutes through plant cell membranes. Thus, AQPs appear to play a role in regulating dynamic changes of root, stem and leaf hydraulic conductivity and whole plant water usage, especially in response to environmental changes, opening the door to using AQP expression to regulate plant water-use efficiency. We highlight the role of vascular AQPs in regulating leaf hydraulic conductivity and raise questions regarding their role in determining growth rate, fruit yield production and harvest index. The following chapter will discuss the cellular tissue and whole-plant role of AQPs in regulation of plant water balance in response to environmental challenges.

1 Introduction

Over the course of their evolution and as part of their adjustment to terrestrial environmental conditions, vascular plants have evolved complex roots and hydraulic systems to absorb water and minerals from the soil and transport them to the leaves, in order to support biochemical reactions, such as photosynthesis. However, most of the water absorbed by plant roots is transpired via stomata due to the order of magnitude differences between the water potential of the leaf and that of the atmosphere. Thus, soil water availability is one of the major abiotic factors limiting

N. Sade
Department of Plant Sciences, University of California, Davis, CA 95616, USA

M. Moshelion (✉)
Institute of Plant Sciences and Genetics in Agriculture, The Robert H. Smith Faculty of Agriculture, Food and Environment, The Hebrew University of Jerusalem, Rehovot 76100, Israel
e-mail: menachem.moshelion@mail.huji.ac.il

F. Chaumont, S.D. Tyerman (eds.), *Plant Aquaporins*, Signaling and Communication in Plants, DOI 10.1007/978-3-319-49395-4_9

the growth and development of terrestrial plants in most areas of the world. Many types of abiotic stress have fast-acting and substantial effects on the plant's water management behavior, resulting in reduced plant hydraulic and gas conductance (Christmann et al. 2013; Maurel et al. 2015). Aquaporins (AQPs) play a major role in many of the molecular-physiological mechanisms that control plant-water relations, particularly under stressful conditions. The following chapter will discuss the cellular tissue and whole-plant role of AQPs in regulation of plant water balance in response to environmental challenges.

1.1 Regulation of Plant Hydraulic Conductance

Water flow along the soil-plant-atmosphere continuum (SPAC) is driven by hydrostatic negative pressure. Generally, plant hydraulic conductance has two aspects: axial conductance and radial conductance. Axial conductance (from the roots to the leaves) is mainly determined by vascular anatomy (i.e., tracheid diameter, cell wall composition, pit structure, and the presence of embolisms) and does not involve membrane selectivity mechanisms. Radial conductance controls the rate at which water enters the roots (known as root hydraulic conductivity, L_{Pr}; reviewed by Maurel et al. 2010) and the radial water outflux through the leaf, i.e., from xylem vessels toward the evaporation sites on the mesophyll cell walls (known as leaf hydraulic conductance, K_{leaf}) reviewed by (Sack and Holbrook 2006). In addition, parenchymatic cells, old (nonfunctional) xylem vessels, and apoplastic spaces along the length of the stem vascular system may serve as reservoirs of water, supporting the transpiration stream through the radial transfer of water to it.

Rapid and dynamic valve-like behavior has been observed for both K_{leaf} (Cochard et al. 2007; Levin et al. 2007) and L_{Pr} (Maggio and Joly 1995; Carvajal et al. 1996; Clarkson et al. 2000; Tournaire-Roux et al. 2003; Gorska et al. 2008; Bramley et al. 2010). Radial conductance responds rapidly to ambient signals. For example, reducing *Arabidopsis* root water potential to −0.8 MPa leads to a rapid reduction in leaf turgor [the signal transfer rate was higher than 40 cm/min; (Christmann et al. 2013)]. Another study revealed that the opposite direction signal (shoot to root) was also very rapid. In that study, grapevine shoot injury reduced root hydraulic conductance within a few minutes (Vandeleur et al. 2014). Other ambient signals (e.g., light, temperature) and internal signals such as the stress phytohormone abscisic acid (ABA) have also been shown to regulate plant hydraulic conductance (Sack and Holbrook 2006; Shatil-Cohen et al. 2011; Aroca et al. 2012; Pantin et al. 2013; Sade et al. 2015b).

It has been suggested that the leaf bundle sheath (BS) and root endodermis (as well as other parenchymal cells surrounding the xylem) may act as hydraulic control centers in the regulation of K_{leaf} and L_{Pr}, respectively (Sack and Holbrook 2006; Shatil-Cohen et al. 2011; Shatil-Cohen and Moshelion 2012; Prado et al. 2013; Sade et al. 2014c) (see also chapters "Aquaporins and Root Water Uptake" and "Aquaporins and Leaf Water Relations"). It was recently shown that K_{leaf} is dynamically controlled by the permeability of the membranes of BS cells to water, with the osmotic permeability coefficient (P_f) likely reflecting the regulated activity of AQPs in BS cells (Shatil-Cohen et al. 2011).

The endodermis is a layer of root inner cortex cells that tightly encases the stele of vascular plants. Its hydrophobic Casparian strip (Steudle and Peterson 1998), which separates the stele from passive apoplastic diffusion (Moon et al. 1986; Alassimone et al. 2010), has a major effect on the radial transport of water and ions. Thus, the endodermis (together with other xylem-surrounding cells) represents the most critical boundary along the apoplastic route, markedly limiting plant radial water uptake (Alassimone et al. 2010). Many of these studies and others (Tournaire-Roux et al. 2003; Maurel et al. 2008; Maurel et al. 2009) have suggested that AQPs play a major role in hydraulic regulation by determining the membrane's resistance to the flow of water. In addition, it has been suggested that AQPs may play a role in the detection of osmotic and turgor pressure gradients (Hill et al. 2004). Therefore, AQPs are considered key players in the hydraulic sensing-responding mechanism that determines early and late whole-plant hydraulic responses to changes in ambient conditions.

1.2 Early and Late Responses Controlling Plant Hydraulic Conductance

The rapid reduction in the hydraulic and gas conductance of leaves of plants exposed to abiotic stress conditions around the root (Sack and Holbrook 2006) is related to root-produced ABA, which is thought to be transported to the shoot thereby acting as the long-term signal. However, there is new evidence of a hydraulic ABA long-term signal that controls the hydraulic response of leaves in response to stress. The first increase in ABA concentration (~50-fold) following exposure to soil drought stress is seen in the xylem sap (ABA_{xyl}) in the shoot. This is followed by an increase in the concentration of ABA in the guard cells, while ABA levels in the roots increase only several hours later (Christmann et al. 2007). Moreover, root ABA did not affect stomatal closure when the stomata of WT scion (tomato and *Arabidopsis*) closed in response to soil signals although those scions were grafted onto ABA-deficient roots and exposed to low water potential or when the stomata of wetted *Arabidopsis* leaves remained open even though the roots of those plants were dry (Holbrook and Zwieniecki 1999; Christmann et al. 2007). In addition, it was suggested that the reduction in the expression of AQPs in leaf vascular BS or their inactivation in response to a xylem-fed ABA treatment might control the decrease in K_{leaf} (Shatil-Cohen et al. 2011). Additionally, artificial overexpression of a stress-induced AQP gene from tobacco (NtAQP1) in the BS cells reduced the effect of ABA on K_{leaf} (Sade et al. 2015b). Interestingly, both guard cells and roots exhibited a hydraulic response opposite to that observed in the vascular BS. Finally, and despite the fact that the expression levels of many AQPs are reduced under stress (Alexandersson et al. 2005; Boursiac et al. 2005), there have been several reports of increased expression of some root AQPs in response to ABA (Jang et al. 2004; Mahdieh and Mostajeran 2009; Parent et al. 2009). These reports and others [e.g., (Hose et al. 2000)] demonstrate that when ABA concentrations are increased or there is some mild abiotic stress, there is a transient increase in root hydraulic conductance [L_{Pr}; (Singh and Sale 2000; Siemens and Zwiazek 2003, 2004; Aroca et al.

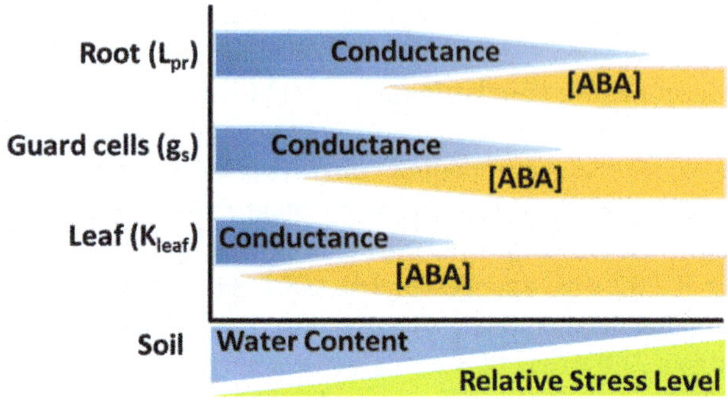

Fig. 1 The whole-plant, stress-induced sequence response model describing the increase in ABA concentration [*ABA*] and the reduction in the hydraulic conductance of the plant under increasingly stressful conditions. A reduction in soil water content increases the plant's stress level, with a differential increase in [*ABA*] and differential changes in the organs' hydraulics. Initially, there is a reduction in the leaf hydraulic conductance, which is followed by an increase in the P_f of the guard cells, which leads to stomatal closure. At this point, root hydraulic conductance is still high; it will be reduced only later and in response to more severe stress

2006; Li et al. 2014)]. Thus, the new hydraulic ABA model suggests a differential and dynamic hydraulic regulation system that balances the whole-plant water status by minimizing water loss and maximizing water uptake (Fig. 1).

This differential sensitivity of different tissues in response to the same signal is most likely associated with different transcriptional, translational, and posttranslational modifications of AQPs, which are unique to each tissue and would be expected based on the hydraulic ABA model. The following sections will discuss the available knowledge and the putative role of tonoplast intrinsic proteins (TIPs), nodulin 26-like intrinsic proteins (NIPs), and plasma membrane intrinsic proteins (PIPs) in plant stress responses. The current understanding of small and basic intrinsic proteins (SIPs), GlpF-like intrinsic proteins (GIPs), hybrid intrinsic proteins (HIPs), and X intrinsic proteins (XIPs) is very limited, particularly with regard to stress, and so those AQPs will not be discussed here.

2 The Role of Tonoplast Intrinsic Proteins (TIPs) in Regulating Plant Stress Behavior

The cellular water-deficit threshold can be defined as the point at which the cytoplasmic water volume cannot continue to support optimal biochemical reactions in the cell. Many types of environmental stress (e.g., drought, salinity, heat, and high vapor pressure deficit (VPD) induce an outflux of water from plant cells, which can result in a significant drop in cell turgor and lead, ultimately, to cell plasmolysis. Therefore, the cytosol, which accounts for 5–10 % of the plant cell volume, may be very sensitive to water deficits. Changes in cytosolic volume can theoretically be avoided if mobilization of water from or into the vacuole is non-limiting at the cost of reductions in the cell volume (plasmolysis) and water potential.

Isohydric water balance behavior involves the maintenance of a constant leaf water potential at midday, which is similar under well-irrigated and drought conditions. In contrast, plants exhibiting anisohydric behavior have markedly decreased water potentials following the evaporative demand experienced during the day. This permits lower leaf water potentials in the presence of drought stress (Tardieu and Simonneau 1998). Recently, it was hypothesized that TIP activity can regulate plant leaf water status and contribute to the difference between isohydric and anisohydric behavior under stress conditions. The constitutive expression of SlTIP2;2 in tomato plants led to greater transpiration and yield under optimal and mild-to-moderate drought conditions, at the cost of lower plant water content, as compared to that observed among the control plants. Interestingly, the transgenic tomatoes had higher yield parameters (including harvest index) when they received relatively small volumes of water (under 50 % irrigation levels), as long as the frequency of irrigation was high (Sade et al. 2009). The fact that overexpression of TIP in isohydric plants led to an increase of the mesophyll cells P_f (which may extend the capacity of the vacuole for osmotic buffering of the cytoplasm under stress conditions) and caused those plants to behave in an anisohydric manner led us to hypothesize that high levels of TIP activity may be part of the cellular mechanism regulating cellular and, therefore, whole-plant water balance (Sade et al. 2009). This risk-taking behavior benefits the plant under conditions of mild-to-moderate abiotic stress (drought and salinity). Interestingly, an anisohydric grapevine cultivar exhibited a high correlation between TIP2 AQP expression and stomatal conductance (Pou et al. 2013). A similar effect was observed when PgTIP1 from *Panax ginseng* was expressed in *Arabidopsis* plants (Peng et al. 2007). Those plants acquired the ability to germinate and grow in the presence of high NaCl concentrations (150 mM). The overexpression of a stress-inducible soybean TIP2 isoform in *Arabidopsis* seedlings resulted in increased transpiration and lower water potential, in agreement with the findings of previous studies (Peng et al. 2007; Sade et al. 2009). However, this phenomenon led to hypersensitivity to abiotic stress (Wang et al. 2011). In contrast, overexpression of a stress-inducible TIP1;2 from *Thellungiella salsuginea* in *Arabidopsis* conferred both drought and salt tolerance, but did so by limiting water loss and thus protecting the deleterious effect caused by the severe stress (Wang et al. 2014). In a different study, heterologous expression of the wheat TaTIP2 compromised the abiotic stress tolerance of *Arabidopsis* (Xu et al. 2013). Recently, expression of the stress-inducible tomato SlTIP2;2 in *Arabidopsis* under the control of its native promoter confirmed its ability to confer salt tolerance (Xin et al. 2014). SlTIP2;2 expression significantly affected the Na and K fluxes from the root meristematic zones and resulted in remarkable changes in the morphology of the pith ray cells in the inflorescence stems of transgenic *Arabidopsis* (Xin et al. 2014). Another recent study in sweet orange revealed strong upregulation of leaf TIPs in response to drought and salt stress, concurrent with downregulation of root TIPs (Martins et al. 2015). Interestingly, this pattern was not observed for PIPs. Altogether these results are in agreement with the idea that TIPs may play a role in adjusting water content of mesophyll cell cytoplasm under stressful conditions (Sade et al. 2009).

The posttranslation regulation of TIPs plays an important role in determining organ water status. A previous study showed that under stress conditions, TIPs are exported from the membrane via a vesicle trafficking mechanism (Vera-Estrella 2004) (see also chapter "Plant Aquaporin Trafficking"). This implies that, under stressful conditions, the permeability of the tonoplast to water should be reduced (Maurel et al. 1993; Ohshima et al. 2001; Sutka et al. 2005). In addition, in salt-treated *Arabidopsis* roots, AtTIP1;1, but not the AtTIP2;1 homolog, is re-localized to vacuolar bulbs (Boursiac et al. 2005). The complexity of this issue increases with the evidence that different stages of vacuole biogenesis and different types of vacuolar compartments can be present in a single cell type and changes can be triggered by environmental conditions (Paris et al. 1996; Marty 1999). This trafficking regulation mechanism not only controls plant cell water homeostasis via the regulation of the tonoplast water permeability but also suggests a protein mass balance mechanism. Accordingly, the balance between the accumulation of TIPs in the tonoplast and the velocity of the trafficking of TIP out of the membrane (in response to a stress signal) will determine the amount of time it takes the tonoplast to reduce its water permeability. Exporting abundant TIP via an endogenous trafficking system will take more time and might result in anisohydric-like behavior.

2.1 Root TIPs Under Abiotic Stress

In the roots, TIPs have a variable pattern of expression, with low abundance isoforms as well as isoforms that are highly abundant (Boursiac et al. 2005). The tissue localization of TIPs in roots is also variable. For example, in maize, high levels of ZmTIP1;1 expression have been found in perivascular root and shoot tissues (Barrieu et al. 1998). Interestingly, an early effect of salt treatment is a reduction in TIP transcript levels that is not accompanied by a reduction in the abundance of TIP proteins, although the root hydraulic conductivity does decrease drastically (Boursiac et al. 2005). A similar trend was observed in roots of radish seedlings treated with salt and ABA (Suga et al. 2002). In agreement with this, a recent study of the relationship between root hydraulic conductivity under conditions of environmental stress and the abundance of TIP protein showed that the abundance of TIP is not a major determinant of L_{Pr} (Di Pietro et al. 2013). On the other hand, exposure of *Arabidopsis* shoots to various stress conditions resulted in a significant increase in TIP2;3 in the roots of those plants, suggesting a role for TIPs in shoot-to-root stress signaling (Levin et al. 2009). Genetic evidence supporting a major role for TIPs in root water transport is currently lacking, and further research is needed to address the role of TIPs in root water transport under stressful conditions.

2.2 TIPs and Reactive Oxygen Species (ROS)

In response to environmental and developmental stimuli, reactive oxygen species (ROS; e.g., H_2O_2) are some of the most important secondary messengers that regulate downstream signaling events and, ultimately, the fate of cells (Steinhorst and

Kudla 2013). H_2O_2 is also involved in mediating cell-to-cell transfer of signals to distal parts of the plants (Steinhorst and Kudla 2013), as well as an important signaling molecule during stress (Neill et al. 2002). In recent years, several AQPs isoforms have been found to transport ROS like H_2O_2 (Bienert and Chaumont 2014). The first direct evidence of TIPs H_2O_2 transport came when the expression of the *Arabidopsis* AtTIP1;1 and AtTIP1;2 in yeast led to markedly reduced growth and cell survival on medium containing H_2O_2. This effect was shown to correlate with higher levels of H_2O_2 inside the yeast cells (Bienert et al. 2007).

AtTIP1;1, which is highly permeable to H_2O_2 (Bienert et al. 2007), is also a very efficient water channel, suggesting that AQPs may possess the same capacity and selectivity toward water and H_2O_2. The involvement of other AQPs in mediating H_2O_2 transport has also been suggested. For example, the seed-specific TIP3 was shown to transport both H_2O_2 and water and to affect the longevity of *Arabidopsis* seeds (Mao and Sun 2015). Another example is the overexpression of a TIP (TsTIP1;2) from *T. salsuginea* that enhances tolerance to multiple stresses, including drought, salinity, and oxidative stress (Wang et al. 2014). The ability of TsTIP1;2 to transport H_2O_2 was also confirmed in the yeast model. Therefore, overexpression of TsTIP1;2 may enhanced the accumulation of stressed-induced H_2O_2 into the vacuoles, resulting in the reduction of cytosolic ROS and alleviation of the injury caused by stress treatment (Wang et al. 2014). The involvement of TIPs in mediating the role of H_2O_2 in stress tolerance needs further research.

2.3 *TIPs Affect the Viability of Pollen and Seed Longevity Under Stressful Conditions*

AQPs supply water and nutrients to pollen and, therefore, are important for reproduction and seed development. *Arabidopsis* pollen grains contain one large vegetative cell and two smaller sperm cells. Two TIPs (AtTIP1;3 and AtTIP5;1) are highly expressed in the pollen cells. The tonoplast localization of AtTIP5;1 was confirmed to be specific to sperm cells. Another TIP isoform, AtTIP1;3, was found to be present in the tonoplast of vegetative cells. Interestingly, the double knockout mutant of AtTIP1;3 and AtTIP5;1 displayed an abnormal rate of barren siliques under conditions of limited water or nutrient supply. These results strongly imply that TIPs are present in two distinct cells, such as vegetative and sperm cells, and interact functionally to maintain pollen viability under adverse environmental conditions (Wudick et al. 2014).

TIPs have also been shown to be involved in seed germination and longevity (Mao and Sun 2015). In embryos of maturing seeds, two TIPs (TIP3;1 and TIP3;2) are uniquely expressed in the tonoplast and plasma membrane (Gattolin et al. 2011). The *tip3;1/tip3;2* double knockout exhibited altered seed longevity (Mao and Sun 2015). This phenomenon was shown to be regulated by the ABA signaling abscisic acid-insensitive 3 (ABI3) gene.

2.4 TIPs and Biotic Stress

Interestingly, bioinformatic analysis also found that TIP AQPs are upregulated in response to biotic stress (Sade et al. 2009). For example, tomato plants that express SlTIP2;2 exhibit resistance to the fungus *Botrytis cinerea* BO5.10 (Sade et al. 2012) and to TYLCV infection (Sade et al. 2014a). Moreover, the *TIP1;1* transcript level was recently reported to be higher in a tomato line that exhibits resistance to tomato yellow leaf curl virus (TYLCV) than in a susceptible line (Eybishtz et al. 2009). Recently, it was suggested that TIP might affect plant water balance indirectly via improved sugar metabolism and/or hormone homeostasis (i.e., ABA/salicylic acid) (Sade et al. 2014a). In agreement with this finding, the SlTIP2;2 tomato TIP was recently shown to interact with a UDP-galactose transporter in vivo (Xin et al. 2014). In addition, a role for salicylic acid (a well-known biotic stress plant hormone) in stimulating the regulation of PIPs was proposed by (Boursiac et al. 2008). It is also worthwhile noting that TIPs have been shown to facilitate the transport of urea in heterologous systems [i.e., yeast and *Xenopus laevis* oocytes; (Liu et al. 2003; Gu et al. 2012)]. Interestingly, urea pathways (both urea transporter transcript levels and urea levels) are upregulated in TYLCV-resistant tomato upon infection (Sade et al. 2015a). This upregulation was not observed in TYLCV-susceptible tomatoes and was correlated with TIP upregulation. It is possible that the upregulation of TIPs might activate a polyamine urea-dependent resistance pathway (Hussain et al. 2011). This pathway seems to be lacking in susceptible tomato plants and appears to be unique to resistance plants (Sade et al. 2015a, b). Interestingly, an analysis of AQP expression in the presence of biotic stress (i.e., *Candidatus Liberibacter asiaticus* infection) in sweet orange revealed strong upregulation of most of the TIPs at a late stage of infection, while little or no change in PIPs was observed (Martins et al. 2015). To date, the nature of the involvement of AQPs in resistance to biotic stress is not well understood, and there are many questions regarding its direct and indirect mechanisms of actions. This seems to be an interesting and promising field for future research.

3 Nodulin 26-Like Intrinsic Proteins (NIPs) and Stress

The NIP subfamily represents a group of multifunctional membrane intrinsic proteins that are named for their sequence similarity to the archetype of the family, soybean nodulin 26 (see chapters "The Nodulin 26 Intrinsic Protein Subfamily" and "Plant Aquaporins and Metalloids"). Functionally, the NIP family constitutes a unique group of major intrinsic proteins with little or no water permeability that are permeable to a wide but defined range of small solutes (Dean et al. 1999; Niemietz and Tyerman 2000; Weig and Jakob 2000; Ma et al. 2006; Takano et al. 2006). Compared to the more widely expressed TIP and PIP genes, NIP genes are generally expressed at low levels (Alexandersson et al. 2005). In addition, they are often expressed only in roots, suggesting that NIP transport activities may be prevalent in a more defined set of cells in the plant. NIPs are subject to posttranslational phosphorylation that increases their permeability. Abiotic stresses such as salinity and drought have been shown to increase phosphorylation (Guenther 2003). There are very few reports suggesting NIP involvement in abiotic stress tolerance. The role of NIPs in plant abiotic stress tolerance has not been widely studied, and further research is needed to determine whether NIPs contribute in this area. However, the high permeability of NIPs to several small molecules (e.g., arsenite, boric acid, silicon) that are potential toxins or nutrients suggests that NIPs may play a unique role in the regulation of plant responses to non-optimal levels of these molecules.

3.1 NIPs and Arsenite Accumulation Stress

Arsenite [As(III)] is highly toxic to many organisms, including plants. As(III) is taken up by the cell, probably via transporters. As(III) transporters have been isolated in several plants and shown to be a part of the NIP family (Bienert and Jahn 2010) (see chapter "Plant Aquaporins and Metalloids"). For example, in *Arabidopsis*, six of the nine members of the NIP subfamily (NIP1;1, NIP1;2,NIP3;1, NIP5;1, NIP6;1, and NIP7;1) are permeable to arsenite (Bienert et al. 2008; Isayenkov and Maathuis 2008; Kamiya et al. 2009). OsNIP2;1, an NIP responsible for As(III) transport in rice, has also been described (Zhao et al. 2010), and an *Osnip2;1* mutant line exhibited tolerance to arsenite stress (Zhao et al. 2010). AtNIP1;1, which has been localized to the root plasma membrane, facilitates the transport of As(III) in plants, and the *nip1;1* mutant lines exhibit strong resistance to As(III) toxicity (Kamiya et al. 2009). Very recently, *Arabidopsis* NIP3;1 was identified as playing an important role in both the uptake and the root-to-shoot distribution of arsenic under arsenite stress conditions (Xu et al. 2015). The *nip3;1* loss-of-function mutants displayed considerable improvements in As(III) tolerance, in terms of aboveground growth, and accumulated less arsenic in their shoots than wild-type plants, whereas the *nip3;1 nip1;1* double mutant showed strong As(III) tolerance and exhibited improved growth of both roots and shoots under As(III) stress conditions, as compared with WT plants (Xu et al. 2015).

3.2 NIPs and Boric Acid Deficiency and Toxicity Stress

Boron (B) is an essential micronutrient for plants. B is often present at low concentrations in the environment; thus B deficiency (as well as toxicity; see below) occurs widely and is a major constraint for cereal and brassica crops (see chapter "Plant Aquaporins and Metalloids"). Reproductive growth is especially sensitive to B deficiency, and substantial crop losses can occur in situations in which no clear vegetative signs of deficiency are observed. In recent years, a role for NIPs in membrane permeability to B has been suggested (Takano et al. 2006; Tanaka et al. 2008). Both *Arabidopsis* NIP5;1 (a root-localized NIP) and NIP6;1 (a leaf-localized NIP) have been shown to be permeable to B and were significantly upregulated in *Arabidopsis* treated with low levels of B (Takano et al. 2006). When grown under low B conditions, AtNIP6;1 knockout lines exhibited reduced leaf expansion and decreased concentrations of B in their young leaves. Loss-of-function mutants of NIP5;1 exhibited severe shoot and root growth retardation and low B contents when grown under low B conditions (Takano et al. 2006). Moreover, enhanced expression of NIP5;1 resulted in improved *Arabidopsis* root elongation under limited B conditions (Kato et al. 2009). Recently, the rice OsNIP3;1 (an AtNIP5;1 homolog) was shown to transport B and to be upregulated under conditions of B deficiency. The growth of *OsNIP3;1* RNAi plants was impaired under limited B conditions, and the distribution of B among their shoot tissues was altered (Hanaoka et al. 2014).

B toxicity can also be a significant limitation to cereal crop production in a number of regions of the world. Tolerance of levels of B that are generally toxic was confirmed in barley through heterologous expression in a yeast system and reduced expression of the rice homolog of HvNIP2;1 (OsNIP2;1) (Schnurbusch et al. 2010). Taken together, NIP manipulation can lead to improved crop performance under conditions of B deficiency and/or toxicity tolerance and carries potential for the regulation of the expression of a mineral nutrient channel gene to improve growth under limited nutrient or toxic conditions.

3.3 NIPs and Silicon Deficiency Stress

It has been suggested that silicon (Si) might support plant resistance to diseases, pests, and lodging through Si depositions in the apoplast and induced resistance, improve the light interception ability of plants, and minimize transpiration losses (Epstein 1994; Ma et al. 2004). Lsil protein (a member of an NIP subfamily) has been identified in cereals including rice, barley, maize, and wheat (Ma 2010). Lsil and its homologs are Si transporters, responsible for the transport of Si from the external solution to root cells (see chapter "Plant Aquaporins and Metalloids").

In rice, an Lsil (NIP2;1) mutant was shown to exhibit reduced uptake of Si and to be susceptible to biotic stress (Ma et al. 2006). Moreover, the grain yield of these mutant plants was 90 % less than the WT (Ma et al. 2006). Although it is not clear whether this reduction in yield is directly related to Si deficiency, the accumulation of Si in rice via NIPs might be related to the maintenance of the vigor of plants exposed to biotic and abiotic stress. However, constitutive expression (35S promoter) of wheat Lsil in *Arabidopsis* caused deleterious symptoms (Montpetit et al. 2012), raising some questions regarding the role of Si in stress resistance of dicots. This might be due to the ectopic expression and/or the wrong protein localization as cell specificity and membrane polarity are important for the channel function (see chapters "Plant Aquaporin Trafficking" and "Plant Aquaporins and Metalloids"). In contrast, specific expression in *Arabidopsis* roots (under a root-specific promoter) resulted in plants that grew normally and did not exhibit any visible symptoms (Montpetit et al. 2012). These results demonstrate that TaLsi1 expression must be confined to root cells for healthy plant development. At this point, further research is needed to address the physiological importance of Si uptake via NIPs and its relationship with stress responses.

4 Plasma Membrane Intrinsic Proteins (PIPs) and Stress

Traditionally, PIPs have been divided into two groups, PIP1 and PIP2, based on their sequence similarities (Johanson et al. 2001). Under abiotic stress, *Arabidopsis* PIP transcripts are generally downregulated or inactivated by various posttranslational mechanisms (Alexandersson et al. 2005; Boursiac et al. 2008; Hachez et al. 2014), with the exception of PIP1;4 and PIP2;5, which are upregulated (Alexandersson et al. 2005). Indeed, the genes for these two proteins have stress response factors in their promoter region (Alexandersson et al. 2010). In addition, in other plant species (e.g., maize, wheat, poplar, tobacco, soybean), some PIPs are upregulated under stressful conditions (Zhu et al. 2005; Mahdieh et al. 2008; Ayadi et al. 2011; Bae et al. 2011; Zhou et al. 2014). Moreover, different genotypes of rice (e.g., upland vs. lowland) exhibit different PIP expression patterns in response to ABA treatment and osmotic stress (Lian et al. 2006). These and other studies have demonstrated that PIPs are responsive to environmental stress. Nevertheless, their complex pattern of expression in different tissues and different parts of the plant (Prado et al. 2013) and the possibility of functional redundancy complicate our understanding of the role(s) of these proteins in the whole-plant stress response. An additional aspect of PIP complexity is the ability of these proteins to increase or decrease plant stress tolerance depending on the organism, isoform, promoter, and tissue in question (see Table 1).

4.1 The Role of PIPs in Whole-Plant Water Budget Regulation

It has also been suggested that PIPs may control isohydric/anisohydric behavior and root hydraulics under drought stress and recovery from that stress (Vandeleur et al. 2009). In grapevine, the tighter control of stomatal conductance in the more isohydric cultivar grenache as compared with the cultivar chardonnay (anisohydric) under drought conditions was also reflected in root hydraulic behavior over the course of the day. The lower L_{Pr} of cv. grenache as compared to cv. chardonnay under drought stress is associated with a lower rate of transpiration. "Chardonnay" seems to compensate with increased expression of VvPIP1;1, which correlates with increased cortical cell water permeability. Interestingly, cv. grenache does not display this behavior (Vandeleur et al. 2009).

Table 1 Representative studies addressing the effects of overexpression or silencing of PIPs on the stress responses of genetically modified plants. Transgenic plants reported to exhibit increased or decreased stress tolerance are labeled in green and yellow, respectively

AQP isoform	Expression level	Stress response	Genetically modified plant	Tissue studied	Source
NtAQP1	Overexpression	Salt tolerant	Arabidopsis	Shoot	Sade et al. (2010)
NtAQP1		Salt tolerant	Tomato	Shoot and root	Sade et al. (2010)
TdPIP1;1		Drought and salt tolerant	Tobacco	Shoot and root	Ayadi et al. (2011)
TdPIP2;1		Drought and salt tolerant	Tobacco	Shoot and root	Ayadi et al. (2011)
TaAQP8		Salt tolerant	Tobacco	Shoot and root	Hu et al. (2012)
AtPIP2;5		Cold tolerant	*Arabidopsis*	Shoot and root	Lee et al. (2012)
BnPIP1		Drought tolerant	Tobacco	Shoot	Yu et al. (2005)
OsPIP1;1		Salt tolerant	Rice	Shoot and root	Liu et al. (2013)
OsPIP1		Drought tolerant	Rice	Shoot	Lian et al. (2004)
OsPIP1;1		Drought and salt tolerant	*Arabidopsis*	Shoot and root	Guo et al. (2006)
OsPIP2;2		Drought and salt tolerant	*Arabidopsis*	Shoot and root	Guo et al. (2006)
MaPIP1;1		Drought and salt tolerant	*Arabidopsis*	Shoot and root	Xu et al. (2014)
GmPIP1;6		Salt tolerant	Soybean	Shoot and root	Zhou et al. (2014)
AtPIP1;2		Drought sensitive	Tobacco	Shoot	Aharon et al. (2003)
AtPIP1;4		Drought sensitive	Tobacco, *Arabidopsis*	Shoot	Jang et al. (2007)
AtPIP2;5		Drought sensitive	Tobacco, *Arabidopsis*	Shoot	Jang et al. (2007)
HvPIP2;1		Salt sensitive	Rice	Shoot and root	Katsuhara et al. (2003)
GaPIP1		Drought sensitive	*Arabidopsis*	Shoot and root	Li et al. (2015)
NtAQP1	Downregulation	Drought sensitive	Tobacco	Shoot and root	Siefritz et al. (2002)
AtPIPs		Drought sensitive	*Arabidopsis*	Shoot and root	Martre et al. (2002)
BnPIP1		Drought sensitive	Tobacco	Shoot	Yu et al. (2005)
PoptrPIP1s		Drought sensitive	Poplar	Shoot	Secchi and Zwieniecki (2014)

Different roles have been suggested for AQPs in the regulation of plant hydraulic conductance (both roots and shoots, see above). Several studies have suggested the existence of tissue-specific mechanisms involving PIPs [(Sade et al. 2010; Shatil-Cohen et al. 2011; Shatil-Cohen and Moshelion 2012; Di Pietro et al. 2013; Prado et al. 2013; Grondin et al. 2015); see also the part on the regulation of plant hydraulic conductance)]. High levels of PIP expression have been observed in the BS cells and xylem parenchyma of several species (Kirch et al. 2000; Otto and Kaldenhoff 2000; Oliviusson et al. 2001; Hachez et al. 2008). This site of expression may be crucial for the radial cell-to-cell movement of water as it exits the xylem vessels (Prado et al. 2013) and for the osmotically driven loading of water into xylem vessels during embolism refilling (Sakr et al. 2003; Secchi and Zwieniecki 2010), suggesting a role for PIPs in the maintenance of vascular tissue function under drought stress (Montalvo-Hernandez et al. 2008). A good example is AtPIP2;1, which was shown to have a major role in the movement of water out of the xylem vessels and an independent role in the stomata as an ABA-dependent regulator of stomatal closure (Prado et al. 2013; Grondin et al. 2015) (see also chapter "Roles of Aquaporins in Stomata"). Similarly, NtAQP1 was shown to be involved in maintaining the root hydraulics of tobacco and tomato under stressful conditions (Siefritz et al. 2002; Sade et al. 2010) and independently affecting leaf photosynthesis (Flexas et al. 2006; Sade et al. 2010, 2014b). In addition, specific expression of NtAQP1 in root endodermis and leaf BS contributed to gas exchange under salt stress, whereas expression of this protein in stomata did not had any affect (Sade et al. 2014b). All of these studies emphasized the necessity of combining tissue-specific transcriptomic, proteomic, and posttranslational regulation data for AQPs under stressful conditions and correlating that data with physiological responses.

PIPs take a major role in the regulation of root hydraulics in general and in particular in response to various environmental stimuli (reviewed by Aroca et al. 2012 and papers within). This role was shown to be strongly affected by both transcriptional and posttranslational regulation (Boursiac et al. 2005; Di Pietro et al. 2013) and to vary significantly between plants and isoforms. Indeed, looking at different studies, it is hard to find a common response of root PIP expression and PIP protein abundance under drought conditions [reviewed in (Aroca et al. 2012)]. These findings may also imply that different PIPs play different roles in different parts of the root under stressful conditions. Indeed, PIP1s and PIP2s have different patterns of expression in *Arabidopsis* roots (Maurel et al. 2015) and also respond differently to different forces [i.e., osmotic and hydrostatic; (Javot 2003; Postaire et al. 2010)]. This state of affairs further emphasizes the necessity for tissue-specific data regarding AQP expression and function.

5 A Hydraulic Analogy Model

Keeping the leaf relative, water content above a certain threshold is a fundamental necessity for plant survival. Terrestrial plants use different strategies for managing their water budgets to keep them above their own critical threshold, under varying environmental conditions. In this chapter, we have summarized the involvement of AQPs in the regulation of plant water status and their role in determining the water flux balance within and between key tissues along the SPAC. We have emphasized the role of TIPs in cellular water homeostasis as regulated by the vacuole and the role of PIPs in the regulation of plant radial hydraulic conductance. We have also emphasized the fact that water status is very dynamic in the face of fluctuations in soil water availability and atmospheric demand for water, both for specific tissues and between different tissues along the SPAC.

In conclusion, we would like to suggest a hydraulic circuit analogy model that emphasizes the role of plant AQPs at the key hydraulic junctions, which regulate whole-plant water balance under changing ambient conditions. This hydraulic analogy is suggested as an alternative to the popular electrical circuit analogy for the movement of water through the SPAC, which includes resistors, a current, and a driving electrical potential (e.g., Steudle 2000; Sack and Holbrook 2006). In the electrical circuit analogy, the potential difference between the soil and atmosphere is analogous to the driving electrical potential in an electrical circuit, and the water flux is analogous to the electron flux based on a steady-state assumption. The basic definition of steady state is that inputs and outputs are equal and thus net balance is even and there is minimal fluctuation in the plant water status (relative water content and water potential). Using an electrical analogy, one can still incorporate this problem by using electric capacitors (Bramley et al. 2007). Nevertheless, the fact that water status at noon [when transpiration is at its peak; (e.g., Attia et al. 2015)] is lower than that seen during the morning and that under stress conditions even greater reductions in water status are observed suggests that the steady-state assumption is incorrect. The water status reduction is likely due to more water flowing out than in during the morning hours (and during stress vs. nonstress periods). Therefore, the regulation of the flow of water has to be supported with an additional water capacitor to buffer the deficits and provide a supply of water to support the transpiration outflow. The vacuoles and the storage of water in stems may serve as a capacitor (Dainty 1976), which should be taken into consideration in the SPAC steady-state flow assumption.

Our alternative hydraulic analogy (presented in Fig. 2) is comprised of dynamic hydraulic resistances that vary nonlinearly with water potential and are controlled by PIPs. Moreover, the stem and vacuole water reservoirs retain and release water, thereby serving as capacitors buffering cytoplasmic water volume. The regulation of the osmotic water permeability of the tonoplast, via its TIPs, might play a role in the compensating mechanism of the cytoplasm volume in the presence of a turgor decrease.

Leaf water balance is determined by the relationship between the flow of water into the leaf and the flow of water out of the leaf. Water balance is based on the counterbalancing of stomatal gas conductance (g_s), which controls the rate at which water vapor is lost from leaves during transpiration) E), and the outward flow of water through the leaf toward the evaporation sites on the mesophyll cell walls.

As depicted in Fig. 2a, under conditions of high soil water availability and relative low atmospheric water demand (e.g., early morning), the plant experiences low

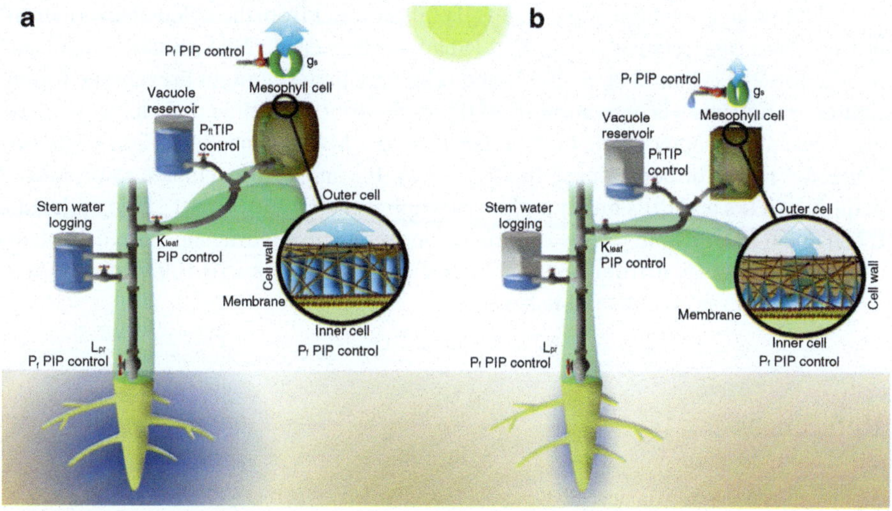

Fig. 2 A hypothetical model for the role of AQPs in controlling the dynamic soil-plant-atmosphere water balance under (**a**) well irrigated and (**b**) drought condition. See text for explanations

hydraulic tension, which results in a positive water balance and full leaf turgor. In this situation, all key hydraulic tissues along the root-leaf-mesophyll (except for guard cells, due to inactivity of their AQPs; Grondin et al. 2015) exhibit high levels of hydraulic conductance due to the activity of PIPs and TIPs (depicted as open taps in Fig. 2a) (Cochard et al. 2007; Levin et al. 2007; Maggio and Joly 1995; Carvajal et al. 1996; Clarkson et al. 2000; Tournaire-Roux et al. 2003; Gorska et al. 2008; Bramley et al. 2010; Tournaire-Roux et al. 2003; Maurel et al. 2009; Shatil-Cohen et al. 2011; Shatil-Cohen and Moshelion 2012), including high mesophyll osmotic water permeability (P_f) (Morillon and Chrispeels 2001).

As depicted in Fig. 2b, as soil water availability decreases and the hydraulic tension on the whole plant increases, the plant water balance becomes negative and leaves lose turgor. At this point, K_{leaf} and g_s are reduced due to increasing ABA levels that decrease the P_f of the BS cells and increase the P_f of the guard cells (see Fig. 1). Nevertheless, we do not know how exactly the hydraulic tension signal is translated into an ABA signal. One possibility is that the negative water potential generated within the leaf (due to the difference between leaf water demand and the vascular system supply) results in a rapid reduction in the mesophyll cell wall hydration level. This reduction in the water content of the microcapillary structure of the cell walls generates a negative water potential within the microcapillary structure of the cell walls (Nobel 2009). The close proximity of the mesophyll plasma membrane to the dehydrated cell wall might put the cytoplasm at risk of water loss. We propose that, in addition to stomatal closure, this risk is averted by (i) maintaining the low conductivity of the mesophyll plasma membrane (P_f) that separates these two compartments and (ii) the buffering effect of the vacuole, based on the water conductivity of the tonoplast, P_{ft}. In parallel and in accordance with the cohesion-tension theory, the decreasing water potential of the cell wall pulls in water from the xylem. Nevertheless, at this stage, the relatively low K_{leaf} (Shatil-Cohen et al. 2011; Shatil-Cohen and Moshelion 2012; Pantin et al. 2013) and relatively high L_{Pr} might prevent the formation of embolisms via activation of AQPs in the root (Secchi and Zwieniecki 2010, 2014; see also chapter "Role of Aquaporins in the Maintenance of Xylem Hydraulic Capacity"). We further propose that higher P_f, P_{ft}, and K_{leaf} levels due to prolonged activity of TIPs and PIPs might prolong transpiration (E) and be part of the molecular mechanism responsible for the difference between anisohydric and isohydric plants (image was modified from Sade and Moshelion (2014) doi:10.1093/treephys/tpu-070with permission of the publisher).

References

Aharon R, Shahak Y, Wininger S, Bendov R, Kapulnik Y, Galili G (2003) Overexpression of a plasma membrane aquaporin in transgenic tobacco improves plant vigor under favorable growth conditions but not under drought or salt stress. Plant Cell 15:439–447

Alassimone J, Naseer S, Geldner N (2010) A developmental framework for endodermal differentiation and polarity. Proc Natl Acad Sci 107:5214–5219

Alexandersson E, Fraysse L, Sjovall-Larsen S, Gustavsson S, Fellert M, Karlsson M, Johanson U, Kjellbom P (2005) Whole gene family expression and drought stress regulation of aquaporins. Plant Mol Biol 59:469–484

Alexandersson E, Danielson JÅH, Råde J, Moparthi VK, Fontes M, Kjellbom P, Johanson U (2010) Transcriptional regulation of aquaporins in accessions of arabidopsis in response to drought stress. Plant J 61:650–660

Aroca R, Ferrante A, Vernieri P, Chrispeels MJ (2006) Drought, abscisic acid and transpiration rate effects on the regulation of pip aquaporin gene expression and abundance in *Phaseolus vulgaris* plants. Ann Bot 98:1301–1310

Aroca R, Porcel R, Manuel Ruiz-Lozano J (2012) Regulation of root water uptake under abiotic stress conditions. J Exp Bot 63:43–57

Attia Z, Domec J-C, Oren R, Way DA, Moshelion M (2015) Growth and physiological responses of isohydric and anisohydric poplars to drought. J Exp Bot 66:4373–4381

Ayadi M, Cavez D, Miled N, Chaumont F, Masmoudi K (2011) Identification and characterization of two plasma membrane aquaporins in durum wheat (*Triticum turgidum* L. subsp. durum) and their role in abiotic stress tolerance. Plant Physiol Biochem 49:1029–1039

Bae E-K, Lee H, Lee J-S, Noh E-W (2011) Drought, salt and wounding stress induce the expression of the plasma membrane intrinsic protein 1 gene in poplar (*Populus alba* × *P. tremula* var. glandulosa). Gene 483:43–48

Barrieu F, Chaumont F, Chrispeels MJ (1998) High expression of the tonoplast aquaporin ZmTIP1 in epidermal and conducting tissues of maize. Plant Physiol 117:1153–1163

Bienert GP, Chaumont F (2014) Aquaporin-facilitated transmembrane diffusion of hydrogen peroxide. Biochim Biophys Acta-Gen Subj 1840:1596–1604

Bienert GP, Jahn TP (2010) Major intrinsic proteins and arsenic transport in plants: new players and their potential role. In: Jahn TP, Bienert GP (eds) Mips and their role in the exchange of metalloids, Landes Bioscience Publishers Advances in Experimental Medicine and Biology, vol 679, New York, pp 111–125

Bienert GP, Moller ALB, Kristiansen KA, Schulz A, Moller IM, Schjoerring JK, Jahn TP (2007) Specific aquaporins facilitate the diffusion of hydrogen peroxide across membranes. J Biol Chem 282:1183–1192

Bienert GP, Thorsen M, Schussler MD, Nilsson HR, Wagner A, Tamas MJ, Jahn TP (2008) A subgroup of plant aquaporins facilitate the bi-directional diffusion of As(OH)(3) and Sb(OH)(3) across membranes. BMC Biol 6:26

Boursiac Y, Chen S, Luu DT, Sorieul M, van den Dries N, Maurel C (2005) Early effects of salinity on water transport in arabidopsis roots. Molecular and cellular features of aquaporin expression. Plant Physiol 139:790–805

Boursiac Y, Boudet J, Postaire O, Luu D-T, Tournaire-Roux C, Maurel C (2008) Stimulus-induced downregulation of root water transport involves reactive oxygen species-activated cell signalling and plasma membrane intrinsic protein internalization. Plant J 56:207–218

Bramley H, Turner NC, Turner DW, Tyerman SD (2007) Comparison between gradient-dependent hydraulic conductivities of roots using the root pressure probe: the role of pressure propagations and implications for the relative roles of parallel radial pathways. Plant Cell Environ 30:861–874

Bramley H, Turner NC, Turner DW, Tyerman SD (2010) The contrasting influence of short-term hypoxia on the hydraulic properties of cells and roots of wheat and lupin. Funct Plant Biol 37:183–193

Carvajal M, Cooke D, Clarkson D (1996) Responses of wheat plants to nutrient deprivation may involve the regulation of water-channel function. Planta 199:372–381

Christmann A, Weiler EW, Steudle E, Grill E (2007) A hydraulic signal in root-to-shoot signalling of water shortage. Plant J 52:167–174

Christmann A, Grill E, Huang J (2013) Hydraulic signals in long-distance signaling. Curr Opin Plant Biol 16:293–300

Clarkson DT, Carvajal M, Henzler T, Waterhouse RN, Smyth AJ, Cooke DT, Steudle E (2000) Root hydraulic conductance: diurnal aquaporin expression and the effects of nutrient stress. J Exp Bot 51:61–70

Cochard H, Venisse J-S, Barigah TS, Brunel N, Herbette S, Guilliot A, Tyree MT, Sakr S (2007) Putative role of aquaporins in variable hydraulic conductance of leaves in response to light. Plant Physiol 143:122–133

Dainty J (1976) Water relations of plant cells. In: Gottingen AP, Zimmermann MH (eds) Encyclopedia of plant physiology, New Series, Part A, vol 2. Springer, Berlin, pp 12–35

Dean RM, Rivers RL, Zeidel ML, Roberts DM (1999) Purification and functional reconstitution of soybean nodulin 26. An aquaporin with water and glycerol transport properties. Biochemical 38:347–353

Di Pietro M, Vialaret J, Li G-W, Hem S, Prado K, Rossignol M, Maurel C, Santoni V (2013) Coordinated post-translational responses of aquaporins to abiotic and nutritional stimuli in arabidopsis roots. Mol Cell Proteomics 12:3886–3897

Epstein E (1994) The anomaly of silicon in plant biology. Proc Natl Acad Sci U S A 91:11–17

Eybishtz A, Peretz Y, Sade D, Akad F, Czosnek H (2009) Silencing of a single gene in tomato plants resistant to tomato yellow leaf curl virus renders them susceptible to the virus. Plant Mol Biol 71:157–171

Flexas J, Ribas-Carbó M, Hanson DT, Bota J, Otto B, Cifre J, McDowell N, Medrano H, Kaldenhoff R (2006) Tobacco aquaporin NtAQP1 is involved in mesophyll conductance to CO_2 in vivo. Plant J 48:427–439

Gattolin S, Sorieul M, Frigerio L (2011) Mapping of tonoplast intrinsic proteins in maturing and germinating arabidopsis seeds reveals dual localization of embryonic tips to the tonoplast and plasma membrane. Mol Plant 4:180–189

Gorska A, Zwieniecka A, Michele Holbrook N, Zwieniecki M (2008) Nitrate induction of root hydraulic conductivity in maize is not correlated with aquaporin expression. Planta 228:989–998

Grondin A, Rodrigues O, Verdoucq L, Merlot S, Leonhardt N, Maurel C (2015) Aquaporins contribute to ABA-triggered stomatal closure through OST1-mediated phosphorylation. Plant Cell 27:1945–1954

Gu R, Chen X, Zhou Y, Yuan L (2012) Isolation and characterization of three maize aquaporin genes, ZmNIP2;1, ZmNIP2;4 and ZmTIP4;4 involved in urea transport. BMB Rep 45:96–101

Guenther JF (2003) Phosphorylation of soybean nodulin 26 on serine 262 enhances water permeability and is regulated developmentally and by osmotic signals. Plant Cell 15:981–991

Guo L, Wang ZY, Lin H, Cui WE, Chen J, Liu MH, Chen ZL, Qu LJ, Gu HY (2006) Expression and functional analysis of the rice plasma-membrane intrinsic protein gene family. Cell Res 16:277–286

Hachez C, Heinen RB, Draye X, Chaumont F (2008) The expression pattern of plasma membrane aquaporins in maize leaf highlights their role in hydraulic regulation. Plant Mol Biol 68:337–353

Hachez C, Veljanovski V, Reinhardt H, Guillaumot D, Vanhee C, Chaumont F, Batoko H (2014) The arabidopsis abiotic stress-induced TSPO-related protein reduces cell-surface expression of the aquaporin PIP2;7 through protein-protein interactions and autophagic degradation. Plant Cell 26:4974–4990

Hanaoka H, Uraguchi S, Takano J, Tanaka M, Fujiwara T (2014) OsNIP3;1, a rice boric acid channel, regulates boron distribution and is essential for growth under boron-deficient conditions. Plant J 78:890–902

Hill AE, Shachar-Hill B, Shachar-Hill Y (2004) What are aquaporins for? J Membr Biol 197:1–32

Holbrook NM, Zwieniecki MA (1999) Embolism repair and xylem tension: do we need a miracle? Plant Physiol 120:7–10

Hose E, Steudle E, Hartung W (2000) Abscisic acid and hydraulic conductivity of maize roots: a study using cell- and root-pressure probes. Planta 211:874–882

Hu W, Yuan Q, Wang Y, Cai R, Deng X, Wang J, Zhou S, Chen M, Chen L, Huang C, Ma Z, Yang G, He G (2012) Overexpression of a wheat aquaporin gene, TaAQP8, enhances salt stress tolerance in transgenic tobacco. Plant Cell Physiol 53:2127–2141

Hussain SS, Iqbal MT, Arif MA, Amjad M (2011) Beyond osmolytes and transcription factors: drought tolerance in plants via protective proteins and aquaporins. Biol Plant 55:401–413

Isayenkov SV, Maathuis FJM (2008) The *Arabidopsis thaliana* aquaglyceroporin AtNIP7;1 is a pathway for arsenite uptake. FEBS Lett 582:1625–1628

Jang JY, Kim DG, Kim YO, Kim JS, Kang HS (2004) An expression analysis of a gene family encoding plasma membrane aquaporins in response to abiotic stresses in *Arabidopsis thaliana*. Plant Mol Biol 54:713–725

Jang JY, Lee SH, Rhee JY, Chung GC, Ahn SJ, Kang H (2007) Transgenic arabidopsis and tobacco plants overexpressing an aquaporin respond differently to various abiotic stresses. Plant Mol Biol 64:621–632

Javot H (2003) Role of a single aquaporin isoform in root water uptake. Plant Cell Online 15:509–522

Johanson U, Karlsson M, Johansson I, Gustavsson S, Sjovall S, Fraysse L, Weig AR, Kjellbom P (2001) The complete set of genes encoding major intrinsic proteins in arabidopsis provides a framework for a new nomenclature for major intrinsic proteins in plants. Plant Physiol 126:1358–1369

Kamiya T, Tanaka M, Mitani N, Ma JF, Maeshima M, Fujiwara T (2009) NIP1;1, an aquaporin homolog, determines the arsenite sensitivity of *Arabidopsis thaliana*. J Biol Chem 284:2114–2120

Kato Y, Miwa K, Takano J, Wada M, Fujiwara T (2009) Highly boron deficiency-tolerant plants generated by enhanced expression of NIP5;1, a boric acid channel. Plant Cell Physiol 50:58–66

Katsuhara M, Koshio K, Shibasaka M, Hayashi Y, Hayakawa T, Kasamo K (2003) Over-expression of a barley aquaporin increased the shoot/root ratio and raised salt sensitivity in transgenic rice plants. Plant Cell Physiol 44:1378–1383

Kirch HH, Vera-Estrella R, Golldack D, Quigley F, Michalowski CB, Barkla BJ, Bohnert HJ (2000) Expression of water channel proteins in *Mesembryanthemum crystallinum*. Plant Physiol 123:111–124

Lee SH, Chung GC, Jang JY, Ahn SJ, Zwiazek JJ (2012) Overexpression of PIP2;5 aquaporin alleviates effects of low root temperature on cell hydraulic conductivity and growth in arabidopsis (vol 159, pg 479, 2012). Plant Physiol 159:1291–1291

Levin M, Lemcoff JH, Cohen S, Kapulnik Y (2007) Low air humidity increases leaf-specific hydraulic conductance of *Arabidopsis thaliana* (L.) Heynh (Brassicaceae). J Exp Bot 58:3711

Levin M, Resnick N, Rosianskey Y, Kolotilin I, Wininger S, Lemcoff JH, Cohen S, Galili G, Koltai H, Kapulnik Y (2009) Transcriptional profiling of *Arabidopsis thaliana* plants' response to low relative humidity suggests a shoot-root communication. Plant Sci 177:450–459

Li G, Santoni V, Maurel C (2014) Plant aquaporins: roles in plant physiology. Biochim Biophys Acta-Gen Subj 1840:1574–1582
Li J, Ban L, Wen H, Wang Z, Dzyubenko N, Chapurin V, Gao H, Wang X (2015) An aquaporin protein is associated with drought stress tolerance. Biochem Biophys Res Commun 459:208–213
Lian HL, Yu X, Ye Q, Ding XS, Kitagawa Y, Kwak SS, Su WA, Tang ZC (2004) The role of aquaporin RWC3 in drought avoidance in rice. Plant Cell Physiol 45:481–489
Lian HL, Yu X, Lane D, Sun WN, Tang ZC, Su WA (2006) Upland rice and lowland rice exhibited different PIP expression under water deficit and ABA treatment. Cell Res 16:651–660
Liu LH, Ludewig U, Gassert B, Frommer WB, von Wiren N (2003) Urea transport by nitrogen-regulated tonoplast intrinsic proteins in arabidopsis. Plant Physiol 133:1220–1228
Liu C, Fukumoto T, Matsumoto T, Gena P, Frascaria D, Kaneko T, Katsuhara M, Zhong S, Sun X, Zhu Y, Iwasaki I, Ding X, Calamita G, Kitagawa Y (2013) Aquaporin OsPIP1;1 promotes rice salt resistance and seed germination. Plant Physiol Biochem 63:151–158
Ma JF (2010) Silicon transporters in higher plants. In: Jahn TP, Bienert GP (eds) Mips and their role in the exchange of metalloids, Adv Exp Med Biol 679:99–109
Ma JF, Mitani N, Nagao S, Konishi S, Tamai K, Iwashita T, Yano M (2004) Characterization of the silicon uptake system and molecular mapping of the silicon transporter gene in rice. Plant Physiol 136:3284–3289
Ma JF, Tamai K, Yamaji N, Mitani N, Konishi S, Katsuhara M, Ishiguro M, Murata Y, Yano M (2006) A silicon transporter in rice. Nature 440:688–691
Maggio A, Joly RJ (1995) Effects of mercuric-chloride on the hydraulic conductivity of tomato root systems – evidence for a channel-mediated water pathway. Plant Physiol 109:331
Mahdieh M, Mostajeran A (2009) Abscisic acid regulates root hydraulic conductance via aquaporin expression modulation in *Nicotiana tabacum*. J Plant Physiol 166:1993–2003
Mahdieh M, Mostajeran A, Horie T, Katsuhara M (2008) Drought stress alters water relations and expression of PIP-type aquaporin genes in *Nicotiana tabacum* plants. Plant Cell Physiol 49:801–813
Mao Z, Sun W (2015) Arabidopsis seed-specific vacuolar aquaporins are involved in maintaining seed longevity under the control of ABSCISIC ACID INSENSITIVE 3. J Exp Bot 66:4781–4794
Martins CDPS, Pedrosa AM, Du D, Goncalves LP, Yu Q, Gmitter FG Jr, Costa MGC (2015) Genomewide characterization and expression analysis of major intrinsic proteins during abiotic and biotic stresses in sweet orange (Citrus sinensis L. Osb.). PLoS ONE 10:e0138786
Martre P, Morillon R, Barrieu F, North GB, Nobel PS, Chrispeels MJ (2002) Plasma membrane aquaporins play a significant role during recovery from water deficit. Plant Physiol 130:2101–2110
Marty F (1999) Plant vacuoles. Plant Cell 11:587–599
Maurel C, Reizer J, Schroeder JI, Chrispeels MJ (1993) The vacuolar membrane-protein gamma-tip creates water specific channels in xenopus-oocytes. EMBO J 12:2241–2247
Maurel C, Verdoucq L, Luu D-T, Santoni V (2008) Plant aquaporins: membrane channels with multiple integrated functions. Annu Rev Plant Biol 59:595–624
Maurel C, Santoni V, Luu D-T, Wudick MM, Verdoucq L (2009) The cellular dynamics of plant aquaporin expression and functions. Curr Opin Plant Biol 12:690–698
Maurel C, Simonneau T, Sutka M (2010) The significance of roots as hydraulic rheostats. J Exp Bot 61:3191–3198
Maurel C, Boursiac Y, Luu D-T, Santoni V, Shahzad Z, Verdoucq L (2015) Aquaporins in plants. Physiol Rev 95:1321–1358
Montalvo-Hernández L, Piedra-Ibarra E, Gómez-Silva L, Lira-Carmona R, Acosta-Gallegos J, Vazquez-Medrano J, Xoconostle-Cázares B, Ruíz-Medrano R (2008) Differential accumulation of mRNAs in drought-tolerant and susceptible common bean cultivars in response to water deficit. New Phytol 177:102–113
Montpetit J, Vivancos J, Mitani-Ueno N, Yamaji N, Remus-Borel W, Belzile F, Ma JF, Belanger RR (2012) Cloning, functional characterization and heterologous expression of TaLsi1, a wheat silicon transporter gene. Plant Mol Biol 79:35–46

Moon GJ, Clough BF, Peterson CA, Allaway WG (1986) Apoplastic and symplastic pathways in avicennia-marina (Forsk) vierh roots revealed by fluorescent tracer dyes. Aust J Plant Physiol 13:637–648

Morillon R, Chrispeels MJ (2001) The role of ABA and the transpiration stream in the regulation of the osmotic water permeability of leaf cells. Proc Natl Acad Sci U S A 98:14138–14143

Neill N, Desikan R, Hancock J (2002) Hydrogen peroxide signaling. Curr Opin Plant Biol 5:388–395

Niemietz CM, Tyerman SD (2000) Channel-mediated permeation of ammonia gas through the peribacteroid membrane of soybean nodules. FEBS Lett 465:110–114

Nobel P (2009) Physicochemical and environmental plant physiology. Elsevier, New York

Ohshima Y, Iwasaki I, Suga S, Murakami M, Inoue K, Maeshima M (2001) Low aquaporin content and low osmotic water permeability of the plasma and vacuolar membranes of a CAM plant graptopetalum paraguayense: comparison with radish. Plant Cell Physiol 42:1119–1129

Oliviusson P, Salaj J, Hakman I (2001) Expression pattern of transcripts encoding water channel-like proteins in Norway spruce (*Picea abies*). Plant Mol Biol 46:289–299

Otto B, Kaldenhoff R (2000) Cell-specific expression of the mercury-insensitive plasma-membrane aquaporin NtAQP1 from *Nicotiana tabacum*. Planta 211:167–172

Pantin F, Monnet F, Jannaud D, Costa JM, Renaud J, Muller B, Simonneau T, Genty B (2013) The dual effect of abscisic acid on stomata. New Phytol 197:65–72

Parent B, Hachez C, Redondo E, Simonneau T, Chaumont F, Tardieu F (2009) Drought and abscisic acid effects on aquaporin content translate into changes in hydraulic conductivity and leaf growth rate: a trans-scale approach. Plant Physiol 149:2000–2012

Paris N, Stanley CM, Jones RL, Rogers JC (1996) Plant cells contain two functionally distinct vacuolar compartments. Cell 85:563–572

Peng Y, Lin W, Cai W, Arora R (2007) Overexpression of a *Panax ginseng* tonoplast aquaporin alters salt tolerance, drought tolerance and cold acclimation ability in transgenic arabidopsis plants. Planta 226:729–740

Postaire O, Tournaire-Roux C, Grondin A, Boursiac Y, Morillon R, Schaeffner AR, Maurel C (2010) A PIP1 aquaporin contributes to hydrostatic pressure-induced water transport in both the root and rosette of arabidopsis. Plant Physiol 152:1418–1430

Pou A, Medrano H, Flexas J, Tyerman SD (2013) A putative role for TIP and PIP aquaporins in dynamics of leaf hydraulic and stomatal conductances in grapevine under water stress and re-watering. Plant Cell Environ 36:828–843

Prado K, Boursiac Y, Tournaire-Roux C, Monneuse JM, Postaire O, Da Ines O, Schaffner AR, Hem S, Santoni V, Maurel C (2013) Regulation of arabidopsis leaf hydraulics involves light-dependent phosphorylation of aquaporins in veins. Plant Cell 25:1029–1039

Sack L, Holbrook NM (2006) Leaf hydraulics. Annu Rev Plant Biol 57:361

Sade N, Moshelion M (2014) The dynamic isohydric-anisohydric behavior of plants upon fruit development: taking a risk for the next generation. Tree Physiol. doi:10.1093/treephys/tpu070

Sade N, Vinocur BJ, Diber A, Shatil A, Ronen G, Nissan H, Wallach R, Karchi H, Moshelion M (2009) Improving plant stress tolerance and yield production: is the tonoplast aquaporin SlTIP2;2 a key to isohydric to anisohydric conversion? New Phytol 181:651–661

Sade N, Gebretsadik M, Seligmann R, Schwartz A, Wallach R, Moshelion M (2010) The role of tobacco aquaporin1 in improving water use efficiency, hydraulic conductivity, and yield production under salt stress. Plant Physiol 152:245–254

Sade N, Gebremedhin A, Moshelion M (2012) Risk-taking plants: anisohydric behavior as a stress-resistance trait. Plant Signal Behav 7:767–770

Sade D, Sade N, Shriki O, Lerner S, Gebremedhin A, Karavani A, Brotman Y, Osorio S, Fernie AR, Willmitzer L, Czosnek H, Moshelion M (2014a) Water balance, hormone homeostasis, and sugar signaling are all involved in tomato resistance to tomato yellow leaf curl virus. Plant Physiol 165:1684–1697

Sade N, Galle A, Flexas J, Lerner S, Peleg G, Yaaran A, Moshelion M (2014b) Differential tissue-specific expression of NtAQP1 in *Arabidopsis thaliana* reveals a role for this protein in stomatal and mesophyll conductance of CO2 under standard and salt-stress conditions. Planta 239:357–366

Sade N, Shatil-Cohen A, Attia Z, Maurel C, Boursiac Y, Kelly G, Granot D, Yaaran A, Lerner S, Moshelion M (2014c) The role of plasma membrane aquaporins in regulating the bundle sheath-mesophyll continuum and leaf hydraulics. Plant Physiol 166:1609

Sade D, Shriki O, Cuadros-Inostroza A, Tohge T, Semel Y, Haviv Y, Willmitzer L, Fernie AR, Czosnek H, Brotman Y (2015a) Comparative metabolomics and transcriptomics of plant response to tomato yellow leaf curl virus infection in resistant and susceptible tomato cultivars. Metabolomics 11:81–97

Sade N, Shatil-Cohen A, Moshelion M (2015b) Bundle-sheath aquaporins play a role in controlling arabidopsis leaf hydraulic conductivity. Plant Signal Behav 10:e1017177–e1017177

Sakr S, Alves G, Morillon RL, Maurel K, Decourteix M, Guilliot A, Fleurat-Lessard P, Julien JL, Chrispeels MJ (2003) Plasma membrane aquaporins are involved in winter embolism recovery in walnut tree. Plant Physiol 133:630

Schnurbusch T, Hayes J, Hrmova M, Baumann U, Ramesh SA, Tyerman SD, Langridge P, Sutton T (2010) Boron toxicity tolerance in barley through reduced expression of the multifunctional aquaporin HvNIP2;1. Plant Physiol 153:1706–1715

Secchi F, Zwieniecki MA (2010) Patterns of PIP gene expression in *Populus trichocarpa* during recovery from xylem embolism suggest a major role for the PIP1 aquaporin subfamily as moderators of refilling process. Plant Cell Environ 33:1285–1297

Secchi F, Zwieniecki MA (2014) Down-regulation of plasma intrinsic protein1 aquaporin in poplar trees is detrimental to recovery from embolism. Plant Physiol 164:1789–1799

Shatil-Cohen A, Moshelion M (2012) Smart pipes: the bundle sheath role as xylem-mesophyll barrier. Plant Signal Behav 7:1088–1091

Shatil-Cohen A, Attia Z, Moshelion M (2011) Bundle-sheath cell regulation of xylem-mesophyll water transport via aquaporins under drought stress: a target of xylem-borne ABA? Plant J 67:72–80

Siefritz F, Tyree MT, Lovisolo C, Schubert A, Kaldenhoff R (2002) PIP1 plasma membrane aquaporins in tobacco: from cellular effects to function in plants. Plant Cell 14:869–876

Siemens JA, Zwiazek JJ (2003) Effects of water deficit stress and recovery on the root water relations of trembling aspen (*Populus tremuloides*) seedlings. Plant Sci 165:113–120

Siemens JA, Zwiazek JJ (2004) Changes in root water flow properties of solution culture-grown trembling aspen (*Populus tremuloides*) seedlings under different intensities of water-deficit stress. Physiol Plant 121:44–49

Singh DK, Sale PWG (2000) Growth and potential conductivity of white clover roots in dry soil with increasing phosphorus supply and defoliation frequency. Agron J 92:868–874

Steinhorst L, Kudla J (2013) Calcium and reactive oxygen species rule the waves of signaling. Plant Physiol 163:471–485

Steudle E (2000) Water uptake by roots: effects of water deficit. J Exp Bot 51:1531–1542

Steudle E, Peterson CA (1998) How does water get through roots? J Exp Bot 49:775–788

Suga S, Komatsu S, Maeshima M (2002) Aquaporin isoforms responsive to salt and water stresses and phytohormones in radish seedlings. Plant Cell Physiol 43:1229–1237

Sutka M, Alleva K, Parisi M, Amodeo G (2005) Tonoplast vesicles of *Beta vulgaris* storage root show functional aquaporins regulated by protons. Biol Cell 97:837–846

Takano J, Wada M, Ludewig U, Schaaf G, von Wiren N, Fujiwara T (2006) The arabidopsis major intrinsic protein NIP5;1 is essential for efficient boron uptake and plant development under boron limitation. Plant Cell 18:1498–1509

Tanaka M, Wallace IS, Takano J, Roberts DM, Fujiwara T (2008) NIP6;1 is a boric acid channel for preferential transport of boron to growing shoot tissues in arabidopsis. Plant Cell 20:2860–2875

Tardieu F, Simonneau T (1998) Variability among species of stomatal control under fluctuating soil water status and evaporative demand: modelling isohydric and anisohydric behaviours. J Exp Bot 49:419–432

Tournaire-Roux C, Sutka M, Javot H, Gout E, Gerbeau P, Luu D-T, Bligny R, Maurel C (2003) Cytosolic pH regulates root water transport during anoxic stress through gating of aquaporins. Nature 425:393–397

Vandeleur RK, Mayo G, Shelden MC, Gilliham M, Kaiser BN, Tyerman SD (2009) The role of plasma membrane intrinsic protein aquaporins in water transport through roots: diurnal and drought stress responses reveal different strategies between isohydric and anisohydric cultivars of grapevine. Plant Physiol 149:445–460

Vandeleur RK, Sullivan W, Athman A, Jordans C, Gilliham M, Kaiser BN, Tyerman SD (2014) Rapid shoot-to-root signalling regulates root hydraulic conductance via aquaporins. Plant Cell Environ 37:520–538

Vera-Estrella R (2004) Novel regulation of aquaporins during osmotic stress. Plant Physiol 135:2318–2329

Wang X, Li Y, Ji W, Bai X, Cai H, Zhu D, Sun X-L, Chen L-J, Zhu Y-M (2011) A novel *Glycine soja* tonoplast intrinsic protein gene responds to abiotic stress and depresses salt and dehydration tolerance in transgenic *Arabidopsis thaliana*. J Plant Physiol 168:1241–1248

Wang L-L, Chen A-P, Zhong N-Q, Liu N, Wu X-M, Wang F, Yang C-L, Romero MF, Xia G-X (2014) The *Thellungiella salsuginea* tonoplast aquaporin TsTIP1;2 functions in protection against multiple abiotic stresses. Plant Cell Physiol 55:148–161

Weig AR, Jakob C (2000) Functional identification of the glycerol permease activity of *Arabidopsis thaliana* NLM1 and NLM2 proteins by heterologous expression in *Saccharomyces cerevisiae*. FEBS Lett 481:293–298

Wudick MM, Doan-Trung L, Tournaire-Roux C, Sakamoto W, Maurel C (2014) Vegetative and sperm cell-specific aquaporins of arabidopsis highlight the vacuolar equipment of pollen and contribute to plant reproduction. Plant Physiol 164:1697–1706

Xin S, Yu G, Sun L, Qiang X, Xu N, Cheng X (2014) Expression of tomato SlTIP2;2 enhances the tolerance to salt stress in the transgenic arabidopsis and interacts with target proteins. J Plant Res 127:695–708

Xu C, Wang M, Zhou L, Quan T, Xia G (2013) Heterologous expression of the wheat aquaporin gene TaTIP2;2 compromises the abiotic stress tolerance of *Arabidopsis thaliana*. Plos ONE 8:e79618

Xu Y, Hu W, Liu J, Zhang J, Jia C, Miao H, Xu B, Jin Z (2014) A banana aquaporin gene, MaPIP1;1, is involved in tolerance to drought and salt stresses. Bmc Plant Biol 14:59

Xu W, Dai W, Yan H, Li S, Shen H, Chen Y, Xu H, Sun Y, He Z, Ma M (2015) Arabidopsis NIP3;1 plays an important role in arsenic uptake and root-to-shoot translocation under arsenite stress conditions. Mol Plant 8:722–733

Yu QJ, Hu YL, Li JF, Wu Q, Lin ZP (2005) Sense and antisense expression of plasma membrane aquaporin BnPIP1 from *Brassica napus* in tobacco and its effects on plant drought resistance. Plant Sci 169:647–656

Zhao XQ, Mitani N, Yamaji N, Shen RF, Ma JF (2010) Involvement of silicon influx transporter OsNIP2;1 in selenite uptake in rice. Plant Physiol 153:1871–1877

Zhou L, Wang C, Liu R, Han Q, Vandeleur RK, Du J, Tyerman S, Shou H (2014) Constitutive overexpression of soybean plasma membrane intrinsic protein GmPIP1;6 confers salt tolerance. BMC Plant Biol 14:181

Zhu CF, Schraut D, Hartung W, Schaffner AR (2005) Differential responses of maize MIP genes to salt stress and ABA. J Exp Bot 56:2971–2981

Root Hydraulic and Aquaporin Responses to N Availability

Stephen D. Tyerman*, **Jonathan A. Wignes***, **and Brent N. Kaiser**

Abstract Nitrogen (N) uptake in most plants is positively correlated to water flow in and through roots. This allows transpiration to drive convection of mobile N sources in the soil to the root surface. Generally N starvation suppresses root hydraulic conductivity (Lp_r) while resupply stimulates. However, ammonium and nitrate can give different responses depending on the species, and this may be associated with the form of N that different species prefer to transport and assimilate. Responses in Lp_r in the short to medium term are largely explained by aquaporin regulation at both the transcript and post-translational level. Local and systemic signalling is indicated in the regulation of aquaporins. A direct role for NO_3 sensing has been shown as well as a role of a high-affinity NO_3^- transporter (NRT2.1), but the mode of action in post-translational modification is not known. Transcripts of the aquaporins are also altered by N treatments, yet the regulations of these transcripts and the function of the proteins in N transport remain unclear. Further research is required to uncover the signalling and regulation of aquaporins in relation to N transport since modifications to this process may improve N use efficiency.

1 Introduction

Nitrogen (N) is essential for plant survival, being vital for cell chemistry and processes, and is the macronutrient taken up in the greatest quantity (80 % of total ion uptake) with the greatest energy requirement for transport and assimilation (Marschner 1995). While 78 % of the atmosphere is nitrogen gas, it is unavailable

*Equal contribution

S.D. Tyerman (✉) • J.A. Wignes
Australian Research Council Centre of Excellence in Plant Energy Biology, Department of Plant Science, School of Agriculture Food and Wine, The University of Adelaide, Glen Osmond, SA 5064, Australia
e-mail: steve.tyerman@adelaide.edu.au

B.N. Kaiser
Centre for Carbon Water and Food, University of Sydney, Camden, NSW 2570, Australia

F. Chaumont, S.D. Tyerman (eds.), *Plant Aquaporins*, Signaling and Communication in Plants, DOI 10.1007/978-3-319-49395-4_10

for direct use by most plants except those that enter into symbiosis with N_2-fixing bacteria, where nitrogen gas (N_2) is converted to inorganic ammonium (NH_4^+). Once reduced, NH_4^+ can be assimilated into organic (amino acids, organic molecules) and inorganic (nitrate NO_3^-) forms by plants and soil microbes, respectively. These forms of N often present within the root rhizosphere have large effects on plant growth and development (Miller and Cramer 2004). In this chapter, the root hydraulic responses to N concentration and form across many species are analysed and discussed in terms of the roles played by aquaporins in this process. The possible sensors involved in NO_3^- detection and signal transduction to regulate aquaporins will be examined, and we will attempt to integrate the various signals and responses that have been shown to regulate N transport and aquaporins.

2 Forms of N and the Water Connection

The two major forms of available N in the soil are NO_3^- and NH_4^+, the availability of which depends on soil structure and composition, microbial activity, temperature and water (Glass 2003; Warren 2009). Under conditions of poor aeration, low temperatures and/or acidity, nitrifying bacteria become inactivated and the less mobile NH_4^+ increases relative to that of the more mobile NO_3^- (Marschner 1995; Laanbroek 1990). In contrast, in well-aerated pH-neutral soils, activity of nitrifying bacteria ensures NO_3^- is the predominant form of available N (Marschner 1995). Most plants have a preference for NO_3^- despite the overall energy costs of NO_3^- assimilation relative to that of NH_4^+ being higher (5–12 %) taking into account pH regulation (Raven 1985). The water costs per mole of N assimilated are also higher for NO_3^- by up to 12 % depending on where the NO_3^- is reduced and how pH is regulated (Raven 1985). There is a strong interaction between N uptake and water content of the soil, since this can have diverse effects on N availability through its effect on the degree of aeration and in the dissolution and transport of available N to the root surface (discussed below) (Moyano et al. 2013; Gonzalez-Dugo et al. 2010).

2.1 Soil N Availability and Mass Flow

N availability has interacting effects on many physiological and developmental processes in plants including water relations, and there is cross talk in the regulation of N and water relations (Easlon and Bloom 2013). The link between N and water movement is also important in biogeochemical models that influence predictions from global climate models (Simunek and Hopmans 2009; Seneviratne et al. 2010). Optimisation of water and NO_3^- uptake by root structure and architecture are also closely linked (Lynch 2013). In addition to supplying water for growth and nutrient movement in the xylem, water flow through plants has a role in nutrient acquisition and delivery from the soil.

Mass flow in the soil has been long observed (Barber 1962) and is the process by which water flow to the root brings soluble and mobile molecules to the root surface. In the absence of mass flow or for nutrients that are poorly mobile in soil (e.g. inorganic phosphate), uptake of nutrients by the root cells reduces the surface concentration and creates a localised diffusion gradient to the root so that the rate of uptake may be limited by the rate of diffusion to the root surface (or hyphae of mycorrhizae) (Nye 1977). Mass flow of water during transpiration can convey (via convection) dissolved and mobile nutrients to the surface of the root (Cramer et al. 2008; Chapman et al. 2012) thus reducing or possibly reversing the concentration gradient and assisting the membrane transport systems by increasing the external concentration at the sites of uptake. The degree of "assistance" will depend on the flow rate to the root surface; concentration kinetics of the transport systems, i.e. high, low or dual affinity (HATS, LATS, dual); and the type of transport systems engaged (i.e. AMTs and AMFs for NH_4^+ (Ludewig et al. 2007; Chiasson et al. 2014) or NRT2 or NPF (NRT1) for NO_3^- (Tsay et al. 2007)), which in turn depends on the preference for different forms of N by the plant and the relative availability of these N forms in the soil solution. NH_4^+ and NO_3^- uptake kinetics are complicated since there are also inducible components, e.g. iHATS. Most of the components recognised from physiological studies can be accounted for by specific transporters in *Arabidopsis*, e.g. AtNRT2.1, AtNRT2.2, AtNRT2.4 and AtNRT2.5 as HATs NO_3^- transporters (Li et al. 2007; O'Brien et al. 2016) and AtNRT1.1 and AtNRT1.2 as LATS NO_3^- transporters (Léran et al. 2014). The NRT2s interact with a nitrate assimilation- related protein (AtNAR2.1) to be active (Kotur et al. 2012).

A whole plant model for the acquisition of NO_3^- by mass flow has been proposed (Cramer et al. 2009b). By increasing total water flow through the plant when NO_3^- is sensed, more NO_3^- may be brought to the root surface. NO_3^- reduction to NH_4^+ is shifted from the root to the shoot as NO_3^- acquisition rises and closes stomata as it is reduced, providing a feedback mechanism to slow water uptake when NO_3^- levels are sufficient. NO_3^- at concentrations above 2 mM in the xylem reduces stomatal conductance in maize in an ABA-dependent manner (Wilkinson et al. 2007). The Cramer et al. (2009b) model necessitates a root NO_3^- sensor and a hydraulic response triggered by NO_3^-, but not by NH_4^+.

2.2 *Transpiration and Stomatal Control*

Transpiration is considered to be an essential driver of various forms of N uptake by roots by increasing both mass flow and diffusive fluxes of N to the root surface (Oyewole et al. 2014). Stomatal control of transpiration is regulated by N availability (Matimati et al. 2014; Cramer et al. 2008), and NO_3^- in the xylem in concert with ABA and xylem pH is a signal from roots that regulates stomatal conductance (Wilkinson et al. 2007). By inference from recent studies, stomatal regulation in response to NO_3^- is likely to involve internal hydraulic conductivities in the leaf bundle sheath regulated by plasma membrane intrinsic proteins (PIP) aquaporins

(Shatil-Cohen et al. 2011) as well as particular PIPs in the guard cell membrane (Wang et al. 2016; Grondin et al. 2015). In turn it is likely that transpiration itself regulates root hydraulic conductivity via root aquaporins (Vandeleur et al. 2014; Laur and Hacke 2013; Tardieu et al. 2015) thus providing a rather complex feedback regulation in concert with N availability. These factors add to the plethora of factors that must be integrated by the plant in regulating transpiration and nutrient uptake (Raven 2008).

3 Linking N and Water Movement: Basic Root Characteristics and Environmental Effects

3.1 Measurement of Root Hydraulic Properties

Root hydraulic properties can be measured using several methods, and these are important in the interpretation of the data obtained and the links with N transport. They apply to different root dimensions, tissue scales and rapidity of measurement at the level of single cells; root segments; whole roots in soil, intact or excised; and direction of flow. Each has advantages and disadvantages in interpretation of physiological significance. The water carrying capacity of the roots is generally measured as the hydraulic conductance variously normalised to account for root size (Lo) or hydraulic conductivity (Lp) in the strict sense when normalised to root surface area or root length. In this chapter we will use the term root hydraulic conductivity (Lp_r) rather loosely to indicate that conductance has been normalised to some component of root size.

Total flow of water through plants can be determined by observing the loss of water over time. This method is useful, since it avoids any wounding effects, but is unable to isolate root versus shoot responses or the effects of hydraulic conductance changes in different parts of the plant versus changes in water potential gradients unless these are measured (e.g. Franks et al. 2007). Isolating the effect of the roots can be done by excising the shoot and observing only the root system. This opens the xylem to the atmosphere and may interfere with signals from the shoot (Vandeleur et al. 2014). Simple observations are done by determining the rate and volume of sap exuded over time at atmospheric or applied suction to the root system (e.g. Hachez et al. 2012). Hydraulic conductivity and solute effects are determined from measurement of the osmotic concentration of the exuded sap and that of the nutrient solution supplied. Hydraulic conductivity can also be determined by observing the rate of solution flow through the root at varying applied pressures to the whole root system forcing water to flow in the direction of the shoot (e.g. Gorska et al. 2008a). Alternatively pressure variations can be applied to the excised root system (high-pressure flowmeter, HPFM (Tyree et al. 1995)) or single root segments (root pressure probe, e.g. Lee et al. 2004b)), and from relaxation kinetics or steady-state volume flow, the hydraulic conductance can be determined. These do not always give the same answer when compared and depending on driving force (osmotic or hydrostatic) (Bramley et al. 2007b). Some techniques can give conductances very rapidly after root system excision, for example, the transient method using the HPFM, and although water flow is in the reverse direction to normal transpirational flow, the values obtained correlate well with measurements of flow in the normal direction (Tsuda and Tyree 2000). While many have used excised root systems or roots for measurement of hydraulic properties, it has been reported that shoot excision itself causes a decrease in hydraulic conductivity of grapevine, soybean and maize (Vandeleur et al. 2014). In soybean the reduction after excision is large and rapid (half-time $\approx$ 5 min).

Root hydraulics can also be observed on a cellular basis using the cell pressure probe (Tomos and Leigh 1999). The hydraulic conductivity of individual plant cells can be determined from the elasticity of the cell wall and the half-time of water flow from pressure pulse relaxations after the cell volume is rapidly changed (Hüsken et al. 1978). This is useful to attempt to reconstruct the components of radial flow across the root and to determine potential cells that could govern whole root hydraulic conductance (Bramley et al. 2009). It also allows a more direct link between root hydraulic conductance and the molecular components that determine cell membrane water permeability, i.e. the aquaporins.

3.2 Root Components That Influence Water Movement

Like N and other inorganic nutrients, water acquisition occurs via the root in nearly all plants. In the root, water is absorbed from the soil by flowing down its free energy gradient, the components of which in the plant can be modified through changing solute concentration and the rate of transpiration (Kramer and Boyer 1995). The free energy gradient (water potential gradient) for water to flow to roots through the soil, radially across roots, and then up to the shoot via the xylem, is generated by hydrostatic pressure and osmotic gradients, osmotic gradients only being effective when across a semipermeable membrane. Osmotic gradients may dominate when there is no transpiration, while pressure gradients established in the soil-root-xylem via the cohesion tension mechanism dominate during transpiration (Steudle 2001; Sperry et al. 2002). Given the relatively large energy requirements for the uptake of NO_3^- compared to other nutrients (Cannell and Thornley 2000; Poorter et al. 1991), it is not surprising that there may have been selection pressure to evolve mechanisms to assist the process through coupling with the energetically downhill flow of water in the soil-plant-atmosphere continuum.

Plants can modify the rate that water flows down the water potential gradient via the hydraulic conductivity in the flow pathways. In roots, water can enter the root through the apoplast and a parallel cell-to-cell pathway to reach the xylem (Steudle and Peterson 1998). In many systems, barriers interrupt the apoplastic water flow, forcing water to use the cell-to-cell pathway (Steudle 2001). The *sgn3* mutant of *Arabidopsis* that has compromised development of the Casparian strip due to the lack of deposition of lignin has increased Lp_r but only shows signs of compromised nutrient transport for potassium (Pfister et al. 2014). Suberin is deposited later in development to surround the endodermal cells in *Arabidopsis*, and its deposition is dependent (reversibly) on nutrient and ionic stress (Barberon et al. 2016). This is mediated by ethylene (reduces suberisation) and ABA (increases suberisation), but of the nutrients examined by Barberon et al. (2016), NO_3^- or NH_4^+ were not included and should be examined in the future. Root ABA and ethylene are reported to be involved in N signalling (Kudoyarova et al. 2015; Ondzighi-Assoume et al. 2016; O'Brien et al. 2016), so it is possible that N nutrition could regulate suberisation of the endodermis and hence influence both water and N transport into the pericycle or the movement of a diffusible chemical signal from the pericycle out to the cortex and epidermis.

3.3 Aquaporins

Membrane water permeability of roots is modulated by the activity of aquaporins, certainly in the plasma membrane of certain cell layers, but probably also in the tonoplast, such that, depending on species, more than 50 % of water flow through roots may occur via aquaporins as determined by inhibitor effects (Chaumont and Tyerman 2014; Maurel et al. 2015) (see also chapter "Aquaporins and Root Water Uptake"). Higher plants have many aquaporin isoforms that can be classified into five main types (see also chapter "Structural Basis of the Permeation Function of Plant Aquaporins"). The plasma membrane intrinsic proteins (PIPs) will be largely discussed in this chapter related to root water movement, but tonoplast intrinsic proteins (TIPS) and nodulin intrinsic proteins (NIPs, see also chapter "Plant Aquaporins and Metalloids") are also relevant to interactions with N transport.

The PIPs are regulated on many levels (transcription, protein amount, protein location and by gating (Boursiac et al. 2008; Chaumont and Tyerman 2014; Maurel et al. 2015) (see also chapters "Structural Basis of the Permeation Function of Plant Aquaporins", "Heteromerization of Plant Aquaporins", "Plant Aquaporin Trafficking" and "Plant Aquaporin Posttranslational Regulation"), and this can account for considerable variation in hydraulic conductivity, sometimes by an order of magnitude in roots depending on species (Vandeleur et al. 2014). In maize roots, aquaporin expression has been seen to change diurnally and circadially (Lopez et al. 2003; Caldeira et al. 2014), in different developmental stages (Hachez et al. 2006) and during water stress (Hachez et al. 2012). Aquaporin protein activity can be modified in several ways including protein-protein interactions (Fetter et al. 2004; Zelazny et al. 2007), pH gating (Hedfalk et al. 2006; Tornroth-Horsefield et al. 2006; Nyblom et al. 2009), blocking with certain metals (Niemietz and Tyerman 2002; Verdoucq et al. 2008), ubiquitination (Lee et al. 2009), amidation (Pietro et al. 2013), phosphorylation (Van Wilder et al. 2008; Grondin et al. 2015) and membrane trafficking (Chevalier and Chaumont 2015) (see also chapters "Structural Basis of the Permeation Function of Plant Aquaporins", "Heteromerization of Plant Aquaporins", "Plant Aquaporin

Trafficking" and "Plant Aquaporin Posttranslational Regulation"). Any one or a combination of these including different combinations of isoforms could be involved in regulating water flow through roots to influence N uptake.

Apart from water, certain aquaporins also facilitate passive transport of small solutes, including N-containing molecules. Some members of the *Arabidopsis* TIP family have been shown to be permeable to urea in yeast complementation assays, and their expression can increase during N starvation (Liu et al. 2003). Other *Arabidopsis* TIP members, maize NIP and TIP, and a cucumber NIP, have been found to complement yeast for growth on urea (Klebl et al. 2003; Gu et al. 2012). ZmPIP1;5-b is induced with NO_3^- application and has been suggested to be water and urea permeable when expressed in *Xenopus* oocytes for 5 days (Gaspar et al. 2003). Wheat TaTIP2;1 and TaTIP2;2 have NH_4^+ permeability in yeast or *Xenopus* (Jahn et al. 2004; Holm et al. 2005; Bertl and Kaldenhoff 2007), and AtTIP2;1 protein structure indicates a unique side pore through which protons can be extracted to deprotonate NH_4^+ and allow NH_3 to permeate (Kirscht et al. 2016). Currently no plant aquaporins have been found to be NO_3^- permeable, but a NO_3^- permeable mammalian aquaporin is known – AQP6 (Yasui et al. 1999; Hazama et al. 2002; Liu et al. 2005).

3.4 Diurnal Variations

Root hydraulic conductivity can change diurnally with higher conductivity observed in the daytime, which has been correlated with expression patterns of PIP aquaporins (Henzler et al. 1999). This has been shown for a number of species (Vandeleur et al. 2009; Almeida-Rodriguez et al. 2011; Hachez et al. 2008) and in some cases related to circadian oscillation in aquaporin activity or expression (Caldeira et al. 2014; Takase et al. 2011; Lopez et al. 2003) and additionally via a transpiration signal from the shoots to the root (Sakurai-Ishikawa et al. 2011; Almeida-Rodriguez et al. 2011; Vandeleur et al. 2014). Though contrary to this, transition from shade to light for several tropical tree species over a long period caused a reduction in Lp_r despite increased transpirational demand (Shimizu et al. 2005). This was correlated with a reduced inhibition by mercury indicating that root aquaporins had a diminished role after the transition to light and further growth.

It is not surprising that NO_3^- or NH_4^+ transport by roots also varies diurnally with higher rates observed in the day or light period (e.g. in soybean (Delhon et al. 1995) and citrus (Camanes et al. 2007)). In citrus the expression of a high-affinity NH_4^+ transporter in the root is regulated by sugar, independently of a circadian control (Camanes et al. 2007). In *Arabidopsis* both NH_4^+ and NO_3^- uptake by roots appears to be regulated via sugar from photosynthesis. In addition to the NH_4^+ transporters (*AMT1.1, 1.2, 1.3*), *NRT1.1* (a NO_3^- transceptor, now called NPF6.3 (O'Brien et al. 2016)) and the high-affinity NO_3^- transporter and transceptor AtNRT2.1 (see below) are both sugar (sucrose and glucose) inducible, and deletion of AtNRT2.1 results in the loss of the regulation of root NO_3^- uptake by light and sugar (Lejay et al. 2003).

In rice NRT transcripts are also upregulated by light and sugar (Feng et al. 2011). Aquaporins can be upregulated by a protein kinase sugar sensor in seedlings (see below) (Wu et al. 2013) that could lead to entrainment of root water movement to N transport and shoot photosynthesis in the light. Recently a shoot-to-root mobile bZIP transcription factor activated by light (HY5) that interacts with sugar signalling in *Arabidopsis* increased expression of *AtNRT2.1* in roots (Chen et al. 2016). It is not known if this changes expression of any aquaporins.

3.5 Temperature

Root hydraulic conductivity is also sensitive to temperature. Cucumber roots experience a decline in both root pressure and hydraulic conductivity in response to temperature reductions below 25 °C (Lee et al. 2004a). In temperature-resistant rice (Ahamed et al. 2012) and maize (Aroca et al. 2005), aquaporin activity is linked with maintenance of hydraulic conductivity and recovery. In cucumber roots exposed to lowered temperature, there was a larger reduction in the osmotic component of water flow compared with the hydrostatic component. This was associated with an affect on root aquaporins (specific type not identified), while the reduction in root pressure was associated with reduced activity of the plasma membrane proton pump (Lee et al. 2004b). The kinetics of NH_4^+ uptake across the plasma membrane of rice roots also differs with temperature where the high-affinity transport system (HATS) is more strongly reduced by low temperature than that of the low-affinity transport system (LATS) (Wang et al. 1993). The relative uptake of NH_4^+ and NO_3^- also changes at low temperature due to different temperature effects on NO_3^- and NH_4^+ transport in ryegrass (Clarkson and Warner 1979). The interaction between soil temperature, water movement and N transport is poorly understood (Pregitzer et al. 2000; Kreuzwieser and Gessler 2010), and this is important in the context of global warming and predictions of biogeochemical fluxes (Davidson et al. 2006).

3.6 Anoxia and Hypoxia

Anoxia or hypoxia in the soil as a result of soil water logging also influences water flux through roots, leading to an immediate decrease in water flux and Lp_r (Vandeleur et al. 2005), in a species-dependent way (Bramley et al. 2010), and this again can have ramifications for predications of fluxes at ecosystem scales (Shaw et al. 2013). Anoxia reduced the hydraulic conductivity of *Arabidopsis* roots, which coincided with cytosolic acidification (Tournaire-Roux et al. 2003). This acidification caused protonation of a histidine on loop D of PIP proteins, which led to a conformational change that reduced water movement. Anoxia greatly reduced the water flow of NO_3^- fed tomato, cucumber (Gorska et al. 2008a), maize (Gorska et al. 2008b) and sunflower indicating that aquaporins are important in hydraulic responses to N in some species. Increase in reactive oxygen species under hypoxia (van Dongen and Licausi 2015; Shabala et al. 2014) may also inhibit aquaporin activity (Wudick et al. 2015). Several aquaporin transcripts are increased under anoxia/hypoxia (Bramley et al. 2007a) that may be linked to tolerance mechanisms related to gas transport (CO_2 or O_2) or signalling (Bramley and Tyerman 2010) or release of a fermentation product (Choi and Roberts 2007).

Despite being passive transporters, it is interesting to note how dependent aquaporin function is on cell energy status as indicated by one of the best inhibitors of aquaporins, azide (Postaire et al. 2010; Grondin et al. 2016), which inhibits mitochondrial respiration (Hodges and Elzam 1967). Since the uptake of NH_4^+ and NO_3^- are highly energy-dependent processes requiring a proton electrochemical gradient established by the H^+-ATPase to drive their influx, it is not surprising that anoxia and soil water logging have profound effects on N uptake, e.g. wheat (Herzog et al. 2016; Robertson et al. 2009), probably due to collapse of pH gradients (Felle 2005). There are however differences between species in N uptake under flooding and in the amount of NH_4^+ versus NO_3^- taken up (Kreuzwieser et al. 2002) partly related to NO_3^- becoming less available relative to NH_4^+ in water logged soils due to decreased nitrification (Laanbroek 1990). An acidification of root cell cytoplasm, that can result from anoxia (Felle 2005), could increase NO_3^- efflux via the AtNAXT NO_3^- transporter (Segonzac et al. 2007) exacerbating the reduced availability in the soil. NO_3^- is particularly interesting in relation to the integration of water and N fluxes because of its stimulatory effect on root water movement (examined below) (Gloser et al. 2007) via aquaporins (Gorska et al. 2008b) and because of the ameliorative effects of NO_3^- and nitrite on cytosolic pH regulation under anoxia (Libourel et al. 2006). Recently it was shown that the *Arabidopsis* NRT1.1 is required for acid tolerance, and when knocked out, the roots were unable to increase the pH of the external medium when acidic. This was linked to NO_3^- uptake rather than NO_3^- sensing (Fang et al. 2016).

3.7 *Energy and Nutrient Signalling*

Energy and nutrient sensing in plant cells is also an area of potential overlap between NO_3^- transport and assimilation and aquaporin activity, especially since PIP aquaporins seem to be energy dependent. The TOR (target of rapamycin) and SnRK1 (sucrose non-fermenting-1 (SNF1)-related kinase 1) signalling pathways interact to regulate metabolism and development in response to energy supply and nutrition (Robaglia et al. 2012). SnRK1 is a central energy sensor important for metabolic homeostasis in plants (Baena-Gonzalez et al. 2007; Baena-González and Sheen 2008) and is partly controlled via trehalose-6-phosphate sensing among other signals (Jossier et al. 2009; Emanuelle et al. 2016). It is activated under energy-deficient conditions, e.g. by hypoxia and carbon starvation (Cho et al. 2016; O'Hara et al. 2013), and inhibits NO_3^- reductase activity via phosphorylation (Sugden et al. 1999). Out of the 1,021 genes transcripts reported to be upregulated by SnRK1.1 in *Arabidopsis*, only 16 are channels or transporters, and three of these are aquaporins (AtTIP2;1, AtTIP1;3 and AtPIP1;3) (Baena-Gonzalez et al. 2007). AtTIP2;1 is of special interest because of its ability to facilitate the diffusion of NH_4^+ directly or indirectly as NH_3 (Holm et al. 2005; Kirscht et al. 2016) and its correlation to stomatal conductance (Pou et al. 2013). The TOR protein kinase on the other hand is a regulatory hub involved in nutrient sensing (Dobrenel et al. 2016). The TOR pathway is proposed to increase N assimilation under energy and nutrient-replete conditions (Dobrenel et al. 2016; Lillo et al. 2014). Protein phosphatase 2A is a component of the TOR pathway via the regulatory subunit (Tap46). When Tap46 is overexpressed in *Arabidopsis*, this results in growth stimulation, increased NO_3^- reductase activity and induction of transcription of genes involved in N transport including AtNRT1.1 (AtNPF6.3) and AtNRT2.1 (Ahn et al. 2015). Both are components of the NO_3^- transceptor signalling cascade, which regulates the primary nitrate response in plant roots (O'Brien et al. 2016) with AtNRT2.1 proposed to have a role in influencing root aquaporin activity in response to NO_3^- (see below) (Li et al. 2016).

Sugar, and hence energy supply, also regulates aquaporins in *Arabidopsis* via a sucrose-activated receptor kinase (SIRK1) in whole seedling plasma membrane, able to phosphorylate aquaporins and thereby positively affecting their activity (Wu et al. 2013). Several PIP aquaporins were shown to interact with SIRK1 including those that have been shown to be associated with Lp_r (e.g. AtPIP2;1) (Wudick et al. 2015; Sutka et al. 2011). Sugar signalling in plants is linked to C:N balance and root/shoot coordination with cross talk with hormone signalling (Wang and Ruan 2016). Altogether these data suggest that aquaporins could be added as an important component of nutrient signalling pathways, but location in roots needs to be determined for SIRK1.

3.8 Co-location of Water and N Uptake

For mass flow to have an effect on NO_3^- transport to the root, it would be expected that the location of NO_3^- influx along the axis of the root or in different root classes should correspond to the location of the maximum volume-flux density of water inflow (Gorska et al. 2008b). It is not a trivial exercise to determine where along the axis of roots or the type of roots where water inflow is maximal. Work with lupin and wheat has indicated contrasting locations for water flow based on the scaling of root hydraulic conductance with root length (Bramley et al. 2009). Wheat tends to have a greater concentration of water flow near the root tip and is more dependent on aquaporins than that for lupins (Bramley et al. 2009). Using neutron radiography to image deuterated water movement in and through roots in soil during transpiration with a model to separate convection from diffusion, Zarebanadkouki et al. (2014) concluded for lupin roots that radial flow into roots was maximal 12–16 cm from the root tip of lateral roots. For maize roots lateral roots show a greater uptake of water relative to that of primary and seminal roots as determined by the same technique (Ahmed et al. 2016). It was concluded that lateral roots were the primary sites of water absorption and that primary and seminal roots functioned to axially transport water to the shoot. Since whole maize roots have been shown to have a positive response of water movement to NO_3^- (Gorska et al. 2008b), it would be expected from the results of Ahmed et al. (2016) that NO_3^- uptake would occur mainly through lateral roots. NO_3^- uptake kinetics measured for different root classes at two plant ages and in response to NO_3^- deprivation showed large differences, with lateral roots giving the highest affinity (lowest Michaelis-Menten constant, K_m) and highest maximum rate of uptake, which depended on plant age (York et al. 2016). Interestingly linear uptake kinetics indicating LATS transport of NO_3^- were not observed for laterals but occurred in all other root classes. Modelling of plant growth based on these kinetics and taking into account mass flow to the roots indicated that the maximum rate of influx was important in determining plant growth rather than the concentration kinetics (York et al. 2016).

Ion sensitive electrodes have been used to scan NH_4^+ and NO_3^- fluxes at finer scale along roots. In young maize roots, NO_3^- influx reached a maximum at 3.5–8.5 mm from the root tip and then remained more or less constant with distance to 21–61 mm (limit of measurements) (Taylor and Bloom 1998). NH_4^+ influx was high at the root tip and increased with distance but was about 1/3 that of NO_3^-. It was concluded that NH_4^+ was most actively taken up at the root tip reflecting the energy advantage over NO_3^-. In fine wheat roots, highest net fluxes of NH_4^+ and NO_3^- occurred at 20–25 mm from the root tip (Zhong et al. 2014), which would correspond to maximum hydraulic conductivity measured by Bramley et al. (2009). In rice, maximum net influx of NH_4^+ and NO_3^- occurred at 10–20 mm from the apex and then declined (Colmer and Bloom 1998). This was different from maize since maize did not show a decline with distance. It was concluded that sclerenchymatous fibres on the outer side of the cortex of older parts of the rice root could impede influx. However, later measurements of water flow across this region indicated that it was largely determined by apoplastic flow (Ranathunge et al. 2004). Examination of NO_3^- uptake kinetics, expression of *ZmNRT2.1* and H^+-ATPase activity along maize primary roots using a compartmental system showed that the root tip (0–4 cm) had higher capacity to take up NO_3^- due to higher maximum influx and more rapid induction of a high-affinity transport system (Sorgona et al. 2011). This did not correlate with the expression of *ZmNRT2.1*, which was highest at 4–6 cm from the root tip. To date there are no studies that we are aware of that have attempted to define water and NO_3^- or NH_4^+ fluxes simultaneously with root position or root type, and this data is required to better understand the link between water and N uptake.

In *Arabidopsis* specific NRT2-HATS transporters are associated with root development, plant age and specific cell types (Kiba and Krapp 2016). AtNRT2.4 and AtNRT2.5 are expressed in a polarised manner in epidermal cells (facing the external medium) with a greater level of expression of AtNRT2.4 in young plants and AtNRT2.5 in mature plants (Lezhneva et al. 2014; Kiba et al. 2012). *AtNRT2.4* is lowly expressed with high NO_3^- and is only increased to levels equivalent to *AtNRT2.1* without NO_3^- in young plants (Kiba et al. 2012). In mature plants

AtNRT2.1 has higher expression than both *AtNRT2.4* and *AtNRT2.5* after 1 day of N starvation but then declines with *AtNRT2.5* and *AtNRT2.4* expression increasing to dominate by 10 days (Lezhneva et al. 2014). Many experiments investigating the link between root water transport and NO_3^- transport initially starve plants of N or have them grown in low NO_3^- for various periods, then resupply with NO_3^- to examine the response in water transport. In this case HATS NRT2s would be induced and may account for a portion of the NO_3^- uptake when NO_3^- is introduced to the roots. However, the concentrations of NO_3^- used are generally well above the HATS concentration range and would require NRT1-LATs transporters or anion channels to carry the load of NO_3^- uptake. In roots of *Arabidopsis*, these would be AtNRT1.1, AtNRT1.2 (AtNPF4.6) and AtNAXT1 (AtNPF2.7) (O'Brien et al. 2016) or possibly an anion channel (initially before cytoplasmic NO_3^- increases) (Skerrett and Tyerman 1994). The location of the inducible LATS transporter *AtNRT1.1* mRNA is in the epidermis near the root tip and in the cortex or endodermis in more mature regions (Huang et al. 1996). Likewise the constitutively expressed *AtNRT1:2* (*AtNP4.6*) is located in root hairs and epidermis in mature regions of roots.

The location of AtNRT2.1 as a NO_3^- sensor that regulates PIP activity in *Arabidopsis* (Li et al. 2016) makes sense since mass flow of water to the root would be expected to build NO_3^- concentration in the apoplast of cortex cells due to endodermal barriers blocking apoplastic movement. However, being a high-affinity transporter with saturation occurring around 0.5 mM NO_3^-, one would expect that mass flow under conditions of NO_3^- resupply (often several millimolar NO_3^-) would require a LATS transport or dual-affinity transporter to do most of the influx, i.e. AtNRT1.1 and AtNRT1.2. The locations of these transporters (discussed above) would match with our present view of high water transport intensity, particularly in the endodermis and epidermis.

4 N Form and Amount Affects Root Hydraulic Properties

4.1 Growth and Development

Over the long term, N affects both root growth and development (Wilson 1988; Robinson 1994; Lopez-Bucio et al. 2003; Guo et al. 2007c; Desnos 2008). In *Arabidopsis* local patches of NO_3^- promote lateral root outgrowth via NO_3^--inducible MADS-box gene (ANR1) and coordinated by a systemic signal dependent on the plant's N status (Zhang and Forde 2000). N responses on root growth can affect different properties in different species. Górska et al. (2010) showed that cucumber, tomato, maize, *Arabidopsis* and *Festuca arundinacea* have increased relative root growth on 2 mM NO_3^- compared to 0.2 mM, whereas *Populus trichocarpa* and *Nephrolepis exaltata* do not have a relative root growth response. Tomato, maize, *Arabidopsis* and *Festuca arundinacea* also have increased root/shoot ratios on low NO_3^- nutrition, whereas cucumber, *Populus trichocarpa* and *Nephrolepis exaltata* do not.

The form of N available to roots impacts growth and development as well. French bean grown in NO_3^- as sole N source has larger roots and shoots than when grown on NH_4^+ (Guo et al. 2002). These results were repeated with split root experiments where one half of the root was grown on NO_3^- and the other NH_4^+ (Guo et al. 2007b). This demonstrated a repression of growth of roots grown in NH_4^+. In experiments where half the roots were grown on N (NH_4^+ or NO_3^-) and the other half starved of N, either form resulted in increased growth. Still NO_3^--grown roots had increased growth. In tomato, roots grown in NH_4^+ have stagnated growth over 3 weeks compared to plants grown in NO_3^- (Pill and Lambeth 1977).

Stimulation of lateral root growth in response to NO_3^- is regulated by multiple pathways related to initiation of the lateral root, early development and elongation of lateral root primordia (Forde 2014). There is an interaction between auxin and NO_3^- signalling in this process. Aquaporins also play a role in lateral root outgrowth. Auxin-mediated repression of AtPIP2;1 accounts for reduced polarised water flow into overlying tissue for optimum extension (Peret et al. 2012), while the tonoplast located aquaporins AtTIP1;1, AtTIP1;2 and AtTIP2;1 are also involved in the development of lateral root primordia in a complex manner, each isoform having different spatio-temporal expression patterns (Reinhardt et al. 2016). It is yet unknown if these aquaporins also play similar roles in lateral root outgrowth under NO_3^- stimulation. AtTIP2;1 is particularly interesting given that it complemented the triple *tip1;1/1;2/2;1* knockouts (Reinhardt et al. 2016) and that its expression can be linked to N and energy signalling (Baena-Gonzalez et al. 2007). However, neither *AtTIP2;1* nor *AtPIP2;1* comes up as regulated genes in local NO_3^- stimulation in wild-type *Arabidopsis* or the NO_3^- signalling mutant of NRG2 (Xu et al. 2016), while *AtTIP1;1, 2;2* and *2;3* are subject to changes in expression under local NO_3^- supply (Li et al. 2014; Ruffel et al. 2011).

Maize cultivars show large differences in the response to NO_3^- of lateral root outgrowth both in terms of homogeneous and locally high NO_3^- (Yu et al. 2016). In this study pericycle cell transcriptomes were examined for all root classes and compared between locally high NO_3^- and uniform low NO_3^-. The transcript responses were root-class dependent with brace roots showing the greatest response of some 2,740 NO_3^--regulated genes with unique cell-cycle control genes being strongly represented. Brace roots would be worthy of further physiological studies to examine the water: NO_3^- link.

4.2 Root Hydraulics

Root hydraulic properties are sensitive to N form and concentration and are dependent on species and length of treatment. For many species, an increase in water flow and hydraulic conductivity are observed. For example, in wheat, plants grown on NO_3^- 0.5–16 mM, de-topped root exudation rates are increased in a dose-dependent manner with increasing NO_3^- (Barthes et al. 1996). When starved of NO_3^- for 5 days, wheat plants showed a decrease in hydraulic conductivity and water flow

(Carvajal et al. 1996a). Resupplying NO_3^- to N-starved wheat caused quick increase in flow and hydraulic conductivity, resulting in a return to control values by 24 h (Carvajal et al. 1996b). A rapid stimulation of Lp_r by NO_3^- is not always observed (Górska et al. 2010), for example, in the tree *Eucalyptus grandis*, no root response was observed, though in this species there was a longer-term increase in whole plant hydraulic conductance when gown on high N (Graciano et al. 2016).

It may be that the root cellular concentration of NO_3^- plays an important role in these responses. Tomato and cucumber both increased water flow through roots in response to NO_3^- (Gorska et al. 2008a). This was observed at the cellular level as well using a cell pressure probe to measure the half-time of water flow to applied pressure relaxations. The half-time fell when roots were exposed to NO_3^-, indicating water flowed with smaller resistance in these cells. In the presence of tungstate, a NO_3^- reductase inhibitor, tomato root responses at the cellular and whole root levels were unaffected, but the response of cucumber roots and cells was eliminated. It was shown that tomato roots maintained NO_3^- uptake, but tungstate treatment inhibited uptake of NO_3^- by cucumber roots. Using the cell pressure probe, injecting NO_3^- directly into cucumber root cells treated with NO_3^- reductase recovered the cellular response. In maize, a delay is observed between NO_3^- application and increase in hydraulic flow (Gorska et al. 2008b). Tungstate application to maize roots results in a decreased NO_3^- uptake rate over 4 h and also eliminates the hydraulic response to NO_3^-. Tungstate application greatly inhibited *PIP* expression not seen in starved plants, indicating tungstate may disrupt many cellular pathways (Gorska et al. 2008b). In this case, diurnal patterns in maize root sap flow and osmolality (Lopez et al. 2003) or de-topping effects may have confounded the hydraulic response to increased cellular NO_3^- levels.

Observations over several species show that a species' NO_3^- uptake rate is linearly correlated with the magnitude of the hydraulic response (Górska et al. 2010). In these experiments, species with high NO_3^- uptake rates had large increases in water flow after NO_3^- application, while species with low NO_3^- uptake rates lacked a hydraulic response to NO_3^- (Górska et al. 2010). Cellular NO_3^- levels may have been important in the hydraulic response to NO_3^- in these plants. The link between NO_3^- uptake rate and hydraulic response also supports the mass flow hypothesis for the basic hydraulic response described in Cramer et al. (2009a). Larger NO_3^- uptake rates are quicker to deplete the root zone of NO_3^-, and mass flow could be a mechanism to quickly resupply NO_3^- pools near the root surface when NO_3^- is present (Gorska et al. 2008b).

4.3 NO_3^- Versus NH_4^+ in Regulation of Root Hydraulics

The form of N supplied affects root hydraulic responses and is species dependent. In French bean, roots exposed to NO_3^- (whole root or in split root) have increased water uptake compared to roots exposed to NH_4^+. Water uptake in root systems split between NO_3^- and zero N has more total uptake from the NO_3^- fed roots, but when normalised by root volume, starved roots absorb more water per root volume. Under

NH_4^+ or starvation, roots absorb more water and more water per unit root volume on the starvation side. Root volume and dry mass are larger when in NO_3^- or NH_4^+ compared to starvation and was more pronounced during the day than at night (Guo et al. 2002, 2007a, b).

Ammonium nutrition has mixed results in other species. Like French bean, maize NO_3^- resupply led to an increase in water flow through the root in N starved plants, whereas urea and NH_4^+ had slightly negative effects on water flow (Gorska et al. 2008b). In tomato, application of either NO_3^- , NH_4^+ or $NH_4^+ + NO_3^-$ increased the total exudate from de-topped plants, with NH_4^+ having a slightly larger effect than NO_3^- (Minshall 1964). Clover plants grown in 5 mM NO_3^- had a slight increase in Lp_r compared to 0.5 mM, but no increase was seen with NH_4^+. Contrasting to this, ryegrass had a 300 % increase to either 5 mM NO_3^- or NH_4^+ compared to 0.5 mM (Belastegui-Macadam et al. 2007).

In rice, NO_3^- and NH_4^+ have mixed effects on plant hydraulics. In cultivar "Shanyou63" hybrid *indica*, root dry weights between NO_3^- and NH_4^+ fed plants are not different (Li et al. 2009; Yang et al. 2012), whereas root fresh weight is significantly greater for NO_3^- grown plants (Gao et al. 2010). When normalised to fresh weight, the water uptake was found to be significantly greater for NH_4^+ grown plants in two out of three of the studies (Li et al. 2009; Yang et al. 2012) and that total root uptake of water (not normalised) was greatest for $NH_4^+ + NO_3^-$ fed plants, intermediate for NO_3^- and least with NH_4^+ (Gao et al. 2010). All three of these studies show that for NO_3^- fed plants, xylem water flow and total water uptake (normalised or not) decrease by greater amounts during water stress (PEG). These studies used the same rice cultivar and were treated with different N forms for up to 2 weeks potentially masking the hydraulic effects behind growth changes. In mixed N (NH_4^+ NO_3^-) treatment, lowering N from 20 to 0.2 ppm led to a decrease in xylem sap flow and hydraulic conductivity over 3 days (Ishikawa-Sakurai et al. 2014). Upon resupply, xylem sap flow rate and hydraulic conductivity were raised. Interestingly, the initial response for starvation or resupply was opposite from the final response. The first 6 h of starvation saw an increase in the hydraulic conductivity and xylem flow rate in N-starved plants. At 24 h, rates began to fall compared to plants in full N treatment. Between 0 and 6 h, N-fed plants showed a slight increase in xylem sap flow but a decrease in Lp_r. By the next light period, a large increase in both root xylem sap flow and Lp_r were observed. This highlights that the magnitude of hydraulic response to N is time dependent and suggests that the direction of change may be time dependent as well probably linked to circadian and diurnal regulation.

Further study on rice comparing two cultivars that have different water relations responses to N supply during growth showed a strong increase in Lp_r with increasing N supply (as equal NH_4^+ and NO_3^-) with larger responses observed for Yangdao6 than Shanyou63 (Ren et al. 2015). This difference was reflected in the response of leaf water potential, which became more negative in Shanyou63 and was constant in Yangdao6 with increasing N. It was concluded that the increase in Lp_r was due to a combination of reduced aerenchyma and lignin as well as increased PIP aquaporin expression in roots (see below). The greater response of Yangdao6 was linked to aquaporin expression and more reduced aerenchyma (Ren et al. 2015).

4.4 Transcriptional Responses of Aquaporins in Response to N

Aquaporins undoubtedly play a role in the N responses as indicated above. Transcriptional responses of aquaporins to changes in external N status, if they occur, tend to take days to become evident. Caution must be used, as transcription and protein amount are not always correlated (Hachez et al. 2012). In tomato, increased hydraulic conductivity to N is seen quickly (Gorska et al. 2008a, 2010), but aquaporin gene responses were not seen before 48 h of treatment (Wang et al. 2001). Exposing *Arabidopsis* to NO_3^- from NH_4^+ growth for 20 min only repressed *AtNIP2;1* expression (see also chapter "Plant Aquaporins and Metalloids") but left all other aquaporin levels the same (Wang et al. 2003). In *Arabidopsis* resupply of NO_3^- to N-stressed plants strongly induced a *TIP* member, and several others weakly (Scheible et al. 2004). In maize, NO_3^- addition was not reported to change *PIP* gene expression within 4 h, while tungstate treatment greatly inhibited the expression of most *PIP* genes (Gorska et al. 2008b). Eight hours after NO_3^- addition, *ZmPIP1-5b* transcript level was strongly induced (Gaspar et al. 2003). In French bean, NO_3^--induced *PIP1* expression occurred around 4 days after addition, but no *PIP1* gene induction was seen with NH_4^+ (Guo et al. 2007a). In rice, switching from 10 ppm NH_4^+ + NO_3^- to 0.5 ppm led to a slight repression of *OsPIP1;1*, *OsPIP2;3*, *OsTIP1;1* and *OsTIP2;2* expression. Major reduction in expression was seen for *OsPIP2;4* and *OsPIP2;5* with induction for *OsTIP2;1* and *OsPIP2;6*. During resupply, *OsPIP2;4*, *OsPIP2;5* and *OsTIP2;2* gene transcription was largely induced. Several other genes were slightly induced. This correlated with decreased hydraulic conductivity during starvation and increased during resupply (Ishikawa-Sakurai et al. 2014).

5 Post-translational Regulation of Aquaporins in Response to N

Aquaporin blockers result in reduced hydraulic conductivity under N treatments of wheat (Carvajal et al. 1996b), barley (Ruggiero and Angelino 2007) and figleaf gourd (Rhee et al. 2011). An investigation of the link between Lp_r and aquaporin protein abundance and post-translational responses in *Arabidopsis* revealed that AtPIP2 protein amount correlated with Lp_r responses to various treatments (Pietro et al. 2013). A higher correlation was observed with the PIP phosphorylated peptides. In contrast PIP1 and TIP abundance did not correlate with Lp_r. N starvation tended to have the largest impact on PIP protein abundance. They also observed differences in deamination in PIP2 protein fragments, suggesting a role in post-secondary modification of aquaporins during the response. Largely, however, these pathways, including pH-mediated gating, localization and phosphorylation, have remained unexplored in N regulation of plant hydraulics (Pietro et al. 2013). Other connections in this regard between NO_3^- signalling and factors that regulate aquaporins include elevation of ABA in certain cells (Ondzighi-Assoume et al. 2016) and transient increased cytosolic Ca^{2+} concentration (Riveras et al. 2015) in response to NO_3^-. The Ca^{2+} response was abolished in mutants of NRT1.1 (NPF6.3). Both cytosolic Ca^{2+} and ABA are known to regulate aquaporins (Wan et al. 2004; Alleva et al. 2006).

Global protein phosphorylation patterns have been examined in N-starved *Arabidopsis* seedlings when resupplied with either NO_3^- or NH_4^+ compared with potassium chloride after 30 min (Engelsberger and Schulze 2012). Distinct phosphorylation patterns were observed for proteins that have signalling functions and transporters/channels. Phosphopeptides from aquaporins were identified, and these were rapidly dephosphorylated in response to KCl and NH_4Cl, but KNO_3 produced a weaker effect, and there was a subsequent increase in phosphorylation over time.

Recently a significant advance was made in understanding the regulation by NO_3^- of Lp_r and aquaporins in *Arabidopsis* (Li et al. 2016). Examination of the Lp_r of NO_3^- transporter mutants including the transceptors NRT1.1 (NPF6.3) (dual HATS LATS) and NRT2.1 (iHATS) showed that only the NRT2.1 knockout had reduced Lp_r. Though Lp_r was lowered in the NRT2.1 mutant, it still responded to reduced NO_3^- supply. Overall there was a better correlation between Lp_r and shoot NO_3^- concentration than root NO_3^- across all mutants and WT. Although NRT1.1 mutant was not considered further in the work, it did appear that the correlation between Lp_r and shoot NO_3^- was depressed in this mutant compared to WT. Transcripts of both *PIP1;1, PIP1;2, PIP2;1* and *PIP2;3* correlated with changes in Lp_r for NRT2.1 mutants under different NO_3^- concentrations, and both PIP1 and PIP2 protein abundance were also correlated with Lp_r. In WT, however, the regulation appeared to be mostly at the post-translational level. The conclusion from this study was that root aquaporins were regulated by a shoot-to-root signal communicating shoot NO_3^- status as well as the function of NRT2.1 (Fig. 1). It would be worthwhile examining a split root protocol to distinguish between local and systemic signals to test this model, especially since NO_3^- initially upregulates the level of the *NRT2.1* transcript after starvation and split root experiments indicate that a shoot-to-root signal is involved in transmitting shoot N-demand to changes in *NRT2.1* expression (Kudoyarova et al. 2015).

6 Conclusions and Research Questions

While not uniform, plant hydraulics are modified depending on N form and amount. This involves changes in membrane water permeability mediated by PIP aquaporins as well as root architectural modifications to facilitate N foraging in the soil. Cellular mechanisms of signalling of NO_3^- and transduction to aquaporin activity to regulate Lp_r are emerging and indicate that NO_3^- concentration in cells is a signal rather than the assimilated products. Both local and systemic (shoot-to-root) signalling are involved, but the identity of the systemic signal is unknown. NRT2.1

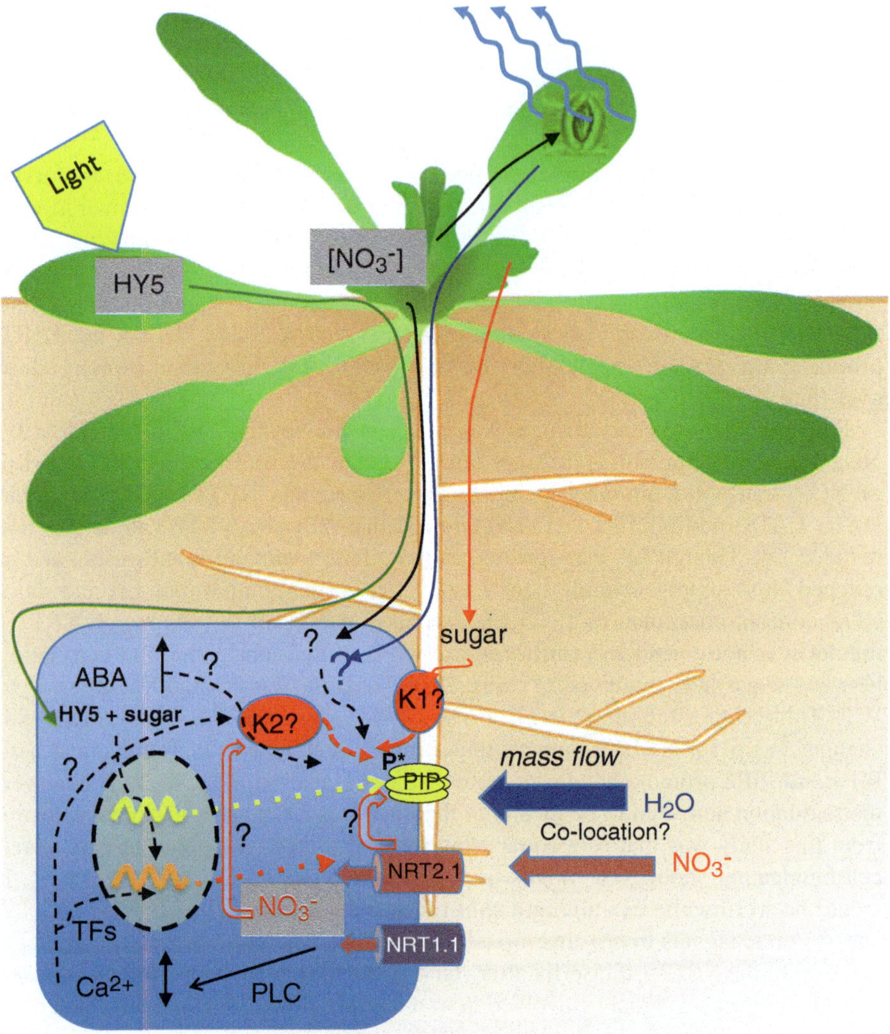

appears to be an important component in the signalling of NO_3^- to aquaporin responses resulting in transcriptional changes in aquaporin genes, but mainly post-translational changes affecting aquaporin activity (Fig. 1). There are parallel regulation paths via translocated sugar from the shoot on both NO_3^- transporters and probably aquaporins. If mass flow of NO_3^- to the root surface is the functional outcome of increased aquaporin activity, then major NO_3^- transporters and aquaporins involved need to be co-located in the root. This co-location need not be in the same cell type but in the relevant root class (i.e. laterals) and similar position along the root axis. Single-cell transcriptomics combined with sensitive techniques to resolve fine spatial and kinetic responses in NO_3^- and NH_4^+ transport and water

Fig. 1 Regulation of root PIP aquaporins involves specific and probably parallel pathways of local and systemic signalling. Local: The presence of NRT2.1 is required for maintaining a high Lp_r in *Arabidopsis* (Li et al. 2016), and while the signalling cascade to PIP aquaporins is unknown, it is likely to be via phosphorylation by an unidentified kinase(s) (denoted K2) (Engelsberger and Schulze 2012). NRT1.1 is also required for nitrate signalling and regulation of expression of NRT2.1 (Bouguyon et al. 2015) (short-term upregulation and long-term downregulation) via a cytosolic Ca^{2+} signal involving phospholipase C (PLC), and various transcription factors are activated (TFs) (O'Brien et al. 2016; Riveras et al. 2015). Although Ca^{2+} is known to directly regulate PIP aquaporins via gating (Verdoucq et al. 2008), the presence of NRT1.1 was not required for Lp_r responses to NO_3^- in *Arabidopsis* (Li et al. 2016). Abscisic acid (ABA) is also elevated in certain cells (endodermis) of the root tip of *Arabidopsis* in response to NO_3^- (Ondzighi-Assoume et al. 2016). ABA has been shown to have a positive effect on root cell water permeability by an unknown but rapid mechanism (Wan et al. 2004; Olaetxea et al. 2015). Transcription and translation to increase protein abundance are indicated as *dotted arrows* to PIPs and NRTs. *Systemic*: Shoot NO_3^- positively correlates with Lp_r in *Arabidopsis*, and unknown shoot-to-root signal was suggested (*black arrow*) (Li et al. 2016). Sugar has been shown to activate a kinase (K1? = SIRK1), which phosphorylates PIP aquaporins to activate them in *Arabidopsis* seedlings (Wu et al. 2013), and it is assumed that this occurs in roots in the diagram (red arrow, K1?). Sugar signals also regulate root nitrate uptake and NRT2.1 (Wang and Ruan 2016). A shoot-to-root mobile bZIP transcription factor activated by light (HY5) that interacts with sugar signalling in *Arabidopsis* increases expression of *AtNRT2.1* in roots (*green arrow*) (Chen et al. 2016). NO_3^- is also a signal to control stomatal aperture (Wilkinson et al. 2007), which will impact on transpiration and thence mass flow to the root. Also an unidentified shoot-to-root signal (*blue arrow*) conveying transpiration or leaf water potential regulates Lp_r in rice (Sakurai-Ishikawa et al. 2011) and probably other plants (Vandeleur et al. 2014); this could be ABA in wheat (Kudoyarova et al. 2011). Not shown in the diagram is the probable signalling involving NH_4^+, the role of certain aquaporins in lateral root initiation in response to NO_3^- and outgrowth that is important for NO_3^- foraging in soil, nor the relative expression patterns of NO_3^- transporters in the root that may influence the efficiency of coupling of water flow with PIP-activated increase in Lp_r. Other hormones involved in N signalling are also not included, and some of these, e.g. auxin, have also been demonstrated to have effects on aquaporin regulation (Peret et al. 2012)

movement will potentially resolve the location issue and genes involved in signalling. Research questions include:

- How is the cellular NO_3^- signal perceived and transduced to give relatively rapid changes in kinase and phosphatases that target PIP aquaporins?
- Is there a direct effect of NO_3^- and/or NH_4^+ on PIP and TIP aquaporins? Given that an animal aquaporin can transport NO_3^- and one TIP aquaporin (AtTIP2;1) has a structure to facilitate NH_4^+ transport, it would be wise to consider that both NH_4^+ and NO_3^- may have direct effects to regulate water flow through some aquaporins that could be tested in heterologous expression systems. Responses via changes in cytosolic pH have also not been thoroughly explored.
- What is the systemic signal that regulates aquaporin response to shoot N? For example, do shoot-derived microRNAs such as miR172c (Wang et al. 2014) or shoot-derived transcription factors such as HY5 (Chen et al. 2016) have a role in root aquaporin expression?
- Is there a control via NO_3^- signalling on the key PIP and TIP involved in lateral root growth in response to local NO_3^- concentration?

- Fine scale mapping is required of water flow along a root that could be used in conjunction with ion flux measurement and gene expression to correlate with NO_3^- and NH_4^+ fluxes. Even more interesting would be to map this with energy consumption during the process of N and water uptake.
- The role of the root rhizosphere including its physical, chemical and microbial characteristics needs to be investigated in relation to N and aquaporin signalling and the effective delivery of both water and solutes to the root surface.

References

Ahamed A, Murai-Hatano M, Ishikawa-Sakurai J, Hayashi H, Kawamura Y, Uemura M (2012) Cold stress-induced acclimation in rice is mediated by root-specific aquaporins. Plant Cell Physiol 53(8):1445–1456

Ahmed MA, Zarebanadkouki M, Kaestner A, Carminati A (2016) Measurements of water uptake of maize roots: the key function of lateral roots. Plant Soil 398(1–2):59–77. doi:10.1007/s11104-015-2639-6

Ahn CS, Ahn HK, Pai HS (2015) Overexpression of the PP2A regulatory subunit Tap46 leads to enhanced plant growth through stimulation of the TOR signalling pathway. J Exp Bot 66(3):827–840. doi:10.1093/jxb/eru438

Alleva K, Niemietz CM, Maurel C, Parisi M, Tyerman SD, Amodeo G (2006) Plasma membrane of *Beta vulgaris* storage root shows high water channel activity regulated by cytoplasmic pH and a dual range of calcium concentrations. J Exp Bot 57(3):609–621. doi:10.1093/Jxb/Erj046

Almeida-Rodriguez AM, Hacke UG, Laur J (2011) Influence of evaporative demand on aquaporin expression and root hydraulics of hybrid poplar. Plant Cell Environ 34(8):1318–1331. doi:10.1111/j.1365-3040.2011.02331.x

Aroca R, Amodeo G, Fernández-Illescas S, Herman EM, Chaumont F, Chrispeels MJ (2005) The role of aquaporins and membrane damage in chilling and hydrogen peroxide induced changes in the hydraulic conductance of maize roots. Plant Physiol 137:341–353

Baena-González E, Sheen J (2008) Convergent energy and stress signaling. Trends Plant Sci 13(9):474–482. doi:10.1016/j.tplants.2008.06.006

Baena-Gonzalez E, Rolland F, Thevelein JM, Sheen J (2007) A central integrator of transcription networks in plant stress and energy signalling. Nature 448(7156):938–U910. doi:10.1038/nature06069

Barber SA (1962) A diffusion and mass-flow concept of soil nutrient availability. Soil Sci 93:39–49

Barberon M, Vermeer JEM, De Bellis D, Wang P, Naseer S, Andersen TG, Humbel BM, Nawrath C, Takano J, Salt DE, Geldner N (2016) Adaptation of root function by nutrient-induced plasticity of endodermal differentiation. Cell 164:447–459

Barthes L, Deleens E, Bouseer A, Hoarau J, Prior J-L (1996) Xylem exudation is regulated to nitrate assimilation pathway in detopped maize seedlings: use of nitrate reductase and glutamine synthetase inhibitors as tools. J Exp Bot 47(297):485–495

Belastegui-Macadam XM, Estavillo JM, Garcia-Mina JM, Gonzalez A, Bastias E, Gonzalez-Murua C (2007) Clover and ryegrass are tolerant species to ammonium nutrition. J Plant Physiol 164(12):1583–1594. doi:10.1016/j.jplph.2006.11.013

Bertl A, Kaldenhoff R (2007) Function of a separate NH_3 pore in Aquaporin TIP2;2 from wheat. FEBS Lett 581:5413–5417

Bouguyon E, Brun F, Meynard D, Kubes M, Pervent M, Leran S, Lacombe B, Krouk G, Guiderdoni E, Zazimalova E, Hoyerova K, Nacry P, Gojon A (2015) Multiple mechanisms of nitrate sensing by Arabidopsis nitrate transceptor NRT1.1. Nat Plants 1(3):8. doi:10.1038/nplants.2015.15

Boursiac Y, Boudet J, Postaire O, Luu DT, Tournaire-Roux C, Maurel C (2008) Stimulus-induced downregulation of root water transport involves reactive oxygen species-activated cell signalling and plasma membrane intrinsic protein internalization. Plant J 56(2):207–218. doi:10.1111/J.1365-313x.2008.03594.X

Bramley H, Tyerman SD (2010) Root water transport under waterlogged conditions and the roles of aquaporins in transport of water and other solutes relevant to oxygen deficient conditions. In: Mancuso S, Shabala S (eds) Waterlogging signalling and tolerance in plants. Springer, Heidelberg, pp 151–180. doi:10.1007/978-3-642-10305-6

Bramley H, Turner DW, Tyerman SD, Turner NC (2007a) Water flow in the roots of crop species: the influence of root structure, aquaporin activity, and waterlogging. In: Sparks DL (ed) Advances in agronomy, vol 96. Elsevier Academic Press, San Diego, pp 133–196. doi:10.1016/s0065-2113(07)96002-2

Bramley H, Turner NC, Turner DW, Tyerman SD (2007b) Comparison between gradient-dependent hydraulic conductivities of roots using the root pressure probe: the role of pressure propagations and implications for the relative roles of parallel radial pathways. Plant Cell Environ 30(7):861–874. doi:10.1111/J.1365-3040.2007.01678.X

Bramley H, Turner NC, Turner DW, Tyerman SD (2009) Roles of morphology, anatomy, and aquaporins in determining contrasting hydraulic behavior of roots. Plant Physiol 150(1):348–364. doi:10.1104/Pp.108.134098

Bramley H, Turner NC, Turner DW, Tyerman SD (2010) The contrasting influence of short-term hypoxia on the hydraulic properties of cells and roots of wheat and lupin. Funct Plant Biol 37(3):183–193. doi:10.1071/Fp09172

Caldeira CF, Jeanguenin L, Chaumont F, Tardieu F (2014) Circadian rhythms of hydraulic conductance and growth are enhanced by drought and improve plant performance. Nat Commun:5. doi:10.1038/ncomms6365

Camanes G, Cerezo M, Primo-Millo E, Gojon A, Garcia-Agustin P (2007) Ammonium transport and CitAMT1 expression are regulated by light and sucrose in Citrus plants. J Exp Bot 58(11):2811–2825. doi:10.1093/jxb/erm135

Cannell MGR, Thornley JHM (2000) Modelling the components of plant respiration: some guiding principles. Ann Bot 85(1):45–54. doi:10.1006/anbo.1999.0996

Carvajal M, Cooke DT, Clarkson DT (1996b) Plasma membrane fluidity and hydraulic conductance in wheat roots: interactions between root temperature and nitrate or phosphate deprivation. Plant Cell Environ 19(9):1110–1114. doi:10.1111/j.1365-3040.1996.tb00219.x

Carvajal M, Cooke DT, Clarkson DT (1996a) Responses of wheat plants to nutrient deprivation may involve the regulation of water-channel function. Planta 199(3):372–381

Chapman N, Miller AJ, Lindsey K, Whalley WR (2012) Roots, water, and nutrient acquisition: let's get physical. Trends Plant Sci 17(12):701–710

Chaumont F, Tyerman SD (2014) Aquaporins: highly regulated channels controlling plant water relations. Plant Physiol 164:1600–1618

Chen XB, Yao QF, Gao XH, Jiang CF, Harberd NP, Fu XD (2016) Shoot-to-root mobile transcription factor HY5 coordinates plant carbon and nitrogen acquisition. Curr Biol 26(5):640–646. doi:10.1016/j.cub.2015.12.066

Chevalier AS, Chaumont F (2015) Trafficking of plant plasma membrane aquaporins: multiple regulation levels and complex sorting signals. Plant Cell Physiol 56(5):819–829. doi:10.1093/pcp/pcu203

Chiasson DM, Loughlin PC, Mazurkiewicz D, Mohammadidehcheshmeh M, Fedorova EE, Okamoto M, McLean E, Glass ADM, Smith SE, Bisseling T, Tyerman SD, Day DA, Kaiser BN (2014) Soybean SAT1 (Symbiotic Ammonium Transporter 1) encodes a bHLH transcription factor involved in nodule growth and NH_4^+ transport. Proc Natl Acad Sci U S A 111(13):4814–4819. doi:10.1073/pnas.1312801111

Cho HY, Wen TN, Wang YT, Shih MC (2016) Quantitative phosphoproteomics of protein kinase SnRK1 regulated protein phosphorylation in Arabidopsis under submergence. J Exp Bot 67(9):2745–2760. doi:10.1093/jxb/erw107

Choi WG, Roberts DM (2007) Arabidopsis NIP2;1, a major intrinsic protein transporter of lactic acid induced by anoxic stress. J Biol Chem 282(33):24209–24218. doi:10.1074/jbc.M700982200

Clarkson DT, Warner AJ (1979) Relationships between root temperature and the transport of ammonium and nitrate ions by Italian and perennial ryegrass (Lolium-multiflorum and Lolium-perenne). Plant Physiol 64(4):557–561. doi:10.1104/pp.64.4.557

Colmer TD, Bloom AJ (1998) A comparison of NH4+ and NO3- net fluxes along roots of rice and maize. Plant Cell Environ 21(2):240–246. doi:10.1046/j.1365-3040.1998.00261.x

Cramer MD, Hoffmann V, Verboom GA (2008) Nutrient availability moderates transpiration in *Ehrharta calycina*. New Phytol 179:1048–1057

Cramer MD, Hawkins H-J, Verboom GA (2009a) The importance of nutritional regulation of plant water flux. Oecologia 161(1):15–24. doi:10.1007/s00442-009-1364-3

Cramer MD, Hawkins H-J, Verboom GA (2009b) The importance of nutritional regulation of plant water flux. Oecologia 161:15–24

Davidson EA, Janssens IA, Luo YQ (2006) On the variability of respiration in terrestrial ecosystems: moving beyond Q(10). Glob Chang Biol 12(2):154–164. doi:10.1111/j.1365-2486.2005.01065.x

Delhon P, Gojon A, Tillard P, Passama L (1995) Diurnal regulation of NO_3^- uptake in soybean plants.1. Changes in NO_3^- influx, efflux, and N utilization in the plant during the day-night cycle. J Exp Bot 46(291):1585–1594. doi:10.1093/jxb/46.10.1585

Desnos T (2008) Root branching responses to phosphate and nitrate. Curr Opin Plant Biol 11(1):82–87. doi:10.1016/j.pbi.2007.10.003

Dobrenel T, Caldana C, Hanson J, Robaglia C, Vincentz M, Veit B, Meyer C (2016) TOR signaling and nutrient sensing. Annu Rev Plant Biol 67:261–285. doi:10.1146/annurev-arplant-043014-114648

Easlon HM, Bloom AJ (2013) The effects of rising atmospheric carbon dioxide on shoot-root nitrogen and water signaling. Front Plant Sci 4:8. doi:10.3389/fpls.2013.00304

Emanuelle S, Doblin MS, Stapleton DI, Bacic A, Gooley PR (2016) Molecular insights into the enigmatic metabolic regulator, SnRK1. Trends Plant Sci 21(4):341–353. doi:10.1016/j.tplants.2015.11.001

Engelsberger WR, Schulze WX (2012) Nitrate and ammonium lead to distinct global dynamic phosphorylation patterns when resupplied to nitrogen-starved Arabidopsis seedlings. Plant J 69(6):978–995. doi:10.1111/j.1365-313X.2011.04848.x

Fang XZ, Tian WH, Liu XX, Lin XY, Jin CW, Zheng SJ (2016) Alleviation of proton toxicity by nitrate uptake specifically depends on nitrate transporter 1.1 in Arabidopsis. New Phytol 211(1):149–158. doi:10.1111/nph.13892

Felle HH (2005) pH regulation in anoxic plants. Ann Bot 96(4):519–532. doi:10.1093/aob/mci207

Feng HM, Yan M, Fan XR, Li BZ, Shen QR, Miller AJ, Xu GH (2011) Spatial expression and regulation of rice high-affinity nitrate transporters by nitrogen and carbon status. J Exp Bot 62(7):2319–2332. doi:10.1093/jxb/erq403

Fetter K, Van Wilder V, Moshelion M, Chaumont F (2004) Interactions between plasma membrane aquaporins modulate their water channel activity. Plant Cell 16(1):215–228

Forde BG (2014) Nitrogen signalling pathways shaping root system architecture: an update. Curr Opin Plant Biol 21:30–36. doi:10.1016/j.pbi.2014.06.004

Franks PJ, Drake PL, Froend RH (2007) Anisohydric but isohydrodynamic: seasonally constant plant water potential gradient explained by a stomatal control mechanism incorporating variable plant hydraulic conductance. Plant Cell Environ 30(1):19–30. doi:10.1111/J.1365-3040.01600.X

Gao Y, Li Y, Yang X, Li H, Shen Q, Guo S (2010) Ammonium nutrition increases water absorption in rice seedlings (*Oryza sativa L.*) under water stress. Plant Soil 331(1–2):193–201. doi:10.1007/s11104-009-0245-1

Gaspar M, Bousser A, Sissoeff I, Roche O, Hoarau J, Mahe A (2003) Cloning and characterization of ZmPIP1-5b, an aquaporin transporting water and urea. Plant Sci 165:21–31

Glass ADM (2003) Nitrogen use efficiency of crop plants: physiological constraints upon nitrogen absorption. Crit Rev Plant Sci 22(5):453–470. doi:10.1080/07352680390243512

Gloser V, Zwieniecki MA, Orians CM, Holbrook NM (2007) Dynamic changes in root hydraulic properties in response to nitrate availability. J Exp Bot 58(10):2409–2415

Gonzalez-Dugo V, Durand JL, Gastal F (2010) Water deficit and nitrogen nutrition of crops. A review. Agron Sustain Dev 30(3):529–544. doi:10.1051/agro/2009059

Gorska A, Ye Q, Holbrook NM, Zwieniecki MA (2008a) Nitrate control of root hydraulic properties in plants: translating local information to whole plant response. Plant Physiol 148:1159–1167

Gorska A, Zwieniecka A, Michele Holbrook N, Zwieniecki MA (2008b) Nitrate induction of root hydraulic conductivity in maize is not correlated with aquaporin expression. Planta 228:989–998

Górska A, Lazor JW, Zwieniecka AK, Benway C, Zwieniecki MA (2010) The capacity for nitrate regulation of root hydraulic properties correlates with species' nitrate uptake rates. Plant Soil 337:447–455

Graciano C, Faustino LI, Zwieniecki MA (2016) Hydraulic properties of *Eucalyptus grandis* in response to nitrate and phosphate deficiency and sudden changes in their availability. J Plant Nutr Soil Sci 179(2):303–309. doi:10.1002/jpln.201500207

Grondin A, Rodrigues O, Verdoucq L, Merlot S, Leonhardt N, Maurel C (2015) Aquaporins contribute to ABA-triggered stomatal closure through OST1-mediated phosphorylation. Plant Cell 27:1945–1954

Grondin A, Mauleon R, Vadez V, Henry A (2016) Root aquaporins contribute to whole plant water fluxes under drought stress in rice (*Oryza sativa L.*). Plant Cell Environ 39(2):347–365. doi:10.1111/pce.12616

Gu R, Chen X, Zhou Y, Yuan L (2012) Isolation and characterization of three maize aquaporin genes, ZmNIP2;1, ZmNIP2;1 and ZmTIP4;4 involved in urea transport. BMB Rep 45(2):96–101

Guo S, Bruck H, Sattelmacher B (2002) Effects of supplied nitrogen form on growth and water uptake of French bean (*Phaseolus vulgaris L.*) plants – nitrogen form and water uptake. Plant Soil 239(2):267–275. doi:10.1023/a:1015014417018

Guo S, Kaldenhoff R, Uehlein N, Sattelmacher B, Brueck H (2007a) Relationship between water and nitrogen uptake in nitrate- and ammonium-supplied *Phaseolus vulgaris L. plants*. J Plant Nutr Soil Sci 170:73–80. doi:10.1002/jpln.200625073

Guo S, Shen Q, Brueck H (2007b) Effects of local nitrogen supply on water uptake of bean plants in a split root system. J Integr Plant Biol 49(4):472–480. doi:10.1111/j.1744-7909.2007.00436.x

Guo S, Zhou Y, Shen Q, Zhang F (2007c) Effect of ammonium and nitrate nutrition on some physiological processes in higher plants - Growth, photosynthesis, photorespiration, and water relations. Plant Biol 9(1):21–29. doi:10.1055/s-2006-924541

Hachez C, Moshelion M, Zelazny E, Cavez D, Chaumont F (2006) Localization and quantification of plasma membrane aquaporin expression in maize primary root: a clue to understanding their role as cellular plumbers. Plant Mol Biol 62(1–2):305–323. doi:10.1007/S11103-006-9022-1

Hachez C, Heinen RB, Draye X, Chaumont Fc, ois (2008) The expression pattern of plasma membrane aquaporins in maize leaf highlights their role in hydraulic regulation. Plant Mol Biol 68:337–353

Hachez C, Veselov D, Ye Q, Reinhardt H, Knipfer T, Fricke W, Chaumont Fc, ois (2012) Short-term control of maize cell and root water permeability through plasma membrane aquaporin isoforms. Plant Cell Environ 35:185–198

Hazama A, Kozono D, Guggino WB, Agre P, Yasui M (2002) Ion permeation of AQP6 water channel protein. J Biol Chem 277(32):29224–29230

Hedfalk K, Törnroth-Horsefield S, Nyblom M, Johanson U, Kjellbom P, Neutze R (2006) Aquaporin gating. Curr Opin Struct Biol 16:447–456

Henzler T, Waterhouse RN, Smyth AJ, Carvajal M, Cooke DT, Schaffner AR, Steudle E, DT C (1999) Diurnal variations in hydraulic conductivity and root pressure can be correlated with the expression of putative aquaporins in the roots of Lotus japonicus. Planta 210:50–60

Herzog M, Striker GG, Colmer TD, Pedersen O (2016) Mechanisms of waterlogging tolerance in wheat – a review of root and shoot physiology. Plant Cell Environ 39(5):1068–1086. doi:10.1111/pce.12676

Hodges TK, Elzam OE (1967) Effect of azide and oligomycin on transport of calcium ions in corn mitochondria. Nature 215:970–972. doi:10.1038/215970a0

Holm LM, Jahn TP, Moller ALB, Schjoerring JK, Ferri D, Klaerke DA, Zeuthen T (2005) NH3 and NH4+ permeability in aquaporin-expressing Xenopus oocytes. Pflugers Arch 450(6):415–428. doi:10.1007/s00424-005-1399-1

Huang NC, Chiang CS, Crawford NM, Tsay YF (1996) CHL1 encodes a component of the low-affinity nitrate uptake system in Arabidopsis and shows cell type-specific expression in roots. Plant Cell 8(12):2183–2191

Hüsken D, Steudle E, Zimmermann U (1978) Pressure probe technique for measuring water relations of cells in higher plants. Plant Physiol 61:158–163

Ishikawa-Sakurai J, Hayashi H, Murai-Hatano M (2014) Nitrogen availability affects hydraulic conductivity of rice roots, possibly through changes in aquaporin gene expression. Plant Soil 379:289–300. doi:10.1007/s11104-014-2070-4

Jahn TP, Moller ALB, Zeuthen T, LM H, Klaerke DA, Mohsin B, Kühlbrandt W, Schjoerring JK (2004) Aquaporin homologues in plants and mammals transport ammonia. FEBS Lett 574:13–36

Jossier M, Bouly JP, Meimoun P, Arjmand A, Lessard P, Hawley S, Grahame Hardie D, Thomas M (2009) SnRK1 (SNF1-related kinase 1) has a central role in sugar and ABA signalling in *Arabidopsis thaliana*. Plant J 59(2):316–328. doi:10.1111/j.1365-313X.2009.03871.x

Kiba T, Krapp A (2016) Plant nitrogen acquisition under low availability: regulation of uptake and root architecture. Plant Cell Physiol. doi:10.1093/pcp/pcw052

Kiba T, Feria-Bourrellier AB, Lafouge F, Lezhneva L, Boutet-Mercey S, Orsel M, Brehaut V, Miller A, Daniel-Vedele F, Sakakibara H, Krapp A (2012) The arabidopsis nitrate transporter NRT2.4 plays a double role in roots and shoots of nitrogen-straved plants. Plant Cell 24(1):245–258. doi:10.1105/tpc.111.092221

Kirscht A, Kaptan SS, Bienert GP, Chaumont F, Nissen P, de Groot BL, Kjellbom P, Gourdon P, Johanson U (2016) Crystal structure of an ammonia-permeable aquaporin. PLoS Biol 14(3):19. doi:10.1371/journal.pbio.1002411

Klebl F, Wolf M, Sauer N (2003) A defect in the yeast plasma membrane urea transporter Dur3p is complemented by CpNIP1, a Nod26-like protein from zucchini *(Cucurbita pepo L.*), and by *Arabidopsis thaliana* δ-TIP or γ-TIP. FEBS Lett 547:69–74

Kotur Z, Mackenzie N, Ramesh S, Tyerman SD, Kaiser BN, Glass ADM (2012) Nitrate transport capacity of the *Arabidopsis thaliana* NRT2 family members and their interactions with AtNAR2.1. New Phytol 194(3):724–731. doi:10.1111/j.1469-8137.2012.04094.x

Kramer PJ, Boyer JS (1995) Water relations of plants and soils. Academic, San Diego

Kreuzwieser J, Gessler A (2010) Global climate change and tree nutrition: influence of water availability. Tree Physiol 30(9):1221–1234. doi:10.1093/treephys/tpq055

Kreuzwieser J, Furniss S, Rennenberg H (2002) Impact of waterlogging on the N-metabolism of flood tolerant and non-tolerant tree species. Plant Cell Environ 25(8):1039–1049. doi:10.1046/j.1365-3040.2002.00886.x

Kudoyarova G, Veselova S, Hartung W, Farhutdinov R, Veselov D, Sharipova G (2011) Involvement of root ABA and hydraulic conductivity in the control of water relations in wheat plants exposed to increased evaporative demand. Planta 233(1):87–94. doi:10.1007/S00425-010-1286-7

Kudoyarova GR, Dodd IC, Veselov DS, Rothwell SA, Veselov SY (2015) Common and specific responses to availability of mineral nutrients and water. J Exp Bot 66(8):2133–2144. doi:10.1093/jxb/erv017

Laanbroek HJ (1990) Bacterial cycling of minerals that affect plant-growth in waterlogged soils – a review. Aquat Bot 38(1):109–125. doi:10.1016/0304-3770(90)90101-p

Laur J, Hacke UG (2013) Transpirational demand affects aquaporin expression in poplar roots. J Exp Bot 64(8):2283–2293. doi:10.1093/jxb/ert096

Lee SH, Singh AP, Chung GC (2004a) Rapid accumulation of hydrogen peroxide in cucumber roots due to exposure to low temperature appears to mediate decreases in water transport. J Exp Bot 55(403):1733–1741

Lee SH, Singh AP, Chung GC, Ahn SJ, Noh EK, Steudle E (2004b) Exposure of roots of cucumber (*Cucumis sativus*) to low temperature severely reduces root pressure, hydraulic conductivity and active transport of nutrients. Physiol Plant 120(3):413–420. doi:10.1111/j.0031-9317.2004.00248.x

Lee HK, Cho SK, Son O, Xu Z, Hwang I, Kim WT (2009) Drought stress-induced Rma1H1, a RING membrane-anchor E3 ubiquitin ligase homolog, regulates aquaporin levels via ubiquitination in transgenic Arabidopsis plants. Plant Cell 21:622–641

Lejay L, Gansel X, Cerezo M, Tillard P, Muller C, Krapp A, von Wiren N, Daniel-Vedele F, Gojon A (2003) Regulation of root ion transporters by photosynthesis: functional importance and relation with hexokinase. Plant Cell 15(9):2218–2232. doi:10.1105/tpc.013516

Léran S, Varala K, Boyer JC, Chiurazzi M, Crawford N, Daniel-Vedele F, David L, Dickstein R, Fernandez E, Forde B, Gassmann W, Geiger D, Gojon A, Gong JM, Halkier BA, Harris JM, Hedrich R, Limami AM, Rentsch D, Seo M, Tsay YF, Zhang M, Coruzzi G, Lacombe B (2014) A unified nomenclature of nitrate transporter 1/peptide transporter family members in plants. Trends Pant Sci 19(1):5–9

Lezhneva L, Kiba T, Feria-Bourrellier AB, Lafouge F, Boutet-Mercey S, Zoufan P, Sakakibara H, Daniel-Vedele F, Krapp A (2014) The Arabidopsis nitrate transporter NRT2.5 plays a role in nitrate acquisition and remobilization in nitrogen-starved plants. Plant J 80(2):230–241. doi:10.1111/tpj.12626

Li WB, Wang Y, Okamoto M, Crawford NM, Siddiqi MY, Glass ADM (2007) Dissection of the AtNRT2.1: AtNRT2.2 inducible high-affinity nitrate transporter gene cluster. Plant Physiol 143(1):425–433. doi:10.1104/pp.106.091223

Li Y, Gao Y, Ding L, Shen Q, Guo S (2009) Ammonium enhances the tolerance of rice seedlings (*Oryza sativa L.*) to drought condition. Agric Water Manag 96(12):1746–1750. doi:10.1016/j.agwat.2009.07.008

Li Y, Krouk G, Coruzzi GM, Ruffel S (2014) Finding a nitrogen niche: a systems integration of local and systemic nitrogen signalling in plants. J Exp Bot 65(19):5601–5610. doi:10.1093/jxb/eru263

Li G, Tillard P, Gojon A, Maurel C (2016) Dual regulation of root hydraulic conductivity and plasma membrane aquaporins by plant nitrate accumulation and high-affinity nitrate transporter NRT2.1. Plant Cell Physiol 57(4):733–742. doi:10.1093/pcp/pcw022

Libourel IGL, van Bodegom PM, Fricker MD, Ratcliffe RG (2006) Nitrite reduces cytoplasmic acidosis under anoxia. Plant Physiol 142(4):1710–1717. doi:10.1104/pp.106.088898

Lillo C, Kataya ARA, Heidari B, Creighton MT, Nemie-Feyissa D, Ginbot Z, Jonassen EM (2014) Protein phosphatases PP2A, PP4 and PP6: mediators and regulators in development and responses to environmental cues. Plant Cell Environ 37(12):2631–2648. doi:10.1111/pce.12364

Liu L-H, Ludewig U, Gassert BF, Wolf B, von Wirén N (2003) Urea transport by nitrogen-regulated tonoplast intrinsic proteins in Arabidopsis. Plant Physiol 133:1220–1228

Liu K, Kozono D, Kato Y, Agre P, Hazama A, Yasui M (2005) Conversion of aquaporin 6 from an anion channel to a water-selective channel by a single amino acid substitution. Proc Natl Acad Sci 102:2192–2197

Lopez M, Bousser AS, Sissoeff I, Gaspar M, Lachaise B, Hoarau J, Mahe A (2003) Diurnal regulation of water transport and aquaporin gene expression in maize roots: Contribution of PIP2 proteins. Plant Cell Physiol 44(12):1384–1395

Lopez-Bucio J, Cruz-Ramirez A, Herrera-Estrella L (2003) The role of nutrient availability in regulating root architecture. Curr Opin Plant Biol 6(3):280–287. doi:10.1016/s1369-5266(03)00035-9

Ludewig U, Neuhduser B, Dynowski M (2007) Molecular mechanisms of ammonium transport and accumulation in plants. FEBS Lett 581(12):2301–2308. doi:10.1016/j.febslet.2007.03.034

Lynch JP (2013) Steep, cheap and deep: an ideotype to optimize water and N acquisition by maize root systems. Ann Bot 112(2):347–357. doi:10.1093/aob/mcs293

Marschner H (1995) Mineral nutrition of higher plants, 2nd edn. Academic, 889 pp Academic Press 24–28. Oval Road, London NW1 7DX

Matimati I, Verboom GA, Cramer MD (2014) Nitrogen regulation of transpiration controls mass-flow acquisition of nutrients. J Exp Bot 65(1):159–168. doi:10.1093/jxb/ert367

Maurel C, Boursiac Y, Luu DT, Santoni V, Shahzad Z, Verdoucq L (2015) Aquaporins in plants. Physiol Rev 95(4):1321–1358. doi:10.1152/physrev.00008.2015

Pietro M, Vialaret J, Li G, Hem S, Prado K, Rossignol M, Maurel C, Santoni V (2013) Coordinated post-translational responses of aquaporins to abiotic and nutritional stimuli in Arabidopsis roots. Mol Cell Proteomics 12(12):M113.028241–M028113.028241

Miller AJ, Cramer MD (2004) Root nitrogen acquisition and assimilation. Plant Soil 274:1–36

Minshall WH (1964) Effect of nitrogen-containing nutrients on the exudation from detopped tomato plants. Nature 202(4935):925–926

Moyano FE, Manzoni S, Chenu C (2013) Responses of soil heterotrophic respiration to moisture availability: an exploration of processes and models. Soil Biol Biochem 59:72–85. doi:10.1016/j.soilbio.2013.01.002

Niemietz CM, Tyerman SD (2002) New potent inhibitors of aquaporins: silver and gold compounds inhibit aquaporins of plant and human origin. FEBS Lett 531(3):443–447

Nyblom M, Frick A, Wang Y, Ekvall M, Hallgren K, Hedfalk K, Neutze R, Tajkhorshid E, Törnroth-Horsefield S (2009) Structural and functional analysis of SoPIP2;1 mutants adds insight into plant aquaporin gating. J Mol Biol 387:653–668

Nye PH (1977) The rate-limiting step in plant nutrient absorption from soil. Soil Sci 123(5):292–297

O'Brien JA, Vega A, Bouguyon E, Krouk G, Gojon A, Coruzzi G, Gutierrez RA (2016) Nitrate transport, sensing, and responses in plants. Mol Plant 9(6):837–856. doi:10.1016/j.molp.2016.05.004

O'Hara LE, Paul MJ, Wingler A (2013) How do sugars regulate plant growth and development? New insight into the role of trehalose-6-phosphate. Mol Plant 6(2):261–274. doi:10.1093/mp/sss120

Olaetxea M, Mora V, Bacaicoa E, Garnica M, Fuentes M, Casanova E, Zamarreno AM, Iriarte JC, Etayo D, Ederra I, Gonzalo R, Baigorri R, Garcia-Mina JM (2015) Abscisic acid regulation of root hydraulic conductivity and aquaporin gene expression is crucial to the plant shoot growth enhancement caused by rhizosphere humic acids. Plant Physiol 169(4):2587–2596. doi:10.1104/pp.15.00596

Ondzighi-Assoume CA, Chakraborty S, Harris JM (2016) Environmental nitrate stimulates abscisic acid accumulation in arabidopsis root tips by releasing it from inactive stores. Plant Cell 28(3):729–745. doi:10.1105/tpc.15.00946

Oyewole OA, Inselsbacher E, Nasholm T (2014) Direct estimation of mass flow and diffusion of nitrogen compounds in solution and soil. New Phytol 201(3):1056–1064. doi:10.1111/nph.12553

Peret B, Li GW, Zhao J, Band LR, Voss U, Postaire O, Luu DT, Da Ines O, Casimiro I, Lucas M, Wells DM, Lazzerini L, Nacry P, King JR, Jensen OE, Schaffner AR, Maurel C, Bennett MJ (2012) Auxin regulates aquaporin function to facilitate lateral root emergence. Nat Cell Biol 14(10):991. doi:10.1038/ncb2573

Pfister A, Barberon M, Alassimone J, Kalmbach L, Lee Y, Vermeer JEM, Yamazaki M, Li G, Maurel C, Takano J, Kamiya T, Salt DE, Roppolo D, Geldner N (2014) A receptor-like kinase mutant with absent endodermal diffusion barrier displays selective nutrient homeostasis defects. eLIFE 3(e03115):1–20. doi:10.7554/eLife.03115

Pill WG, Lambeth VN (1977) Effects of NH_4^+ and NO_3^- nutrition with and without pH adjustment on tomato growth, ion composition, and water relations. J Am Soc Hortic Sci 102(1):78–81

Poorter H, Vanderwerf A, Atkin OK, Lambers H (1991) Respiratory energy-requirements of roots vary with the potential growth-rate of a plant-species. Physiol Plant 83(3):469–475. doi:10.1034/j.1399-3054.1991.830321.x

Postaire O, Tournaire-Roux C, Grondin A, Boursiac Y, Morillon R, Schaffner AR, Maurel C (2010) A PIP1 aquaporin contributes to hydrostatic pressure-induced water transport in both the root and rosette of arabidopsis. Plant Physiol 152(3):1418–1430. doi:10.1104/pp.109.145326

Pou A, Medrano H, Flexas J, Tyerman SD (2013) A putative role for TIP and PIP aquaporins in dynamics of leaf hydraulic and stomatal conductances in grapevine under water stress and re-watering. Plant Cell Environ 36(4):828–843. doi:10.1111/pce.12019

Pregitzer KS, King JA, Burton AJ, Brown SE (2000) Responses of tree fine roots to temperature. New Phytol 147(1):105–115. doi:10.1046/j.1469-8137.2000.00689.x

Ranathunge K, Kotula L, Steudle E, Lafitte R (2004) Water permeability and reflection coefficient of the outer part of young rice roots are differently affected by closure of water channels (aquaporins) or blockage of apoplastic pores. J Exp Bot 55(396):433–447. doi:10.1093/jxb/erh041

Raven JA (1985) Regulation of pH and generation of osmolarity in vascular plants – a cost-benefit analysis in relation to efficiency of use of energy, nitrogen and water. New Phytol 101(1):25–77. doi:10.1111/j.1469-8137.1985.tb02816.x

Raven JA (2008) Transpiration: how many functions? New Phytol 179(4):905–907

Reinhardt H, Hachez C, Bienert MD, Beebo A, Swarup K, Voss U, Bouhidel K, Frigerio L, Schjoerring JK, Bennett MJ, Chaumont F (2016) Tonoplast aquaporins facilitate lateral root emergence. Plant Physiol 170(3):1640–1654. doi:10.1104/pp.15.01635
Ren BB, Wang M, Chen YP, Sun GM, Li Y, Shen QR, Guo SW (2015) Water absorption is affected by the nitrogen supply to rice plants. Plant Soil 396(1–2):397–410. doi:10.1007/s11104-015-2603-5
Rhee JY, Chung GC, Katsuhara M, Ahn S-J (2011) Effect of nutrient deficiencies on the water transport properties in figleaf gourd plants. Hortic Environ Biotechnol 52(6):629–634. doi:10.1007/s13580-011-0046-3
Riveras E, Alvarez JM, Vidal EA, Oses C, Vega A, Gutierrez RA (2015) The calcium ion is a second messenger in the nitrate signaling pathway of arabidopsis. Plant Physiol 169(2):1397–1404. doi:10.1104/pp.15.00961
Robaglia C, Thomas M, Meyer C (2012) Sensing nutrient and energy status by SnRK1 and TOR kinases. Curr Opin Plant Biol 15(3):301–307. doi:10.1016/j.pbi.2012.01.012
Robertson D, Zhang HP, Palta JA, Colmer T, Turner NC (2009) Waterlogging affects the growth, development of tillers, and yield of wheat through a severe, but transient, N deficiency. Crop Pasture Sci 60(6):578–586. doi:10.1071/cp08440
Robinson D (1994) The responses of plants to nonuniform supplies of nutrients. New Phytol 127(4):635–674. doi:10.1111/j.1469-8137.1994.tb02969.x
Ruffel S, Krouk G, Ristova D, Shasha D, Birnbaum KD, Coruzzi GM (2011) Nitrogen economics of root foraging: transitive closure of the nitrate-cytokinin relay and distinct systemic signaling for N supply vs. demand. Proc Natl Acad Sci U S A 108(45):18524–18529. doi:10.1073/pnas.1108684108
Ruggiero C, Angelino G (2007) Changes of root hydraulic conductivity and root/shoot ratio of durum wheat and barley in relation to nitrogen availability and mercury exposure. Ital J Agron 2(3):281–290
Sakurai-Ishikawa J, Murai-Hatano M, Hayashi H, Ahamed A, Fukushi K, Matsumoto T, Kitagawa Y (2011) Transpiration from shoots triggers diurnal changes in root aquaporin expression. Plant Cell Environ 34(7):1150–1163. doi:10.1111/J.1365-3040.2011.02313.X
Scheible W-R, Morcuende R, Czechowski T, Fritz C, Osuna D, Palacios-Rojas N, Schindelasch D, Thimm O, Udvardi MK, Stitt M (2004) Genome-wide reprogramming of primary and secondary metabolism, protein synthesis, cellular growth processes, and the regulatory infrastructure of Arabidopsis in response to nitrogen. Plant Physiol 136:2483–2499
Segonzac C, Boyer JC, Ipotesi E, Szponarski W, Tillard P, Touraine B, Sommerer N, Rossignol M, Gibrat R (2007) Nitrate efflux at the root plasma membrane: Identification of an Arabidopsis excretion transporter. Plant Cell 19:3760–3777
Seneviratne SI, Corti T, Davin EL, Hirschi M, Jaeger EB, Lehner I, Orlowsky B, Teuling AJ (2010) Investigating soil moisture-climate interactions in a changing climate: a review. Earth-Sci Rev 99(3–4):125–161. doi:10.1016/j.earscirev.2010.02.004
Shabala S, Shabala L, Barcelo J, Poschenrieder C (2014) Membrane transporters mediating root signalling and adaptive responses to oxygen deprivation and soil flooding. Plant Cell Environ 37(10):2216–2233. doi:10.1111/pce.12339
Shatil-Cohen A, Attia Z, Moshelion M (2011) Bundle-sheath cell regulation of xylem-mesophyll water transport via aquaporins under drought stress: a target of xylem-borne ABA? Plant J 67(1):72–80. doi:10.1111/j.1365-313X.2011.04576.x
Shaw RE, Meyer WS, McNeill A, Tyerman SD (2013) Waterlogging in Australian agricultural landscapes: a review of plant responses and crop models. Crop Pasture Sci 64(6):549–562. doi:10.1071/cp13080
Shimizu M, Ishida A, Hogetsu T (2005) Root hydraulic conductivity and whole-plants water balance in tropical saplings following a shade-to-sun transfer. Oecologia 143:189–197
Simunek J, Hopmans JW (2009) Modeling compensated root water and nutrient uptake. Ecol Model 220(4):505–521. doi:10.1016/j.ecolmodel.2008.11.004
Skerrett M, Tyerman SD (1994) A channel that allows inwardly directed fluxes of anions in protoplasts derived from wheat roots. Planta 192(3):295–305

Sorgona A, Lupini A, Mercati F, Di Dio L, Sunseri F, Abenavoli MR (2011) Nitrate uptake along the maize primary root: an integrated physiological and molecular approach. Plant Cell Environ 34(7):1127–1140. doi:10.1111/j.1365-3040.2011.02311.x

Sperry JS, Hacke UG, Oren R, Comstock JP (2002) Water deficits and hydraulic limits to leaf water supply. Plant Cell Environ 25(2):251–263. doi:10.1046/j.0016-8025.2001.00799.x

Steudle E (2001) The cohesion-tension mechanism and the acquisition of water by plant roots. Annu Rev Plant Physiol Plant Mol Biol 52:847–875

Steudle E, Peterson CA (1998) How does water get through roots? J Exp Bot 49(322):775–788. doi:10.1093/Jexbot/49.322.775

Sugden C, Donaghy PG, Halford NG, Hardie DG (1999) Two SNF1-Related protein kinases from spinach leaf phosphorylate and inactivate 3-hydroxy-3-methylglutaryl-coenzyme A reductase, nitrate reductase, and sucrose phosphate synthase in vitro. Plant Physiol 120(1):257–274. doi:10.1104/pp.120.1.257

Sutka M, Li GW, Boudet J, Boursiac Y, Doumas P, Maurel C (2011) Natural variation of root hydraulics in arabidopsis grown in normal and salt-stressed conditions. Plant Physiol 155(3):1264–1276. doi:10.1104/Pp.110.163113

Takase T, Ishikawa H, Murakami H, Kikuchi J, Sato-Nara K, Suzuki H (2011) The circadian clock modulates water dynamics and aquaporin expression in arabidopsis roots. Plant Cell Physiol 52(2):373–383. doi:10.1093/pcp/pcq198

Tardieu F, Simonneau T, Parent B (2015) Modelling the coordination of the controls of stomatal aperture, transpiration, leaf growth, and abscisic acid: update and extension of the Tardieu-Davies model. J Exp Bot 66(8):2227–2237. doi:10.1093/jxb/erv039

Taylor AR, Bloom AJ (1998) Ammonium, nitrate, and proton fluxes along the maize root. Plant Cell Environ 21(12):1255–1263. doi:10.1046/j.1365-3040.1998.00357.x

Tomos AD, Leigh RA (1999) The pressure probe: a versatile tool in plant cell physiology. Annu Rev Plant Physiol Plant Mol Biol 50:447–472. doi:10.1146/annurev.arplant.50.1.447

Tornroth-Horsefield S, Wang Y, Hedfalk K, Johanson U, Karlsson M, Tajkhorshid E, Neutze R, Kjellbom P (2006) Structural mechanism of plant aquaporin gating. Nature 439(7077): 688–694

Tournaire-Roux C, Sutka M, Javot H, Gout E, Gerbeau P, Luu D-T, Bligny R, Maurel C (2003) Cytosolic pH regulates root water transport during anoxic stress through gating of aquaporins. Nature 425:393–397

Tsay YF, Chiu CC, Tsai CB, Ho CH, Hsu PK (2007) Nitrate transporters and peptide transporters. FEBS Lett 581(12):2290–2300. doi:10.1016/j.febslet.2007.04.047

Tsuda M, Tyree MT (2000) Plant hydraulic conductance measured by the high pressure flow meter in crop plants. J Exp Bot 51(345):823–828. doi:10.1093/jexbot/51.345.823

Tyree MT, Patiño S, Bennink J, Alexander J (1995) Dynamic measurements of root hydraulic conductance using a high-pressure flowmeter in the laboratory and field. J Exp Bot 46(282):83–94

van Dongen JT, Licausi F (2015) Oxygen sensing and signaling. Annu Rev Plant Biol 66:345–367. doi:10.1146/annurev-arplant-043014-114813

van Wilder V, rie Miecielica U, Degand H, Derua R, Waelkens E, ois, Chaumont F (2008) Maize plasma membrane aquaporins belonging to the PIP1 and PIP2 subgroups are in vivo phosphorylated. Plant Cell Physiol 49:1364–1377

Vandeleur R, Niemietz C, Tilbrook J, Tyerman SD (2005) Roles of aquaporins in root responses to irrigation. Plant Soil 274(1–2):141–161. doi:10.1007/S11104–004-8070-Z

Vandeleur RK, Mayo G, Shelden MC, Gilliham M, Kaiser BN, Tyerman SD (2009) The role of plasma membrane intrinsic protein aquaporins in water transport through roots: diurnal and drought stress responses reveal different strategies between isohydric and anisohydric cultivars of grapevine. Plant Physiol 149(1):445–460. doi:10.1104/pp.108.128645

Vandeleur RK, Sullivan W, Athman A, Jordans C, Gilliham M, Kaiser BN, Tyerman SD (2014) Rapid shoot-to-root signalling regulates root hydraulic conductance via aquaporins. Plant Cell Environ 37(2):520–538. doi:10.1111/pce.12175

Verdoucq L, Grondin A, Maurel C (2008) Structure-function analysis of plant aquaporin AtPIP2;1 gating by divalent cations and protons. Biochem J 415:409–416

Wan XC, Steudle E, Hartung W (2004) Gating of water channels (aquaporins) in cortical cells of young corn roots by mechanical stimuli (pressure pulses): effects of ABA and of $HgCl_2$. J Exp Bot 55(396):411–422. doi:10.1093/Jxb/Erh051

Wang L, Ruan YL (2016) Shoot-root carbon allocation, sugar signalling and their coupling with nitrogen uptake and assimilation. Funct Plant Biol 43(2):105–113. doi:10.1071/fp15249

Wang MY, Siddiqi MY, Ruth TJ, Glass ADM (1993) Ammonium uptake by rice roots.2. kinetics of NH_4^+-N-13 influx across the plasmalemma. Plant Physiol 103(4):1259–1267

Wang Y-H, Garvin DF, Kochian LV (2001) Nitrate-induced genes in tomato roots. Array analysis reveals novel genes that may play a role in nitrogen nutrition. Plant Physiol 127:345–359

Wang R, Okamoto M, Xing X, Crawford NM (2003) Microarray analysis of the nitrate response in Arabidopsis roots and shoots reveals over 1,000 rapidly responding genes and new linkages to glucose, trehalose-6-phosphate, iron, and sulfate metabolism. Plant Physiol 132:556–567

Wang YN, Wang LX, Zou YM, Chen L, Cai ZM, Zhang SL, Zhao F, Tian YP, Jiang Q, Ferguson BJ, Gresshoff PM, Li X (2014) Soybean miR172c targets the repressive AP2 transcription factor NNC1 to activate ENOD40 expression and regulate nodule initiation. Plant Cell 26(12):4782–4801. doi:10.1105/tpc.114.131607

Wang C, Hu HH, Qin X, Zeise B, Xu DY, Rappel WJ, Boron WF, Schroeder JI (2016) Reconstitution of CO2 regulation of SLAC1 anion channel and function of CO_2-permeable PIP2;1 aquaporin as CARBONIC ANHYDRASE4 interactor. Plant Cell 28(2):568–582. doi:10.1105/tpc.15.00637

Warren CR (2009) Why does temperature affect relative uptake rates of nitrate, ammonium and glycine: a test with *Eucalyptus pauciflora*. Soil Biol Biochem 41(4):778–784. doi:10.1016/j.soilbio.2009.01.012

Wilkinson S, Bacon MA, Davies WJ (2007) Nitrate signalling to stomata and growing leaves: interactions with soil drying, ABA, and xylem sap pH in maize. J Exp Bot 58(7):1705–1716

Wilson JB (1988) A review of evidence on the control of shoot-root ratio, in relation to models. Ann Bot 61(4):433–449

Wu XN, Rodriguez CS, Pertl-Obermeyer H, Obermeyer G, Schulze WX (2013) Sucrose-induced receptor Kinase SIRK1 regulates a plasma membrane aquaporin in arabidopsis. Mol Cell Proteomics 12(10):2856–2873. doi:10.1074/mcp.M113.029579

Wudick MM, Li XJ, Valentini V, Geldner N, Chory J, Lin JX, Maurel C, Luu DT (2015) Subcellular redistribution of root aquaporins induced by hydrogen peroxide. Mol Plant 8(7):1103–1114. doi:10.1016/j.molp.2015.02.017

Xu N, Wang RC, Zhao LF, Zhang CF, Li ZH, Lei Z, Liu F, Guan PZ, Chu ZH, Crawford NM, Wang Y (2016) The arabidopsis NRG2 protein mediates nitrate signaling and interacts with and regulates key nitrate regulators. Plant Cell 28(2):485–504. doi:10.1105/tpc.15.00567

Yang X, Li Y, Ren B, Ding L, Gao C, Shen Q, Guo S (2012) Drought-induced root aerenchyma formation restricts water uptake in rice seedlings supplied with nitrate. Plant Cell Physiol 53(3):495–504. doi:10.1093/pcp/pcs003

Yasui M, Hazama A, Kwon T-H, Nielsen S, Guggino WB, Agre P (1999) Rapid gating and anion permeability of an intracellular aquaporin. Nature 402:184–187

York LM, Silberbush M, Lynch JP (2016) Spatiotemporal variation of nitrate uptake kinetics within the maize (*Zea mays L.*) root system is associated with greater nitrate uptake and interactions with architectural phenes. J Exp Bot 67(12):3763–3775. doi:10.1093/jxb/erw133

Yu P, Baldauf JA, Lithio A, Marcon C, Nettleton D, Li CJ, Hochholdinger F (2016) Root type-specific reprogramming of maize pericycle transcriptomes by local high nitrate results in disparate lateral root branching patterns. Plant Physiol 170(3):1783–1798. doi:10.1104/pp.15.01885

Zarebanadkouki M, Kroener E, Kaestner A, Carminati A (2014) Visualization of root water uptake: quantification of deuterated water transport in roots using neutron radiography and numerical modeling. Plant Physiol 166(2):487–499. doi:10.1104/pp.114.243212

Zelazny E, Borst JW, Muylaert M, Batoko H, Hemminga MA, Chaumont F (2007) FRET imaging in living maize cells reveals that plasma membrane aquaporins interact to regulate their subcellular localization. Proc Natl Acad Sci 104:12359–12364

Zhang HM, Forde BG (2000) Regulation of Arabidopsis root development by nitrate availability. J Exp Bot 51(342):51–59. doi:10.1093/jexbot/51.342.51

Zhong YQW, Yan WM, Chen J, Shangguan ZP (2014) Net ammonium and nitrate fluxes in wheat roots under different environmental conditions as assessed by scanning ion-selective electrode technique. Sci Rep 4:9. doi:10.1038/srep07223

Role of Aquaporins in the Maintenance of Xylem Hydraulic Capacity

Maciej A. Zwieniecki and Francesca Secchi

Abstract Terrestrial plants' well-being depends upon an uninterrupted supply of water from roots to leaves. Water stress or high transpirational demand results in an increase of water tension in the xylem, followed by an increased likelihood of embolism formation and reduction of xylem capacity to conduct water. The prolonged presence of xylem hydraulic dysfunction caused by embolism can have dramatic short- and long-term effects on plant function including the decrease of photosynthetic capacity, reduced vitality, or plant death. As the presence of embolisms is a negative trait, plants have evolved several strategies to prevent and/or mitigate the effects of hydraulic failure and restore xylem transport capacity. Recovery process requires a set of physiological activities that promote water flow into embolized conduits to restore its transport function. As hydraulic repair necessitates movement of water across xylem parenchyma cell membranes, an understanding of xylem-specific aquaporin expression patterns, their localization and activity are essential for the development of biological models describing embolism recovery process in woody plants. In this chapter, we provide an overview of aquaporin distributions and activity during development of drought stress, formation of embolism, and subsequent recovery from stress that result in restoration of xylem hydraulic capacity.

1 Water Transport and Embolism Formation

Terrestrial plants depend upon an uninterrupted supply of water from roots to photosynthetic tissue (Sperry 2003). This supply is guaranteed, in part, via the apoplastic axial transport of water through the lumens of interconnected dead cells characterized by thick, lignified walls. In the case of non-angiosperms, these cells

M.A. Zwieniecki (✉)
Department of Plant Sciences, UC Davis, Davis, CA 95616, USA
e-mail: mzwienie@ucdavis.edu

F. Secchi
Department of Agricultural, Forest and Food Sciences (DISAFA), University of Turin, Grugliasco 10095, Italy

F. Chaumont, S.D. Tyerman (eds.), *Plant Aquaporins*, Signaling and Communication in Plants, DOI 10.1007/978-3-319-49395-4_11

are typically uniform in shape and length and connected one to the next via specialized bordered pits that include structures called the torus and margo (Fig. 1). Angiosperms, on the other hand, possess water transport conduits called vessels that are formed from continuous linear files of cells (vessel elements) with large diameters and are separated within a vessel by partially or completely digested walls referred to as perforation plates. One vessel is connected to the next by bordered pits, but these bordered pits do not include the torus and margo as they do in non-angiosperms (Fig. 1). The flow of a plant's water supply is driven by a decrease in water potential from soil to air and described using electrical analogs. Coupling this pressure-driven flow with the fact that the cellular conduits for flow are dead, we are led to consider the transport capacity of xylem in a purely physical context. Such a non-biological perspective on water transport has predominantly focused xylem research on the anatomy and morphology that protect transport from failure, however, while largely ignoring the role of living cells.

What is transport failure? Under drought stress or high transpirational demands, water tension (often described in plant literature as negative pressure) in the xylem increases, increasing the likelihood of embolism formation. When xylem tension

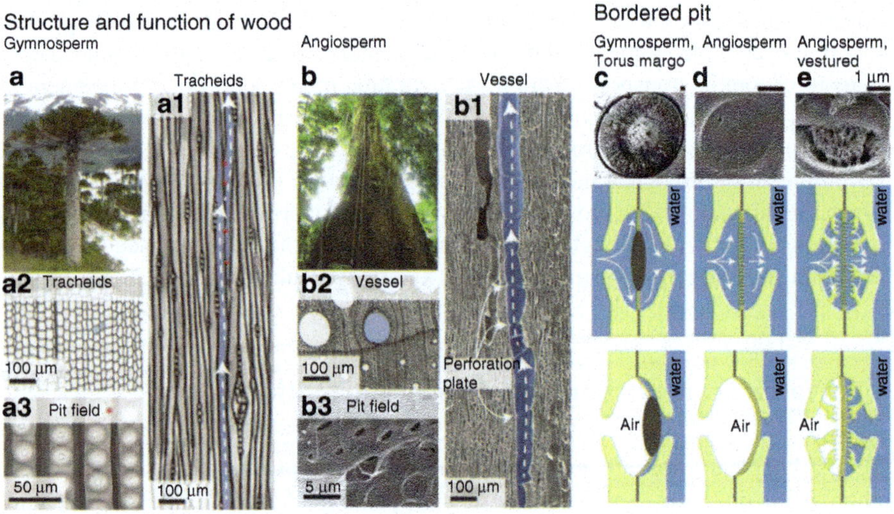

Fig. 1 Principles of wood structure in gymnosperms (**a**) and angiosperms (**b**) (**a1**) and (**b1**) depict typical features of conductive wood (xylem). In gymnosperms, the longitudinal conducting elements (tracheids (**a1**), cross-sectional view (**a2**)) are connected via linearly aligned bordered pits (**a3**). In angiosperm, wood vessels are composed from large cells that are vertically aligned and joined via fully or partially digested end walls (**b1**); vessels are also seen as large ovals in cross section (**b2**). Vessels are connected via fields composed from tens to hundreds of bordered pits (**b3**). The structure and function of xylem bordered pit pores vary between species (**c–e**). (**c**) Bordered pits of gymnosperm with torus and margo, (**d**) typical angiosperm pit, and (**e**) angiosperm with vestured pits. Drawings below (**c–e**) represent cross sections of bordered pits under normal nonembolized conditions and their potential function as protection from gas spread (Jensen et.al. 2016)

forces the radii of the air-water interface beyond a critical threshold, the tensile strength of water is overcome. This happens either as (1) air aspirates through the bordered pit membranes separating adjacent conduits or (2) preexisting gas bubbles spontaneously expand (Tyree and Zimmermann 2002). Embolism formation is considered to be a spatially and temporally unpredictable phenomenon related to the degree of tension in the xylem, the thermal environment, the physical properties of the xylem, the chemical properties of water, and a plant's previous embolism activity (Holbrook and Zwieniecki 1999; Hacke et al. 2001; Stiller and Sperry 2002; Tyree and Zimmermann 2002). As a consequence of embolism, a plant's water continuum is broken and transport is blocked via the vacuum or air-filled tracheid or vessel. Thus, the presence of embolism reduces a stem's capacity to transport water and can magnify leaf water stress, forcing stomatal closure and reducing leaf photosynthetic activity (Brodribb and Jordan 2008). In the event of a severe overload of the water transport system (when water loss exceeds transport capacity of xylem, runaway cavitation may occur resulting in plant death (Sperry et al. 1998). Therefore, the capacity of a plant to reduce the detrimental effects of embolism is an important trait for growth and survival (Tyree and Ewers 1991; Pockman et al. 1995; Choat et al. 2012; Barigah et al. 2013).

2 Why Do Plants Need to Remove Embolism?

The prolonged presence of xylem hydraulic dysfunction caused by embolism can have dramatic short- and long-term effects on plant function including the reduction of photosynthetic capacity, reduced vitality, or death. As the prolonged presence of embolisms is a negative trait, plants have evolved several strategies to prevent and/or mitigate the effects of hydraulic failure and restore xylem transport capacity post embolism. Although embolism formation is a purely physical process (Brenner 1995; Tyree and Zimmermann 2002), embolism removal requires that empty vessels fill with water against existing energy gradients as the bulk of water in the xylem remains under tension. Thus, recovery from embolism cannot happen spontaneously and necessitates some physiological activities that promote water flow into embolized conduits. The restoration of xylem capacity can be divided in two sets of strategies:

1. Strategies requiring both relief from water stress/transpiration and a prolonged period of time. This group includes shedding leaves or small branches (shrubs) to lower evaporative demand followed by the growth of new shoots, generating root pressure (small herbaceous plants) to refill embolized conduits, or growing new vessels or tracheids (radial xylem growth) to replace lost capacity with a new transport system (Sperry et al. 1987; Stiller and Sperry 2002). However, as these strategies depend on plant growth, they are slow and may result in the temporary loss of species competitiveness in a highly variable environment.
2. Strategies requiring cellular activities to dynamically repair embolized conduits and relieve tension. These strategies may be fast (minutes to hours) and thus

allow for greater flexibility in response to water stress (Zwieniecki and Holbrook 2009). They also provide protection from temporary reductions in photosynthetic capacity that might reduce competitiveness. Whether or not this type of refilling can occur in the presence of xylem tension has been difficult to prove although proposed conceptual theories attempt to reconcile experimental data with our physical understanding of xylem function (Tyree et al. 1999; Holbrook and Zwieniecki 1999).

Because the second group of strategies requires physiological activity in the xylem to maintain or restore transport function, it also requires that the xylem tissue is not dead. Indeed, even in woody plants, living cells constitute at least a few percent of the xylem and up to more than 80 % in baobabs where water storage is exceptional (Chapotin et al. 2006). The majority of living cells in the xylem are located in parenchyma rays – radially extending files of cells produced by the cambium alongside water conduits and often remaining in direct contact with vessels or tracheids. In many angiosperms, vessels are in contact with multiple parenchymal rays that link the vertical water transport system into an intricate network of interconnected pathways. At the extreme, vessels are fully surrounded by living cells – as in the case of the black locust (Robinia) (Fromard et al. 1995), multiple palm trees (Tomlinson et al. 2001; Tomlinson and Spangler 2002), and even maize roots (Barrieu et al. 1998). These parenchymal cells are often connected with vessels via simple pits with narrow straight walls. The role of living axial and radial parenchyma cells in the xylem remains ambiguous. They have been shown to store carbohydrates in the form of starch that may be used to support spring bloom or bud growth (Lebon et al. 2005, 2008; Sperling et al. 2015). In some cases, these cells are responsible for the formation of tyloses – vascular occlusions formed by the ingrowth of cells into the vessel through the pits. These ingrowths usually occur in winter (Cochard and Tyree 1990), in response to infection by pathogens (Beckman and Talboys 1981; Davison and Tay 1985) and/or in response to wounding (Sun et al. 2006, 2008) and completely cease the transport function of occluded vessels. Yet another potential function might be related to radial redistribution of water among functional vessels, cambium, and phloem, which may provide both water and energy to redistribute solutes and actively refill the embolized conduits.

Interestingly, both angiosperm and gymnosperm species transport water only in conduits adjoined to living parenchymal cells. The death of parenchyma cells inevitably is linked to the loss of water transport capacity and formation of heartwood. Cell death is most likely caused by a decrease in oxygen concentration rather than a loss of xylem water transport capacity (Spicer and Holbrook 2007). Thus the dependence of water transport on living parenchymal cells further suggests that cellular activity is the key aspect of xylem function maintenance over long time periods.

The major interruption of xylem water transport is embolism formation. The close association of viable xylem conduits and xylem parenchymal cells suggests that these living support cells are involved in xylem recovery from embolism, possibly enabling the mobilization of water against existing energy gradients. Visual evidence from cryo-scanning electron microscopy studies, magnetic resonance imaging observations, and computed tomography scans shows that vessels indeed fill up with water during recovery (Holbrook et al. 2001; Clearwater and Goldstein 2005; Scheenen et al. 2007) and water droplets preferentially form and grow until the lumen completely refills on the vessel walls in contact with living parenchymal cells (Brodersen et al. 2010; Holbrook et al. 2001). However, these observational studies do not provide any indication of sources and pathways involved in moving the water required for recovery. As processes related to water transport across the cellular membrane involve the activity of specific water channels named aquaporins (AQP), the role of those and in particular the involvement of the plasma intrinsic proteins (PIPs) must be considered when contemplating how plants recover from embolism formation.

3 Aquaporins in the Vascular Tissue

The tissue-specific localization of AQP expression, with consideration of specific isoforms and rates of expression, can provide clues about the physiological roles of aquaporins and their temporal activity. The localization of AQPs is well described for leaves and roots of angiosperms, where they are expressed in the leaf sheath cells, in/around vascular bundles (*Arabidopsis*) and apoplastic barriers of roots (exodermis and endodermis), suggesting a crucial role in transmembrane water diffusion/control in the barriers separating the plant from its environment (Gambetta et al. 2013; Kirch et al. 2000; Perrone et al. 2012a; Chaumont and Tyerman 2014; Schaffner 1998; Suga et al. 2003; Hachez et al. 2006, 2008, 2012; Shatil-Cohen et al. 2011; Vandeleur et al. 2009; Prado et al. 2013). If xylem parenchyma cells have to supply a significant fraction of the water required for refilling embolized vessels, water must pass through a cellular membrane, and therefore, the flow must be facilitated by aquaporins and can be controlled by the number, activity, and localization of these proteins. Despite reported observations of AQP abundance in stem tissues, only a few studies have focused on the xylem. Molecular and microscopic studies have revealed that AQPs are highly expressed in the xylem parenchymal cells. For instance, the ZmTIP1;1 is expressed in tonoplast of cells surrounding the mature xylem vessels of roots and stems and in the phloem companion cells of maize plants (Barrieu et al. 1998, Fig. 2). In spinach, the SoPIP1;2 is highly expressed in the phloem sieve elements of leaves, roots, and petioles while SoPIP1;1 is present in stomatal guard cells (Fraysse et al. 2005). A detailed description of tobacco NtAQP1 localization reports that younger stems express the protein in developing xylem vessels and internal phloem cells, while older stems accumulate the protein in the outer xylem border and internal phloem (Otto and Kaldenhoff 2000). This specific localization of AQP isoforms in or around conduits implicates their role in permitting a transcellular water transfer between xylem conduits and, potentially, phloem via xylem parenchyma cells. It also reflects geometry of water transport, as flux density is highest near the narrow conduits.

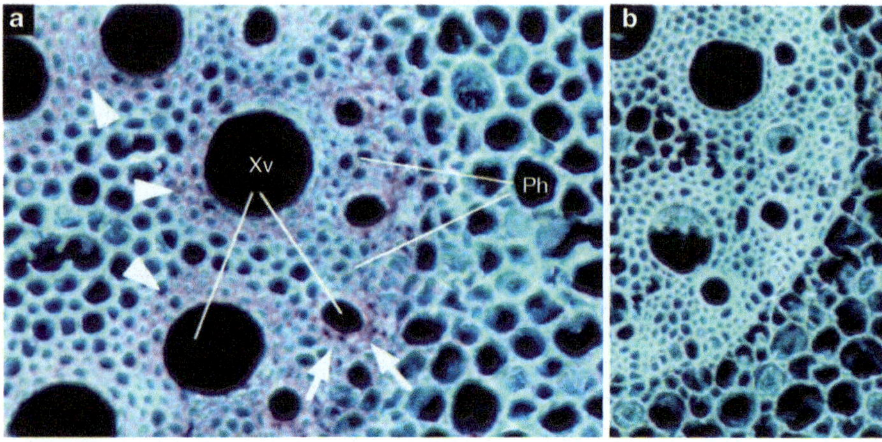

Fig. 2 *In situ* localization of ZmTIP1 mRNA in mature maize root. Transverse sections of the root (10–12 cm from the tip) were hybridized with ZmTIP1 antisense (**a**) or sense (**b**) digoxigenin-labeled RNA probes and photographed under dark-field conditions. The transcript signal is *red*. (**a**) Expression of ZmTIP1 in the parenchyma cells of early (*arrows*) and late (*arrowheads*) xylem vessels. *Xv* Xylem vessels, *Ph* phloem strand. (**b**) Control section hybridized with a ZmTIP1 sense probe (Barrieu et al. 1998)

Most studies on AQPs have applied bulk tissue analysis to herbaceous plants (i.e., total xylem and/or bark), while the stems of trees have received much less attention. However, detailed information for trees showing the localization of AQP expression in particular cells and tissue types is reported for the stems of a hybrid poplar (Almeida-Rodriguez and Hacke 2012). There, the greatest accumulation of expression occurred in the cambial region and adjacent xylem-phloem cells. Aquaporin accumulation was also detected in ray cells. Interestingly, the cells connected with vessels through pits, or contact cells, exhibited particularly high AQP protein expression, suggesting an increased potential for water exchange between apoplast and symplast. The ray cells not in contact with vessels, or isolation cells, accumulated water channels to varying degrees. Additional studies on walnut (*Juglans regia*) showed a higher expression of two aquaporin proteins (JrPIP2.1 and JrPIP2.2) in specific vessel-associated parenchyma cells (VACs), which are living cells in direct contact with vessels (Sakr et al. 2003, Fig. 3). In a study with another woody perennial, the aquaporin transcript profile was examined on VACs isolated from petioles (by laser microdissection) and on whole petioles of grapevine, confirming their specificity. While some of the *VvPIP1*- and *VvPIP2*-tested genes were activated by stress and subsequent recovery in whole petioles, some aquaporin genes *VvPIP1;1* and *VvPIP2;4N* were exclusively expressed in VACs (Chitarra et al. 2014).

Another omission in the AQP localization literature applies to non-angiosperm plant groups like ferns and gymnosperms. Available data on the expression of aquaporins in the needles of *Picea glauca* show that for drought-stressed trees, expression is abundant in the endodermis-like bundle sheath, in phloem cells, and in

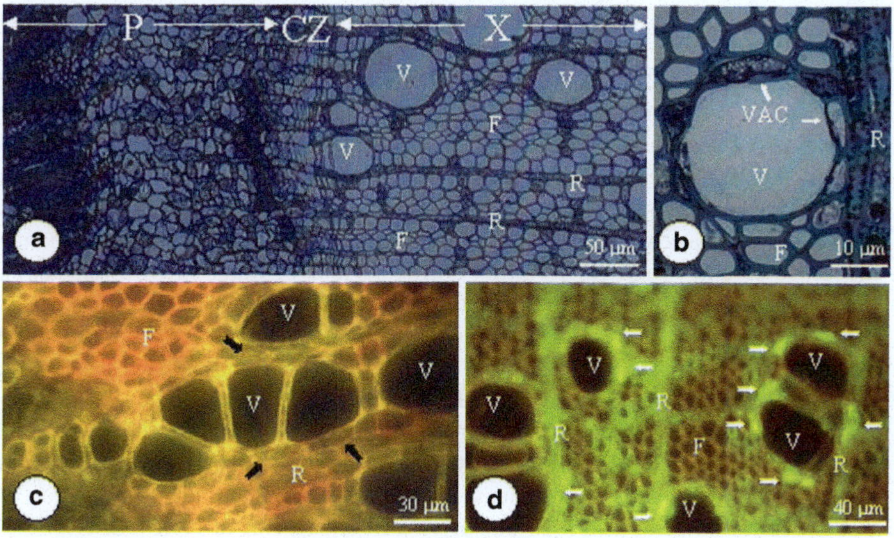

Fig. 3 Transversal section showing the general organization of the stem in walnut tree. (**a**) Cambium zone appears at the phloem-xylem interface. Xylem contains lignified xylem vessels and fibers (died cells) associated with parenchyma cells (living cells). The support tissue is constituted of fibers emptied of their content. (**b**) Cell types in walnut xylem. VACs (*white arrows*) are specialized cells that surround the xylem vessels. From their localization in the tissue, VACs have the ability to control nutrient exchanges between the parenchyma cells and the xylem vessels. Localization of aquaporins in xylem tissue of walnut tree sampled in February (winter period). (**c**) No recognition in control (AtPIP2 antiserum saturated by the purified JrPIP2,1 protein). *Black arrows* indicated the localization of VACs. (**d**) Section showing extensive green immunofluorescence (*white arrows*) in VACs. *X* xylem, *P* phloem, *CZ* cambial zone, *V* xylem vessel, *F* fiber, *R* ray cell (Sakr et al. 2003)

transfusion parenchyma tissue, further suggesting that water channels are localized in vascular tissue (Laur and Hacke 2014a). Similar results were reported by Mayr et al. 2014, showing higher amounts of PIP1 and PIP2 proteins in the endodermis and phloem cells of the needles of Norway spruce (*Picea abies*). In *Cheilanthes lanosa*, a xerophytic fern, it was shown that a PIP1 might have a key role in water balance mainly in the gametophyte stages (Diamond et al. 2012).

4 Aquaporins in Stems Under Water Stress and Embolism Formation

Water stress has a strong influence on AQP gene expression (see also chapter "Plant Aquaporins and Abiotic Stress"). However, studies attempting to relate physiological water stress responses to the expression patterns of different aquaporins have led to contrasting results. Upregulation, downregulation, and no change have all been reported (Baiges et al. 2002). Variation in the range of transcriptional responses might be species or tissue specific, associated with stress level and duration, or dependent upon the specific physiological role of each AQP gene isoform (Alexandersson et al. 2005; Kaldenhoff et al. 2008; Galmes et al. 2007). Furthermore, differences between drought-adapted and nonadapted varieties can affect aquaporin expression (Lian et al. 2004). Thus, it is currently very difficult to provide a general pattern of AQP gene expression in response to water stress; aquaporin upregulation is thought to increase membrane permeability to water transport when water is less available (Yamada et al. 1997), but the downregulation of AQP gene expression may encourage cellular water conservation during periods of water stress (Smart et al. 2001; Li et al. 2004). It is probable that in order to maintain a suitable water status under abiotic stress, both increased water transport via AQP in some tissues and reduced water transport in other tissues are required (Jang et al. 2004).

The effects of drought treatment on the expression of *AQP* genes have been studied in numerous species, and the downregulation has been frequently observed. *PIP* and *TIP* genes are downregulated in the leaves, shoots, and roots of *Nicotiana glauca* (Smart et al. 2001). In the leaves of *Arabidopsis*, the gradual imposition of drought stress downregulated 10 out of 13 PIP aquaporins at both transcript and protein levels. Of the three remaining, one of the isoforms (*AtPIP2;6*) was maintained at the same expression level and two (*AtPIP1;4* and *AtPIP2;5*) were upregulated (Alexandersson et al. 2005). The strong downregulation of *PIP* gene transcription under drought stress was also observed in the roots and twigs of olives (Secchi et al. 2007a, b) as well as in tobacco roots (Mahdieh et al. 2008) and in the leaves of *Populus trichocarpa* (Laur and Hacke 2014b).

Evidence for the downregulation of AQPs in response to drought may be contrasted with data suggesting that some tissue-specific AQP isoforms show increased expression in response to drought. For example, the *VvPIP1;1* gene in the roots of grapevines was upregulated by drought stress in an anisohydric but not in an isohydric cultivar (Vandeleur et al. 2009). In the stems of *P. trichocarpa*, expression levels of the PIP2 subfamily did not change in response to water stress or embolism presence, while some genes from the PIP1 subfamily were highly upregulated (Secchi and Zwieniecki 2010). Similar results were found by Chitarra et al. 2014 showing that two aquaporin genes (*VvPIP2;1* and *VvPIP2;4N*) were activated upon stress in petioles. Two other studies performed on different rootstocks of *Vitis* sp. showed similar results where drought treatments resulted in significant variations (both up- and downregulation) in leaf aquaporin gene expression over time (Galmes et al. 2007; Pou et al. 2013).

It is difficult to isolate response to embolism specifically from response to water stress in general and, consequently, analyses of embolism-induced AQP expression may often be confounded by water-stress-induced AQP expression. As mentioned earlier, the expression of aquaporins in *J. regia* was induced in VACs in response to water stress, while xylem parenchyma cells not in contact with vessels varied in their AQP expression due to water stress (Sakr et al. 2003). Considering this pattern and after imposing stress levels large enough to cause widespread embolism in the xylem of *J. regia*, it was inferred that the presence of embolism is associated with the expression of at least two aquaporins (JrPIP2.1 and JrPIP2.2). The progression of recovery in both stem water potential and stem water conductance did not immediately reduce the expression suggesting that embolism presence or rather a lack of vessel functionality in terms of water transport under tension was required to maintain this elevated expression level of studied AQPs in VACs. It is worth noting that this over expression of AQPs in VACs was ubiquitously occurring along the xylem vessels and continued to be present during recovery suggesting that the two studied proteins may play a role in the regulation of water flux between VACs and adjacent vessels (Sakr et al. 2003) by redistributing water between functional and embolized vessels.

The induction of water stress is usually a slow process that in natural conditions can extend for days or even weeks. The prolonged implementation of the stress might be the basis for the high variability in AQP expression observed in response to drought treatments. Embolism formation is, on the other hand, a very fast event that results in both the evacuation of vessels and the cessation of water movement due to the blocking of conduits by embolisms in distal locations. Can plants exclusively respond to embolism formation without stress-related changes? To answer this question, a study involving the induction of embolism in the stem of *P. trichocarpa* plants and the determination of concurrent expressions of AQPs was performed (Secchi et al. 2011; Secchi and Zwieniecki 2010). In these studies, embolism was induced by forcing air into the stem of non-stressed plants. Both a genome-wide analysis and the specific analysis of selected genes showed that the expression of some AQPs from the PIP1 subfamily (*PtPIP1.1* and *PtPIP1.3*) in poplar stems increased due to embolism presence alone without changes in stem water potential. This expression change occurred in less than half an hour following embolism formation (Secchi and Zwieniecki 2010; Secchi et al. 2011). These studies suggest that plants can possibly sense the formation of embolism (or its presence) separately from water stress.

There is no direct evidence that embolism formation independent of water stress can induce AQP expression in conifers. However, conifers growing at high elevations are often subjected to winter embolism (Sparks and Black 2000; Mayr et al. 2002, 2006), and some of these species were found to recover from hydraulic failure in late winter and spring when the snow on branch surfaces started to melt (Sparks et al. 2001; Limm et al. 2009). In such situations, water stress is minimal as humidity and water availability are very high despite the fact that conduits are mostly embolized. In these conditions, increased amounts of PIP1 and PIP2 proteins in the needle endodermis and phloem cells were detected (Mayr et al. 2014) (providing support that conifers also may detect embolism and respond to its presence with the upregulation of a few specific AQPs).

5 The Role of Aquaporins in the Maintenance of Xylem Water Transport

Observed changes in AQP expression during the onset of water stress may be interpreted and tested at the sites of water exchange between plant and environment. For example, an increase in AQP expression or activity might aim to reduce the resistance to water flow at distally located water uptake sites such as roots in order to beneficially reduce stress. Conversely, decreasing AQP expression and activity and so increasing resistance to water movement across living cell membranes might be beneficial at sites of water loss like leaves. In fact, the roles of root and leaf aquaporins in relation to changes in hydraulic resistance and plant susceptibility to stress have been tested using genetic manipulation approaches. For example, tobacco plants (*Nicotiana tabacum)* with downregulated aquaporin 1 (NtAQP1) show reduced root hydraulic conductivity and lower water stress resistance (Siefritz et al. 2002). *Arabidopsis thaliana* plants expressing *PIP* antisense genes exhibited an impaired ability to recover from water stress (Martre et al. 2002), and *Arabidopsis* knockout mutants were characterized by reduced leaf hydraulic conductivity (Da Ines et al. 2010) and root hydrostatic hydraulic conductivity (Postaire et al. 2010). Plants with silenced *PIP1* genes demonstrated decreased transcript and protein levels and decreased mesophyll and bundle sheath osmotic water permeability among many other physiological parameters (Sade et al. 2014). On the other hand, *Arabidopsis* expressing an AQP (PIP1) from *Vicia faba* exhibit a faster growth rate, a lower transpiration rate, and a greater drought tolerance compared to control plants (Cui et al. 2008). A better tolerance to several abiotic stresses was also displayed in transgenic banana plants overexpressing an isoform of the PIP1 subfamily (Sreedharan et al. 2013). All of these studies have focused on distal locations (roots, leaves), or ubiquitous changes in expression throughout the entire plant, however, and so do not provide insight into the physiological role of AQPs expressed in vascular tissue.

Interestingly, applications of genetic manipulation technologies to elucidate the roles of particular aquaporins in woody plants have not been as successful as equivalent studies in herbaceous plants. For example, the overexpression of PIP2;4 root-specific aquaporin enhanced water transport in transformed *Vitis* spp. under well-watered conditions, but not under water stress (Perrone et al. 2012a), and *Eucalyptus* spp. hybrid clones overexpressing two *Raphanus sativus* genes (*RsPIP1;1* and *RsPIP2;1*) did not display any increase in drought tolerance (Tsuchihira et al. 2010). The downregulation of the entire PIP1 family in *Populus tremula x alba* by 80 % also had a very minimal effect on the majority of physiological functions prior to the onset of water stress (Secchi and Zwieniecki 2013). These ambiguous results might be related to the fact that in woody plants the xylem contributes significantly to a plant's stress response only. Considering that xylem functionality likely dominates hydraulic resistance under stress due to dynamic changes in embolism level, a tree's overall response might not depend on the distal locations of water exchange with the environment so much as AQPs that maintain xylem hydraulic capacity.

The indication that AQPs are indeed involved in the maintenance of xylem hydraulic function involves the earlier described localization of AQPs in xylem parenchymal cells, including VACs, observed increases in expression with the onset of embolism formation, and the fact that water droplets grow, form, and expand on walls in contact with living cells. This deductive interpretation of evidence is supported by observations of transgenic poplar trees characterized by the dramatic downregulation of multiple isoforms belonging to the PIP1 subfamily (Secchi and Zwieniecki 2014). As different poplar PIP1 isoforms in the stem were upregulated in response to the induction of embolism in the presence of water stress and were downregulated soon after full recovery occurred (Secchi and Zwieniecki 2010), the downregulation of AQPs was expected to delay the hydraulic restoration process. Indeed, transgenic plants with significantly reduced amounts of various AQP isoform transcripts in *P. trichocarpa* leaves (Laur and Hacke 2014b) and stems (Secchi and Zwieniecki 2014) significantly delayed the restoration of leaf hydraulic conductance and xylem functionality upon recovery from water stress. In addition to the delay in recovery, an unexpected finding was that the downregulation of PIP1 expression resulted in a significant shift in the susceptibility of *P. tremula x alba* xylem to embolism formation; the transgenic poplars were found indeed to be more sensitive to imposed water stress resulting in increased vulnerability to embolism formation (Secchi and Zwieniecki 2014). This shift in susceptibility to embolism formation holds important clues to both the role of aquaporins in xylem responses to embolism and the recovery process itself. As embolism formation is considered a function of cellular wall chemistry and xylem anatomical features, aquaporins should not affect susceptibility to embolism. Thus, the observed shift must be the result of physiological processes happening in the xylem during its normal function and should be referred to as an "apparent susceptibility." As we currently understand it, apparent susceptibility to embolism is a balance between the rates of embolism formation (functionally linked to water stress) and the capacity of living parenchyma cells to refill (inversely linked to water stress) (Secchi and Zwieniecki 2012, 2014). Because the latter is related to the parenchymal capacity to redistribute water between living cells and the xylem apoplast, it is thus a function of aquaporin activity. In such a context, it is easy to see that the downregulation of AQPs may affect this balance by both reducing the capacity to refill and shifting apparent susceptibility.

The radial redistribution of stem water during the process of hydraulic recovery may only be a secondary role while the primary role of stem AQPs is related to the absorption of water from the environment. Some recent studies suggest that many plant species have the capacity to absorb rain, melting snow, or fog water directly into their leaves and even through bark. This strategy of water uptake provides a means to relieve localized disruptions to hydraulic conductivity and to reduce tracheid embolization (Limm et al. 2009; Mayr et al. 2014; Laur and Hacke 2014a; Earles et al. 2015). In such a case, stem AQPs may behave similarly to roots and leaves, following the typical function/expression patterns observed in distal locations, i.e., activation during times of water availability (wet bark) and deactivation during drought. For example, increased aquaporin expression in the endodermis-like bundle sheath, phloem cells, and transfusion parenchyma of drought-stressed *P. glauca* needles was observed when the same needles were exposed to high relative humidity, supporting the idea that there is a role for AQPs in transferring absorbed water to vascular tissue (Laur and Hacke 2014a). The reconciliation of observed increases in AQP expression in the stems of drought-stressed trees and the above-mentioned role of AQPs in the transfer of water from wet bark requires an analysis of the temporal and spatial aspects of expression distribution.

In general, drought results in the induction of AQP expression in the xylem but not necessarily in the outer part of the stem. Large concurrent changes in the expression of other gene groups, however, may complete the pathway. In poplar xylem the upregulation of ion transport, additional aquaporins, and carbon metabolism has been detected (Secchi et al. 2011). Carbon metabolism and aquaporin expression were also strongly upregulated in drought-stressed grapevine petioles (Perrone et al. 2012b). This type of drought upregulation may be an indication of the stem priming for recovery. Under natural conditions, prolonged drought should eventually end with a significant rain event and the recovery of stem function. Even if drought leads to the loss of leaves, a rain event would wet the entire tree, including bark, providing easily accessible water in a relatively short period of time. Uptake would be facilitated if the hydraulic resistance of the path is low, and the energy gradient favors the flow of water into the stem. The cambium with its limited apoplastic path and multiple layers of membrane serves as a significant barrier to water flow, and the reduction of resistance requires the presence of aquaporin (Barrowclough et al. 2000). In addition, if the plant can manage to direct flow into embolized vessels with the specific localized expression of AQPs in VACs (see Chitarra et al. 2014), then such global crown wetting may indeed provide a mechanism for the recovery of functional xylem transport.

6 Aquaporin Upregulation in the Xylem: Signaling

Very little is known about embolism-related regulation of aquaporins in the xylem. There are currently two, not mutually exclusive, experimentally supported views on the topic. The first idea about signaling relies on the sudden transition between high tension when water is in its liquid state to zero tension/pressure when water under tension converts to vapor, releasing mechanical stresses in the cellular wall as well as energy in the form of sound (Tyree and Sperry 1989; Nardini et al. 2011; Tyree and Dixon 1983; Johnson et al. 2009; Zweifel and Zeugin 2008). Such spatially specific events are thought to interact with mechanosensors in VAC membranes and trigger embolism-specific expression patterns that are upregulated until the restoration of tension is achieved (Salleo et al. 2000). This idea has some experimental support, as there have been experiments aimed to test generation of mechanosensor triggering without the implementation of water stress (see more details in chapter "Root Hydraulic and Aquaporin Responses to N Availability"). We do, however, know that induction of embolism in non-stressed plants (when mechanical stresses are at their minimum) also results in the triggering of AQP expression (Secchi and Zwieniecki 2010, 2011).

The second signaling path relates to the fact that embolism formation stops water transport around VAC cells, dramatically changing surrounding mass flow/diffusional paths including the rate of diffusion of respirational CO_2, thus facilitating its efflux to the void and changes in apoplastic pH, concentration of ions and sugars, and the cell vicinity by cessation of the washout effect of the transpirational stream (Nardini et al.

2011). All of these changes are known to be involved in multiple signaling paths including expression of AQPs. Recent experiments aimed at testing if the presence of sucrose in the stem would trigger a similar expression response as in the case of embolism only showed that indeed multiple gene ontology (GO) groups including ion transporters, AQPs, and reactive oxygen species responded similarly, suggesting that there might be some shared signaling pathways between sugar concentration and embolism presence (Secchi and Zwieniecki 2011). Despite these efforts, the question on what signaling paths lead to embolism-specific expression changes is wide open.

7 Conclusions

The role of aquaporins in the maintenance of xylem hydraulic function remains an active research field. The spatial distribution of AQPs in xylem parenchyma cells, the dynamics of expression in response to the development of water stress, and during the recovery from water stress only indirectly point to their role in facilitating recovery from embolism. With exception of few studies using chemical inhibitors of aquaporins (Lovisolo and Schubert 2006; Voicu and Zwiazek 2010), most of the observations aimed at testing radial water transport in stems and leaves in relation to AQPs were correlative and not manipulative and still need direct experimental proof of concept. The genetic manipulation of expression level designed specifically to test stem hydraulic recovery was only performed on one species (*P. tremula x alba*) and only for one subfamily of AQPs genes (PIP1) (Secchi and Zwieniecki 2014). These results proved that while the susceptibility of xylem hydraulic capacity to water stress was affected by the downregulation of AQPs, recovery was only marginally affected. This suggests that our understanding of the embolism formation-recovery cycle is the main obstacle to progress in our knowledge of the physiological role of xylem aquaporins.

References

Alexandersson E, Fraysse L, Sjovall-Larsen S, Gustavsson S, Fellert M, Karlsson M, Johanson U, Kjellbom P (2005) Whole gene family expression and drought stress regulation of aquaporins. Plant Mol Biol 59:469–484

Almeida-Rodriguez AM, Hacke UG (2012) Cellular localization of aquaporin mRNA in hybrid poplar stems. Am J Bot 99(7):1249–1254. doi:10.3732/ajb.1200088

Baiges I, Schaffner AR, Affenzeller MJ, Mas A (2002) Plant aquaporins. Physiol Plant 115(2): 175–182. doi:10.1034/j.1399-3054.2002.1150201.x

Barigah TS, Charrier O, Douris M, Bonhomme M, Herbette S, Ameglio T, Fichot R, Brignolas F, Cochard H (2013) Water stress-induced xylem hydraulic failure is a causal factor of tree mortality in beech and poplar. Ann Bot 112(7):1431–1437. doi:10.1093/aob/mct204

Barrieu F, Chaumont F, Chrispeels MJ (1998) High expression of the tonoplast aquaporin ZmTIP1 in epidermal and conducting tissues of maize. Plant Physiol 117(4):1153–1163. doi:10.1104/pp.117.4.1153

Barrowclough DE, Peterson CA, Steudle E (2000) Radial hydraulic conductivity along developing onion roots. J Exp Bot 51(344):547–557. doi:10.1093/jexbot/51.344.547

Beckman CH, Talboys PW (1981) Anatomy of resistance. In: Mace ME, Bell AA, Beckman CH (eds) Fungal wilt diseases of plants. Academic, New York, pp 487–521

Brenner CE (1995) Cavitation and bubble dynamics, Oxford Engineering Science Series, vol 44. Oxford University Press, Oxford

Brodersen CR, McElrone AJ, Choat B, Matthews MA, Shackel KA (2010) The dynamics of embolism repair in xylem: in vivo visualizations using high-resolution computed tomography. Plant Physiol 154(3):1088–1095. doi:10.1104/pp.110.162396

Brodribb TJ, Jordan GJ (2008) Internal coordination between hydraulics and stomatal control in leaves. Plant Cell Environ 31(11):1557–1564. doi:10.1111/j.1365-3040.2008.01865.x

Chapotin SM, Razanameharizaka JH, Holbrook NM (2006) Abiomechanical perspective on the role of large stem volume and high water content in baobab trees (Adansonia spp.; Bombacaceae). Am J Bot 93(9):1251–1264. doi:10.3732/ajb.93.9.1251

Chaumont F, Tyerman SD (2014) Aquaporins: highly regulated channels controlling plant water relations. Plant Physiol 164(4):1600–1618. doi:10.1104/pp.113.233791

Chitarra W, Balestrini R, Vitali M, Pagliarani C, Perrone I, Schubert A, Lovisolo C (2014) Gene expression in vessel-associated cells upon xylem embolism repair in *Vitis vinifera* L. petioles. Planta 239(4):887–899. doi:10.1007/s00425-013-2017-7

Choat B, Jansen S, Brodribb TJ, Cochard H, Delzon S, Bhaskar R, Bucci SJ, Feild TS, Gleason SM, Hacke UG, Jacobsen AL, Lens F, Maherali H, Martinez-Vilalta J, Mayr S, Mencuccini M, Mitchell PJ, Nardini A, Pittermann J, Pratt RB, Sperry JS, Westoby M, Wright IJ, Zanne AE (2012) Global convergence in the vulnerability of forests to drought. Nature 491(7426): 752–755. doi:10.1038/nature11688

Clearwater M, Goldstein G (2005) Embolism repair and long distance transport. In: Holbrook NM, Zwieniecki MA (eds) Vascular transport in plants. Elsevier, Amsterdam, pp 201–220

Cochard H, Tyree MT (1990) Xylem dysfunction in Quercus vessel sizes, tyloses, cavitation and seasonal-changes in embolism. Tree Physiol 6(4):393–407

Cui XH, Hao FS, Chen H, Chen J, Wang XC (2008) Expression of the *Vicia faba* VfPIP1 gene in *Arabidopsis thaliana* plants improves their drought resistance. J Plant Res 121(2):207–214. doi:10.1007/s10265-007-0130-z

Da Ines O, Graf W, Franck KI, Albert A, Winkler JB, Scherb H, Stichler W, Schaffner AR (2010) Kinetic analyses of plant water relocation using deuterium as tracer – reduced water flux of Arabidopsis pip2 aquaporin knockout mutants. Plant Biol 12:129–139. doi:10.1111/j.1438-8677.2010.00385.x

Davison EM, Tay FCS (1985) The effect of waterlogging on seedlings of *Eucalyptus marginata*. New Phytol 101(4):743–753. doi:10.1111/j.1469-8137.1985.tb02879.x

Diamond HL, Jones HR, Swatzell LJ (2012) The role of aquaporins in water balance in *Cheilanthes lanosa* (Adiantaceae) gametophytes. Am Fern J 102(1):11–31

Earles JM, Sperling O, Silva LC, McElrone A, Brodersen C, North M, Zwieniecki M (2015) Bark water uptake promotes localized hydraulic recovery in coastal redwood crown. Plant Cell Environ. doi:10.1111/pce.12612

Fraysse LC, Wells B, McCann MC, Kjellbom P (2005) Specific plasma membrane aquaporins of the PIP1 subfamily are expressed in sieve elements and guard cells. Biol Cell 97(7):519–534

Fromard L, Babin V, Fleuratlessard P, Fromont JC, Serrano R, Bonnemain JL (1995) Control of vascular sap pH by the vessel-associated cells in woody species – physiological and immunological studies. Plant Physiol 108(3):913–918

Galmes J, Pou A, Alsina MM, Tomas M, Medrano H, Flexas J (2007) Aquaporin expression in response to different water stress intensities and recovery in Richter-110 (Vitis sp.): relationship with ecophysiological status. Planta 226(3):671–681. doi:10.1007/s00425-007-0515-1

Gambetta GA, Fei J, Rost TL, Knipfer T, Matthews MA, Shackel KA, Walker MA, McElrone AJ (2013) Water uptake along the length of grapevine fine roots: developmental anatomy, tissue-specific aquaporin expression, and pathways of water transport. Plant Physiol 163(3): 1254–1265. doi:10.1104/pp.113.221283

Hachez C, Moshelion M, Zelazny E, Cavez D, Chaumont F (2006) Localization and quantification of plasma membrane aquaporin expression in maize primary root: a clue to understanding their role as cellular plumbers. Plant Mol Biol 62(1–2):305–323. doi:10.1007/s11103-006-9022-1

Hachez C, Heinen RB, Draye X, Chaumont F (2008) The expression pattern of plasma membrane aquaporins in maize leaf highlights their role in hydraulic regulation. Plant Mol Biol 68(4–5):337–353. doi:10.1007/s11103-008-9373-x

Hachez C, Veselov D, Ye Q, Reinhardt H, Knipfer T, Fricke W, Chaumont F (2012) Short-term control of maize cell and root water permeability through plasma membrane aquaporin isoforms. Plant Cell Environ 35(1):185–198. doi:10.1111/j.1365-3040.2011.02429.x

Hacke UG, Stiller V, Sperry JS, Pittermann J, McCulloh KA (2001) Cavitation fatigue. Embolism and refilling cycles can weaken the cavitation resistance of xylem. Plant Physiol 125(2):779–786. doi:10.1104/pp.125.2.779

Holbrook NM, Zwieniecki MA (1999) Embolism repair and xylem tension: do we need a miracle? Plant Physiol 120(1):7–10

Holbrook NM, Ahrens ET, Burns MJ, Zwieniecki MA (2001) In vivo observation of cavitation and embolism repair using magnetic resonance imaging. Plant Physiol 126(1):27–31

Jang JY, Kim DG, Kim YO, Kim JS, Kang HS (2004) An expression analysis of a gene family encoding plasma membrane aquaporins in response to abiotic stresses in *Arabidopsis thaliana*. Plant Mol Biol 54(5):713–725. doi:10.1023/B:PLAN.0000040900.61345.a6

Jensen KH, Berg-Sørensen K, Bruus H, Holbrook NM, Liesche J, Schulz A, Zwieniecki MA, Bohr T (2016) Sap flow and sugar transport in plants. Reviews of modern physics (in press)

Johnson DM, Meinzer FC, Woodruff DR, McCulloh KA (2009) Leaf xylem embolism, detected acoustically and by cryo-SEM, corresponds to decreases in leaf hydraulic conductance in four evergreen species. Plant Cell Environ 32(7):828–836. doi:10.1111/j.1365-3040.2009.01961.x

Kaldenhoff R, Ribas-Carbo M, Flexas J, Lovisolo C, Heckwolf M, Uehlein N (2008) Aquaporins and plant water balance. Plant Cell Environ 31(5):658–666. doi:10.1111/j.1365-3040.2008.01792.x

Kirch HH, Vera-Estrella R, Golldack D, Quigley F, Michalowski CB, Barkla BJ, Bohnert HJ (2000) Expression of water channel proteins in *Mesembryanthemum crystallinum*. Plant Physiol 123(1):111–124. doi:10.1104/pp.123.1.111

Laur J, Hacke UG (2014a) Exploring *Picea glauca* aquaporins in the context of needle water uptake and xylem refilling. New Phytol 203(2):388–400. doi:10.1111/nph.12806

Laur J, Hacke UG (2014b) The role of water channel proteins in facilitating recovery of leaf hydraulic conductance from water stress in *Populus trichocarpa*. PLoS One 9(11):e111751. doi:10.1371/journal.pone.0111751

Lebon G, Duchene E, Brun O, Clement C (2005) Phenology of flowering and starch accumulation in grape (*Vitis vinifera* L.) cuttings and vines. Ann Bot 95(6):943–948. doi:10.1093/aob/mci108

Lebon G, Wojnarowiez G, Holzapfel B, Fontaine F, Vaillant-Gaveau N, Clement C (2008) Sugars and flowering in the grapevine (*Vitis vinifera* L.). J Exp Bot 59(10):2565–2578. doi:10.1093/jxb/ern135

Li Y, Wang GX, Xin M, Yang HM, Wu XJ, Li T (2004) The parameters of guard cell calcium oscillation encodes stomatal oscillation and closure in *Vicia faba*. Plant Sci 166(2):415–421. doi:10.1016/j.plantsci.2003.10.008

Lian HL, Yu X, Ye Q, Ding XS, Kitagawa Y, Kwak SS, Su WA, Tang ZC (2004) The role of aquaporin RWC3 in drought avoidance in rice. Plant Cell Physiol 45(4):481–489. doi:10.1093/pcp/pch058

Limm EB, Simonin KA, Bothman AG, Dawson TE (2009) Foliar water uptake: a common water acquisition strategy for plants of the redwood forest. Oecologia 161(3):449–459. doi:10.1007/s00442-009-1400-3

Lovisolo C, Schubert A (2006) Mercury hinders recovery of shoot hydraulic conductivity during grapevine rehydration: evidence from a whole-plant approach. New Phytol 172(3):469–478. doi:10.1111/j.1469-8137.2006.01852.x

Mahdieh M, Mostajeran A, Horie T, Katsuhara M (2008) Drought stress alters water relations and expression of PIP-type aquaporin genes in *Nicotiana tabacum* plants. Plant Cell Physiol 49(5):801–813. doi:10.1093/pcp/pcn054

Martre P, Morillon R, Barrieu F, North GB, Nobel PS, Chrispeels MJ (2002) Plasma membrane Aquaporins play a significant role during recovery from water deficit. Plant Physiol 130(4):2101–2110. doi:10.1104/pp.009019

Mayr S, Wolfschwenger M, Bauer H (2002) Winter-drought induced embolism in Norway spruce (*Picea abies*) at the alpine timberline. Physiol Plant 115(1):74–80. doi:10.1034/j.1399-3054.2002.1150108.x

Mayr S, Hacke U, Schmid P, Schwienbacher F, Gruber A (2006) Frost drought in conifers at the alpine timberline: xylem dysfunction and adaptations. Ecology 87(12):3175–3185. doi:10.1890/0012-9658(2006)87[3175:fdicat]2.0.co;2

Mayr S, Schmid P, Laur J, Rosner S, Charra-Vaskou K, Damon B, Hacke UG (2014) Uptake of water via branches helps timberline conifers refill embolized xylem in late winter. Plant Physiol 164(4):1731–1740. doi:10.1104/pp.114.236646

Nardini A, Lo Gullo MA, Salleo S (2011) Refilling embolized xylem conduits: is it a matter of phloem unloading? Plant Sci 180:604–611

Otto B, Kaldenhoff R (2000) Cell-specific expression of the mercury insensitive plasma-membrane aquaporin NtAQP1 from *Nicotiana tabacum*. Planta 211:167–172

Perrone I, Gambino G, Chitarra W, Vitali M, Pagliarani C, Riccomagno N, Balestrini R, Kaldenhoff R, Uehlein N, Gribaudo I, Schubert A, Lovisolo C (2012a) The grapevine root-specific aquaporin VvPIP2;4 N controls root hydraulic conductance and leaf gas exchange under well-watered conditions but not under water stress. Plant Physiol 160(2):965–977. doi:10.1104/pp.112.203455

Perrone I, Pagliarini C, Lovisolo C, Chitarra W, Roman F, Schubert A (2012b) Recovery from water stress affects grape leaf petiole transcriptome. Planta 235(6):1383–1396

Pockman WT, Sperry JS, Oleary JW (1995) Sustained and significant negative water-pressure in xylem. Nature 378:715–716

Postaire O, Tournaire-Roux C, Grondin A, Boursiac Y, Morillon R, Schaffner AR, Maurel C (2010) A PIP1 aquaporin contributes to hydrostatic pressure-induced water transport in both the root and rosette of Arabidopsis. Plant Physiol 152(3):1418–1430. doi:10.1104/pp.109.145326

Pou A, Medrano H, Flexas J, Tyerman SD (2013) A putative role for TIP and PIP aquaporins in dynamics of leaf hydraulic and stomatal conductances in grapevine under water stress and re-watering. Plant Cell Environ 36(4):828–843. doi:10.1111/pce.12019

Prado K, Boursiac Y, Tournaire-Roux C, Monneuse J-M, Postaire O, Da Ines O, Schaeffner AR, Hem S, Santoni V, Maurel C (2013) Regulation of Arabidopsis Leaf Hydraulics involves light-dependent phosphorylation of aquaporins in veins. Plant Cell 25(3):1029–1039. doi:10.1105/tpc.112.108456

Sade N, Shatil-Cohen A, Attia Z, Maurel C, Boursiac Y, Kelly G, Granot D, Yaaran A, Lerner S, Moshelion M (2014) The role of plasma membrane aquaporins in regulating the bundle sheath-mesophyll continuum and leaf hydraulics. Plant Physiol 166(3):1609–1620. doi:10.1104/pp.114.248633

Sakr S, Alves G, Morillon RL, Maurel K, Decourteix M, Guilliot A, Fleurat-Lessard P, Julien JL, Chrispeels MJ (2003) Plasma membrane aquaporins are involved in winter embolism recovery in walnut tree. Plant Physiol 133:630–641

Salleo S, Nardini A, Pitt F, Lo Gullo MA (2000) Xylem cavitation and hydraulic control of stomatal conductance in Laurel (*Laurus nobilis* L.). Plant Cell Environ 23(1):71–79. doi:10.1046/j.1365-3040.2000.00516.x

Schaffner AR (1998) Aquaporin function, structure, and expression: are there more surprises to surface in water relations? Planta 204(2):131–139. doi:10.1007/s004250050239

Scheenen TWJ, Vergeldt FJ, Heemskerk AM, Van As H (2007) Intact plant magnetic resonance imaging to study dynamics in long-distance sap flow and flow-conducting surface area. Plant Physiol 144(2):1157–1165. doi:10.1104/pp.106.089250

Secchi F, Zwieniecki MA (2010) Patterns of PIP gene expression in *Populus trichocarpa* during recovery from xylem embolism suggest a major role for the PIP1 aquaporin subfamily as moderators of refilling process. Plant Cell Environ 33(8):1285–1297

Secchi F, Zwieniecki MA (2011) Sensing embolism in xylem vessels: the role of sucrose as a trigger for refilling. Plant Cell Environ 34(3):514–524

Secchi F, Zwieniecki MA (2012) Analysis of xylem sap from functional (Nonembolized) and nonfunctional (Embolized) vessels of *Populus nigra*: chemistry of refilling. Plant Physiol 160(2):955–964. doi:10.1104/pp.112.200824

Secchi F, Zwieniecki MA (2013) The physiological response of *Populus tremula* x alba leaves to the down-regulation of PIP1 aquaporin gene expression under no water stress. Front Plant Sci 4:507. doi:10.3389/fpls.2013.00507

Secchi F, Zwieniecki MA (2014) Down-regulation of plasma intrinsic protein1 aquaporin in poplar trees is detrimental to recovery from embolism. Plant Physiol 164(4):1789–1799. doi:10.1104/pp.114.237511

Secchi F, Lovisolo C, Schubert A (2007a) Expression of OePIP2.1 aquaporin gene and water relations of *Olea europaea* twigs during drought stress and recovery. Ann Appl Biol 150(2):163–167. doi:10.1111/j.1744-7348.2007.00118.x

Secchi F, Lovisolo C, Uehlein N, Kaldenhoff R, Schubert A (2007b) Isolation and functional characterization of three aquaporins from olive (*Olea europaea* L.). Planta 225(2):381–392. doi:10.1007/s00425-006-0365-2

Secchi F, Gilbert ME, Zwieniecki MA (2011) Transcriptome response to embolism formation in stems of *Populus trichocarpa* provides insight into signaling and the biology of refilling. Plant Physiol 157:1419–1429

Shatil-Cohen A, Attia Z, Moshelion M (2011) Bundle-sheath cell regulation of xylem-mesophyll water transport via aquaporins under drought stress: a target of xylem-borne ABA? Plant J 67(1):72–80. doi:10.1111/j.1365-313X.2011.04576.x

Siefritz F, Tyree MT, Lovisolo C, Schubert A, Kaldenhoff R (2002) PIP1 plasma membrane aquaporins in tobacco: from cellular effects to function in plants. Plant Cell 14(4):869–876. doi:10.1105/tpc.000901

Smart LB, Moskal WA, Cameron KD, Bennett AB (2001) MIP genes are down-regulated under drought stress in *Nicotiana glauca*. Plant Cell Physiol 42(7):686–693. doi:10.1093/pcp/pce085

Sparks JP, Black RA (2000) Winter hydraulic conductivity end xylem cavitation in coniferous trees from upper and lower treeline. Arct Antarct Alp Res 32(4):397–403. doi:10.2307/1552388

Sparks JP, Campbell GS, Black RA (2001) Water content, hydraulic conductivity, and ice formation in winter stems of *Pinus contorta*: a TDR case study. Oecologia 127(4):468–475. doi:10.1007/s004420000587

Sperling O, Earles JM, Secchi F, Godfrey J, Zwieniecki MA (2015) Frost induces respiration and accelerates carbon depletion in trees. PLoS One 10(12):e0144124. doi:10.1371/journal.pone.0144124

Sperry JS (2003) Evolution of water transport and xylem structure. Int J Plant Sci 164(3):S115–S127

Sperry JS, Holbrook NM, Zimmermann MH, Tyree MT (1987) Spring filling of xylem vessels in wild grapevine. Plant Physiol 83(2):414–417

Sperry JS, Adler FR, Campbell GS, Comstock JP (1998) Limitation of plant water use by rhizosphere and xylem conductance: result from the model. Plant Cell Environ 21:347–359

Spicer R, Holbrook NM (2007) Effects of carbon dioxide and oxygen on sapwood respiration in five temperate tree species. J Exp Bot 58(6):1313–1320. doi:10.1093/jxb/erl296

Sreedharan S, Shekhawat UKS, Ganapathi TR (2013) Transgenic banana plants overexpressing a native plasma membrane aquaporin MusaPIP1;2 display high tolerance levels to different abiotic stresses. Plant Biotechnol J 8:942–952. doi:10.1111/pbi.12086

Stiller V, Sperry JS (2002) Cavitation fatigue and its reversal in sunflower (*Helianthus annuus* L.). J Exp Bot 53(371):1155–1161. doi:10.1093/jexbot/53.371.1155

Suga S, Murai M, Kuwagata T, Maeshima M (2003) Differences in aquaporin levels among cell types of radish and measurement of osmotic water permeability of individual protoplasts. Plant Cell Physiol 44(3):277–286. doi:10.1093/pcp/pcg032

Sun Q, Rost TL, Matthews MA (2006) Pruning-induced tylose development in stems of current-year shoots of *Vitis vinifera* (*Vitaceae*). Am J Bot 93(11):1567–1576. doi:10.3732/ajb.93.11.1567

Sun Q, Rost TL, Matthews MA (2008) Wound-induced vascular occlusions in *Vitis vinifera* (*Vitaceae*): tyloses in summer and gels in winter. Am J Bot 95(12):1498–1505. doi:10.3732/ajb.0800061

Tomlinson PB, Spangler R (2002) Developmental features of the discontinuous stem vascular system in the rattan palm Calamus (Arecaceae-Calamoideae-Calamineae). Am J Bot 89(7):1128–1141. doi:10.3732/ajb.89.7.1128

Tomlinson PB, Fisher JB, Spangler RE, Richer RA (2001) Stem vascular architecture in the rattan palm Calamus (Arecaceae-Calamoideae-Galaminae). Am J Bot 88(5):797–809. doi:10.2307/2657032

Tsuchihira A, Hanba YT, Kato N, Doi T, Kawazu T, Maeshima M (2010) Effect of overexpression of radish plasma membrane aquaporins on water-use efficiency, photosynthesis and growth of eucalyptus trees. Tree Physiol 30(3):417–430. doi:10.1093/treephys/tpp127

Tyree MT, Dixon MA (1983) Cavitation events in *Thuja occidentalis* L.?: utrasonic acoustic emissions from the sapwood can be measured. Plant Physiol 72(4):1094–1099. doi:10.1104/pp.72.4.1094

Tyree MT, Ewers FW (1991) The hydraulic architecture of trees and other woody-plants. New Phytol 119(3):345–360. doi:10.1111/j.1469-8137.1991.tb00035.x

Tyree MT, Sperry JS (1989) Characterization and propagation of acoustic-emission signals in woody-plants: towards an improved acoustic emission counter. Plant Cell Environ 12(4):371–382. doi:10.1111/j.1365-3040.1989.tb01953.x

Tyree MT, Zimmermann MH (2002) Xylem structure and the ascent of sap, 2nd edn. Springer, New York

Tyree MT, Salleo S, Nardini A, Lo Gullo MA, Mosca R (1999) Refilling of embolized vessels in young stems of Laurel. Do we need a new paradigm? Plant Physiol 120:11–21

Vandeleur RK, Mayo G, Shelden MC, Gilliham M, Kaiser BN, Tyerman SD (2009) The role of plasma membrane intrinsic protein aquaporins in water transport through roots: diurnal and drought stress responses reveal different strategies between isohydric and anisohydric cultivars of grapevine. Plant Physiol 149(1):445–460. doi:10.1104/pp.108.128645

Voicu MC, Zwiazek JJ (2010) Inhibitor studies of leaf lamina hydraulic conductance in trembling aspen (*Populus tremuloides* Michx.) leaves. Tree Physiol 30(2):193–204. doi:10.1093/treephys/tpp112

Yamada S, Nelson DE, Ley E, Marquez S, Bohnert HJ (1997) The expression of an aquaporin promoter from *Mesembryanthemum crystallinum* in tobacco. Plant Cell Physiol 38(12):1326–1332

Zweifel R, Zeugin F (2008) Ultrasonic acoustic emissions in drought-stressed trees – more than signals from cavitation? New Phytol 179(4):1070–1079. doi:10.1111/j.1469-8137.2008.02521.x

Zwieniecki MA, Holbrook NM (2009) Confronting Maxwell's demon: biophysics of xylem embolism repair. Trends Plant Sci 14(10):530–534

Plant Aquaporins and CO_2

Norbert Uehlein, Lei Kai, and Ralf Kaldenhoff

Abstract Aquaporins in plants show more abundant and greater diversity than aquaporins in bacteria and animals. This unique characteristic provided versatile tool boxes for plants, dealing with environmental changes, which overcome the disadvantage of immobility. Aquaporins were first known for their function as water channel proteins. Later on, more and more studies showed that other small solutes, i.e., ammonia, glycerol, urea, hydrogen peroxide and metalloids, can also pass through the channel of various aquaporins. Moreover, the function of aquaporins as CO_2 gas channels was studied by several groups (Nakhoul et al. Am J Physiol Cell Physiol 43(2):C543–C548, 1998; Yang et al. J Biol Chem 275(4):2686–2692, 2000; Tholen and Zhu Plant Physiol 156(1):90–105, 2011). In parallel, studies on model reconstituted membranes claim that no such type of channel would be needed due to the high permeability of those model membranes (Missner et al. Proc Natl Acad Sci USA 105(52):E123, 2008a; J Biol Chem 283(37):25340–25347, 2008b). However, experimental data showed the physiological significance of CO_2-conducting channels, particularly in plants. It is generally accepted that plant science presented the first evidence for the physiological relevance and importance of aquaporins as CO_2 transport facilitators (Boron Exp Physiol 95(12):1107–1130, 2010; Terashima and Ono Plant Cell Physiol 43(1):70–78, 2002; Uehlein et al. Nature 425 (6959):734–737, 2003; Heckwolf et al. Plant J 67 (5):795–804, 2011; Uehlein et al. Plant Cell 20(3):648–657, 2008). In this chapter, we discuss the CO_2 diffusion across membranes and the role of plant aquaporins during this process.

N. Uehlein • R. Kaldenhoff (✉)
Department of Biology, Applied Plant Sciences, Technische Universität Darmstadt, Schnittspahnstrasse 10, 64287 Darmstadt, Germany
e-mail: Kaldenhoff@bio.tu-darmstadt.de

L. Kai
Department of Cellular and Molecular Biophysics, Max-Planck-Institute of Biochemistry, Am Klopferspitz 18, 82152 Martinsried, Germany

F. Chaumont, S.D. Tyerman (eds.), *Plant Aquaporins*, Signaling and Communication in Plants, DOI 10.1007/978-3-319-49395-4_12

1 Membrane Permeability According to Meyer and Overton

More than a hundred years ago, H. Meyer and C.E. Overton independently worked on the lipid theory of anaesthetic action (Meyer 1899; Overton 1901). They found out that the higher the lipid solubility of an anaesthetic molecule is, the higher is its anaesthetic activity. Basically, these studies showed a direct correlation between the ability of a molecule to dissolve in lipid and its membrane permeability, and consequently, membranes should not impose resistance to diffusion of small hydrophobic molecules. This dependency has been called solubility-diffusion model or Meyer-Overton correlation. The results of transport studies on lipid bilayers are still valid today, and there are only rare exceptions (Missner et al. 2008a). The collection of substances that should comply with the solubility-diffusion model and possess very high membrane permeability contains among others also biologically important gasses like CO_2, O_2, NH_3 and NO (Missner and Pohl 2009). Insertion of special transport proteins would not be able to increase the diffusion rate because it would not be able to reduce the overall resistance of the membrane any further. In addition, the main resistance limiting permeation of small hydrophobic molecules should come from unstirred water layers on both sides of the membrane, drastically reducing the transmembrane concentration (Missner et al. 2008a). The concept of membrane diffusion based on Meyer and Overton has been presented in textbooks ever since.

Summing up, this model can explain how most gasses diffuse across numerous types of membranes. However, some biological membranes evidentially have a very low or no gas permeability at all and challenge the general validity of the solubility-diffusion model.

2 Experimental Findings That Support or Contradict Meyer and Overton

Studies on membrane permeability to various gasses and solutes complying with the Meyer-Overton correlation generally used artificial lipid bilayer membranes as experimental systems (Antonenko et al. 1993; Mathai et al. 2009; Missner et al. 2008a, b; Missner and Pohl 2009). J. Gutknecht and co-workers could show that an artificial bilayer consisting of egg lecithin and cholesterol does not constitute a substantial barrier to the diffusion of CO_2 (permeability coefficient $P_{CO2} \approx 0.35$ cm/s) (Gutknecht et al. 1977). The CO_2 molecule is very hydrophobic and according to Meyer and Overton has high membrane permeability. A comparable situation using a colorimetric approach was found for NH_3 ($P_{NH3} \approx 0.13$ cm/s) (Walter and Gutknecht 1986). P. Pohl and co-workers analyzed the diffusion of NH_3 (Antonenko et al. 1997) and acetate (Antonenko et al. 1993) as well as H_2S (Mathai et al. 2009), across lipid bilayer membranes employing a pH microelectrode technique and again confirmed the solubility-diffusion model.

However, studies that do not comply with the solubility-diffusion model have been performed on biological membranes rather than artificial lipid bilayers. We were able to show that membrane CO_2 permeability of tobacco chloroplast envelopes is around fivefold lower than that of plasma membranes and the variation is even more pronounced when the membrane integral protein composition is altered (Uehlein et al. 2008). The latter finding by itself is not consistent with the model as it shows that membranes can impose resistance to CO_2 diffusion and that this resistance can depend on protein components. In addition the values determined for plant membranes are roughly by a factor of 10^2–10^3 lower than what has been measured on artificial lipid bilayer membranes (Gutknecht et al. 1977; Missner et al. 2008b) and about 10^4-fold lower than what is expected on the basis of theoretical considerations (Missner et al. 2008b; Missner and Pohl 2009).

In mammals, membranes exhibiting unusually low gas permeability were identified, obviously contradicting the common view of gas permeable membranes. Fermentation of non-absorbed nutrients in the colon generates high concentrations of NH_4^+ in the colonic lumen, which is in equilibrium with ammonia (NH_3). NH_3 can easily permeate lipid bilayer membranes (Antonenko et al. 1997; Walter and Gutknecht 1986) and affect transmembrane pH gradients. However, in isolated colonic crypts, it has been observed that cytosolic pH of the epithelial cells does not increase when the NH_3 concentration in the colon lumen is raised but does so immediately when NH_3 concentration is increased on the basolateral side (Singh et al. 1995). Thus, biomembrane permeability to ammonia can vary even within individual cells.

In the colon lumen, some bacteria generate very high CO_2 partial pressures (Rasmussen et al. 1999, 2002). An extreme CO_2 partial pressure can seriously affect the pH of colon epithelial cells, if the CO_2 molecules can freely permeate across the plasma membrane into the cytosol. However, apical membranes of gastrointestinal endothelia exhibit very low CO_2 permeability (Endeward and Gros 2005; Waisbren et al. 1994). Consequently, it was concluded that there must be a permeability barrier to gasses in certain types of membranes. It appears to be quite reasonable from a physiological point of view, if some membranes build up a gas barrier to protect the cells. By comparison, in organs and tissues important for gas exchange, e.g., in lung alveolar endothelia and red blood cells (Endeward et al. 2006b) or photosynthetic cells, membrane CO_2 permeability is very high, in order to guarantee a quick removal of the gaseous waste product or uptake of the substrate.

3 Physiological Significance of CO_2 Diffusion in Plants

The physiological reactions of CO_2 exchange are unequivocally important for most organisms. CO_2 is a waste product from biochemical reactions, while for plants it is the substrate for sugar synthesis (Kaldenhoff 2012). In addition, the concentration gradient between plant cells taking up CO_2 and the atmosphere is less pronounced than it is the case for animals releasing CO_2. Therefore, the physiological study of

CO_2 diffusion in photosynthetic active plants could be more beneficial (Kaldenhoff 2012). Under light-saturating conditions, the photosynthetic capacity is limited by the availability of CO_2 in the chloroplast. Therefore, reduction of the barriers of CO_2 diffusion is becoming the focus of many studies, especially regarding the regulation of stomata and the biochemistry of CO_2 reactions. The complete cellular CO_2 diffusion path from the atmosphere to the site of chemical fixation in the chloroplast stroma is restricted by resistances with different ranges. After diffusing through the stomata and the leaf internal air spaces, atmospheric CO_2 has to pass the cell wall, the plasma membrane, diffuse a short distance through the cytosol, cross the chloroplast envelope and diffuse in the chloroplast stroma (Evans et al. 2009). All these barriers contribute to the so-called mesophyll conductance (g_m) or the mesophyll resistance, which is significant enough to decrease the CO_2 concentrations in the chloroplast and becomes a major limitation in net photosynthesis (Harley et al. 1992; Kaldenhoff 2012).

4 Aquaporin-Facilitated CO_2 Diffusion

During the past 15 years, proteins facilitating the diffusion of CO_2 across membranes have been identified. It has been shown in heterologous model systems that certain aquaporins can facilitate membrane gas transport. First evidence came from research on the human aquaporin 1 (Nakhoul et al. 1998). Since then, hints have accumulated that also other aquaporins may be involved in the facilitation of membrane transport of gasses like CO_2 or NH_3 and maybe O_2. Studies on HsAQP1 showed that it is highly expressed in cells or tissues that require particularly high gas permeability like the lung or erythrocytes (Cooper et al. 2002; Preston and Agre 1991; Verkman et al. 2000). Besides aquaporins, also rhesus blood group antigen-associated glycoproteins (RhAG) that are present in the red blood cell membrane have been shown to transport NH_3 and CO_2 (Endeward et al. 2008; Ripoche et al. 2004).

In the past, *Xenopus* oocytes served as a useful tool to analyze functional properties of aquaporins, as their plasma membranes generally exhibit low water and also low gas permeability (Cooper et al. 2002; Nakhoul et al. 1995). Using them as an expression system, it was shown that HsAQP1 considerably increases the oocyte membrane permeability to both CO_2 and NH_3 (Cooper and Boron 1998; Nakhoul et al. 1998, 2001; Musa-Aziz et al. 2009). The plant aquaporin NtAQP1 in heterologous expression systems as well as in in-vitro systems allows only very low water transport rates. However, it shares the functionally important amino acid residues with HsAQP1 on the basis of sequence comparison and, like HsAQP1, facilitates uptake of CO_2 into *Xenopus* oocytes (Uehlein et al. 2003). Also other aquaporins from other organisms were shown to increase the gas permeability of oocyte membranes (Jahn et al. 2004; Holm et al. 2005; Musa-Aziz et al. 2009). Using *Saccharomyces cerevisiae* as another heterologous expression system, facilitation of CO_2 or NH_3 membrane transport was confirmed for aquaporins from tobacco, maize and wheat (Bertl and Kaldenhoff 2007; Jahn et al. 2004; Loque et al. 2005; Heinen et al. 2014).

Also in native systems, an effect of aquaporin expression on CO_2 membrane diffusion has been shown. Expression of human AQP1 has turned out to facilitate membrane CO_2 transport in red blood cells (Endeward et al. 2006a, b). The tobacco PIP1 aquaporin NtAQP1 is highly expressed in photosynthetically active tissue (Otto and Kaldenhoff 2000). Manipulating its expression level in the plant has a substantial effect on the rate of photosynthesis. Under antisense limited expression of NtAQP1, net photosynthesis is reduced; when NtAQP1 is overexpressed, the photosynthetic performance increases, in each case by around 30–40 % (Uehlein et al. 2003). *Arabidopsis* mutants with a specific knockout of the NtAQP1 ortholog AtPIP1;2 are available. It can be assumed that in these plants the expression of sequence related aquaporin genes is not affected, as it can be the case when the antisense or RNAi technique is applied. Molecular and physiological analyses supported this assumption. Using those mutant plants, the specific importance of AtPIP1;2 for plant physiology can be analyzed, and hints on its molecular function can be deduced. The studies that were performed in this respect indeed confirmed a contribution of the aquaporin in membrane transport of CO_2. Cellular transport of CO_2 was not limited by unstirred layer effects but was dependent on the expression of AtPIP1;2 (Uehlein et al. 2012b). It could be shown that a knockout of AtPIP1;2 expression led to about 40 % reduction in mesophyll conductance to CO_2 and eventually limited photosynthesis. Introduction of the unmutated gene fully complemented the mutated phenotype (Heckwolf et al. 2011).

A similar situation has been observed for membrane diffusion of ammonia. According to the Meyer-Overton rule, NH_3 should easily pass lipid bilayers down its concentration gradient. However, facilitation of ammonia membrane transport in heterologous and native systems has been reported for aquaporins (Nakhoul et al. 2001; Bertl and Kaldenhoff 2007; Holm et al. 2005) as well as special ammonium transporters (Ludewig et al. 2003). Recently the crystal structure of the water- and ammonia-permeable aquaporin AtTIP2;1 revealed a special selectivity filter containing a conserved arginine residue that helps to understand the molecular mechanism of ammonia permeation (Kirscht et al. 2016). In addition it has been shown in zebrafish larvae that an aquaporin can facilitate the membrane diffusion of CO_2, ammonia and water in a physiologically relevant fashion (Talbot et al. 2015).

5 Molecular Mechanism of Aquaporin-Facilitated CO_2 Diffusion

Atomic structure of aquaporins shows high similarity and unique structural elements (see chapter "Structural Basis of the Permeation Function of Plant Aquaporins"). In general, aquaporins consist of six membrane-spanning helical domains with N- and C-termini heading towards the cytosol. Conserved motifs called 'NPA' in loop B and E from opposite sides located inside the membrane form the water-conducting channel (Murata et al. 2000). Experimental data showed that the water channel forms in a single aquaporin monomer; however, aquaporins form tetramers in the membrane (de Groot et al. 2003). In general, membrane proteins tend to form oligomers not only for stabilization but also for functionality of the protein (Ali and Imperiali 2005). However, the reason why aquaporins tend to form tetramers is still not clear (Strand et al. 2009).

HsAQP1 is an aquaporin that has been closely examined with regard to its function in CO_2 transport. Experimental data gave clear evidence that HsAQP1 can facilitate CO_2 transport in *Xenopus laevis* oocytes (Uehlein et al. 2003; Endeward et al. 2006b). Atomic molecular simulation data based on HsAQP1 crystal structures show that HsAQP1-mediated CO_2 transport via the monomer pores can be expected in membranes with low intrinsic CO_2 permeability (Hub and de Groot 2006). However, it is more likely that CO_2 diffusion is mediated through the central pore formed by the tetramer (Hub and de Groot 2006), because it is lined by hydrophobic amino acid residues and therefore is an ideal path for hydrophobic CO_2 molecules (Wang et al. 2007). Later on, experimental data based on the artificial homo- and heterotetramers of NtPIP1;2 and NtPIP2;1 showed that maximum CO_2 transport rates were obtained when the tetramer consisted of NtPIP1;2 units only. Substitution of two PIP1 by two PIP2 units completely abolished the CO_2 transport capacity. In conclusion, the data showed that tetramer formation is necessary for CO_2 transportation and that a joint structure built by all four units in the tetramer is responsible for this function. It is most likely that the central pore is the respective structure (Otto et al. 2010).

6 In Vitro Studies of CO_2 Diffusion Across Reconstituted Model Membranes

The diffusion of CO_2 across membranes was studied in different systems using different measuring approaches. In early studies, overexpression of aquaporins was done in oocytes, and the measurement of CO_2 diffusion was done with the detection of acidification of the oocytes cytosol in the presence of carbonic anhydrase when subjected to a CO_2 gradient (Nakhoul et al. 1998). Later on, the stopped-flow method was used to detect CO_2 diffusion across yeast cell membranes, in which a CO_2 permeable aquaporin was overexpressed. To guarantee that the membrane transport of CO_2 is the rate limiting step, rather than the conversion reaction of CO_2 to carbonic acid, the yeast cells expressed a carbonic anhydrase. NtAQP1 or NtPIP2;1 was expressed in yeast, and the cells were subjected to functional analysis for intracellular acidification in response to CO_2 uptake. Intracellular acidification was monitored via detecting changes of fluorescein fluorescence. Yeast cells expressing NtAQP1 in addition show faster intracellular decrease in pH compared to control cells and cells expressing NtPIP2;1 (Otto et al. 2010). Similarly, experiments were done using reconstituted proteoliposomes (Prasad et al. 1998; Yang et al. 2000). However, these studies, which employed the stopped-flow technique, were questioned regarding the ability to resolve the rapid kinetics of CO_2 diffusion across the membrane (Boron et al. 2011). In addition, an ^{18}O exchange mass spectrometric technique was applied to measure the CO_2 diffusion across red blood cells (Itel et al. 2012). Recently, a new method for monitoring CO_2 transport through a lipid bilayer using a micro pH electrode was developed. CO_2 transportation through a two-chamber system that was separated by a lipid bilayer resulted in an acidified region close to the membrane. A micro pH electrode attached to a micromanipulator device was used to monitor this acidification (Uehlein et al. 2012a; Missner et al. 2008b). This tool offered a new instrument to study the steady-state CO_2 diffusion through membranes independent of resolving rapid kinetics and just reliant on the diffusion rate of CO_2. In addition, using non-CO_2-permeable triblock-copolymer membranes, it was possible to reduce the background diffusion, which is quite significant in artificial lipid bilayers, to close to zero (Uehlein et al. 2012a). Further studies examining the effect of sterols as well as non-CO_2-permeable aquaporins on CO_2 diffusion were done using the same method. The development of techniques used for measuring CO_2 diffusion could shed some light on the general principles. Several independent studies revealed that sterols and membrane proteins that are major components of a biological membrane also contribute to the limitation of membrane CO_2 permeability (Itel et al. 2012; Kai and Kaldenhoff 2014; Tsiavaliaris et al. 2015).

7 Conclusions and Perspectives

Taken together, experimental data on membrane gas permeability are available that are not consistent with the Meyer-Overton rule and with theoretical considerations, or the process of CO_2 diffusion across biological membranes is not suitable to apply the Meyer-Overton rule. One possible reason could be that our current view of biological membranes and comparison of their permeability properties to well-established experimental test systems like pure lipid bilayers are not quite correct. Biological membranes are considerably more complex than is reflected by the well-known fluid mosaic model (Singer and Nicolson 1972). They contain integral and associated membrane proteins, as well as membrane regions tightly connected to the cytoskeleton. The protein occupancy in highly specialized membranes reaches up to 75 % (Dupuy and Engelman 2008; Engelman 2005). It ranges between ~23 % for red blood cell membranes (Dupuy and Engelman 2008) and ~70–80 % for chloroplast membranes (Kirchhoff et al. 2008). Therefore, it is not surprising that permeability properties of biological membranes differ greatly from that of artificial lipid bilayers. It may be misleading to expect that findings obtained on pure lipid bilayers and in silico studies directly apply to complex biological membranes. However, one should keep in mind that molecules cannot cross the lipid matrix when the lipid matrix is either replaced by integral membrane proteins or covered by membrane-associated proteins and therefore not accessible (Boron 2010).

The literature shows that membranes differ in CO_2 permeability and that certain aquaporins as well as RhAG proteins are able to transport CO_2 when the background permeability of the membrane in which they reside is low or, in case of special block copolymer membranes, even close to zero. Furthermore, studies show that the gas permeability of biological membranes may decrease with increasing protein content, thus making special transport proteins necessary in cases where high gas permeability is required. There are more physiologically important gasses, namely, NH_3, O_2 and NO, which have to cross membranes in order to be taken up or to be released. Membrane diffusion of CO_2 and NH_3 can be detected by measurement of pH changes, for O_2 and NO carbon fibre electrodes can be used. Understanding the mechanism of gas transport across biological membranes will help understand the physiology of organisms and will provide new arguments for the controversial debate on membrane gas transport. Eventually it will help to understand how CO_2 transport in plants can be improved, and this in turn will help to advance photosynthesis and growth of crop plants. In addition, it will give rise to broader impacts like strategies towards reduction of atmospheric CO_2 concentration and development of technical applications like highly gas-selective membranes for gas purification or sensor technology.

References

Ali MH, Imperiali B (2005) Protein oligomerization: how and why. Bioorg Med Chem 13(17):5013–5020. doi:10.1016/j.bmc.2005.05.037

Antonenko YN, Denisov GA, Pohl P (1993) Weak acid transport across bilayer lipid membrane in the presence of buffers. Theoretical and experimental pH profiles in the unstirred layers. Biophys J 64(6):1701–1710

Antonenko YN, Pohl P, Denisov GA (1997) Permeation of ammonia across bilayer lipid membranes studied by ammonium ion selective microelectrodes. Biophys J 72(5):2187–2195

Bertl A, Kaldenhoff R (2007) Function of a separate NH(3)-pore in Aquaporin TIP2;2 from wheat. FEBS Lett 581(28):5413–5417

Boron WF (2010) Sharpey-Schafer lecture: gas channels. Exp Physiol 95(12):1107–1130. doi:10.1113/expphysiol.2010.055244

Boron W, Endeward V, Gros G, Musa-Aziz R, Pohl P (2011) Intrinsic CO2 permeability of cell membranes and potential biological relevance of CO2 channels. Chemphyschem: Eur J Chem Phys Phys Chem 12(5):1017–1019. doi:10.1002/cphc.201100034

Cooper GJ, Boron WF (1998) Effect of PCMBS on CO2 permeability of Xenopus oocytes expressing aquaporin 1 or its C189S mutant. Am J Phys Cell Phys 44(6):C1481–C1486

Cooper GJ, Zhou YH, Bouyer P, Grichtchenko II, Boron WF (2002) Transport of volatile solutes through AQP1. J Physiol 542(1):17–29

de Groot BL, Engel A, Grubmuller H (2003) The structure of the aquaporin-1 water channel: a comparison between cryo-electron microscopy and X-ray crystallography. J Mol Biol 325(3):485–493

Dupuy AD, Engelman DM (2008) Protein area occupancy at the center of the red blood cell membrane. Proc Natl Acad Sci U S A 105(8):2848–2852. doi:10.1073/pnas.0712379105

Endeward V, Gros G (2005) Low carbon dioxide permeability of the apical epithelial membrane of guinea-pig colon. J Physiol 567(Pt 1):253–265

Endeward V, Cartron JP, Ripoche P, Gros G (2006a) Red cell membrane CO2 permeability in normal human blood and in blood deficient in various blood groups, and effect of DIDS. Transfus Clin Biol 13(1–2):123–127. doi:10.1016/j.tracli.2006.02.007

Endeward V, Musa-Aziz R, Cooper GJ, Chen LM, Pelletier MF, Virkki LV, Supuran CT, King LS, Boron WF, Gros G (2006b) Evidence that aquaporin 1 is a major pathway for CO2 transport across the human erythrocyte membrane. FASEB J 20(12):1974–1981

Endeward V, Cartron JP, Ripoche P, Gros G (2008) RhAG protein of the Rhesus complex is a CO2 channel in the human red cell membrane. FASEB J 22(1):64–73

Engelman DM (2005) Membranes are more mosaic than fluid. Nature 438(7068):578–580

Evans JR, Kaldenhoff R, Genty B, Terashima I (2009) Resistances along the CO2 diffusion pathway inside leaves. J Exp Bot 60(8):2235–2248. doi:10.1093/jxb/erp117

Gutknecht J, Bisson MA, Tosteson FC (1977) Diffusion of carbon dioxide through lipid bilayer membranes. Effects of carbonic anhydrase, bicarbonate, and unstirred layers. J Gen Physiol 69(6):779–794

Harley PC, Loreto F, Di Marco G, Sharkey TD (1992) Theoretical considerations when estimating the mesophyll conductance to CO(2) flux by analysis of the response of photosynthesis to CO(2). Plant Physiol 98(4):1429–1436

Heckwolf M, Pater D, Hanson DT, Kaldenhoff R (2011) The *Arabidopsis thaliana* aquaporin AtPIP1;2 is a physiologically relevant CO(2) transport facilitator. Plant J 67(5):795–804. doi:10.1111/j.1365-313X.2011.04634.x

Heinen RB, Bienert GP, Cohen D, Chevalier AS, Uehlein N, Hachez C, Kaldenhoff R, Le Thiec D, Chaumont F (2014) Expression and characterization of plasma membrane aquaporins in stomatal complexes of *Zea mays*. Plant Mol Biol 86(3):335–350. doi:10.1007/s11103-014-0232-7

Holm LM, Jahn TP, Moller AL, Schjoerring JK, Ferri D, Klaerke DA, Zeuthen T (2005) NH3 and NH4+ permeability in aquaporin-expressing Xenopus oocytes. Pflugers Arch 450(6):415–428

Hub J, de Groot B (2006) Does CO2 permeate through aquaporin-1? Biophys J 91(3):842–848. doi:10.1529/biophysj.106.081406

Itel F, Al-Samir S, Oberg F, Chami M, Kumar M, Supuran CT, Deen PM, Meier W, Hedfalk K, Gros G, Endeward V (2012) CO2 permeability of cell membranes is regulated by membrane cholesterol and protein gas channels. FASEB J: Off Publ Fed Am Soc Exp Biol 26(12):5182–5191. doi:10.1096/fj.12-209916

Jahn TP, Moller ALB, Zeuthen T, Holm LM, Klaerke DA, Mohsin B, Kuhlbrandt W, Schjoerring JK (2004) Aquaporin homologues in plants and mammals transport ammonia. FEBS Lett 574(1–3):31–36

Kai L, Kaldenhoff R (2014) A refined model of water and CO(2) membrane diffusion: effects and contribution of sterols and proteins. Sci Rep 4:6665. doi:10.1038/srep06665

Kaldenhoff R (2012) Mechanisms underlying CO2 diffusion in leaves. Curr Opin Plant Biol 15(3):276–281. doi:10.1016/j.pbi.2012.01.011

Kirchhoff H, Haferkamp S, Allen JF, Epstein DB, Mullineaux CW (2008) Protein diffusion and macromolecular crowding in thylakoid membranes. Plant Physiol 146(4):1571–1578. doi:10.1104/pp.107.115170

Kirscht A, Kaptan SS, Bienert GP, Chaumont F, Nissen P, de Groot BL, Kjellbom P, Gourdon P, Johanson U (2016) Crystal structure of an ammonia-permeable aquaporin. PLoS Biol 14(3):e1002411. doi:10.1371/journal.pbio.1002411

Loque D, Ludewig U, Yuan L, von Wiren N (2005) Tonoplast intrinsic proteins AtTIP2;1 and AtTIP2;3 facilitate NH3 transport into the vacuole. Plant Physiol 137(2):671–680

Ludewig U, Wilken S, Wu BH, Jost W, Obrdlik P, El Bakkoury M, Marini AM, Andre B, Hamacher T, Boles E, von Wiren N, Frommer WB (2003) Homo- and hetero-oligomerization of ammonium transporter-1 NH4+ uniporters. J Biol Chem 278(46):45603–45610

Mathai JC, Missner A, Kugler P, Saparov SM, Zeidel ML, Lee JK, Pohl P (2009) No facilitator required for membrane transport of hydrogen sulfide. Proc Natl Acad Sci USA 106(39):16633–16638

Meyer H (1899) Zur Theorie der Alkoholnarkose. Arch Exp Pathol Pharmakol 42:109–118

Missner A, Pohl P (2009) 110 years of the Meyer-Overton rule: predicting membrane permeability of gases and other small compounds. Chem Phys Chem 10(9–10):1405–1414

Missner A, Kugler P, Antonenko YN, Pohl P (2008a) Passive transport across bilayer lipid membranes: Overton continues to rule. Proc Natl Acad Sci U S A 105(52):E123

Missner A, Kugler P, Saparov SM, Sommer K, Mathai JC, Zeidel ML, Pohl P (2008b) Carbon dioxide transport through membranes. J Biol Chem 283(37):25340–25347. doi:10.1074/jbc.M800096200

Murata K, Mitsuoka K, Hirai T, Walz T, Agre P, Heymann JB, Engel A, Fujiyoshi Y (2000) Structural determinants of water permeation through aquaporin-1. Nature 407(6804):599–605

Musa-Aziz R, Chen LM, Pelletier MF, Boron WF (2009) Relative CO2/NH3 selectivities of AQP1, AQP4, AQP5, AmtB, and RhAG. Proc Natl Acad Sci U S A 106:5406

Nakhoul NL, Romero MF, Davis BA, Bron WF (1995) Expression of Chip28 (Aquaporin-1) in Xenopus oocytes accelerates the CO2-induced decrease of intracellular Ph (Ph(I)). J Am Soc Nephrol 6(3):312–312

Nakhoul NL, Davis BA, Romero MF, Boron WF (1998) Effect of expressing the water channel aquaporin-1 on the CO2 permeability of Xenopus oocytes. Am J Physiol Cell Physiol 43(2):C543–C548

Nakhoul NL, Hering-Smith KS, Abdulnour-Nakhoul SM, Hamm LL (2001) Transport of NH3/NH4+ in oocytes expressing aquaporin-1. Am J Physiol-Renal Physiol 281(2):F255–F263

Otto B, Kaldenhoff R (2000) Cell-specific expression of the mercury-insensitive plasma-membrane aquaporin NtAQP1 from *Nicotiana tabacum*. Planta 211(2):167–172

Otto B, Uehlein N, Sdorra S, Fischer M, Ayaz M, Belastegui-Macadam X, Heckwolf M, Lachnit M, Pede N, Priem N, Reinhard A, Siegfart S, Urban M, Kaldenhoff R (2010) Aquaporin tetramer composition modifies the function of tobacco aquaporins. J Biol Chem 285(41):31253–31260

Overton E (1901) Studien über die Narkose. Gustav Fischer, Jena

Prasad GV, Coury LA, Finn F, Zeidel ML (1998) Reconstituted aquaporin 1 water channels transport CO2 across membranes. J Biol Chem 273(50):33123–33126

Preston GM, Agre P (1991) Isolation of the cDNA for erythrocyte integral membrane protein of 28 kilodaltons: member of an ancient channel family. Proc Natl Acad Sci U S A 88(24):11110–11114

Rasmussen H, Kvarstein G, Johnsen H, Dirven H, Midtvedt T, Mirtaheri P, Tonnessen TI (1999) Gas supersaturation in the cecal wall of mice due to bacterial CO2 production. J Appl Physiol 86(4):1311–1318

Rasmussen H, Mirtaheri P, Dirven H, Johnsen H, Kvarstein G, Tonnessen TI, Midtvedt T (2002) PCO2 in the large intestine of mice, rats, guinea pigs, and dogs and effects of the dietary substrate. J Appl Physiol 92(1):219–224

Ripoche P, Bertrand O, Gane P, Birkenmeier C, Colin Y, Cartron JP (2004) Human rhesus-associated glycoprotein mediates facilitated transport of NH(3) into red blood cells. Proc Natl Acad Sci U S A 101 (49):17222–17227. doi: 0403704101 [pii] 10.1073/pnas.0403704101

Singer SJ, Nicolson GL (1972) The fluid mosaic model of the structure of cell membranes. Science 175(23):720–731

Singh SK, Binder HJ, Geibel JP, Boron WF (1995) An apical permeability barrier to NH3/NH4+ in isolated, perfused colonic crypts. Proc Natl Acad Sci U S A 92(25):11573–11577

Strand L, Moe SE, Solbu TT, Vaadal M, Holen T (2009) Roles of aquaporin-4 isoforms and amino acids in square array assembly. Biochemistry 48(25):5785–5793. doi:10.1021/bi802231q

Talbot K, Kwong RW, Gilmour KM, Perry SF (2015) The water channel aquaporin-1a1 facilitates movement of CO2 and ammonia in zebrafish (*Danio rerio*) larvae. J Exp Biol 218(Pt 24):3931–3940. doi:10.1242/jeb.129759

Terashima I, Ono K (2002) Effects of HgCl(2) on CO(2) dependence of leaf photosynthesis: evidence indicating involvement of aquaporins in CO(2) diffusion across the plasma membrane. Plant Cell Physiol 43(1):70–78

Tholen D, Zhu XG (2011) The mechanistic basis of internal conductance: a theoretical analysis of mesophyll cell photosynthesis and CO2 diffusion. Plant Physiol 156(1):90–105. doi:10.1104/pp.111.172346

Tsiavaliaris G, Itel F, Hedfalk K, Al-Samir S, Meier W, Gros G, Endeward V (2015) Low CO2 permeability of cholesterol-containing liposomes detected by stopped-flow fluorescence spectroscopy. FASEB J: Off Publ Fed Am Soc Exp Biol 29(5):1780–1793. doi:10.1096/fj.14-263988

Uehlein N, Lovisolo C, Siefritz F, Kaldenhoff R (2003) The tobacco aquaporin NtAQP1 is a membrane CO2 pore with physiological functions. Nature 425(6959):734–737

Uehlein N, Otto B, Hanson DT, Fischer M, McDowell N, Kaldenhoff R (2008) Function of *Nicotiana tabacum* aquaporins as chloroplast gas pores challenges the concept of membrane CO2 permeability. Plant Cell 20(3):648–657. doi:10.1105/tpc.107.054023

Uehlein N, Otto B, Eilingsfeld A, Itel F, Meier W, Kaldenhoff R (2012a) Gas-tight triblock-copolymer membranes are converted to CO(2) permeable by insertion of plant aquaporins. Sci Rep 2:538. doi:10.1038/srep00538

Uehlein N, Sperling H, Heckwolf M, Kaldenhoff R (2012b) The Arabidopsis aquaporin PIP1;2 rules cellular CO(2) uptake. Plant Cell Environ 35(6):1077–1083. doi:10.1111/j.1365-3040.2011.02473.x

Verkman AS, Matthay MA, Song Y (2000) Aquaporin water channels and lung physiology. Am J Physiol Lung Cell Mol Physiol 278(5):L867–L879

Waisbren SJ, Geibel JP, Modlin IM, Boron WF (1994) Unusual permeability properties of gastric gland cells. Nature 368(6469):332–335

Walter A, Gutknecht J (1986) Permeability of small nonelectrolytes through lipid bilayer membranes. J Membr Biol 90(3):207–217

Wang Y, Cohen J, Boron WF, Schulten K, Tajkhorshid E (2007) Exploring gas permeability of cellular membranes and membrane channels with molecular dynamics. J Struct Biol 157(3):534–544

Yang B, Fukuda N, van Hoek A, Matthay MA, Ma T, Verkman AS (2000) Carbon dioxide permeability of aquaporin-1 measured in erythrocytes and lung of aquaporin-1 null mice and in reconstituted proteoliposomes. J Biol Chem 275(4):2686–2692

The Nodulin 26 Intrinsic Protein Subfamily

Daniel M. Roberts and Pratyush Routray

Abstract Nodulin intrinsic proteins (NIPs) represent a land plant-specific subfamily of the major intrinsic protein/aquaporin superfamily. NIPs are named for the first member of the family discovered, soybean nodulin 26 of symbiotic nitrogen-fixing root nodules. Evolutionarily, NIPs appear in early nonvascular and vascular land plant lineages, with the family undergoing substantial diversification and subfunctionalization during subsequent evolution of seed plants. Structurally, most NIPs can be divided into three "pore" families based on the composition of amino acids comprising the predicted aromatic-arginine selectivity region of the channel pore. Functionally, two of these families (NIP II and NIP III) serve as channels for metalloid nutrients (boric acid and silicic acid respectively), while the biological role of NIP I channels remains more open. Biochemical functions for NIP proteins are diverse, with transport selectivities ranging from metalloid hydroxides to glycerol, lactic acid, urea, and hydrogen peroxide. Some NIPs retain their aquaporin function, while others have lost this signature activity of the aquaporin family. In the present chapter, the evolutionary origins, structural and functional properties, and potential biological functions, particularly beyond their roles as metalloid facilitators, are reviewed.

1 Nodulin 26: A Historical Perspective

The beginning of the NIP subfamily narrative starts with early work on the development of legume-rhizobial nitrogen-fixing symbioses. The establishment of root nodules upon rhizobial infection results in the generation of modified organellar

D.M. Roberts (✉)
Department of Biochemistry and Cellular and Molecular Biology, The University of Tennessee, Knoxville, TN 37996, USA

Program in Genome Science and Technology, The University of Tennessee, Knoxville, TN 37996, USA
e-mail: drobert2@utk.edu

P. Routray
Department of Biochemistry and Cellular and Molecular Biology, The University of Tennessee, Knoxville, TN 37996, USA

F. Chaumont, S.D. Tyerman (eds.), *Plant Aquaporins*, Signaling and Communication in Plants, DOI 10.1007/978-3-319-49395-4_13

structures known as "symbiosomes" (Emerich and Krishnan 2014; Ivanov et al. 2010; Roth and Stacey 1989) that house endosymbiotic rhizobia bacteria. Symbiosomes mediate the metabolic exchange that supports the energetically expensive process of nitrogen fixation (uptake of a reduced carbon energy source to support N_2 reduction and export of fixed NH_3 for cytosolic assimilation (Clarke et al. 2014; Day et al. 2001; Udvardi and Poole 2013)). Early work by Verma and colleagues identified a subclass of symbiosome membrane "nodulin" genes that encode proteins that support this process (Fortin et al. 1985). Among these was an integral polytopic membrane protein, referred to as nodulin 26 (Fortin et al. 1987) that is the major protein component of the mature soybean symbiosome membrane (Dean et al. 1999; Fortin et al. 1987; Rivers et al. 1997; Weaver et al. 1991). Its similarity to the "major intrinsic protein" of bovine lens fibers (later renamed aquaporin 0) was noted (Sandal and Marcker 1988), and it is among the first members of the plant aquaporin superfamily discovered.

Since these seminal observations, a wealth of information has been generated on the NIP gene family, including its wide distribution in land plants, its unusual broad selectivity as a transporter of multiple uncharged solutes and water, and its biological function in plant nutrition and stress responses have been generated. The picture that has emerged is that NIPs are a land plant-specific collection of unique aquaporin-like proteins that have conserved the basic aquaporin structural paradigm but have evolved to acquire unique biochemical and biological functions in plant physiology.

2 Nodulin Intrinsic Proteins: Evolution and Diversification

2.1 The Evolutionary Origins of the NIP Subfamily

The availability of genomic information across all biological kingdoms has provided a detailed picture of the phylogeny of the major intrinsic protein/aquaporin superfamily (Abascal et al. 2014; Danielson and Johanson 2010; Finn and Cerda 2015; Johanson et al. 2001; Perez Di Giorgio et al. 2014). In general, a major branch point between two ancient ancestral paralogous subfamilies, the "glycerol transporter/aquaglyceroporin" (GLP) and "aquaporin" (AQP) subgroups, is proposed with further duplication, diversification, and sub-functionalization (particularly within the AQP group) occurring during evolution of higher eukaryotes (reviewed in Abascal et al. 2014; Zardoya 2005). Within this larger phylogeny, the NIPs form a monophyletic group within the AQP subfamily. Phylogenetic analysis shows that NIP family is quite distinct from other plant MIP clades (e.g., PIP, SIP, TIP, and XIP subfamilies) and shows a closer evolutionary alignment with bacterial and archaeal AQPs (Abascal et al. 2014; Danielson and Johanson 2010; Finn and Cerda 2015; Zardoya 2005; Zardoya et al. 2002). This has led to an initial hypothesis that plants acquired NIP proteins through a horizontal gene transfer from the bacterial AQP-Z family with a common ancestor gene dating to 1,200 million years ago, correlating with the divergence and origin of the kingdom Plantae (Zardoya 2005; Zardoya et al. 2002). Further, given the glyceroporin activities commonly found in NIPs (see Sect. 3) and the lack of GLP-like proteins in seed plants, it was proposed that this transfer and subsequent convergent evolution of the pore selectivity amino acid residues provided an alternative mechanism for the acquisition of glycerol permease activity in land plants (Zardoya 2005; Zardoya et al. 2002).

More recently, other models for the origin of NIPs have also been proposed based on phylogenic analyses of a multitude of plant, bacterial, and archaeal genomes (Anderberg et al. 2011, 2012; Danielson and Johanson 2008, 2010; Finn and Cerda 2015; Gustavsson et al. 2005; Perez Di Giorgio et al. 2014). For example, representative archaeal and cyanobacterial species possess genes encoding NIP-like proteins that show the closest phylogenetic relationships to plant NIPs (Abascal et al. 2014; Danielson and Johanson 2010; Finn and Cerda 2015). This suggests that horizontal transfer of an ancestral gene from this prokaryotic NIP-like group could have given rise to plant NIPs rather than a transfer event of bacterial AQP-Z aquaporins (Danielson and Johanson 2010; Finn and Cerda 2015). More recently, a third model of NIP evolution proposes the vertical decent of plant NIPs and animal aquaglyceroporins of the AQP3 family from a common eukaryotic ancestor (Perez Di Giorgio et al. 2014). While plausible, this model would have to account for the absence of NIP genes in ancient plant lineages that predate the evolution of land plants (Anderberg et al. 2011).

2.2 *Divergence and Sub-functionalization of NIPs*

Although multiple schemes for the origin of NIPs have been proposed, a general consensus is that these genes were acquired, diversified, and sub-functionalized during the evolutionary emergence of the Embryophyta. Plant NIPs are represented in all seed plant genomes (c.f., Chaumont et al. 2001; Diehn et al. 2015; Fouquet et al. 2008; Gupta and Sankararamakrishnan 2009; Hove et al. 2015; Johanson et al. 2001; Quigley et al. 2002; Reuscher et al. 2013; Sakurai et al. 2005; Zhang et al. 2013) as well as in early bryophyte, lycopod, and monilophyte plant lineages (e.g., *Physcomitrella patens* (Danielson and Johanson 2008), *Selaginella moellendorffii* (Anderberg et al. 2012), and *Equisetum arvense* (Gregoire et al. 2012)). In contrast, NIPs are absent in algal genomes (Anderberg et al. 2011), supporting the acquisition of this plant MIP subfamily (as well as others including TIPs and SIPs) upon plant colonization of land.

Further phylogenetic, structural modeling and transport function analysis have revealed substantial diversification and sub-functionalization of NIPs during plant evolution. Based on the early phylogenetic analyses of *Arabidopsis* MIPs (Johanson and Gustavsson 2002; Quigley et al. 2002), and parallel analyses in maize (Chaumont et al. 2001), a nomenclature for assigning MIP protein designations was adopted with the subfamily (e.g., SIP, PIP, TIP, and NIP) followed by a numbering scheme based on monophyletic groups within each subfamily as well as the number of the individual gene within these monophyletic groups. In the case of *Arabidopsis* NIPs, nine genes divided into seven groups were identified (AtNIP1;1 and 1;2, AtNIP2;1, AtNIP3;1, AtNIP4;1 and 4;2, AtNIP5;1, AtNIP6;1, and AtNIP7;1) (Johanson et al. 2001).

Since these early analyses, the availability of a large number of NIPs representing a diverse array of land plant lineages reveals a more complex NIP phylogeny, and broader segregation into distinct paralog groups has been noted in several analyses (Abascal et al. 2014; Anderberg et al. 2012; Chaumont et al. 2001; Danielson and Johanson 2008, 2010; Sakurai et al. 2005). From a recent analysis (Abascal et al. 2014), based on a global MIP phylogenetic tree constructed with 1714 MIP/aquaporin sequences, seed plant NIPs can be divided into four paralogous clades (NIP-1 to NIP-4, Fig. 1) (See also chapter "Plant Aquaporins and Metalloids").

NIP-1 proteins are represented by the family archetype, soybean nodulin 26. NIP-1 proteins are found exclusively in angiosperms and were originally characterized as channels for water and glycerol (Wallace et al. 2006), although they are permeated by other solutes as well (discussed in the next section). NIP-2 proteins were first described in monocot genomes and were first functionally characterized as silicic acid transporters associated with accumulation of silicon nutrients in rice (Ma et al. 2006). NIP-2 proteins are widely distributed in monocots, and are also found in some dicots (although notably absent in *Arabidopsis*) and gymnosperms (Anderberg et al. 2012). NIP-3 proteins are present in all land plant genomes. Functionally, NIP3 channels are permeated by multiple metalloid hydroxide substrates (Pommerrenig et al. 2015) and glycerol (Wallace et al. 2006). Biologically, they mediate the uptake and homeostasis of boron nutrients under limiting conditions (Miwa and Fujiwara 2010; Takano et al. 2006; Tanaka et al. 2008). A fourth paralogous group (NIP-4) is more distal from the other NIP groups, and shows similarity to *Arabidopsis* AtNIP7;1 (Abascal et al. 2014). In addition, a subgroup found exclusively in earlier plants species (designated as NIP-5 by Anderberg et al. 2012; Danielson and Johanson 2008, 2010) are proposed to be phylogenetically associated with the NIP-2 group (Abascal et al. 2014).

2.3 Phylogenetic Versus "Pore" Classification of NIPs

In addition to these phylogenetic designations based on evolutionary relationships of nucleotide and amino acid sequences, a second nomenclature and grouping of seed plant NIPs into "pore families" based on selectivity filter analysis from homology modeling of the NIP subfamilies has also been generated (Mitani et al. 2008; Wallace and Roberts 2004). Using the nomenclature of (Danielson and Johanson 2010), phylogenetic clades are designated by Arabic numerals (e.g., NIP1-5) and "pore families" by Roman numerals (e.g., NIP I–III). An analysis of the NIP pore families will be taken up in the next section on NIP structure and function.

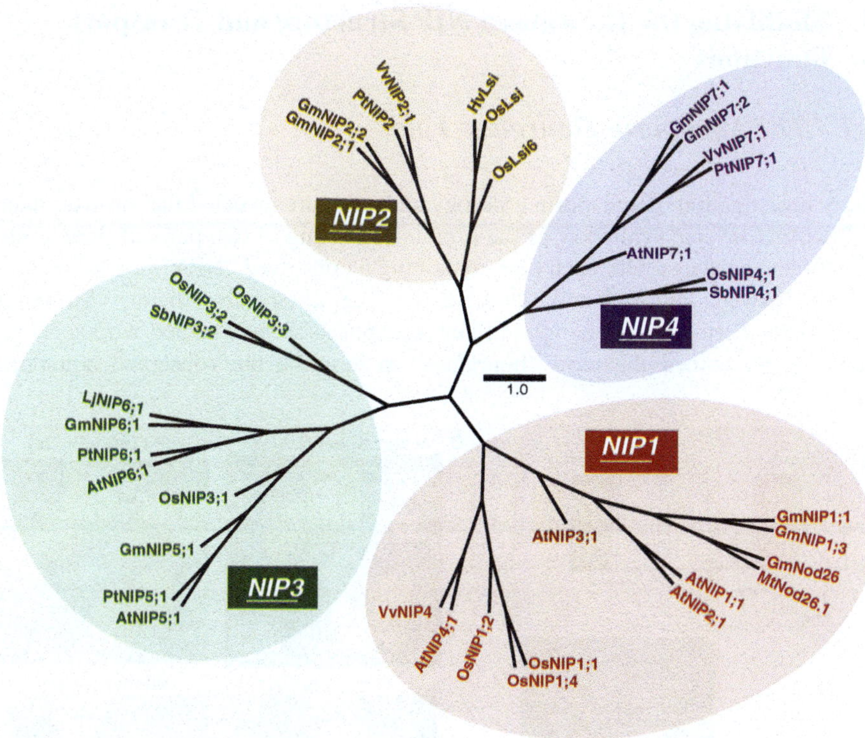

Fig. 1 Phylogenic divisions within representative angiosperm NIP proteins. Selected NIP proteins were aligned by using the ClustalW multiple sequence alignment program, and an unrooted phylogenetic tree was generated using the MegAlign software in the DNASTAR Lasergene Suite. The scale bar represents amino acid substitutions per ten residues. Phylogenetic clades NIP1 to NIP4 using the nomenclature of (Abascal et al. 2014) are shown. The protein sequences correspond to the following: (1) *Arabidopsis* NIPs correspond to the proteins encoded by the indicated genes based on the nomenclature of Johanson et al. (2001); (2) soybean NIPs correspond to the protein nomenclature of Zhang et al. (2013) with the exception of GmNod26 (designated GmNIP1;5 by Zhang et al. 2013) which represents the original archetype of the family and MtNod26.1 (Medtr8g087710.1) which represents the nodulin 26 ortholog in *Medicago truncatula* based on expression data (Benedito et al. 2008); (3) *Oryza sativa* NIPs correspond to the nomenclature of Sakurai et al. (2005) with the exception of OsLsi1 and OsLsi6 (represented by OsNIP2;1 and OsNIP2;2 in Sakurai et al. 2005)) which are named based on their silicic acid transport properties as described in Ma et al. (2006) and Yamaji et al. (2008); the barley ortholog HvLsi is from Chiba et al. (2009); (4) *Vitis vinifera* NIPs correspond to the nomenclature in Fouquet et al. (2008); (5) *Sorghum bicolor* NIPs correspond to the nomenclature in Azad et al. (2016); (6) LjNIP6;1 corresponds to the metalloid transporter from *Lotus japonicus* described in Bienert et al. (2008); (7) The *Populus trichocarpa* NIPs correspond to the following gene IDs: PtNIP5;1 (POPTR_0001s45920), PtNIP6;1 (POPTR_0001s14850), PtNIP7;1 (POPTR_0008s20750), and PtNIP2 (POPTR_0017s11960)

3 Modifying the Hourglass: NIP Structure and Transport Functions

3.1 The Aquaporin Hourglass Fold

Early investigation of aquaporin topology and structure revealed the intrinsic internal twofold inverse symmetry of the protein consisting of two pairs of three transmembrane α-helices with each pair containing a conserved "NPA" box in the loop regions between helices 2 and 3 and 5 and 6 (Fig. 2) (see also chapter "Structural Basis of the Permeation Function of Plant Aquaporins"). Agre and co-workers (Jung et al. 1994) coined the term "hourglass" to describe the conserved aquaporin

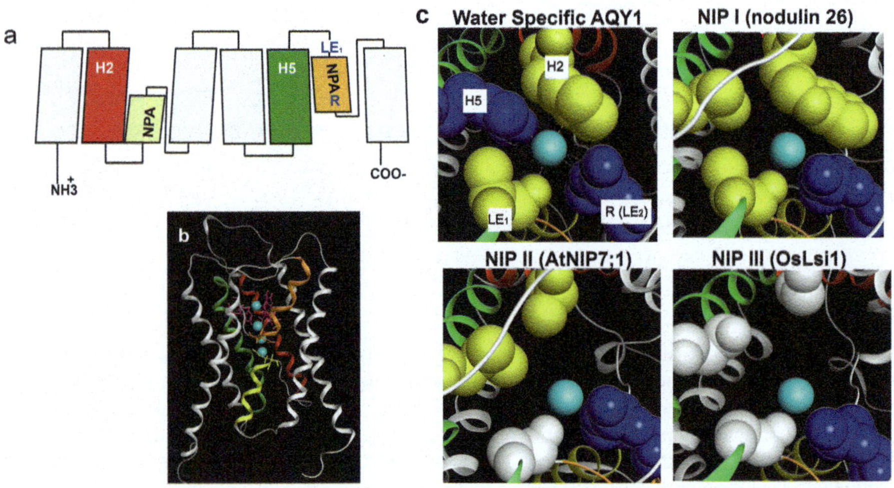

Fig. 2 The aquaporin fold and NIP pore families. (**a**) Diagrammatic representation of the conserved MIP membrane topology. The positions of the four amino acids from helix 2 (H2), helix 5 (H5), and the extracellular NPA loop (LE_1 and the conserved arginine at LE_2) are shown. (**b**) A backbone ribbon depiction of the sub-angstrom structure the *Pichia* aquaporin AQY1 (pdb 3ZOJ Kosinska Eriksson et al. 2013) is shown with the same color scheme for the NPA boxes and helices 2 and 5 as in panel A. The ar/R side chains (magenta) and the NPA asparagines (white and yellow) are shown as well as space filling representations of four water molecules in the transport channel (aqua). (**c**) The ar/R regions of AQY1 compared to the ar/R regions (from homology models) of representative NIP I (soybean nodulin 26), NIP II (AtNIP7;1), and NIP III (OsLsi1) are shown viewed down the pore axis from the extracellular face of each protein. The positions of the four ar/R positions are indicated with the side chain colors indicating chemical properties (yellow, hydrophobic; white, uncharged hydrophilic; blue, basic). The position of the bound water from the AQY1 structure (aqua sphere) is docked into each structure. Hydrogen bond contacts (yellow) and hydrogen bond-π contacts (orange) are indicated

topology even before the solution of the first crystal structures. This term was prescient since the solution of the first crystal structures of aquaporin/MIP family members [the *Escherichia coli* glycerol facilitator (Fu et al. 2000), and mammalian aquaporin 1 (Sui et al. 2001)] confirmed the twofold symmetry and the role of the NPA boxes in folding into the pore to form a seventh pseudo-transmembrane helix (Fig. 2). Further, they identified the nature of the hourglass constriction that serves as the selectivity filter of the pore. This selectivity constriction is formed by the confluence of four amino acids, one residue from two of the transmembrane α-helices (helix 2 and helix 5, referred to as the **H2** and **H5** positions) and two residues from the second NPA box (referred to as the $\mathbf{LE_1}$ and $\mathbf{LE_2}$ positions). Due to the presence of a nearly invariant arginine at the $\mathbf{LE_2}$ position, and the prevalence of aromatic residues at the other positions, the term "aromatic/arginine" region or ar/R was coined for this selectivity filter constriction (de Groot and Grubmuller 2001). The composition of the ar/R in representative GLP and AQP structures showed how substitutions of the side chains at the $\mathbf{LE_1}$, **H2**, and **H5** positions control specificity for water or glycerol transport through alterations in the size of the ar/R aperture and the hydrophobicity and hydrogen bond potential of the ar/R side chains (Hub et al. 2009; Walz et al. 2009). The NPA boxes, on the other hand, interact with transported substrates and, in the case of aquaporin transport, disrupt the hydrogen bond chain of waters in the pore to prevent the transport of protons via the Grotthuss mechanism (Kosinska Eriksson et al. 2013; Kreida and Tornroth-Horsefield 2015).

Early homology modeling studies showed that the diversification of the aquaporin gene family in plants was accompanied by alterations in the compositions of the predicted ar/R selectivity filter sequences which transcended the water-specific AQP and glycerol permease paradigms (Ludewig and Dynowski 2009; Wallace and Roberts 2004). Seed plant NIPs can be classified into three "pore families" (NIP I, NIP II, and NIP III, Fig. 2) based on amino acid motifs found within the ar/R selectivity filter of the pore (Liu and Zhu 2010; Mitani et al. 2008; Rouge and Barre 2008; Wallace et al. 2006; Wallace and Roberts 2004). Interestingly, the "pore families" show phylogenetic segregation into groups as well (Fig. 1), although NIPs from early plant species show more diversity within this region (Anderberg et al. 2012; Danielson and Johanson 2008).

3.1.1 The NIP Pore Subfamilies

3.1.2 Soybean Nodulin 26 and NIP I Pores

Homology modeling of soybean nodulin 26 and related proteins in the NIP-1 phylogenetic subgroup revealed a conserved consensus ar/R tetrad consisting of [in order, the $\mathbf{H2\text{-}H5\text{-}LE_1\text{-}LE_2}$ residues are W-V-A-R (Wallace and Roberts 2004; Wallace et al. 2002)], (Fig. 2). The presence of a tryptophan at **H2** and a valine at **H5** results in a more hydrophobic amphipathic pore compared to the water-specific aquaporins (Wallace and Roberts 2004, 2005; Wallace et al. 2002). In addition, the substitution of valine for a histidine at the **H5** position both increases the pore diameter of the ar/R as well as removes the hydrogen bonding potential of the histidine resulting in a reduction in the ability of NIP I ar/R to form hydrogen bonds with transported waters (Wallace and Roberts 2005), (Fig. 2). Consistent with this alteration in the ar/R, NIP I proteins form aquaglyceroporin channels that mediate the flux of glycerol and water (Dean et al. 1999; Rivers et al. 1997; Schuurmans et al. 2003; Weig et al. 1997; Weig and Jakob 2000), but with an aquaporin single channel rate (roughly 0.1 billion/sec) that is an order of magnitude reduced compared to water-specific aquaporins (Dean et al. 1999; Rivers et al. 1997). The conservation of a tryptophan at the **H2** position is characteristic of glyceroporins and aquaglyceroporins (Fu et al. 2000), and substitution of a histidine at this position (characteristic of the plant TIP family) results in the loss of the glycerol permease activity while retaining the aquaporin activity of the channel (Wallace and Roberts 2005; Wallace et al. 2002). In addition to water and glycerol transport, NIP I proteins, such as nodulin 26, also possess an ammonia permease activity (Hwang et al. 2010; Niemietz and Tyerman 2000). The potential role of nodulin 26 as an ammoniaporin during symbiotic nitrogen fixation is discussed below.

Biochemical and biophysical transport analyses reveal that the NIP I pore is also permeable to uncharged hydrophilic "test" substrates with molecular volumes that can be accommodated by the pore. Among these are formamide (Rivers et al. 1997) and hydrogen peroxide (Katsuhara et al. 2014). The *Arabidopsis* NIP2;1 protein presents an unusual case and an outlier function for NIP I proteins. As discussed below in Sect. 4, NIP2;1 is among a small collection of genes that encode the core "anaerobic response polypeptides" in *Arabidopsis* that are acutely induced during flooding and associated anaerobic stress (Choi and Roberts 2007; Mustroph et al. 2009). Functional analysis of NIP2;1 showed that it differed from canonical nodulin 26 and lacked any detectable glycerol or aquaporin channel activities and instead was found to flux the protonated lactic acid, a fermentation end product (Choi and Roberts 2007). A potential role for NIP2;1 in anaerobic stress responses is discussed in Sect. 4.3 below.

NIP2;1 has an ar/R selectivity filter that is identical to other NIP I proteins, and so the reason for its distinct substrate selectivity is likely due to other (as yet uncharacterized) characteristics of the transport channel. Crystal structures of various mammalian and plant water-specific aquaporins with similar ar/R show that other unique features of the channel besides the ar/R can also contribute to transport activity, selectivity and regulation (Gonen et al. 2004; Kreida and Tornroth-Horsefield 2015; Tornroth-Horsefield et al. 2006). In this regard, while the ar/R is a critical determinant of the channel selectivity, other structural determinants outside of this constriction need to be considered in evaluating transport function (Savage et al. 2010).

As discussed in more detail in the chapter "Plant Aquaporins and Metalloids" in this volume, a major role for the NIP family as permeases for uncharged metalloid hydroxides (i.e., "metalloidoporin" activity) has emerged based on substantial biochemical, biophysical, and genetic evidence. These include metalloid nutrients such as boric acid and orthosilicic acid as well as toxic hydroxides of As (III), Ge, and Sb (reviewed in Pommerrenig et al. 2015). Collectively, these compounds form weak Lewis acid structures with trigonal ($B[OH]_3$) or tetrahedral (e.g.,

$As[OH]_3$ and $Si[OH]_4$) geometry that exist in an uncharged state at physiological and soil pH. Quantum calculations and thermodynamic comparisons suggest that among the myriad conformations that glycerol can adapt (over 500), a "retracted" glycerol conformation shows substantial similarity to the more conformationally restrained $As[OH]_3$ and $Sb[OH]_3$ molecules with respect to structure, the disposition of the hydrogen bonding hydroxyl groups, electrostatic potential, and molecular volume (Porquet and Filella 2007). Consistent with this observation, glycerol permeases of the aquaporin superfamily, such as the *E. coli* GlpF channel, the yeast FPS1 protein, as well as a number of animal aquaglyceroporins, are permeated by $Sb[OH]_3$ and $As[OH]_3$ (Liu et al. 2004; Mukhopadhyay et al. 2014; Sanders et al. 1997; Wysocki et al. 2001). NIP proteins of all three pore subclasses are permeated by $Sb[OH]_3$ and $As[OH]_3$ based on toxicity analyses in yeast backgrounds deficient in glycerol or arsenite transport (Bienert et al. 2008; Isayenkov and Maathuis 2008; Kamiya and Fujiwara 2009; Katsuhara et al. 2014; Xu et al. 2015), or by direct uptake when expressed in *Xenopus* oocytes (Kamiya et al. 2009; Katsuhara et al. 2014; Ma et al. 2008).

While other plant MIPs have been proposed to be permeated by $As[OH]_3$ and other metalloids (see chapter "Plant Aquaporins and Metalloids"), strong evidence for a unique role of NIP I proteins in conferring resistance to As III toxicity comes from an unbiased forward genetic screen in *Arabidopsis* that identified the NIP1;1 protein as the determinant for $As[OH]_3$ sensitivity (Kamiya et al. 2009). This may reflect the high expression level of NIP1;1 compared to other NIPs in the roots, as well as its cellular and subcellular localization (Alexandersson et al. 2005; Kamiya et al. 2009). In a subsequent study, another NIP I protein, NIP3;1, was found to work synergistically with NIP 1;1 in As uptake and root to shoot translocation and partitioning (Xu et al. 2015). As discussed recently (Pommerrenig et al. 2015), it is not clear whether $Sb[OH]_3$ and $As[OH]_3$ permeabilities are non-biological side activities based on the structural similarity of these toxic compounds to actual biological substrates (e.g., metalloid nutrients or glycerol) or whether NIPs are bona fide physiological channels involved in As detoxification. As discussed in the next two sections, there is strong physiological, genetic, and biochemical evidence for NIP II and NIP III proteins as channels for metalloid nutrients (B and Si, respectively). However, a biological role for NIP I as transporters for these metalloids is lacking, and the permeability of NIP I for these compounds is not well demonstrated (Ma et al. 2008; Mitani et al. 2008, Fig. 3).

3.1.3 NIP II Pores: Metalloid and Glycerol Permeases

With respect to channel selectivity, NIP II proteins share the glycerol permease activity of NIP I proteins (Wallace and Roberts 2005), and are permeated by a variety of metalloids (Hanaoka et al. 2014; Isayenkov and Maathuis 2008; Kamiya et al. 2009; Katsuhara et al. 2014; Ma et al. 2008; Takano et al. 2006; Tanaka et al. 2008). Overwhelming genetic evidence suggests that the primary biological function of NIP II channels is to serve as facilitators of boric acid permeability across biological

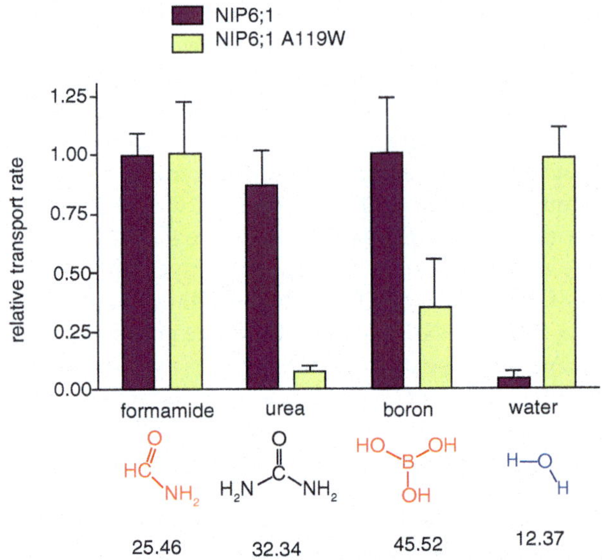

Fig. 3 Transport properties of NIP I and NIP II ar/R. The comparative permeabilities of the NIP II protein AtNIP6;1 for water and several test solutes are shown. Wild-type AtNIP6;1 and AtNIP6;1with the substitution of a tryptophan for an alanine at the H2 position (*NIP6;1 A119W*) were expressed in *Xenopus* oocytes, and solute and water permeability were assayed as described in Wallace and Roberts (2005). For comparison, the relative standardized transport rates are shown, with the maximum rate in each pair set to 1. The structures and molecular volumes (in cm^3/mol) of each test solute are shown below the graph

membranes under limiting conditions of this essential nutrient (Hanaoka et al. 2014; Takano et al. 2006; Tanaka et al. 2008), Miwa et al. 2010, see chapter "Plant Aquaporins and Metalloids").

From a structural and biochemical transport perspective, the NIP II pore differs fundamentally from NIP I proteins by the substitution of a small side chain amino acid (typically alanine) for the bulky aromatic tryptophan at the **H2** position of the ar/R region while retaining an aliphatic branched hydrophobic residue (isoleucine or valine) at position **H5** (Wallace and Roberts 2005). This results in a much wider predicted ar/R selectivity filter, principally from the **H2** alanine substitution (Fig. 2). Transport analyses show that the **H2** position of NIP I and NIP II pores is a critical determinant of which solutes are permeated by each pore class, as well as the aquaporin activity possessed by each (Wallace and Roberts 2005) (Fig. 3). For example, NIP II proteins (e.g., AtNIP6;1 and AtNIP5;1) are readily permeated by urea (Wallace and Roberts 2005; Yang et al. 2015), while NIP I proteins (e.g., nodulin 26) are not (Rivers et al. 1997; Wallace and Roberts 2005). Similarly, the substitution of a tryptophan for the conserved alanine at the **H2** position of AtNIP6;1 results in a reduced flux of boric acid (Fig. 3), suggesting a small amino acid at the H2 position is necessary for optimal transport of this solute. Interestingly, as discussed above, NIP I proteins are readily permeated by other metalloid hydroxides such as

$As[OH]_3$ (Pommerrenig et al. 2015), which has similar or larger molecular volumes than boric acid. The reason for this potential difference remains unresolved but may be due to the different geometry (i.e., planar vs. tetrahedral) of boron and arsenic hydroxides.

Another curiosity regarding NIP II channel activity is the apparent drastic reduction or loss of aquaporin activity in many NIP II proteins compared to their NIP I counterparts (Katsuhara et al. 2014; Li et al. 2011; Takano et al. 2006; Wallace and Roberts 2005). From a size perspective, this is counterintuitive since the predicted width of the NIP II ar/R selectivity region is large enough to accommodate two water molecules (Wallace and Roberts 2005). The **H2** position is again the determinant of NIP aquaporin activity, with reciprocal site-specific substitutions of either tryptophan or alanine in this position conferring aquaporin activity (i.e., tryptophan) or a watertight pore (i.e., alanine) (Fig. 3).

Why does a wider pore result in the loss of aquaporin activity? One possibility, based on molecular dynamics simulations of NIP II homology models (Li 2014), is that the substitution of a smaller alanine side chain in the ar/R stabilizes a rotomeric state of the ar/R arginine in a "down" position which would effectively occlude the pore. In support of this, crystal structures of certain aquaporins (e.g., AQP-Z) have revealed the existence of such a "down" configuration of the ar/R arginine (Jiang et al. 2006), and the fluctuation between the up and down states has been proposed as a potential mechanism to regulate aquaporin activity in mammalian and bacteria aquaporins (Hub et al. 2010; Xin et al. 2011).

Why would a "watertight" pore be desirable for NIP II proteins? These proteins function as boric acid channels at strategic locations on the plasma membranes of selected cells and tissues (Takano et al. 2006, 2010; Tanaka et al. 2008; Uehara et al. 2014). The water permeability of the plant plasma membrane is controlled largely by PM-localized PIP aquaporins, which are subject to tight regulation through multiple mechanisms at a variety of regulatory levels (e.g., genetic, trafficking and pore gating), (Chaumont and Tyerman 2014; Maurel et al. 2015). The plasma membrane possesses an osmotic water permeability (P_f) that is two orders of magnitude lower than internal membranes, which may be necessary for cell volume regulation and buffering (Maurel et al. 1997). In the case of NIP II proteins, the ability to readily transport boric acid in a watertight manner may be necessary to prevent interference with the regulation of plasma membrane P_f.

NIP7;1 and related proteins of the NIP4 clade also possess an ar/R filter characteristic of the NIP II pore (A/G-V-G-R). NIP7;1 has been implicated as a transporter of $Sb[OH]_3$ (Bienert et al. 2008) or $As[OH]_3$ (Isayenkov and Maathuis 2008), and a potential role in boric acid transport was suggested based on its similarity to established NIP II channels. However, despite this similarity, evaluation of transport function in *Xenopus* oocytes showed that AtNIP7;1 exhibits a tenfold lower permeability for boric acid compared to the constitutive AtNIP6;1 boric acid channel (Li et al. 2011). The reason for this discrepancy appears to be the result of a conserved tyrosine residue (Y81) located on helix 2 four residues from the **H2** position (**H2-4** position). Molecular modeling and molecular dynamics simulations suggest that Y81 can exist in two rotomeric states that would open or close the pore and access to the ar/R (Li et al. 2011). In the closed state, Y81 lies perpendicular to the pore axis in a position capable of forming a hydrogen bond between the phenolic hydroxyl group of Y81 and the guanidino group of the ar/R arginine. Substitution of Y81 with either cysteine (characteristic of AtNIP6;1) or a phenylalanine (lacking a side chain hydroxyl) opens the pore and allows boric acid flux. Analysis of NIP7;1 orthologs in other plant species shows the conservation of a tyrosine at the **H2-4** position, whereas NIP6;1 and NIP5;1 orthologs have conserved cysteine or asparagine residues, respectively, at this position (Li et al. 2011). It is acknowledged that boron is both an essential micronutrient and a potential toxin at high concentrations (Camacho-Cristobal et al. 2008), and as a result, boric acid transporters and NIP II channels are subject to regulation at multiple levels (genetic, mRNA, and trafficking; Takano et al. 2005; 2006, 2010; Tanaka et al. 2008, 2011) (See also chapter "Plant Aquaporins and Metalloids"). The conservation of a tyrosine in the pore that could serve as a gate for boric acid transport was suggested to be a potential additional mechanism of regulation for NIP7;1-like proteins. Regulatory mechanisms that could modulate the orientation of this tyrosine residue need to be addressed.

3.1.4 NIP III Pores: Silicic Acid Channels

NIP III proteins are widely distributed among the Gramineae, are found in some dicot species, but are absent in *Arabidopsis*. With respect to channel selectivity, NIP III proteins transport silicic acid ($Si[OH]_4$) and the structurally similar metalloid hydroxide $Ge[OH]_4$ (Chiba et al. 2009; Ma et al. 2006). Overwhelming evidence suggests that the principal biological functions of NIP III proteins are to serve as facilitators for silicon transport in plants (Ma and Yamaji 2015). Silicon is a nonessential nutrient that accumulates in grasses and selected dicots and contributes to optimal growth and development as well as resistance to abiotic and biotic stress (Epstein 1999; Ma and Yamaji 2006). In addition to silicic acid, NIP III proteins are permeated by $As[OH]_3$ and contribute to the arsenic accumulation in rice grain in the presence of this toxic compound (Ma et al. 2008). Transport analyses show that NIP III proteins also flux water, boric acid, and urea (Gu et al. 2012; Mitani et al. 2008).

The ar/R of this class of NIPs retains the invariant arginine but possesses a substitution of a small hydroxylated amino acid (serine) at the **H5** position compared to the common branched aliphatic hydrophobic residues (valine or isoleucine) found in NIP I and NIP II pores (Fig. 2). The NIP III ar/R also possesses glycine residues at positions **H2** and $\mathbf{LE_1}$ (Fig. 2). The overall result is a wider, flexible, and more hydrophilic selectivity pore with an additional hydrogen bond donor at the H5 position. Site-directed mutagenesis of the NIP III silicon transporter OsLsi1 shows that the serine residue at the **H5** position is critical for Si transport, with the substitution of the bulkier, aliphatic isoleucine residue (NIP II-like) abolishing $Si[OH]_4$ transport while retaining the ability to transport $As[OH]_3$ (Mitani-Ueno et al. 2011). Since $Si[OH]_4$ is a larger substrate with an additional hydroxyl group, the **H5** serine may be essential for providing a wider pore and an additional hydrogen bond donor/

acceptor. However, the simple presence of the G-S-G-R ar/R composition is not in itself adequate to confer Si[OH]$_4$ transport (Deshmukh et al. 2015; Mitani-Ueno et al. 2011), and other determinants (e.g., NPA spacing and conformation (Deshmukh et al. 2015)) likely contribute to transport selectivity. Ultimately, validation of these hypotheses, similar to others regarding the nature of the NIP I–III pores, will require the solution of representative NIP structures at atomic resolution bound with substrates.

3.1.5 Ancestral Pore Structures and Evolutionary Origins

Pore models and predicted selectivity ar/R structures from early plant NIPs reveal divergence from the NIP I–III pore paradigm described above. Analysis of *P. patens*, the earliest plant species that possesses NIP genes, shows that NIP ar/R sequences can be classified into three categories (Danielson and Johanson 2008). Two (PpNIP3;1, A-I-A-R and PpNIP6;1, G-V-A-R) are characteristic of NIP II pores, while the remaining three (PpNIP5;1, 5;2, 5;3) have a unique composition consisting of F-A-A-R. This latter ar/R is characteristic of some bacterial and archaeal NIP-like proteins (Abascal et al. 2014; Danielson and Johanson 2010) and thus may represent an evolutionary origin of the NIP selectivity filter (Danielson and Johanson 2010). Since bryophytes do not appear to have a nutritional need for boron and do not accumulate boron within their cell walls (Matsunaga et al. 2004; Wakuta et al. 2015), it is questionable whether these early NIPs play a role in metalloid nutrition. Together with the GIP protein subclass (Gustavsson et al. 2005), these ancient NIPs may participate in glycerol uptake and distribution in early nonvascular terrestrial plants.

Subsequent analyses of the lycopod *S. moellendorffii* reveal an even greater diversification of NIP genes and pore structures (Anderberg et al. 2012). The greatest diversity was found within the NIP-3 clade (five genes), supporting the proposal that this is among the oldest phylogenetic NIP groups (Danielson and Johanson 2010). However, unlike plant NIP-3 proteins, which have the NIP II ar/R signature, a greater diversity of pore selectivity structures was found in *S. moellendorffii*. Some of these are unique among the major intrinsic protein/aquaporin superfamily (e.g., *SmNIP3;2* P-N-A-R and *SmNIP3;3* and *3;4* A-N-A-R). Two additional NIP genes code for proteins characteristic of the *P. patens* NIP-5, and an additional gene encodes a protein (SmNIP7;1) phylogenetically related to AtNIP7;1 (NIP-4 clade) with a similar NIP II-like ar/R composition (A-V-G-R). The transport and biological functions of these proteins have yet to be elucidated. However, *Selaginella* possesses functional borate transporters of the BOR family (Wakuta et al. 2015) and also is a silica-accumulating plant (Trembath-Reichert et al. 2015), and potential functions for *Selaginella* NIPs in boron and silicon metalloid transport are possible.

Analysis of another primitive vascular plant, the horsetail species *E. arvense*, revealed a collection of silicon transporting NIPs (EaNIP3;1–9) that share high (>90 %) sequence identity (Gregoire et al. 2012). Phylogenetically, these NIPs cluster within the NIP-3 clade (Abascal et al. 2014). Unlike silicon transporting NIP-2 proteins from higher plants which possess the NIP III pore signature (G-S-G-R), the *E. arvense* NIPs showed a conserved S-T-A-R ar/R region, suggesting the early evolution of a distinct pore structure capable of fluxing $Si[OH]_4$.

4 Selected Biological Functions and Regulation of NIPs

Since the seminal findings of Takano et al. (Takano et al. 2006) and Ma et al. (Ma et al. 2006) regarding the demonstration of the NIPs as boric acid or silicic acid channels in *Arabidopsis* and rice, respectively, a wealth of information has emerged regarding a primary biological function of NIP channels as metalloidoporins. This is expertly reviewed in the chapter "Plant Aquaporins and Metalloids" in this volume as well as recently by (Pommerrenig et al. 2015). In the final section of the present chapter, selected additional biological functions of NIPs, including potential roles as channels of non-metalloid substrates, will be discussed.

4.1 Nodulin 26 and NIPs in Symbioses

Soybean nodulin 26, the archetype of the NIP subfamily, is a late nodulin which is expressed at a stage of root nodule development associated with a massive induction of endomembrane biosynthesis that precedes the endocytosis of rhizobia bacteria and the formation of the unique nitrogen fixation organelle, the symbiosome (Guenther et al. 2003; Verma and Hong 1996). It remains as the major protein component of the symbiosome membrane throughout the lifetime of nodule through senescence (Guenther et al. 2003). Nodulin 26 orthologs have been found in other legume symbiosome membranes, and thus this protein appears to be a common component of nitrogen-fixing symbiosomes within legume-rhizobia root nodules (Catalano et al. 2004; Clarke et al. 2015; Guenther and Roberts 2000).

Early in vitro studies in planar lipid bilayers suggested a role as a metabolite and ion channel for the protein (Ouyang et al. 1991; Weaver et al. 1994). However, functional analyses of nodulin 26 in *Xenopus* oocytes and in reconstituted proteoliposomes showed the absence of ionic currents and provided overwhelming evidence that nodulin 26 is an aquaglyceroporin (Dean et al. 1999; Rivers et al. 1997) that is responsible for the unusually high osmotic water permeability of the symbiosome membrane (P_f = 0.05 cm/sec) (Rivers et al. 1997). The symbiosome is the major organelle occupying much of the space of the specialized nitrogen-fixing infected cells within the core of the mature soybean nodule. Since these cells lack a central vacuole, symbiosomes may be involved in cell volume regulation and osmotic buffering in response to changing metabolic and environmental conditions, similar to the role proposed for TIPs in vacuolar membranes (Maurel et al. 1997).

Even before its functional characterization, nodulin 26 was first described as a target for phosphorylation by symbiosome membrane-associated calcium-regulated protein kinases of the CDPK (or CPK) family (Weaver et al. 1991; Weaver and Roberts 1992) and was one of the first endogenous targets for these kinases identified in plants. The site of phosphorylation, which is largely conserved within proteins of the NIP-1 clade (Wallace et al. 2006), is serine-262 within the hydrophilic carboxyl terminal domain within the cytosol (Weaver and Roberts 1992). This is a common location for phosphorylation of a number of MIPs/aquaporins from both plants and animals (Hachez and Chaumont 2010; Kreida and Tornroth-Horsefield 2015; Maurel et al. 2015). Phosphorylation of the carboxyl terminal domain of aquaporins regulates activity in a number of ways, including gating of the channel activity (Johansson et al. 1998; Nyblom et al. 2009; Tornroth-Horsefield et al. 2006) as well as by trafficking and targeted localization to specific membranes (Boursiac et al. 2008; Van Balkom et al. 2002). Stopped flow analysis of nodulin 26 vesicles showed that phosphorylation stimulates the intrinsic water permeability activity of nodulin 26 (Guenther et al. 2003), suggesting a potential gating role.

Phosphorylation of the nodulin 26 is developmentally regulated, coinciding with the formation of mature nitrogen-fixing symbiosomes. Phosphorylation is not detected in older nodules entering senescence, or in immature nodules prior to acquisition of nitrogen fixation activity (Guenther et al. 2003). Nodulin 26 phosphorylation is additionally subject to reversible phosphorylation in response to osmotic/drought stress (which stimulates phosphorylation, Guenther et al. 2003) as well as by flooding/hypoxia stress (which triggers rapid dephosphorylation, Hwang 2013). These disparate stresses have a converse effect on an O_2 diffusion pathway within the nodule which exhibits a tight control on the rates of nitrogen fixation (Minchin et al. 2008; Roberts et al. 2010). Regulation of this gas diffusion pathway is proposed to be mediated through aquaporin-mediated reversible movement of water into the interstitial spaces between nodule cells in the cortex and central infected zone that regulates cell shape and apoplastic water content, opening or closing intercellular pathways for O_2 movement (Denison and Kinraide 1995; Minchin et al. 2008; Purcell and Sinclair 1994). Phosphorylation and enhancement of the nodulin 26 aquaporin activity may be part of this coordinated response to environmental regulation of the gas diffusion pathway.

From a metabolic perspective, the fundamental exchange associated with nitrogen fixing symbiosomes is the uptake of organic acids, generally malate, which serve as a carbon source for the nitrogen-fixing rhizobia bacteroids, and the efflux of fixed nitrogen (NH_3/NH_4^+) across the symbiosome membrane to the plant cytosol for assimilation (Udvardi and Poole 2013). A facilitated pathway for the efflux of NH_3 on the symbiosome membrane was first demonstrated by Niemietz and Tyerman (Niemietz and Tyerman 2000) based on stopped flow measurements with isolated symbiosome membrane vesicles. Subsequent analysis of purified nodulin 26 verified its NH_3 permeability and demonstrated that it is the preferred substrate for the channel ($P_{ammonia}$ is fivefold higher than P_f) (Hwang et al. 2010) and is a potential pathway for efflux of NH_3 from the symbiosome. An additional pathway for the efflux of NH_4^+ through a voltage-activated inwardly rectified nonselective cation channel is also present on the symbiosome membrane (Tyerman et al. 1995), and the relative efflux through the two pathways may depend on the pH and voltage gradient across the symbiosome membrane (Hwang et al. 2010; Niemietz and Tyerman 2000).

Additional support for nodulin 26 in the transport of fixed NH_3 comes from the observation that it forms a complex with soybean nodule cytosolic glutamine synthetase (Masalkar et al. 2010). Upon efflux from the symbiosome, ammonia is assimilated into an organic form within the cytosol, principally by the action of ATP-dependent glutamine synthetase (GS), which constitutes a major component of the infected cell cytosol. GS associates with the symbiosome membrane (Clarke et al. 2014; Masalkar et al. 2010) by interacting with the carboxyl terminal domain of nodulin 26 (Masalkar et al. 2010). Interestingly, the carboxyl terminal domains of other aquaporins have been demonstrated to be a common site of binding of a variety of proteins (Kreida and Tornroth-Horsefield 2015; Sjohamn and Hedfalk 2014) that regulate activity [e.g., calmodulin for AQP0 (Reichow et al. 2013)], trafficking and stability [heat shock proteins and LIP5 for AQP2 (Van Balkom et al. 2009)], or assembly of subcellular structures [e.g., cytoskeletal-associated proteins for AQP0 (Lindsey Rose et al. 2006; Wang and Schey 2011)].

Nodulin 26 interaction with GS would localize this critical assimilatory enzyme to the symbiosome surface to facilitate the rapid assimilation of fixed NH_3/NH_4^+. Such an association would not only facilitate rapid promote nitrogen assimilation but could also prevent the accumulation of toxic levels of free ammonia in the cytosol leading to futile cycling (Routray et al. 2015). Since nodulin 26 is an ammonia channel, the possibility of direct substrate channeling to the GS active site through a nodulin 26/GS complex is also possible (Masalkar et al. 2010).

The expression of selected NIP proteins is also induced during the establishment of the symbiosis, between plants and symbiotic arbuscular mycorrhizal (AM) fungi (Abdallah et al. 2014; Giovannetti et al. 2012; Uehlein et al. 2007) (see also chapter "Plant Aquaporins and Mycorrhizae: Their Regulation and Involvement in Plant Physiology and Performance"). Infection of plant roots by AM fungi results in an obligate symbiosis which results in the formation of fungal structures (arbuscules) that invade the host cytosol (Harrison 2012). Arbuscules are surrounded by a specialized membrane (the peri-arbuscular membrane) that mediates the symbiotic exchange of nutrients from the fungal partner (primarily inorganic phosphate but additionally nitrogen in the form of NH_4^+/NH_3) for carbon provided by the plant host (Smith and Smith 2011). In response to AM colonization, plants express transporters [e.g., phosphate transporters (Harrison et al. 2002) and an AMT ammonia transporter (Guether et al. 2009)] localized to the peri-arbuscular membrane to facilitate nutrient delivery from the AM symbiont. Analysis of two model legumes, *Lotus japonicus* (Giovannetti et al. 2012) and *Medicago truncatula* (Uehlein et al. 2007), shows the induction of NIP-1 genes during the formation of AM symbioses. Analysis of the *M. truncatula* gene atlas (Benedito et al. 2008) shows that the *MtNIP1* transcript described in (Uehlein et al. 2007) is identical to the nodulin 26 ortholog that is induced during rhizobia infection and nodulation, suggesting a possible parallel function for the protein in rhizobial as well as AM symbioses. Whether arbuscular NIPs play a role in nutrient delivery (Maurel and Plassard 2011), or are involved in other symbiotic processes, requires further investigation.

4.2 NIPs in Pollen Development and Function

4.2.1 Pollen-Specific NIP4;1 and 4;2

The *Arabidopsis* NIP-1 proteins NIP4;1 and NIP4;2 are paralogous pollen-specific transcripts (Bock et al. 2006; Di Giorgio et al. 2016) that share 88 % amino acid sequence identity, including conservation of ar/R and pore-forming residues. While both show pollen-specific expression, the two genes differ with respect to specific expression patterns and associated biological functions in developing, mature dehydrated pollen, germinating pollen, and growing pollen tubes (Di Giorgio et al. 2016). NIP4;1 appears late in pollen development (tricellular stage) and remains present in mature pollen and germinating pollen/pollen tubes both in vitro and in vivo. In contrast NIP4;2 shows more restricted expression, appearing only after pollen germination and within the plasma membranes of the growing pollen tube (Di Giorgio et al. 2016). Single T-DNA mutants of NIP4;1 and NIP4;2 and a double knockdown mutant of both NIPs using amiRNA technology support distinct roles for the two proteins in plant reproduction and pollen viability and function (Di Giorgio et al. 2016). In general, the mutant lines show reduced fertility of the male gametophyte based on reciprocal crosses with wild-type plants including reductions in seed production and silique size. In the case of *nip4;1* mutants, additional defects and delay in pollen maturation were noted, whereas in the case of *nip4;2* mutants, the major defects were predominantly observed in pollen germination and pollen tube growth (Di Giorgio et al. 2016). The two genes are localized in a tight tandem arrangement on chromosome 5 and likely represent the result of a gene duplication event resulting in sub-functionalization into discreet pollen-specific functions.

Similar to soybean nodulin 26 and consistent with the presence of consensus CDPK phosphorylation sites within the carboxyl terminal domain, NIP4;1 and NIP4;2 are phosphorylated at Ser 267 within this region (Di Giorgio et al. 2016; Sugiyama et al. 2008). This residue is phosphorylated in vitro by selected *Arabidopsis* CPKs (Curran et al. 2011), including an isoform (CPK34) involved in polarized pollen tube growth (Myers et al. 2009). The developmental or environmental conditions that regulate NIP4;1 and NIP4;2 phosphorylation and the effects of phosphorylation on trafficking or activity of the proteins *in planta* remain unresolved.

Similar to soybean nodulin 26, NIP4;1 and 4;2 show aquaporin activities in oocytes (Di Giorgio et al. 2016). Aquaporins have been proposed to be involved in bulk water flow associated with all phases of pollen development and function (Firon et al. 2012; Soto et al. 2008; Wudick et al. 2014). Water homeostasis driven by solute gradients is critical for pollen development and maturation, and the water content of the pollen is spatially and temporally regulated during its development leading to dehydration and pollen release (Firon et al. 2012). Upon landing on a compatible stigma, pollen grains again undergo rapid rehydration and germination, a process that involves rapid transfer of water and nutrients from the stigma to the pollen grain. The cycle of dehydration and rehydration is tightly regulated, and its disruption can lead to male sterility (Johnson and McCormick 2001). Pollen tube growth itself is proposed to be osmotically driven with a potential role for aquaporins as osmotic sensors (Hill et al. 2012; Shachar-Hill et al. 2013). NIP4;1 and/or 4;2 may be part of the network of aquaporins that coordinate direct bulk water flow underpinning these pollen functions.

However, similar to other NIP-1 proteins, the inherent multifunctional nature of NIP I proteins with respect to transport substrates complicates the identification of the specific biochemical activities associated with NIP4;1 and NIP4;2 that lead to developmental phenotypes. In addition to its aquaporin activity, NIP4;1 and NIP4;2 also flux glycerol in oocytes, with additional potential permeabilities for H_2O_2 and NH_3, as well as a weak permeability for boric acid, inferred from yeast-based screens (Di Giorgio et al. 2016). The contributions of these additional activities to NIP4;1 and NIP4;2 function and pollen physiology need to be assessed.

4.2.2 NIP7;1 in Early Pollen Development

Reproductive development in *Arabidopsis* has been divided into 20 stages from the initiation of flower primordia to the fall of seeds from siliques (Smyth et al. 1990). Microarray data (Schmid et al. 2005), Q-PCR (Li et al. 2011), and promoter-reporter analyses (Li 2014) show that *Arabidopsis* NIP7;1 is a floral-specific transcript that is expressed in the anthers of developing flowers within a tight developmental window appearing and peaking at stages 9–10 and declining and disappearing by stage 12. Expression of NIP7;1 protein and transcript within this window occurs both in the tapetum as well as in the developing pollen microspores (Li 2014). During stages 9–11, anther tissues expand, and microsporocytes enter and complete mitotic divisions (Sanders et al. 1999), with the tapetum playing an important role in providing nutrients to developing pollen during this stage. NIP7;1 orthologs are apparent in other dicotyledonous species (Fig. 1), and analysis of the gene expression atlases from these species shows a similar flower-specific expression suggesting a potential conserved role in pollen development.

As discussed above, NIP7;1 belongs to NIP II subfamily with an ar/R identical to the NIP5;1 and NIP6;1 boric acid channels but differs in its intrinsic rate of boric acid transport based on a conserved tyrosine residue in the pore that could serve as a channel gate (Li et al. 2011). Boron plays a significant role in pollen development, and the reproductive organs of the plant show a greater sensitive to boron deficiency than vegetative parts, presumably due to more restricted access of the vascular system to reproductive structures (Dell and Huang 1997). Boron deficiency leads to defective anther and pollen development leading to male sterility (Cheng and Rerkasem 1993; Huang et al. 2000; Rawson 1996). Preliminary analyses suggest that disruption of NIP7;1 expression decreases fertility in a boron-dependent fashion (Li 2014), suggesting that the protein may play a role as a boric acid channel at critical stages of early pollen development.

4.3 *NIPs in Stress Responses*

As part of the network regulating transcellular water flow that controls plant hydraulic conductivity (Steudle and Peterson 1998), aquaporins are subject to extensive regulation at transcriptional, translational, and posttranslational levels in response to multiple abiotic stresses (reviewed in Afzal et al. 2016; Chaumont and Tyerman 2014; Maurel et al. 2015). Considerable emphasis has been placed on the regulation of plasma membrane PIPs because of their central role in this process. Less emphasis has been placed on NIPs as aquaporins involved in this stress responses, likely due to their low intrinsic water permeabilities and overall low expression compared to PIPs and TIPs (Alexandersson et al. 2005). However, the gene expression of some NIPs is induced by drought and salinity stress (Gao et al. 2010; Martins et al. 2015), and in the case of one, the TaNIP of wheat, heterologous expression in *Arabidopsis* confers increased tolerance to salinity (Gao et al. 2010). Similar results have been observed with PIP aquaporins (Hu et al. 2012; Zhou et al. 2012, 2014). It is less clear whether salinity-/drought-induced NIPs participate as aquaporins in osmotic stress responses or play other roles as solute transporters.

With respect to waterlogging stress, the NIP2;1 protein of *Arabidopsis* is clearly a target for regulation. Energy metabolism in higher plants is imperatively dependent on the availability of oxygen. Exposure of plants to low oxygen conditions due to flooding, waterlogging, and poor soil aeration leads to an energy crisis that triggers a variety of adaptation strategies (Voesenek and Bailey-Serres 2015). In *Arabidopsis*, a key adaptation strategy is to reorganize gene expression and translation networks for the limited synthesis of a small collection (~49) of core anaerobic response proteins (ANPs), which include metabolic proteins (e.g., glycolytic and fermentation enzymes) as well as transcription factors, signaling components, and other proteins needed for anaerobic adaptation (Mustroph et al. 2009). NIP2;1 is among these ANPs, and it induced rapidly and acutely (>300-fold) in a root-specific manner in response to anaerobic stress (Choi and Roberts 2007).

In *Xenopus* oocytes, NIP2;1 exhibits specificity for the uncharged, protonated form of lactic acid with little aquaporin or glyceroporin activity detected, suggesting that it might serve as a lactic acid channel during the anaerobic response in *Arabidopsis* (Choi and Roberts 2007). Lactic acid fermentation is necessary, in part, to sustain energy production in the absence of oxygen, but its production can result in cytotoxic levels of lactic acid and acidosis (Felle 2005; Roberts et al. 1984). Experiments with maize roots show that the ability to efflux and release lactic acid enhances survival to hypoxia (Xia and Roberts 1994). In *Arabidopsis* it was proposed that NIP2;1 could participate in the cytosolic release and partitioning of lactic acid produced during fermentation (Choi and Roberts 2007). Interestingly, two distinct subcellular localization profiles have been proposed for NIP2;1, with the protein reported to be associated with the endoplasmic reticulum (Mizutani et al. 2006) as well as the plasma membrane (Choi and Roberts 2007) in *Arabidopsis* cultured cells or mesophyll protoplasts, respectively. It remains to be determined whether, similar to other aquaporins, NIP2;1 is subject to regulation by trafficking to various target membranes, perhaps as part of the hypoxia response.

In addition to NIP2;1, lactic acid permeability has also been described for other members of the aquaporin superfamily (Bienert et al. 2013; Faghiri et al. 2010; Tsukaguchi et al. 1999). For example, *GlpF* genes are associated with the lactate racemization operon (lar) of bacteria in the *Lactobacillales*, which are known to accumulate large quantities of lactic acid (Bienert et al. 2013). *Lactobacillus plantarum GlpF1* and *GlpF4* appear to encode lactic acid permeable channels based on yeast complementation assays, analysis of lactic acid uptake in single and double bacterial mutants, and transport assays of expressed *GlpF4* in *Xenopus* oocytes (Bienert et al. 2013). Lactic acid permeability was not observed in other GlpF isoforms. Unlike *Arabidopsis* NIP2;1 (Choi and Roberts 2007), lactic acid transporting GlpF proteins are multifunctional and are permeable to water, glycerol, H_2O_2, and urea (Bienert et al. 2013). Nevertheless, the presence of *GlpF1/GlpF4* in the lar operon and the observation that a double *glpF1/glpF4* mutation decreases the fitness of *L. plantarum* and enhances its sensitivity to stressful levels of lactic acid (Bienert et al. 2013) argue for a role of these glycerol facilitator isoforms in lactic acid efflux and intracellular pH regulation. A second example of a lactic acid permeable aquaporin is the SmAQP associated with the human parasitic nematode *Schistosoma mansoni* that is essential for excretion of lactic acid generated by fermentation (Faghiri et al. 2010).

5 Summary and Prospects

Since their initial discovery as symbiosis-specific nodulins, NIPs have emerged as a diverse family of plant-specific MIPs with multiple channel activities. The strongest evidence for biological function for these proteins comes from genetic and biochemical evidence supporting a role as facilitators of metalloid hydroxide nutrients (boric acid and silicic acid), particularly under environmental conditions where these nutrients are limiting. However, additional functions as aquaporins, ammoniaporins, and the transport of other biologically relevant substrates (e.g., lactic acid, urea, hydrogen peroxide, and glycerol) are also likely. The inherent multifunctional nature of NIP transport makes the elucidation of biological function difficult on the basis of biochemical activity alone. Moving forward, a number of evolutionary, biochemical, and physiological questions remain to be addressed regarding the NIP family:

- *What is the evolutionary origin and driving force for the diversification of NIPs in land plants?* The details of the evolutionary origin of NIPs, as well as the driving force that lead to the diversification of NIPs upon the colonization of land, are still debated questions. Phylogenetic support for a horizontal gene transfer from a bacterial or archaeal AQP ancestor has been provided, although alternative models have also been advanced. In addition, it is clear that *NIP* genes in early plant lineages were particularly diverse and the encoded proteins show unique pore selectivity regions. Details on the functional properties of NIPs from these lineages, as well as bacterial and archaeal NIP-like proteins, are lacking and could provide insight into early biological roles as metalloid or solute channels.

- *What are the biological functions of the NIP-1 protein group?* While a substantial body of knowledge supports functions for NIP-3 proteins as boric acid channels and NIP-2 proteins as silicic acid channels, the biological function of the NIP-1 group remains less clear. From an evolutionary perspective, the NIP-1 group evolved most recently and are restricted to angiosperms. While it is clear that they participate in arsenite mobility in plants, this is likely not their biological function. A tryptophan substitution within the conserved ar/R region of the protein results in enhanced aquaporin activity and restricted boric acid permeability, and a role in boron nutrition may be less likely. Targeted NIP-1 expression in specific developmental or stress responses (e.g., symbiosis, pollen, and waterlogging stress) provides interesting leads to biological function, but genetic evidence (and possibly bioinformatic- and system-based approaches combined with genetic approaches) is needed to provide more incisive support for these hypothetical roles.
- *What are the structural determinants that guide the substrate selectivity of the three NIP channel classes?* Based on the conserved aquaporin hourglass fold and molecular modeling, angiosperm NIPs have been segregated into three distinct pore families with "signature" amino acids at the proposed ar/R selectivity filter. However, there is considerable variation in biochemical activity even among NIPs with identical ar/R compositions, and it is likely that additional structural and functional determinants control selectivity and regulation. From the multiple structures of aquaporins that have been solved at atomic resolution, it is clear that while the hourglass fold is a conserved feature, each new structure provides unexpected surprises, additional unanticipated pore constrictions, and structural features that confer unique properties on each protein. The solution of an NIP crystal structure is long overdue, and it is essential to provide mechanistic insight beyond simple homology modeling.
- *What is the molecular basis and mechanism of NIP regulation by phosphorylation, other posttranslational modifications, and protein interactions?* Even before its functional properties were known, nodulin 26 was described as an in vivo substrate for calcium-dependent protein kinases. It appears as if NIP proteins are targets for these and possibly other Ser/Thr protein kinases. Potential regulation of transport has been proposed, but a more thorough investigation of the functional effects of phosphorylation on the structure of the channel and potential effects of phosphorylation on other processes including membrane targeting and trafficking [similar to other aquaporins (Kreida and Tornroth-Horsefield 2015; Maurel et al. 2015)] needs to be pursued. Related to this, protein interaction with aquaporins has emerged as another mechanism of regulation, not only

with respect to regulation of activity [e.g., calmodulin and AQP0 (Reichow et al. 2013)] but also controlling membrane localization and targeting [e.g., AQP2 (Van Balkom et al. 2009)]. One interaction target for a NIP (glutamine synthetase) has been described, but it is likely that the collection of NIP-interacting proteins will be more complex and is an underexplored area.

Acknowledgments Supported by National Science Foundation grant MCB-1121465.

References

Abascal F, Irisarri I, Zardoya R (2014) Diversity and evolution of membrane intrinsic proteins. Biochim Biophys Acta 1840(5):1468–1481

Abdallah C, Valot B, Guillier C, Mounier A, Balliau T, Zivy M, Van Tuinen D, Renaut J, Wipf D, Dumas-Gaudot E, Recorbet G (2014) The membrane proteome of *Medicago truncatula* roots displays qualitative and quantitative changes in response to arbuscular mycorrhizal symbiosis. J Proteome 108:354–368

Afzal Z, Howton TC, Sun YL, Mukhtar MS (2016) The roles of aquaporins in plant stress responses. J Dev Biol 4(1):9

Alexandersson E, Fraysse L, Sjovall-Larsen S, Gustavsson S, Fellert M, Karlsson M, Johanson U, Kjellbom P (2005) Whole gene family expression and drought stress regulation of aquaporins. Plant Mol Biol 59(3):469–484

Anderberg HI, Danielson JA, Johanson U (2011) Algal MIPs, high diversity and conserved motifs. BMC Evol Biol 11:110

Anderberg HI, Kjellbom P, Johanson U (2012) Annotation of selaginella moellendorffii major intrinsic proteins and the evolution of the protein family in terrestrial plants. Front Plant Sci 3:33

Azad AK, Ahmed J, Alum MA, Hasan MM, Ishikawa T, Sawa Y, Katsuhara M (2016) Genome-wide characterization of major intrinsic proteins in four grass plants and their non-aqua transport selectivity profiles with comparative perspective. PLoS One 11(6):e0157735

Benedito VA, Torres-Jerez I, Murray JD, Andriankaja A, Allen S, Kakar K, Wandrey M, Verdier J, Zuber H, Ott T, Moreau S, Niebel A, Frickey T, Weiller G, He J, Dai X, Zhao PX, Tang Y, Udvardi MK (2008) A gene expression atlas of the model legume *Medicago truncatula*. Plant J 55(3):504–513

Bienert GP, Desguin B, Chaumont F, Hols P (2013) Channel-mediated lactic acid transport: a novel function for aquaglyceroporins in bacteria. Biochem J 454:559–570

Bienert GP, Thorsen M, Schussler MD, Nilsson HR, Wagner A, Tamas MJ, Jahn TP (2008) A subgroup of plant aquaporins facilitate the bi-directional diffusion of As(OH)(3) and Sb(OH)(3) across membranes. BMC Biol 6:26

Bock KW, Honys D, Ward JM, Padmanaban S, Nawrocki EP, Hirschi KD, Twell D, Sze H (2006) Integrating membrane transport with male gametophyte development and function through transcriptomics. Plant Physiol 140(4):1151–1168

Boursiac Y, Prak S, Boudet J, Postaire O, Luu DT, Tournaire-Roux C, Santoni V, Maurel C (2008) The response of Arabidopsis root water transport to a challenging environment implicates reactive oxygen species- and phosphorylation-dependent internalization of aquaporins. Plant Signal Behav 3(12):1096–1098

Camacho-Cristobal JJ, Rexach J, Gonzalez-Fontes A (2008) Boron in plants: deficiency and toxicity. J Integr Plant Biol 50(10):1247–1255

Catalano CM, Lane WS, Sherrier DJ (2004) Biochemical characterization of symbiosome membrane proteins from *Medicago truncatula* root nodules. Electrophoresis 25(3):519–531

Chaumont F, Tyerman SD (2014) Aquaporins: highly regulated channels controlling plant water relations. Plant Physiol 164(4):1600–1618

Chaumont F, Barrieu F, Wojcik E, Chrispeels MJ, Jung R (2001) Aquaporins constitute a large and highly divergent protein family in maize. Plant Physiol 125(3):1206–1215

Cheng C, Rerkasem B (1993) Effects of boron on pollen viability in wheat. Plant Soil 155:313–315

Chiba Y, Mitani N, Yamaji N, Ma JF (2009) HvLsi1 is a silicon influx transporter in barley. Plant J 57(5):810–818

Choi WG, Roberts DM (2007) Arabidopsis NIP2;1, a major intrinsic protein transporter of lactic acid induced by anoxic stress. J Biol Chem 282(33):24209–24218

Clarke VC, Loughlin PC, Day DA, Smith PMC (2014) Transport processes of the legume symbiosome membrane. Front Plant Sci 5:699

Clarke VC, Loughlin PC, Gavrin A, Chen C, Brear EM, Day DA, Smith PMC (2015) Proteomic analysis of the soybean symbiosome identifies new symbiotic proteins. Mol Cell Proteomics 14(5):1301–1322

Curran A, Chang IF, Chang CL, Garg S, Miguel RM, Barron YD, Li Y, Romanowsky S, Cushman JC, Gribskov M, Harmon AC, Harper JF (2011) Calcium-dependent protein kinases from Arabidopsis show substrate specificity differences in an analysis of 103 substrates. Front Plant Sci 2:36

Danielson JA, Johanson U (2008) Unexpected complexity of the aquaporin gene family in the moss *Physcomitrella patens*. BMC Plant Biol 8:45

Danielson JA, Johanson U (2010) Phylogeny of major intrinsic proteins. Adv Exp Med Biol 679:19–31

Day DA, Poole PS, Tyerman SD, Rosendahl L (2001) Ammonia and amino acid transport across symbiotic membranes in nitrogen-fixing legume nodules. Cell Mol Life Sci 58(1):61–71

De Groot BL, Grubmuller H (2001) Water permeation across biological membranes: mechanism and dynamics of aquaporin-1 and GlpF. Science 294(5550):2353–2357

Dean RM, Rivers RL, Zeidel ML, Roberts DM (1999) Purification and functional reconstitution of soybean nodulin 26. An aquaporin with water and glycerol transport properties. Biochemistry 38(1):347–353

Dell B, Huang LB (1997) Physiological response of plants to low boron. Plant Soil 193(1–2):103–120

Denison RF, Kinraide TB (1995) Oxygen-induced membrane depolarizations in legume root-nodules – possible evidence for an osmoelectrical mechanism controlling nodule gas-permeability. Plant Physiol 108(1):235–240

Deshmukh RK, Vivancos J, Ramakrishnan G, Guerin V, Carpentier G, Sonah H, Labbe C, Isenring P, Belzile FJ, Belanger RR (2015) A precise spacing between the NPA domains of aquaporins is essential for silicon permeability in plants. Plant J 83(3):489–500

Di Giorgio JA, Bienert GP, Ayub ND, Yaneff A, Barberini ML, Mecchia MA, Amodeo G, Soto GC, Muschietti JP (2016) Pollen-specific aquaporins NIP4;1 and NIP4;2 are required for pollen development and pollination in *Arabidopsis thaliana*. Plant Cell 28(5):1053–1077

Diehn TA, Pommerrenig B, Bernhardt N, Hartmann A, Bienert GP (2015) Genome-wide identification of aquaporin encoding genes in *Brassica oleracea* and their phylogenetic sequence comparison to Brassica crops and Arabidopsis. Front Plant Sci 6:166

Emerich DW, Krishnan HB (2014) Symbiosomes: temporary moonlighting organelles. Biochem J 460(1):1–11

Epstein E (1999) Silicon. Annu Rev Plant Physiol Plant Mol Biol 50:641–664

Faghiri Z, Camargo SM, Huggel K, Forster IC, Ndegwa D, Verrey F, Skelly PJ (2010) The tegument of the human parasitic worm *Schistosoma mansoni* as an excretory organ: the surface aquaporin SmAQP is a lactate transporter. PLoS One 5(5):e10451

Felle HH (2005) pH regulation in anoxic plants. Ann Bot 96(4):519–532

Finn RN, Cerda J (2015) Evolution and functional diversity of aquaporins. Biol Bull 229(1):6–23

Firon N, Nepi M, Pacini E (2012) Water status and associated processes mark critical stages in pollen development and functioning. Ann Bot 109(7):1201–1213

Fortin MG, Morrison NA, Verma DP (1987) Nodulin-26, a peribacteroid membrane nodulin is expressed independently of the development of the peribacteroid compartment. Nucleic Acids Res 15(2):813–824

Fortin MG, Zelechowska M, Verma DP (1985) Specific targeting of membrane nodulins to the bacteroid-enclosing compartment in soybean nodules. EMBO J 4(12):3041–3046
Fouquet R, Leon C, Ollat N, Barrieu F (2008) Identification of grapevine aquaporins and expression analysis in developing berries. Plant Cell Rep 27(9):1541–1550
Fu D, Libson A, Miercke LJ, Weitzman C, Nollert P, Krucinski J, Stroud RM (2000) Structure of a glycerol-conducting channel and the basis for its selectivity. Science 290(5491):481–486
Gao ZX, He XL, Zhao BC, Zhou CJ, Liang YZ, Ge RC, Shen YZ, Huang ZJ (2010) Overexpressing a putative aquaporin gene from wheat, TaNIP, enhances salt tolerance in transgenic arabidopsis. Plant Cell Physiol 51(5):767–775
Giovannetti M, Balestrini R, Volpe V, Guether M, Straub D, Costa A, Ludewig U, Bonfante P (2012) Two putative-aquaporin genes are differentially expressed during arbuscular mycorrhizal symbiosis in Lotus japonicus. BMC Plant Biol 12:186
Gonen T, Sliz P, Kistler J, Cheng Y, Walz T (2004) Aquaporin-0 membrane junctions reveal the structure of a closed water pore. Nature 429(6988):193–197
Gregoire C, Remus-Borel W, Vivancos J, Labbe C, Belzile F, Belanger RR (2012) Discovery of a multigene family of aquaporin silicon transporters in the primitive plant *Equisetum arvense*. Plant J 72(2):320–330
Gu R, Chen X, Zhou Y, Yuan L (2012) Isolation and characterization of three maize aquaporin genes, ZmNIP2;1, ZmNIP2;4 and ZmTIP4;4 involved in urea transport. BMB Rep 45(2):96–101
Guenther JF, Roberts DM (2000) Water-selective and multifunctional aquaporins from Lotus japonicus nodules. Planta 210(5):741–748
Guenther JF, Chanmanivone N, Galetovic MP, Wallace IS, Cobb JA, Roberts DM (2003) Phosphorylation of soybean nodulin 26 on serine 262 enhances water permeability and is regulated developmentally and by osmotic signals. Plant Cell 15(4):981–991
Guether M, Neuhauser B, Balestrini R, Dynowski M, Ludewig U, Bonfante P (2009) A mycorrhizal-specific ammonium transporter from Lotus japonicus acquires nitrogen released by arbuscular mycorrhizal fungi. Plant Physiol 150(1):73–83
Gupta AB, Sankararamakrishnan R (2009) Genome-wide analysis of major intrinsic proteins in the tree plant *Populus trichocarpa*: characterization of XIP subfamily of aquaporins from evolutionary perspective. BMC Plant Biol 9:134
Gustavsson S, Lebrun AS, Norden K, Chaumont F, Johanson U (2005) A novel plant major intrinsic protein in *Physcomitrella patens* most similar to bacterial glycerol channels. Plant Physiol 139(1):287–295
Hachez C, Chaumont F (2010) Aquaporins: a family of highly regulated multifunctional channels. Adv Exp Med Biol 679:1–17
Hanaoka H, Uraguchi S, Takano J, Tanaka M, Fujiwara T (2014) OsNIP3;1, a rice boric acid channel, regulates boron distribution and is essential for growth under boron-deficient conditions. Plant J 78(5):890–902
Harrison MJ (2012) Cellular programs for arbuscular mycorrhizal symbiosis. Curr Opin Plant Biol 15(6):691–698
Harrison MJ, Dewbre GR, Liu J (2002) A phosphate transporter from *Medicago truncatula* involved in the acquisition of phosphate released by arbuscular mycorrhizal fungi. Plant Cell 14(10):2413–2429
Hill AE, Shachar-Hill B, Skepper JN, Powell J, Shachar-Hill Y (2012) An osmotic model of the growing pollen tube. PLoS One 7(5):e36585
Hove RM, Ziemann M, Bhave M (2015) Identification and expression analysis of the Barley (*Hordeum vulgare* L.) aquaporin gene family. Plos One 10(6):e0128025
Hu W, Yuan Q, Wang Y, Cai R, Deng X, Wang J, Zhou S, Chen M, Chen L, Huang C, Ma Z, Yang G, He G (2012) Overexpression of a wheat aquaporin gene, TaAQP8, enhances salt stress tolerance in transgenic tobacco. Plant Cell Physiol 53(12):2127–2141
Huang LB, Pant J, Dell B, Bell RW (2000) Effects of boron deficiency on anther development and floret fertility in wheat (*Triticum aestivum* L-'Wilgoyne'). Ann Bot 85(4):493–500
Hub JS, Aponte-Santamaria C, Grubmuller H, De Groot BL (2010) Voltage-regulated water flux through aquaporin channels in silico. Biophys J 99(12):L97–L99

Hub JS, Grubmuller H, De Groot BL (2009) Dynamics and energetics of permeation through aquaporins. What do we learn from molecular dynamics simulations? Handb Exp Pharmacol 90(190):57–76

Hwang JH (2013) Soybean nodulin 26: a channel for water and ammonia at the symbiotic interface of legumes and nitrogen-fixing rhizobia bacteria. Ph.D., The University of Tennessee, Knoxville

Hwang JH, Ellingson SR, Roberts DM (2010) Ammonia permeability of the soybean nodulin 26 channel. FEBS Lett 584(20):4339–4343

Isayenkov SV, Maathuis FJM (2008) The *Arabidopsis thaliana* aquaglyceroporin AtNIP7;1 is a pathway for arsenite uptake. FEBS Lett 582(11):1625–1628

Ivanov S, Fedorova E, Bisseling T (2010) Intracellular plant microbe associations: secretory pathways and the formation of perimicrobial compartments. Curr Opin Plant Biol 13(4): 372–377

Jiang J, Daniels BV, Fu D (2006) Crystal structure of AqpZ tetramer reveals two distinct Arg-189 conformations associated with water permeation through the narrowest constriction of the water-conducting channel. J Biol Chem 281(1):454–460

Johanson U, Gustavsson S (2002) A new subfamily of major intrinsic proteins in plants. Mol Biol Evol 19(4):456–461

Johanson U, Karlsson M, Johansson I, Gustavsson S, Sjovall S, Fraysse L, Weig AR, Kjellbom P (2001) The complete set of genes encoding major intrinsic proteins in Arabidopsis provides a framework for a new nomenclature for major intrinsic proteins in plants. Plant Physiol 126(4):1358–1369

Johansson I, Karlsson M, Shukla VK, Chrispeels MJ, Larsson C, Kjellbom P (1998) Water transport activity of the plasma membrane aquaporin PM28A is regulated by phosphorylation. Plant Cell 10(3):451–459

Johnson SA, Mccormick S (2001) Pollen germinates precociously in the anthers of raring-to-go, an Arabidopsis gametophytic mutant. Plant Physiol 126(2):685–695

Jung JS, Preston GM, Smith BL, Guggino WB, Agre P (1994) Molecular structure of the water channel through aquaporin CHIP. The hourglass model. J Biol Chem 269(20):14648–14654

Kamiya T, Fujiwara T (2009) Arabidopsis NIP1;1 transports antimonite and determines antimonite sensitivity. Plant Cell Physiol 50(11):1977–1981

Kamiya T, Tanaka M, Mitani N, Ma JF, Maeshima M, Fujiwara T (2009) NIP1;1, an aquaporin homolog, determines the arsenite sensitivity of *Arabidopsis thaliana*. J Biol Chem 284(4):2114–2120

Katsuhara M, Sasano S, Horie T, Matsumoto T, Rhee J, Shibasaka M (2014) Functional and molecular characteristics of rice and barley NIP aquaporins transporting water, hydrogen peroxide and arsenite. Plant Biol 31(3):213–U173

Kosinska Eriksson U, Fischer G, Friemann R, Enkavi G, Tajkhorshid E, Neutze R (2013) Subangstrom resolution X-ray structure details aquaporin-water interactions. Science 340(6138):1346–1349

Kreida S, Tornroth-Horsefield S (2015) Structural insights into aquaporin selectivity and regulation. Curr Opin Struct Biol 33:126–134

Li T (2014) Pore selectivity and gating of Arabidopsis nodulin 26 intrinsic proteins and roles in boric acid transport in reproductive growth. Ph.D., The University of Tennessee, Knoxville

Li T, Choi WG, Wallace IS, Baudry J, Roberts DM (2011) *Arabidopsis thaliana* NIP7;1: an anther-specific boric acid transporter of the aquaporin superfamily regulated by an unusual tyrosine in helix 2 of the transport pore. Biochemistry 50(31):6633–6641

Lindsey Rose KM, Gourdie RG, Prescott AR, Quinlan RA, Crouch RK, Schey KL (2006) The C terminus of lens aquaporin 0 interacts with the cytoskeletal proteins filensin and CP49. Invest Ophthalmol Vis Sci 47(4):1562–1570

Liu Q, Zhu Z (2010) Functional divergence of the NIP III subgroup proteins involved altered selective constraints and positive selection. BMC Plant Biol 10:256

Liu Z, Carbrey JM, Agre P, Rosen BP (2004) Arsenic trioxide uptake by human and rat aquaglyceroporins. Biochem Biophys Res Commun 316(4):1178–1185

Ludewig U, Dynowski M (2009) Plant aquaporin selectivity: where transport assays, computer simulations and physiology meet. Cell Mol Life Sci 66(19):3161–3175

Ma JF, Yamaji N (2006) Silicon uptake and accumulation in higher plants. Trends Plant Sci 11(8):392–397

Ma JF, Yamaji N (2015) A cooperative system of silicon transport in plants. Trends Plant Sci 20(7):435–442

Ma JF, Tamai K, Yamaji N, Mitani N, Konishi S, Katsuhara M, Ishiguro M, Murata Y, Yano M (2006) A silicon transporter in rice. Nature 440(7084):688–691

Ma JF, Yamaji N, Mitani N, Xu XY, Su YH, Mcgrath SP, Zhao FJ (2008) Transporters of arsenite in rice and their role in arsenic accumulation in rice grain. Proc Natl Acad Sci U S A 105(29):9931–9935

Martins CDS, Pedrosa AM, Du DL, Goncalves LP, Yu Q, Gmitter FG, Costa MGC (2015) Genome-wide characterization and expression analysis of major intrinsic proteins during abiotic and biotic stresses in sweet orange (*Citrus sinensis* L. Osb.). Plos One 10(9):e0138786

Masalkar P, Wallace IS, Hwang JH, Roberts DM (2010) Interaction of cytosolic glutamine synthetase of soybean root nodules with the C-terminal domain of the symbiosome membrane nodulin 26 aquaglyceroporin. J Biol Chem 285(31):23880–23888

Matsunaga T, Ishii T, Matsumoto S, Higuchi M, Darvill A, Albersheim P, O'neill MA (2004) Occurrence of the primary cell wall polysaccharide rhamnogalacturonan II in pteridophytes, lycophytes, and bryophytes. Implications for the evolution of vascular plants. Plant Physiol 134(1):339–351

Maurel C, Plassard C (2011) Aquaporins: for more than water at the plant-fungus interface? New Phytol 190(4):815–817

Maurel C, Boursiac Y, Luu DT, Santoni V, Shahzad Z, Verdoucq L (2015) Aquaporins in plants. Physiol Rev 95(4):1321–1358

Maurel C, Tacnet F, Guclu J, Guern J, Ripoche P (1997) Purified vesicles of tobacco cell vacuolar and plasma membranes exhibit dramatically different water permeability and water channel activity. Proc Natl Acad Sci U S A 94(13):7103–7108

Minchin FR, James EK, Becana M (2008) Oxygen diffusion, production of reactive oxygen and nitrogen species, and antioxidants in legume nodules. In: Dilworth MJ, James EK, Sprent JI, Newton WE (eds) Nitrogen-fixing leguminous symbioses. Springer, Dordrecht

Mitani N, Yamaji N, Ma JF (2008) Characterization of substrate specificity of a rice silicon transporter, Lsi1. Pflugers Arch 456(4):679–686

Mitani-Ueno N, Yamaji N, Zhao FJ, Ma JF (2011) The aromatic/arginine selectivity filter of NIP aquaporins plays a critical role in substrate selectivity for silicon, boron, and arsenic. J Exp Bot 62(12):4391–4398

Miwa K, Fujiwara T (2010) Boron transport in plants: co-ordinated regulation of transporters. Ann Bot 105(7):1103–1108

Miwa K, Tanaka M, Kamiya T, Fujiwara T (2010) Molecular mechanisms of boron transport in plants: involvement of Arabidopsis NIP5;1 and NIP6;1. Mips Exch Metalloids 679:83–96

Mizutani M, Watanabe S, Nakagawa T, Maeshima M (2006) Aquaporin NIP2;1 is mainly localized to the ER membrane and shows root-specific accumulation in *Arabidopsis thaliana*. Plant Cell Physiol 47(10):1420–1426

Mukhopadhyay R, Bhattacharjee H, Rosen BP (2014) Aquaglyceroporins: generalized metalloid channels. Biochim Biophys Acta 1840(5):1583–1591

Mustroph A, Zanetti ME, Jang CJ, Holtan HE, Repetti PP, Galbraith DW, Girke T, Bailey-Serres J (2009) Profiling translatomes of discrete cell populations resolves altered cellular priorities during hypoxia in Arabidopsis. Proc Natl Acad Sci U S A 106(44):18843–18848

Myers C, Romanowsky SM, Barron YD, Garg S, Azuse CL, Curran A, Davis RM, Hatton J, Harmon AC, Harper JF (2009) Calcium-dependent protein kinases regulate polarized tip growth in pollen tubes. Plant J 59(4):528–539

Niemietz CM, Tyerman SD (2000) Channel-mediated permeation of ammonia gas through the peribacteroid membrane of soybean nodules. FEBS Lett 465(2–3):110–114

Nyblom M, Frick A, Wang Y, Ekvall M, Hallgren K, Hedfalk K, Neutze R, Tajkhorshid E, Tornroth-Horsefield S (2009) Structural and functional analysis of SoPIP2;1 mutants adds insight into plant aquaporin gating. J Mol Biol 387(3):653–668

Ouyang LJ, Whelan J, Weaver CD, Roberts DM, Day DA (1991) Protein phosphorylation stimulates the rate of malate uptake across the peribacteroid membrane of soybean nodules. FEBS Lett 293(1–2):188–190

Perez Di Giorgio J, Soto G, Alleva K, Jozefkowicz C, Amodeo G, Muschietti JP, Ayub ND (2014) Prediction of aquaporin function by integrating evolutionary and functional analyses. J Membr Biol 247(2):107–125

Pommerrenig B, Diehn TA, Bienert GP (2015) Metalloido-porins: essentiality of nodulin 26-like intrinsic proteins in metalloid transport. Plant Sci 238:212–227

Porquet A, Filella M (2007) Structural evidence of the similarity of Sb(OH)3 and As(OH)3 with glycerol: implications for their uptake. Chem Res Toxicol 20(9):1269–1276

Purcell LC, Sinclair TR (1994) An osmotic hypothesis for the regulation of oxygen permeability in soybean nodules. Plant Cell Environ 17(7):837–843

Quigley F, Rosenberg JM, Shachar-Hill Y, Bohnert HJ (2002) From genome to function: the Arabidopsis aquaporins. Genome Biol 3(1):1–17

Rawson HM (1996) The developmental stage during which boron limitation causes sterility in wheat genotypes and the recovery of fertility. Aust J Plant Physiol 23(6):709–717

Reichow SL, Clemens DM, Freites JA, Nemeth-Cahalan KL, Heyden M, Tobias DJ, Hall JE, Gonen T (2013) Allosteric mechanism of water-channel gating by Ca2+–calmodulin. Nat Struct Mol Biol 20(9):1085–1092

Reuscher S, Akiyama M, Mori C, Aoki K, Shibata D, Shiratake K (2013) Genome-wide identification and expression analysis of aquaporins in tomato. PLoS One 8(11):e79052

Rivers RL, Dean RM, Chandy G, Hall JE, Roberts DM, Zeidel ML (1997) Functional analysis of nodulin 26, an aquaporin in soybean root nodule symbiosomes. J Biol Chem 272(26):16256–16261

Roberts JK, Callis J, Wemmer D, Walbot V, Jardetzky O (1984) Mechanisms of cytoplasmic pH regulation in hypoxic maize root tips and its role in survival under hypoxia. Proc Natl Acad Sci U S A 81(11):3379–3383

Roberts DM, Choi WG, Hwang JH (2010) Strategies for adaptation to waterlogging and hypoxia in nitrogen fixing nodules of legumes. Waerlogging signalling and tolerance in plants. S. Mancuso and S. Shabala, eds. Springer-Verlag Berlin Heidelberg, pp. 37–59

Roth LE, Stacey G (1989) Bacterium release into host-cells of nitrogen-fixing soybean nodules – the symbiosome membrane comes from 3 sources. Eur J Cell Biol 49(1):13–23

Rouge P, Barre A (2008) A molecular modeling approach defines a new group of Nodulin 26-like aquaporins in plants. Biochem Biophys Res Commun 367(1):60–66

Routray P, Masalkar PD, Roberts DM (2015) Nodulin intrinsic proteins: facilitators of water and ammonia transport across the symbiosome membrane. In: De Bruijn FJ (ed) Biological nitrogen fixation. Wiley, Hoboken, Volume II, Chapter 69, pp. 695–704.

Sakurai J, Ishikawa F, Yamaguchi T, Uemura M, Maeshima M (2005) Identification of 33 rice aquaporin genes and analysis of their expression and function. Plant Cell Physiol 46(9):1568–1577

Sandal NN, Marcker KA (1988) Soybean nodulin 26 is homologous to the major intrinsic protein of the bovine lens fiber membrane. Nucleic Acids Res 16(19):9347

Sanders PM, Bui AQ, Weterings K, Mcintire KN, Hsu YC, Lee PY, Truong MT, Beals TP, Goldberg RB (1999) Anther developmental defects in *Arabidopsis thaliana* male-sterile mutants. Sex Plant Reprod 11(6):297–322

Sanders OI, Rensing C, Kuroda M, Mitra B, Rosen BP (1997) Antimonite is accumulated by the glycerol facilitator GlpF in *Escherichia coli*. J Bacteriol 179(10):3365–3367

Savage DF, O'Connell JD 3rd, Miercke LJ, Finer-Moore J, Stroud RM (2010) Structural context shapes the aquaporin selectivity filter. Proc Natl Acad Sci U S A 107(40):17164–17169

Schmid M, Davison TS, Henz SR, Pape UJ, Demar M, Vingron M, Scholkopf B, Weigel D, Lohmann JU (2005) A gene expression map of *Arabidopsis thaliana* development. Nat Genet 37(5):501–506
Schuurmans JA, Van Dongen JT, Rutjens BP, Boonman A, Pieterse CM, Borstlap AC (2003) Members of the aquaporin family in the developing pea seed coat include representatives of the PIP, TIP, and NIP subfamilies. Plant Mol Biol 53(5):633–645
Shachar-Hill B, Hill AE, Powell J, Skepper JN, Shachar-Hill Y (2013) Mercury-sensitive water channels as possible sensors of water potentials in pollen. J Exp Bot 64(16):5195–5205
Sjohamn J, Hedfalk K (2014) Unraveling aquaporin interaction partners. Biochim Biophys Acta 1840(5):1614–1623
Smith SE, Smith FA (2011) Roles of arbuscular mycorrhizas in plant nutrition and growth: new paradigms from cellular to ecosystem scales. Annu Rev Plant Biol 62:227–250
Smyth DR, Bowman JL, Meyerowitz EM (1990) Early flower development in Arabidopsis. Plant Cell 2(8):755–767
Soto G, Alleva K, Mazzella MA, Amodeo G, Muschietti JP (2008) AtTIP1;3 and AtTIP5;1, the only highly expressed Arabidopsis pollen-specific aquaporins, transport water and urea. FEBS Lett 582(29):4077–4082
Steudle E, Peterson CA (1998) How does water get through roots? J Exp Bot 49(322):775–788
Sugiyama N, Nakagami H, Mochida K, Daudi A, Tomita M, Shirasu K, Ishihama Y (2008) Large-scale phosphorylation mapping reveals the extent of tyrosine phosphorylation in Arabidopsis. Mol Syst Biol 4:193
Sui H, Han BG, Lee JK, Walian P, Jap BK (2001) Structural basis of water-specific transport through the AQP1 water channel. Nature 414(6866):872–878
Takano J, Miwa K, Yuan L, Von Wiren N, Fujiwara T (2005) Endocytosis and degradation of BOR1, a boron transporter of *Arabidopsis thaliana*, regulated by boron availability. Proc Natl Acad Sci U S A 102(34):12276–12281
Takano J, Tanaka M, Toyoda A, Miwa K, Kasai K, Fuji K, Onouchi H, Naito S, Fujiwara T (2010) Polar localization and degradation of Arabidopsis boron transporters through distinct trafficking pathways. Proc Natl Acad Sci U S A 107(11):5220–5225
Takano J, Wada M, Ludewig U, Schaaf G, Von Wiren N, Fujiwara T (2006) The Arabidopsis major intrinsic protein NIP5;1 is essential for efficient boron uptake and plant development under boron limitation. Plant Cell 18(6):1498–1509
Tanaka M, Takano J, Chiba Y, Lombardo F, Ogasawara Y, Onouchi H, Naito S, Fujiwara T (2011) Boron-dependent degradation of NIP5;1 mRNA for acclimation to excess boron conditions in Arabidopsis. Plant Cell 23(9):3547–3559
Tanaka M, Wallace IS, Takano J, Roberts DM, Fujiwara T (2008) NIP6;1 is a boric acid channel for preferential transport of boron to growing shoot tissues in arabidopsis. Plant Cell 20(10):2860–2875
Tornroth-Horsefield S, Wang Y, Hedfalk K, Johanson U, Karlsson M, Tajkhorshid E, Neutze R, Kjellbom P (2006) Structural mechanism of plant aquaporin gating. Nature 439(7077):688–694
Trembath-Reichert E, Wilson JP, Mcglynn SE, Fischer WW (2015) Four hundred million years of silica biomineralization in land plants. Proc Natl Acad Sci U S A 112(17):5449–5454
Tsukaguchi H, Weremowicz S, Morton CC, Hediger MA (1999) Functional and molecular characterization of the human neutral solute channel aquaporin-9. Am J Phys 277(5 Pt 2):F685–F696
Tyerman SD, Whitehead LF, Day DA (1995) A channel-like transporter for NH_4^+ on the symbiotic interface of N_2-fixing plants. Nature 378(6557):629–632
Udvardi M, Poole PS (2013) Transport and metabolism in legume-rhizobia symbioses. Annu Rev Plant Biol 64(64):781–805
Uehara M, Wang SL, Kamiya T, Shigenobu S, Yamaguchi K, Fujiwara T, Naito S, Takano J (2014) Identification and characterization of an Arabidopsis mutant with altered localization of NIP5;1, a plasma membrane boric acid channel, reveals the requirement for d-Galactose in endomembrane organization. Plant Cell Physiol 55(4):704–714
Uehlein N, Fileschi K, Eckert M, Bienert GP, Bertl A, Kaldenhoff R (2007) Arbuscular mycorrhizal symbiosis and plant aquaporin expression. Phytochemistry 68(1):122–129

Van Balkom BW, Boone M, Hendriks G, Kamsteeg EJ, Robben JH, Stronks HC, Van Der Voorde A, Van Herp F, Van Der S, Deen PM (2009) LIP5 interacts with aquaporin 2 and facilitates its lysosomal degradation. J Am Soc Nephrol 20(5):990–1001

Van Balkom BW, Savelkoul PJ, Markovich D, Hofman E, Nielsen S, Van Der S, Deen PM (2002) The role of putative phosphorylation sites in the targeting and shuttling of the aquaporin-2 water channel. J Biol Chem 277(44):41473–41479

Verma DP, Hong Z (1996) Biogenesis of the peribacteroid membrane in root nodules. Trends Microbiol 4(9):364–368

Voesenek LA, Bailey-Serres J (2015) Flood adaptive traits and processes: an overview. New Phytol 206(1):57–73

Wakuta S, Mineta K, Amano T, Toyoda A, Fujiwara T, Naito S, Takano J (2015) Evolutionary divergence of plant borate exporters and critical amino acid residues for the polar localization and boron-dependent vacuolar sorting of AtBOR1. Plant Cell Physiol 56(5):852–862

Wallace IS, Roberts DM (2004) Homology modeling of representative subfamilies of Arabidopsis major intrinsic proteins. Classification based on the aromatic/arginine selectivity filter. Plant Physiol 135(2):1059–1068

Wallace IS, Roberts DM (2005) Distinct transport selectivity of two structural subclasses of the nodulin-like intrinsic protein family of plant aquaglyceroporin channels. Biochemistry 44(51):16826–16834

Wallace IS, Choi WG, Roberts DM (2006) The structure, function and regulation of the nodulin 26-like intrinsic protein family of plant aquaglyceroporins. Biochim Biophys Acta 1758(8):1165–1175

Wallace IS, Wills DM, Guenther JF, Roberts DM (2002) Functional selectivity for glycerol of the nodulin 26 subfamily of plant membrane intrinsic proteins. FEBS Lett 523(1–3):109–112

Walz T, Fujiyoshi Y, Engel A (2009) The AQP structure and functional implications. Handb Exp Pharmacol 190:31–56

Wang Z, Schey KL (2011) Aquaporin-0 interacts with the FERM domain of ezrin/radixin/moesin proteins in the ocular lens. Invest Ophthalmol Vis Sci 52(8):5079–5087

Weaver CD, Roberts DM (1992) Determination of the site of phosphorylation of nodulin-26 by the calcium-dependent protein-kinase from soybean nodules. Biochemistry 31(37):8954–8959

Weaver CD, Crombie B, Stacey G, Roberts DM (1991) Calcium-dependent phosphorylation of symbiosome membrane-proteins from nitrogen-fixing soybean nodules – evidence for phosphorylation of nodulin-26. Plant Physiol 95(1):222–227

Weaver CD, Shomer NH, Louis CF, Roberts DM (1994) Nodulin-26, a nodule-specific symbiosome membrane-protein from soybean, is an ion-channel. J Biol Chem 269(27):17858–17862

Weig AR, Jakob C (2000) Functional identification of the glycerol permease activity of *Arabidopsis thaliana* NLM1 and NLM2 proteins by heterologous expression in *Saccharomyces cerevisiae*. FEBS Lett 481(3):293–298

Weig A, Deswarte C, Chrispeels MJ (1997) The major intrinsic protein family of Arabidopsis has 23 members that form three distinct groups with functional aquaporins in each group. Plant Physiol 114(4):1347–1357

Wudick MM, Luu DT, Tournaire-Roux C, Sakamoto W, Maurel C (2014) Vegetative and sperm cell-specific aquaporins of arabidopsis highlight the vacuolar equipment of pollen and contribute to plant reproduction. Plant Physiol 164(4):1697–1706

Wysocki R, Chery CC, Wawrzycka D, Van Hulle M, Cornelis R, Thevelein JM, Tamas MJ (2001) The glycerol channel Fps1p mediates the uptake of arsenite and antimonite in *Saccharomyces cerevisiae*. Mol Microbiol 40(6):1391–1401

Xia JH, Roberts J (1994) Improved cytoplasmic pH regulation, increased lactate efflux, and reduced cytoplasmic lactate levels are biochemical traits expressed in root tips of whole maize seedlings acclimated to a low-oxygen environment. Plant Physiol 105(2):651–657

Xin L, Su H, Nielsen CH, Tang C, Torres J, Mu Y (2011) Water permeation dynamics of AqpZ: a tale of two states. Biochim Biophys Acta 1808(6):1581–6.

Xu WZ, Dai WT, Yan HL, Li S, Shen HL, Chen YS, Xu H, Sun YY, He ZY, Ma M (2015) Arabidopsis NIP3;1 plays an important role in arsenic uptake and root-to-shoot translocation under arsenite stress conditions. Mol Plant 8(5):722–733

Yamaji N, Mitatni N, Ma JF (2008) A transporter regulating silicon distribution in rice shoots. Plant Cell 20(5):1381–1389

Yang H, Menz J, Haussermann I, Benz M, Fujiwara T, Ludewig U (2015) High and low affinity urea root uptake: involvement of NIP5;1. Plant Cell Physiol 56(8):1588–1597

Zardoya R (2005) Phylogeny and evolution of the major intrinsic protein family. Biol Cell 97(6):397–414

Zardoya R, Ding X, Kitagawa Y, Chrispeels MJ (2002) Origin of plant glycerol transporters by horizontal gene transfer and functional recruitment. Proc Natl Acad Sci U S A 99(23):14893–14896

Zhang DY, Ali Z, Wang CB, Xu L, Yi JX, Xu ZL, Liu XQ, He XL, Huang YH, Khan IA, Trethowan RM, Ma HX (2013) Genome-wide sequence characterization and expression analysis of major intrinsic proteins in soybean (*Glycine max* L.). Plos One 8(2):e56312

Zhou S, Hu W, Deng X, Ma Z, Chen L, Huang C, Wang C, Wang J, He Y, Yang G, He G (2012) Overexpression of the wheat aquaporin gene, TaAQP7, enhances drought tolerance in transgenic tobacco. PLoS One 7(12):e52439

Zhou L, Wang C, Liu R, Han Q, Vandeleur RK, Du J, Tyerman S, Shou H (2014) Constitutive overexpression of soybean plasma membrane intrinsic protein GmPIP1;6 confers salt tolerance. BMC Plant Biol 14:181

Plant Aquaporins and Metalloids

Manuela Désirée Bienert and Gerd Patrick Bienert

Abstract The metalloids represent a group of physiologically important elements, some of which are essential or at least beneficial (boron and silicon) for plant growth and some of which are toxic (arsenic, antimony and germanium). Exposure to and availability of metalloids can have major effects on plant fitness and yield and can seriously downgrade the end-use quality of certain crop products. Plants have evolved various membrane transport systems to regulate metalloid transport both at the cellular and whole plant level. To date, the channel proteins referred to as aquaporins (AQPs) represent the most favored candidates ensuring metalloid homeostasis. AQPs are found in all living organisms. From bacteria to mammals and also in plants, several distinct AQP subfamilies facilitate the transmembrane diffusion of the set of physiologically and environmentally important metalloids. A subgroup of the Nodulin26-like intrinsic protein AQP subfamily (NIPs) has been designated as functional metalloidoporins. NIPs are the only known transport protein family in the plant kingdom which are essential for the uptake, translocation, or extrusion of various uncharged metalloid species. This chapter describes the various features, and particularly the metalloid transport properties of plant AQPs, and illustrates their physiologically important contributions to metalloid homeostasis. Their intimate involvement in metalloid transport underlines their relevance to plant nutrition, detoxification of toxic mineral elements phytoremediation, phytomining, and biofortification.

1 The Metalloids

The metalloids represent a group of elements whose physical and chemical properties define them as being neither metals nor nonmetals. The six elements falling into this class are boron (B), silicon (Si), arsenic (As), antimony (Sb), germanium (Ge), and tellurium (Te). Selenium (Se), polonium (Po), and astatine (At) also belong to

M.D. Bienert • G.P. Bienert (✉)
Metalloid Transport Group, Department of Physiology and Cell Biology, Leibniz Institute of Plant Genetics and Crop Plant Research, Corrensstrasse 3, 06466 Gatersleben, Germany
e-mail: bienert@ipk-gatersleben.de

F. Chaumont, S.D. Tyerman (eds.), *Plant Aquaporins*, Signaling and Communication in Plants, DOI 10.1007/978-3-319-49395-4_14

pk_{a1}	protonated [metalloid acid]-H		>90% protonated acid	deprotonated [metalloid base]$^-$	
9.25	boric acid	H_3BO_3	pH < 8.30	$[H_4BO_4]^-$	borate
9.51	silicic acid	H_4SiO_4	pH < 8.56	$[H_3SiO_4]^-$	silicate
9.23	arsenous acid	H_3AsO_3	pH < 8.28	$[H_2AsO_3]^-$	arsenite
2.26	arsenic acid	H_3AsO_4	pH < 1.31	$[H_2AsO_4]^-$	arsenate
11.8	antimonous acid	H_3SbO_3	pH < 10.85	$[H_2SbO_3]^-$	antimonite
2.85	antimonic acid	H_3SbO_4	pH < 1.9	$[H_2SbO_4]^-$	antimonate
9.0	germanic acid	H_2GeO_3	pH < 8.05	$[HGeO_3]^-$	germanate
2.57	selenous acid	H_2SeO_3	pH < 1.62	$[HSeO_3]^-$	selenite
1.74	selenic acid	H_2SeO_4	pH < 0.79	$[HSeO_4]^-$	selenate

Fig. 1 pH-dependent acid-base equilibrium of hydroxylated metalloid acids. The *green color* indicates the chemical form and structural formula of the metalloid which predominates at the physiological pH range. Only neutral forms of metalloid acids are channeled by metalloidoporins. pKa values of the metalloid acids and the structural formula are given. The pH range in which more than 90 % of the metalloid acid occurs in its fully protonated acid species is displayed

the group but are less commonly designated as such. The lack of an unambiguous set of defining criteria reflects the dependence of many of their physical and chemical properties on ambient temperature and pressure, as well as on their crystal lattice/ crystal structure. The metalloids have a metallic appearance but are brittle. They are electrical semiconductors, can alloy with metals, and typically form amphoteric to weakly acidic oxides (Fig. 1). Their abundance in the Earth's crust varies from Si – the second most abundant element after oxygen, constituting ~25 % by mass of the Earth's crust (Lombi and Holm 2010) – to At, of which not more than 25 g is present in the total Earth' s crust at any given time (Lombi and Holm 2010).

The biological significance of the metalloids ranges from essential through beneficial to toxic. B is required for plant growth (Marschner 2012); Si is not generally recognized as essential, except for a few algal species and members of the *Equisetaceae* (Epstein 1994), although it is recognized as being beneficial for growth in many species. Se is essential in the human diet and for the growth of some algae, but is not so for plants (Pilon-Smits and Quinn 2010). As, Sb, Ge, and Te are all considered to be (phyto)toxic. The molecular form and the concentration of metalloids are both important in assessing the reaction of a plant to exposure. The impact of beneficial and essential metalloids on a given plant's metabolism can be summarized, *pace* Paracelsus: "the only difference between a nutrient and a poison is the dose."

2 The "Major Intrinsic Proteins" or Aquaporins

The large family of "major intrinsic proteins" comprises transmembrane-spanning channel proteins, found in almost all life forms (the exceptions being certain thermophilic *Archaea* and intracellular bacteria) (Abascal et al. 2014). The term "aquaporin" (AQP) is widely used as a synonym. Despite their sequence variation at the amino acid level, crystal structures acquired to date imply a high degree of conservation. The AQPs form tetramers: each monomer constitutes a functional channel on its own and is composed of six transmembrane-spanning helices (TMHs) with five connecting loops (loop A to loop E) and two cytoplasmic termini (see chapter "Structural Basis of the Permeation Function of Plant Aquaporins"). They define a narrow path across various cellular membranes, including the plasma membrane, the endoplasmic reticulum, the mitochondria, the vacuole, the vesicles involved in the trafficking pathway, the tonoplast, and the chloroplast (Maurel et al. 2015). They facilitate the diffusion of water and small uncharged solutes and have been shown by various means to control water homeostasis. In plants, they function to import water into the root from the soil, to transport it from the root to the shoot, to drive osmotic force-driven growth, and to ensure cytoplasmic osmolarity (Maurel et al. 2015; Chaumont and Tyerman 2014; see chapters "Aquaporins and Root Water Uptake" and "Aquaporins and Leaf Water Relations"). AQPs also have an impact on the uptake, translocation, sequestration, and extrusion of uncharged and physiologically important compounds such as glycerol (Richey and Lin 1972; Luyten et al. 1995), nitric oxide (NO) (Herrera et al. 2006), hydrogen peroxide (H_2O_2) (Bienert et al. 2006, 2007; Dynowski et al. 2008), urea (CH_4N_2O) (Liu et al. 2003), ammonia (NH_3) (Jahn et al. 2004; Loqué et al. 2005), lactic acid (Tsukaguchi et al. 1998; Choi and Roberts 2007; Bienert et al. 2013), and acetic acid (Mollapour and Piper 2007). Of note in the context of this chapter, they also transport arsenous acid (H_3AsO_3) (Bienert et al. 2008a, b; Ma et al. 2008; Kamiya et al. 2009), boric acid (H_3BO_3) (Takano et al. 2006; Tanaka et al. 2008; Hanaoka et al. 2014), silicic acid (H_4SiO_4) (Ma et al. 2006), antimonous acid (H_3SbO_3) (Bienert et al. 2008a; Kamiya et al. 2009), germanic acid (H_4GeO_4) (Ma et al. 2006; Hayes et al. 2013), and selenous acid (H_2SeO_3) (Zhao et al. 2010a, b) (Fig. 2).

AQPs allow the passage of a single continuous file of molecules. While a few ion-mediating AQPs have been identified (reviewed by Yool and Campbell 2012), the consensus, based on chemical species selectivity, is that only noncharged molecules are able to pass through the majority of AQP channels. However, compared to animal AQPs, not many plant AQPs have been assessed for being permeable to ions. The selectivity and transport capacity of each isoform are determined by the identity of the amino acids aligned along the channel pathway (see also chapter "Structural Basis of the Permeation Function of Plant Aquaporins"). The so-called "aromatic/arginine" (ar/R) selective filter, situated on the luminal side of the membrane, comprises four residues (R1–R4), located in TMH2 (R1), TMH5 (R2), and loop E (R3 and R4); this structure

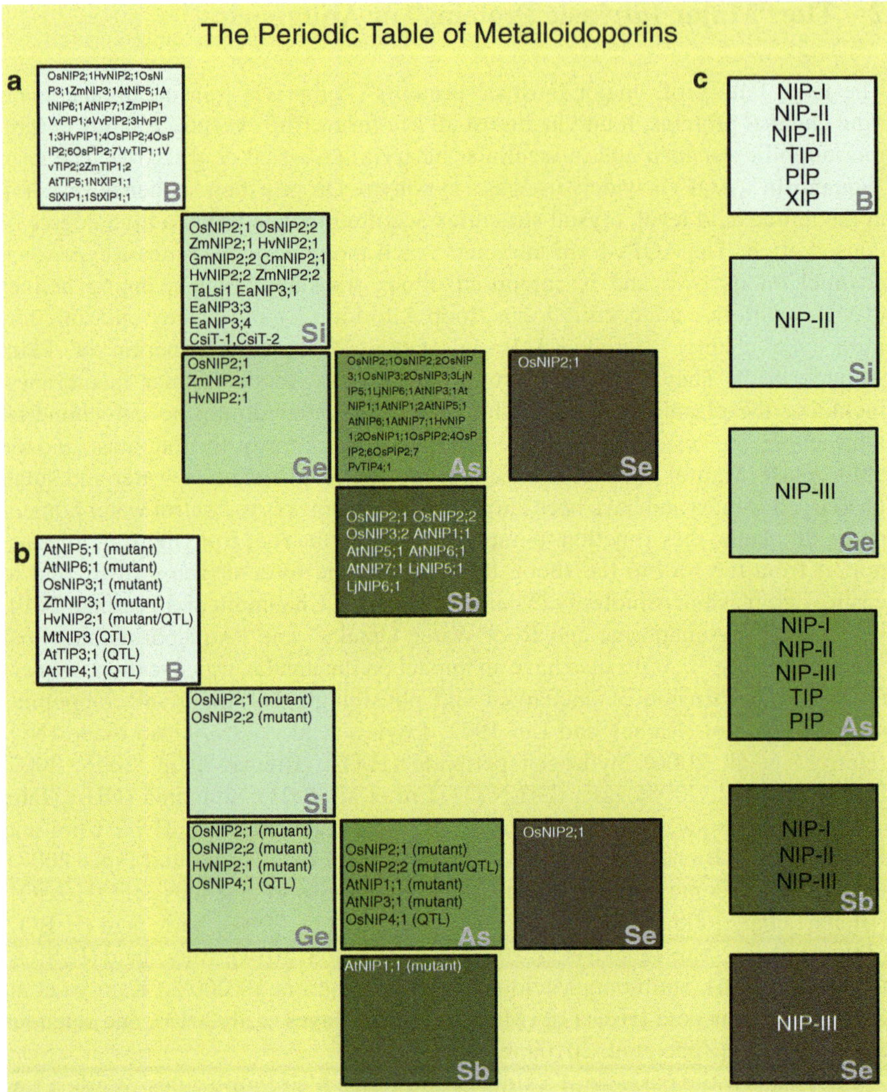

Fig. 2 The periodic table of metalloidoporins. (**a**) Aquaporin channel proteins, which were shown to be permeable to the corresponding metalloid acid in transport assays performed in plants, or heterologous expression systems (i.e., plants, frog oocytes, or yeasts) are listed. (**b**) Listed aquaporins have either been identified to occur in quantitative trait loci genomic regions linked to the tolerance toward toxicity or deficiency of the corresponding metalloid species (indicated by "QTL") or which, when being silenced or knocked out in planta (indicated by "mutant"), caused obvious metalloid deficiency or tolerance phenotypes. (**c**) Phylogenetic or functional plant aquaporin groups which were shown to be permeable to the corresponding metalloid acid in transport assays performed in plants or heterologous expression systems (plants, frog oocytes, or yeasts) are listed

forms a size exclusion barrier and the hydrogen bond environment necessary for the efficient transport of a particular substrate (Murata et al. 2000). A second selectivity filter, the so-called "NPA" motif (asparagine-proline-alanine or variants thereof), is formed by the two membrane-embedded half-helices of loop A and loop E, each containing the conserved AQP signature. The "NPA"motifs meet in the center of the membrane, forming a narrow hydrophilic cavity (Murata et al. 2000) and are responsible for the exclusion of water-mediated proton and ion transport.

A major difference between plants and other organisms is the large number of AQP isoforms encoded by plant genomes (Abascal et al. 2014). While the norm in bacteria, fungi, and mammals is 2–13 genes per genome (Agre and Kozono 2003), the moss *Physcomitrella patens* and the lycophyte *Selaginella moellendorffii* encode, respectively, 23 and 19 *AQP*s (Danielson and Johanson 2008; Anderberg et al. 2012). Higher plant genomes harbor from 30 to 70 isoforms: the number in *Arabidopsis thaliana* is 35 (Johanson et al. 2001), in cabbage (*Brassica oleracea*) 67 (Diehn et al. 2015), in Chinese cabbage (*Brassica rapa*) 57 (Diehn et al. 2015), in poplar (*Populus trichocarpa)* 55 (Gupta and Sankararamakrishnan 2009), in banana (*Musa* sp.) 47 (Hu et al. 2015), in castor bean (*Ricinus communis*) 47 (Zou et al. 2015), in soybean (*Glycine max*) 66 (Zhang et al. 2013), in potato (*Solanum tuberosum*) 41 (Venkatesh et al. 2013), in tomato (*Solanum lycopersicum*) 47 (Sade et al. 2009; Reuscher et al. 2013), in cotton (*Gossypium hirsutum*) 71 (Park et al. 2010), in rice (*Oryza sativa*) 33 (Sakurai et al. 2005), and in maize (*Zea mays*) at least 36 (Chaumont et al. 2001).

Based on their sequence, the AQPs have been classified into two major subgroups, which in both bacteria and mammalians reflect their contrasting functionality: the orthodox AQPs (AQPs) act as channels for water and small solutes such as ammonia or hydrogen peroxide, while the aquaglyceroporins (GLPs) are responsible for the transport of solutes, such as glycerol or urea. In plants, the congruence between phylogeny and functionality is less clear. The sequences present in higher plants cluster phylogenetically with the AQPs and have been arranged into five distinct subfamilies, namely, the nodulin26-like intrinsic proteins (NIPs), the plasma membrane intrinsic proteins (PIPs), the tonoplast intrinsic proteins (TIPs), the small basic intrinsic proteins (SIPs), and the as yet poorly characterized X intrinsic proteins (XIPs) (Chaumont et al. 2001; Johanson et al. 2001; Danielson and Johanson 2008). XIPs are found in many, but not all, species within the section *Magnoliopsida* (they are not present in species belonging to the Brassicaceae), but have not been identified in any section *Liliopsida* species to date (Danielson and Johanson 2008). Analyses of the genomes of lower plants and algae have revealed several mostly not yet functionally characterized but clearly distinct AQP subfamilies (Anderberg et al. 2011; Khabudaev et al. 2014). For some plant AQPs (notably the NIPs and XIPs), certain specific sequence features, along with their functionality, have been taken to suggest a functional equivalence with the GLPs.

3 Non-plant AQP and GLP-Mediated Metalloid Transport

The transport of glycerol mediated by GLPs is an important component of carbon metabolism and osmoregulation in bacteria, *Archaea*, protozoans, and mammals (Hara-Chikuma and Verkman 2006; Laforenza et al. 2015; Ahmadpour et al. 2014). Some GLPs are better described as "metalloidoporins" (Pommerrenig et al. 2015), since they fulfill physiologically important metalloid channel functions, thereby ensuring cellular metalloid homeostasis. Representative examples for such functional metalloidoporin GLPs are isoforms, which are part of As resistance (ars) operons. For example, the As resistance operons in bacteria such as *Escherichia coli* comprise the five genes *arsR*, *arsD*, *arsA*, *arsB*, and *arsC* (Rosen and Tamas 2010). The presence of arsenate ($H_2AsO_4^-$) in the growing medium activates *arsR*, which encodes a regulatory protein; the products of *arsC* and *arsD* are, respectively, an arsenate reductase and an arsenate binding metallochaperone, which together deliver arsenite ($H_2AsO_3^-$) to the ATP-driven extrusion pump encoded by *arsA* and *arsB* (Rosen and Tamas 2010). In the bacterial species *Sinorhizobium meliloti*, *Mesorhizobium loti*, *Caulobacter crescentus*, and *Ralstonia solanacearum,* a gene encoding a GLP aquaporin, which functions as an As-permeable channel, replaces the *arsB*-encoded efflux pump (Yang et al. 2005). These cases demonstrate that certain bacteria have adapted AQPs to handle As efflux and that an inheritable link between AQPs and metalloid transport exist (Yang et al. 2005). A further exciting link between metalloid transport and AQP function is represented in the actinomycete *Salinispora tropica*, where a GLP sequence has been fused to the sequence encoding an arsenate reductase domain, resulting in the translation of a dual function protein (Wu et al. 2010). The N-terminal GLP channel protein shows a greater selectivity for H_3AsO_3 than for either water or glycerol (Mukhopadhyay et al. 2014) and facilitates the efflux of H_3AsO_3 out of the cells directly at its site of production catalyzed by the C terminal arsenate reductase region of the protein (Wu et al. 2010). This spatially identical site of production and transport has the advantage that toxic As species do not pass through the cytoplasm before reaching their efflux site.

In *Saccharomyces cerevisiae*, the GLP FpsI acts normally as an osmoregulator. When the yeast cells are exposed to H_3AsO_3 stress, *FpsI* transcription is downregulated, and the preexisting FpsI in the cell will be inactivated in a phosphorylation-dependent manner (reviewed by Maciaszczyk-Dziubinska et al. 2012). Once inactivated, short-term H_3AsO_3 uptake is prevented; after a longer exposure to the stress, the abundance of *FpsI* transcript rises, which increases the efficiency of H_3AsO_3 efflux. The required concentration gradient is established in the yeast cell via the exudation of glutathione, which enables the exported H_3AsO_3 to be extracellularly chelated (Thorsen et al. 2012). Mammalian GLPs have also been identified as participating in As detoxification. This was demonstrated by the impaired ability of AQP9-null mice and mouse hepatocytes to dispose of As and which therefore suffer an increased severity of toxicity symptoms (Carbrey et al. 2009; Shinkai et al. 2009). Reviews by Mukhopadhyay et al. (Mukhopadhyay et al. 2014) and by

Maciaszczyk-Dziubinska et al. (Maciaszczyk-Dziubinska et al. 2012) have detailed how non-plant AQP channels support the bidirectional cross-membrane movement of metalloids in a range of organisms.

The above-depicted examples of non-plant AQP and GLP-mediated metalloid transport processes are listed to demonstrate that the link between AQPs and metalloid transport is non-incidental in nature and is given across kingdoms. Diverse organisms independently evolved different AQP-employing strategies to regulate the transport and homeostasis of various metalloids. The adaption of AQPs to act as metalloidoporins is, on the one hand, based on the chemical characteristics of the channel and, on the other, on the physicochemical properties of uncharged hydroxylated metalloid species resembling those of glycerol, the suggested original substrate of AQPs. The size of undissociated hydroxy-metalloid acids (Fig. 1) and their volume, dipole moment, surface charge distribution, and ability to form hydrogen bonds are all reminiscent of glycerol. All these attributes are decisive for the efficient passage though the AQP pore and metalloids behave as effective molecular transport mimics of glycerol (Porquet and Filella 2007). The experimental data derived from bacteria to mammals did significantly change the view on how membrane permeability to metalloids might be regulated in planta. The long-held assumption that uncharged metalloids are solely transported across plant membranes via a process of passive nonprotein-facilitated diffusion has had to be reconsidered in the light of the discovery of metalloid-permeable plant AQPs.

The following observations support the view that plant membranes can obstruct the diffusion of metalloids and that plant AQPs, like their GLP counterparts, offer the means to adjust membrane permeability appropriately: (1) concentration gradients across membranes of uncharged metalloid species have been detected (Meharg and Jardine 2003; Dordas et al. 2000; Dordas and Brown 2000), (2) the permeability coefficients for B measured in certain plant vesicles are significantly higher than those measured in synthetic liposomes (Dordas et al. 2000; Dordas and Brown 2000), (3) the transmembrane transport of B and As can be inhibited by potent AQP blockers (Meharg and Jardine 2003; Dordas et al. 2000), while (4) glycerol acts as a competitor for As flux (Meharg and Jardine 2003). As described subsequently, a number of both target-oriented and nontargeted approaches have revealed that certain plant AQPs (and especially members of the NIP subfamily) are physiologically important metalloidoporins.

4 NIP-Mediated Metalloid Transport in Plants

The evolutionary origin of the NIPs is unclear. Phylogenetically, they cluster with bacterial and archaeal NIP-like proteins, forming a basal lineage within the AQPs distinct from the aquaporin Z-like or glycerol uptake facilitator-like proteins (Abascal et al. 2014). Their phylogeny provides support for the notion that plant NIPs were originally acquired via horizontal gene transfer from the prokaryotic chloroplast progenitor (Abascal et al. 2014), but the alternative route of convergent functional evolution cannot be totally excluded. The plant NIPs can be phylogenetically divided into subgroups NIP1 through NIP5, which are remarkably well conserved across species (Danielson and Johanson 2010; see also chapter "The Nodulin 26 Intrinsic Protein Subfamily"). Note that the numerals "1" to "5" designating the five phylogenetically NIP subgroups do not match with the designated numerals designating *NIP* genes within one species. The low level of node support and the various polytomies that arise in *NIP* phylogenies emphasize the uncertain evolutionary relationships obtained between the *NIP* subgroups and isoforms (Danielson and Johanson 2010; Abascal et al. 2014). Based on the amino acid composition of the ar/R constriction region, three functional groups (NIP-I through -III) have been recognized (Wallace and Roberts 2004; Mitani et al. 2008; see chapter "The Nodulin 26 Intrinsic Protein Subfamily"). The three functional NIP subgroups are represented in all higher plants, although NIP-III is largely confined to section *Liliopsida* species (Danielson and Johanson 2010).

The soybean NIP GmNOD26 was the first plant AQP to be described (Fortin et al. 1987; see chapter "The Nodulin 26 Intrinsic Protein Subfamily") and became the eponym of the NIP subfamily. It is the major proteinaceous constituent of the root nodule membranes (Fortin et al. 1987; Dean et al. 1999). Transport assays designed to assess the permeability of diverse functional NIP subgroups have shown that glycerol, NH_3, CH_4N_2O, water, H_2O_2, and metalloids can all be transported via these proteins (Bienert and Chaumont 2011). To date, however, in planta evidence for physiologically relevant non-metalloid transport is lacking. NIPs are not only channel metalloids but are also essentially required for their transport into and within the plant. Evidence gathered from genetic, physiological, and molecular biology experiments argues for them having a major impact on metalloid homeostasis. Indeed, they are the only protein family in plants known to be essential for the uptake, translocation, as well as extrusion of a number of uncharged metalloids (Fig. 2).

4.1 NIP-Mediated Transport of Boron

B has long been recognized as essential for plant growth (Warrington 1923); nevertheless, the only known function of B surrounds the formation of borate ester bridges within the primary cell wall, which serve to crosslink rhamnogalacturonan-II (RG-II) monomers. Dimerized RG-II contributes to the overall network of pectic polysaccharides (Funakawa and Miwa 2015). In a standard plant cell wall, >90 % of RG-II monomers are dimerized, and although the overall proportion of cell wall pectin represented by RG-II is only around 10 %, it is clear that the quantity of free and cross-linked RG-II is critical for cell differentiation and elongation, as well as for plant growth and development (Funakawa and Miwa 2015). Insufficient cross-linking induced by B-deficient growing conditions has a deleterious effect on plant growth and results in dwarfed plants (O'Neill et al. 2001). *Magnoliopsida* species tend to have a higher B demand than those in the class of *Liliopsida*, which correlates with the quantity of RG-II found within the cell wall (Pérez et al. 2003). B deficiency manifests itself in the form of meristematic defects, abnormal cell differentiation, and a compromised expansion of the stem, leaf, and vascular system. Flowering – especially pollen development – and pollen tube growth are also highly sensitive to B deficiency (Marschner 2012). While the molecular roles of B are enigmatic, the last years have provided detailed understanding on B transport mechanisms in plants.

The transcription of NIP-II genes in roots such as *AtNIP5;1* and its orthologs in various plant species responds rapidly to B starvation (Takano et al. 2006; Hanaoka et al. 2014; Zhou et al. 2015). *NIP5;1* transcripts of Arabidopsis, citrus, and rice are strongly upregulated within 24 h after the onset of B-deficient conditions. Reverse genetic approaches in Arabidopsis and rice using NIP-II knockout and silenced plants have shown that B uptake into the roots requires a functional *AtNIP5;1* and *OsNIP3;1*, respectively (Takano et al. 2006; Hanaoka et al. 2014). The heterologous expression of *AtNIP5;1*, *AtNIP6;1*, and *OsNIP3;1* promotes the transport of H_3BO_3 in yeast, frog oocytes, and plants, demonstrating that they are all functional B transporters (Takano et al. 2006; Tanaka et al. 2008; Hanaoka et al. 2014). *Atnip5;1* and *Atnip6;1* knockouts display characteristic symptoms of B deficiency, i.e., reduced stability of the epidermis abolished apical dominance and perturbed cell differentiation (Takano et al. 2006; Tanaka et al. 2008). While *AtNIP5;1* is expressed in the root epidermis and operates to move H_3BO_3 into the root, the *AtNIP6;1* product is deposited in young leaf phloem companion and parenchyma cells, where it presumably is involved in unloading H_3BO_3 from the xylem into the phloem (Takano et al. 2006; Tanaka et al. 2008) (see also chapter "Plant Aquaporin Trafficking").

Under B-deficient conditions, the shoot growth of *Atnip6;1* knockouts is restricted, suggesting that AtNIP6;1 is important for the allocation of B to developing and meristematic tissue (Tanaka et al. 2008). Under such conditions, both *Atnip5;1* and *Atnip6;1* knockouts form largely sterile flowers. In rice, OsNIP3;1 has been shown as responsible for the uptake of B into the root, its translocation into the shoot, and its unloading from the xylem into the phloem in the mature leaf (Hanaoka et al. 2014). Its encoding gene is strongly transcribed in the root exodermis and in the cells surrounding the vascular bundles in both the root and shoot. When the *OsNIP3;1* gene is silenced, neither the total B concentration nor its distribution between the shoot and root is disturbed, provided that the conditions are not B-deficient; however, when the supply of B is limiting, the shoot's B content is significantly decreased. This indicates different regulations of AtNIP5;1 and its ortholog OsNIP3;1. Consistent with this result, a map-based cloning approach targeting the *Dwarf and tiller-enhancing* 1 (*dte-1*) rice mutant identified *OsNIP3;1* as the candidate gene underlying the mutated locus (Liu et al. 2015); the mutant displays B deficiency symptoms when the supply of B is suboptimal (Liu et al. 2015). These results clearly indicate the crucial function of NIPs in plant B homeostasis. *ZmNIP3;1*, the maize ortholog of *OsNIP3;1*, has been similarly identified thanks to its positional cloning to underlie the phenotype of the *tassel-less1* (*tsl-1*) mutant (Durbak et al. 2014). This mutant produces not only an aberrant flower, but its vegetative growth resembles that of a B-deficient maize plant. When expressed heterologously, *ZmNIP3;1* facilitates the uptake of B into both frog oocytes and yeast cells (Durbak et al. 2014). Tissue B content is suboptimal in the *tassel-less1* mutant, and the mutant phenotype can be rescued by providing a source of B. *ZmNIP3;1* transcript is highly abundant in the wild-type silk and (to a lesser extent) in the tassel and root, a distribution which is dissimilar to that shown by its *A. thaliana* and rice orthologs. The *tassel-less 1* mutant carries a gene encoding for a mutated ZmNIP3;1 protein resulting in a nonfunctional channel protein (Durbak et al. 2014, Leonard et al. 2014).

Even though they differ with respect to their spatial transcription profile, each of the *Atnip5;1*, *Atnip6;1*, *Osnip3;1* (*dte-1*), and *Zmnip3;1* (*ts-l1*) loss-of-function mutants expresses a normal phenotype, provided that the supply of B is non-limiting; however, when this is not the case, the plants remain stunted, their apical dominance is compromised, and they suffer from inflorescence defects and reproductive sterility (Takano et al. 2006; Tanaka et al. 2008; Durbak et al. 2014; Hanaoka et al. 2014; Liu et al. 2015). The NIP-II group AQP isoforms are therefore crucial for the uptake and distribution of B within the plant not just in section *Magnoliopsida* species, which have a relatively high B requirement, but also in section *Liliopsida* ones, which do not need as much B for growth (Marschner 2012).

Excessive soil B is phytotoxic. B is taken up in the transpiration stream, so tends to accumulate initially in more mature leaves (Nable et al. 1997). As a result, B toxicity manifests itself as leaf chlorosis/necrosis, spreading from the leaf margin into the center of the leaf (Nable et al. 1997; Shatil-Cohen and Moshelion 2012). Barley (which, like all of the cereals, has a relatively low B requirement) is particularly sensitive to B toxicity (Schnurbusch et al. 2010). The genomic region of barley associated with B tolerance harbors *HvNIP2;1*. The mapping population progeny carrying the *HvNIP2;1* allele that is present in the B tolerant mapping parent (the Algerian landrace Sahara 3771) exhibits a higher level of tolerance and accumulates less B in their leaves than those which carry the alternative allele from cv. Clipper. The level of *HvNIP2;1* transcript increases from the root tip to the basal root region in both parental lines, but its abundance is up to 15-fold higher in the roots of the sensitive parent (Schnurbusch et al. 2010). A sequence comparison of the alternative *HvNIP2;1* coding sequences identified only one base variation, while the predicted encoded proteins are identical. The *HvNIP2;1* upstream sequence (up to −1377 nt) is wholly monomorphic, so the differential transcription of the gene has been concluded to reflect sequence variation even further upstream (Schnurbusch et al. 2010). Based on its H_4SiO_4 permeability both in frog oocytes and *in planta*, and its tissue distribution, the barley protein HvNIP2;1 is also thought to have an impact on the supply of the metalloid Si, even though no correlation could be established between *HvNIP2;1* transcription and the plant's Si uptake capacity (Chiba et al. 2009).

In *Medicago truncatula*, Bogacki et al. (2013) show that 95 % of the phenotypic variation for B tolerance displayed by the progeny of a cross between two contrasting parents could be linked to two microsatellite loci, which flank a cluster of five predicted AQP genes. Among them, only one (*MtNIP3*) is transcribed in the leaf and root. While the transcript levels are low and indistinguishable in the roots of tolerant and sensitive types, a fourfold difference in the leaf is observed, and the leaf B concentration is correlated with the phenotype, suggesting that *MtNIP3* is likely the gene underlying B tolerance (Bogacki et al. 2013). Based on the observed B distribution, it can be excluded that an enhanced B translocation from the roots is responsible for the differential B tolerance between these genotypes. It has been suggested that the redistribution of B from the symplast to the apoplast of leaves and subsequent leaching through rain and/or the removal of B via guttation represents the basis for the observed MtNIP3-dependent tolerance (Bogacki et al. 2013).

These examples demonstrate that the regulation of NIP-metalloidoporin activity and expression are important mechanisms for plants to adapt to either B-deficient or toxic environmental conditions.

NIPs are not the only proteins known to be involved in B transport in plants. The first B transporter to be described was identified from the analysis of an *A. thaliana* mutant in which shoot, but not root growth, was severely inhibited by B deficiency (Takano et al. 2002). The product of the mutated gene *AtBOR1* was shown to mediate the xylem loading of B. BOR proteins share homology with the Slc4 bicarbonate transporters (Parker and Boron 2013) and are predicted to form 14 plasma membrane-spanning helices. Potentially, a secondary active transport process is responsible for the BOR-mediated efflux of B (Parker and Boron 2013). The substrate used by HvBOT1, a sodium-dependent BOR transport protein from barley, was demonstrated to be the borate anion $H_4BO_4^-$ (Nagarajan et al. 2015). $H_4BO_4^-$ represents highly likely also the substrate of other BOR proteins. BOR-type transporters and NIP-II group AQPs cooperatively regulate B influx and efflux in a species-dependent manner. In rice, *OsNIP3;1* – but not *OsBOR1* – is expressed in the stele, while in the exodermis and endodermis, the genes are co-expressed (Nakagawa et al. 2007). In contrast, in *A. thaliana* AtBOR1 and AtNIP5;1 together control the radial transport of B to the vascular system in various cell types together, and are co-expressed in the endodermis (Takano et al. 2008, 2010).

A responsive metalloid transport system is of biological importance because plants can face sudden changes in the availability of these elements. Several *AtNIP5;1* gene homologs, the products of which are both able to channel H_3BO_3 and are known to be important for B uptake, are transcriptionally upregulated when the availability of B is limiting but downregulated when B is in oversupply (Takano et al. 2006; Tanaka et al. 2008; Hanaoka et al. 2014; Zhou et al. 2015; Martínez-Cuenca et al. 2015). The *AtNIP5;1* 5′-UTR is particularly important both for the induction of *AtNIP5;1* transcription and for its mRNA degradation under B-sufficient conditions (Tanaka et al. 2011). A similar regulatory role has been suggested for the almost identical 5′-UTR of *OsNIP3;1* in B-deficient conditions and after B resupply. While the molecular basis for this upregulation is unknown, an 18 bp sequence within the *AtNIP5;1* 5′ UTR has been shown to be responsible for the rapid destabilization of *AtNIP5;1* mRNA when the plants are oversupplied with B, shortening the mRNA's half-life to about 30 % compared to plants grown under B-limiting conditions (Tanaka et al. 2011). This specific 18 bp sequence also influences the abundance of other tested downstream mRNA sequences in a B concentration-dependent manner (Tanaka et al. 2011), leading to the suggestion that a number of genes are regulated via a B-dependent mRNA (de-)stabilization or translational efficiency mechanism. It remains to be shown if the mRNA destabilization is caused by a direct interaction between H_3BO_3 and the ribose sugar component of the RNA, as ribose moieties can chemically interact with H_3BO_3 or via other yet unknown mechanisms.

4.2 NIP-Mediated Transport of Silicon

Si is a non-essential element for most plants, but it does exert some highly beneficial effects on growth and productivity (Ma et al. 2002; Ma and Yamaji 2015). The presence of silica in plant tissue has been associated with an enhancement to certain plants' tolerance to drought, salinity, extreme temperature stress, and nutrient imbalance, as well as providing physical strength to the stem and leaves, thereby increasing lodging resistance in the field (Ma and Yamaji 2015). In addition, small herbivores typically avoid feeding on grasses that deposit significant quantities of silica in their leaves and digest them rather inefficiently. High silica contents also protect plants from fungal pathogens. The element has been designated as quasi-essential for rice (Epstein 1994), and Si fertilizers (the bioavailable form is silicic acid [H_4SiO_4]) are widely used in rice production in various continents (Ma and Yamaji 2015). The tissue concentration of Si in the aerial part of the plant varies across species from 0.1 % to 10 % of dry weight and by 5–10 % from rice cultivar to rice cultivar (Ma and Takahashi 2002).

The first higher plant Si transporter to be identified was OsNIP2;1 (syn. OsLsi1) (Ma et al. 2006). The *low-silicon* (*lsi*) mutant displays severe Si deficiency symptoms; the mutated gene product differs from that of the wild type by a single residue. The substitution of ala132 by thr132 significantly alters the protein conformation, resulting in a loss of its channel functionality. RNAi-induced suppression of *OsNIP2;1* expression in cv. Nipponbare reduces Si uptake considerably, producing a phenotype resembling that of the *lsi1* mutant (Ma et al. 2006). The wild-type gene product localizes to the exodermis and endodermis and to root zones, which are decisive for and intimately associated with Si uptake. The expression of *OsNIP2;1* in frog oocytes results in a de novo capacity to transport H_4SiO_4, but not glycerol or water (Ma et al. 2006). OsNIP2;1 is a NIP-III AQP, a class of protein typically characterized by an ar/R selectivity filter comprising gly, ser, gly, and arg. The small size of these four residues leads to the formation of a pore diameter that is somewhat larger than those produced by NIP-I and -II proteins. Once Si is taken up by rice roots, more than 95 % of it is translocated from the roots to the shoots (Ma and Takahashi 2002). In the shoot, OsNIP2;2 is responsible for the unloading of H_4SiO_4 from the xylem sap into the cytoplasmic leaf space (Yamaji and Ma 2009). This protein is polar-localized to the adaxial side of xylem parenchyma cells in the leaf sheath and blade (Yamaji and Ma 2009). Transpirational water loss drives the gradual polymerization of H_4SiO_4 into amorphous silica, which is deposited as a double layer beneath the cuticle (Ma and Takahashi 2002). In *OsNIP2;2* knockout plants, H_4SiO_4 accumulates in the leaf guttation sap, and an altered pattern of silica deposition in the leaf is observed (Yamaji and Ma 2009). NIP-III isoforms permeable to H_4SiO_4 and important for the uptake and distribution of Si have been identified in barley (HvLsi1 [HvNIP2;1] and HvLsi6 [HvNIP2;2]: Chiba et al. 2009; Yamaji et al. 2012), wheat (TaLsi1: Montpetit et al. 2012), maize (ZmLsi1 [ZmNIP2;1] and ZmLsi6 [ZmNIP2;2]: Mitani et al. 2009a), cucumber (CsiT-1 and CsiT-2: Wang et al. 2015), pumpkin (CmNIP2;1: Mitani et al. 2011), and soybean (GmNIP2;2: Deshmukh et al. 2013).

OsNIP2;1, HvNIP2;1, ZmNIP2;1, and TaLsi1 channels are present mainly in the root and are known to be required both for the uptake of H_4SiO_4 into the plant and for its transport toward the vasculature (Ma et al. 2006; Montpetit et al. 2012; Mitani et al. 2009a; Chiba et al. 2009). *OsNIP2;2*, *HvNIP2;2*, *ZmNIP2;2*, *GmNIP2;1*, *GmNIP2;2*, *CmNIP2;1*, *CSiT1*, and *CSiT2* transcripts are all detectable in both the root and the shoot; the function of their products in rice, barley, and maize is considered to lie in xylem unloading in the leaf sheath and blade (Yamaji and Ma 2009; Yamaji et al. 2012; Mitani et al. 2009a); an additional function in rice is the intervascular transfer of nutrients at the nodes (Yamaji and Ma 2009; Yamaji et al. 2015).

While the abovementioned NIP-IIIs all share a capacity to transport H_4SiO_4, the various orthologs differ from one another with respect to both their spatial expression and their transcriptional response to specific stimuli. For example, *OsNIP2;1*, *OsNIP2;2*, *GmNIP2;1*, and *GmNIP2;2* are all downregulated by the presence of H_4SiO_4 (Ma et al. 2006; Yamaji and Ma 2009; Deshmukh et al. 2013), whereas *ZmNIP2;1*, *TaLsi*1, and *HvNIP2;1* are nonresponsive (Chiba et al. 2009; Mitani et al. 2009a; Montpetit et al. 2012). OsNIP2;1 is abundant in the exodermis and endodermis in primary and lateral roots where casparian strips exist (Ma et al. 2006); both *HvNIP2;1* and *ZmNIP2;1* are active in the epidermis, hypodermis, and cortex (Chiba et al. 2009; Mitani et al. 2009a); CmNIP2;1 is ubiquitous throughout the root (Mitani et al. 2011); OsNIP2;2/Lsi6 homologs in rice, barley, and maize are deposited throughout the root tip and in xylem parenchyma in the leaf (Yamaji et al. 2008; Yamaji et al. 2012; Yamaji and Ma 2009; Mitani et al. 2009a). The herbaceous perennial horsetail (*Equisetum arvense*) is one of the highest accumulators of Si in the plant kingdom (Chen and Lewin 1969). It encodes nine NIPs (EaNIP3;1 through 9), of which EaNIP3;1, EaNIP3;3, and EaNIP3;4 have each been shown to be permeable to H_4SiO_4 and to feature a distinct amino acid residue composition in their selectivity filter, namely, composed of ser, thr, ala, and arg (Grégoire et al. 2012).

The composition of cereal and horsetail Si channel ar/R selectivity filters is too variable for it to be usable as a diagnostic for Si transporters. Nonetheless, an in silico analysis has identified a phenylalanine in TMH6 and a polar serine/threonine residue in TMH5 that are shared by all Si-permeable NIP-III group proteins while being absent from all other NIPs (Pommerrenig et al. 2015). However, whether these residues are indeed critical for H_4SiO_4 selectivity has yet to be experimentally verified.

NIP-III group channels are encoded by the genomes of both *Liliopsida* and *Magnoliopsida* species, including the Gramineae, Arecaceae, Musaceae, Solanaceae, Rosaceae, Cucurbitaceae, Leguminosae, Vitaceae, Rubiaceae, and Rutaceae, as well as in the species *Amborella trichopoda*, which has been placed at, or near the base of, the angiosperm lineage (Ma and Yamaji 2015). *A. thaliana* lacks any *NIP-III* genes. Note that the presence of a *NIP-III* gene(s) does not correlate with an enhanced capacity to accumulate Si. For example, NIP-III group isoforms are produced by tomato, which is a non-accumulator (Mitani and Ma 2005). Thus, NIP-IIIs likely fulfill also other physiological functions – an example is the previously mentioned barley HvNIP2;1 protein, associated with B tolerance (Schnurbusch et al. 2010).

NIP channels are not the only plant proteins able to transport Si. The Lsi2-type transporters have been designated as putative anion transporters (Ma et al. 2007; Mitani et al. 2009b; Mitani-Ueno et al. 2011; Yamaji et al. 2015); they form 11 predicted plasma membrane-spanning helices and remove Si from the cell via a secondary active process driven by the establishment of a proton gradient across the plasma membrane (Ma et al. 2007). In rice, an Lsi2 homolog governs the uptake and transport of $H_3AsO_3/H_2AsO_3^-$ and its translocation into the grain (Ma et al. 2008). Lsi2-type transporters are found in many *Magnoliopsida* (including *A. thaliana*) and *Liliopsida* species (Ma and Yamaji 2015). The function of the *A. thaliana* homolog (encoded by *At1g02260*) is still unknown. The cooperation of Lsi2-type transporters and NIP-III channels is required for cell-to-cell Si transport (reviewed by Ma and Yamaji 2015). In some cases, NIP channels and Lsi2-type efflux transporters are located within the same cell type but with opposite polarity; in other cases, they appear in adjacent cell layers. A mathematical modeling approach has calculated that the polar localization of the two transporter types (NIPs and Lsi2-type transporters) at the exodermis and endodermis is optimal with respect to an energy efficient and high capacity Si uptake into the rice root (Sakurai et al. 2015).

4.3 NIP-Mediated Transport of Arsenic

As is an acutely toxic and carcinogenic though relatively abundant and highly bioavailable metalloid, which can enter the human food chain via contaminated water or plant biomass (mainly via staple crops) (Meharg and Zhao 2012). The most common forms present in soil are $H_2AsO_4^-$ and H_3AsO_3. In well-aerated (oxidative) soils, the former type predominates, while the latter type is associated with hypoxic (reducing) conditions. Both forms are readily taken up by plants (Meharg and Zhao 2012). Arsenate ($H_2AsO_4^-$) and phosphate ($H_2PO_4^-$), the salts of arsenic acid (H_3AsO_4) and phosphoric acid (H_3PO_4), share a similar tetrahedral structure, pK_a, molecular volume, and electrostatic behavior. Thus, being chemical analogs, $H_2AsO_4^-$ can readily replace $H_2PO_4^-$, entering the plant via phosphate transporters (Zangi and Filella 2012). High affinity phosphate transporters are unable to distinguish between the two compounds (Zangi and Filella 2012; Li et al. 2015). Once taken up, $H_2AsO_4^-$ forms As-adducts which are typically short-lived and nonfunctional compared to the physiologically functional P-adducts; an example is the formation and rapid autohydrolysis of $H_2AsO_4^-$-ADP, initiating a futile cycle which uncouples oxidative phosphorylation and interferes with enzymes regulated by phosphorylation (Finnegan and Chen 2012). As most arable soils are not hypoxic, most of the As taken up by plants is in the form $H_2AsO_4^-$. Shortly after entering the root, it is enzymatically or nonenzymatically reduced to $H_2AsO_3^-$ and then protonated to form H_3AsO_3 (Finnegan and Chen 2012). The reduction of $H_2AsO_4^-$ to $H_2AsO_3^-$ is a common detoxification strategy used by most organisms, including plants (Bienert and Jahn 2010b). In the form $H_2AsO_3^-/H_3AsO_3$, As is more easily transported than $H_2AsO_4^-$, but its toxicity is enhanced by its ready reactivity with sulfur groups, thereby inactivating enzymes for which their functionality depends on cysteine residues or dithiol cofactors (Finnegan and Chen 2012). In non-hyperaccumulators, most of the $H_2AsO_3^-/H_3AsO_3$ taken up is chelated by glutathione or a metallothionein and sequestered into root cell vacuoles by the action of ABC transporters; alternatively it can be effluxed out of the cells (Li et al. 2015). The $H_2AsO_3^-/H_3AsO_3$ which is neither compartmentalized nor effluxed is distributed throughout the plant either actively by members of the secondary active Si-transporting Lsi2-type transporter family or passively along a concentration gradient by NIPs which transport Si and B (see elsewhere in this chapter; Pommerrenig et al. 2015; Li et al. 2015).

In bacteria, fungi, fish, and mammals (including humans), H_3AsO_3 is transported by specific GLPs (reviewed in Bienert and Jahn 2010a; Maciaszczyk-Dziubinska et al. 2012; Mukhopadhyay et al. 2014). Evidence supporting the involvement of AQPs in As transport has been obtained from kinetic uptake studies of the rice root (Meharg and Jardine 2003). In particular, when $H_2AsO_3^-$ was supplied to rice roots, As uptake can be partially inhibited by alternative AQP substrates (such as glycerol and antimonite) or by the AQP inhibitor $HgCl_2$ (Meharg and Jardine 2003). Consequently, Meharg and Jardine postulated already in 2003 that H_3AsO_3 is transported across plant plasma membranes via MIPs/AQPs. In 2008, three studies independently and congruently demonstrated in direct uptake experiments that certain plant NIPs are permeable to H_3AsO_3 (Isayenkov and Maathuis 2008; Bienert et al. 2008a; Ma et al. 2008). The effect of exposing plants to $NaAsO_2$ and As trioxide (As_2O_3) on uptake and growth implies strongly that the uncharged H_3AsO_3 molecule permeates plant NIPs ($NaAsO_2$ and As_2O_3 form H_3AsO_3 in aqueous solution). A detailed study has shown that H_3AsO_3 shares several physicochemical and structural characteristics with the canonical NIP substrate glycerol, further supporting the idea that it is transported in planta through AQP channels (Porquet and Filella 2007).

A number of rice (OsNIP2;1, OsNIP2;2, and OsNIP 3;2), *A. thaliana* (AtNIP5;1, AtNIP6;1, and AtNIP7;1), and *Lotus japonicus* (LjNIP5;1 and LjNIP6;1) proteins have been tested for their ability to abolish the As tolerance displayed by certain *S. cerevisiae* yeast strains (Fig. 2); all of them significantly increase the level of sensitivity to $NaAsO_2$ (Bienert et al. 2008a). When the yeast is cultured on a medium containing $H_2AsO_4^-$, the NIP proteins also facilitate the efflux of the H_3AsO_3 generated in vivo through enzymatic reduction of $H_2AsO_4^-$ (Bienert et al. 2008a), clearly demonstrating the bidirectional flux of H_3AsO_3 carried out by plant NIPs.

The physiological consequences of NIP-mediated H_3AsO_3 transport have been revealed by exposing an *Osnip2;1* knockout rice line (defective in Si uptake) to $H_2AsO_3^-$. The accumulation of As in the mutant's shoot and root is reduced by, respectively, 71 % and 53 % compared to that recorded for a wild-type plant grown in a medium lacking H_4SiO_4 (Ma et al. 2008). The presence of H_4SiO_4 reduces As uptake in the wild type but not in the *Osnip2;1* mutant plants, indicating a competitively inhibited flux of the two substrates through the native OsNIP2;1 channel. A short-term uptake assay has demonstrated that As uptake by the mutant is 57 % less than that of the wild type (Ma et al. 2008). The conclusion is that OsNIP2;1 is responsible for H_3AsO_3 uptake in planta (Ma et al. 2008). The addition of $H_2AsO_4^-$ to the growth medium promotes OsNIP2;1-mediated H_3AsO_3 efflux (Zhao et al. 2010b). The suggestion here was that NIPs are able to reinforce As detoxification by effluxing H_3AsO_3 out of the roots after its intracellular formation through the reduction of $H_2AsO_4^-$, provided that the rhizosphere environment is permissive. When challenged with organic (methylated) molecules involving As, the *Osnip2;1* mutant takes up only half the amount of either monomethylarsonic acid (CH_5AsO_3) or dimethylarsinic acid ($C_2H_7AsO_2$) taken up by wild-type plants (Li et al. 2009). The heterologous expression of *OsNIP2;1* in frog oocytes has shown that this NIP facilitates both the influx and efflux of H_3AsO_3, as well as that of CH_5AsO_3 and $C_2H_7AsO_2$ (Ma et al. 2008; Li et al. 2009). The indications are therefore that OsNIP2;1 represents an important bidirectional channel for a range of uncharged As species and represents the major uptake pathway for these species into rice.

A screen of an EMS mutagenized population of *A. thaliana* was used by Kamiya et al. (Kamiya et al. 2009) to identify individuals compromised for root growth in the presence of $H_3AsO_3/H_2AsO_3^-$. The three selected mutants all carry a mutation in the *AtNIP1;1* coding sequence. The heterologous expression of each of the mutant alleles in frog oocytes has shown that they specify a nonfunctional As-impermeable AtNIP1;1 channel. The abundance of wild-type *AtNIP1;1* transcript was 20 times higher in the root than the shoot, and a promoter-GUS fusion analysis showed that the *AtNIP1;1* promoter is active in the stomata, the root-hypocotyl junction, the lateral root tip and stele, and the primary root stele (Kamiya et al. 2009). These data suggest that AtNIP1;1 contributes to As uptake into Arabidopsis roots. Similarly, AtNIP3;1 has been shown to participate in both As uptake and root-to-shoot translocation in plants subjected to H_3AsO_3 stress (Xu et al. 2015). Several independent *Atnip3;1* loss-of-function mutants display a clear improvement in their level of H_3AsO_3 tolerance, as expressed by their aerial growth and their reduced ability to accumulate As in the shoot (Xu et al. 2015). The *Atnip3;1/Atnip1;1* double mutant exhibits a strong degree of H_3AsO_3 tolerance; its root and shoot continue to grow even in the presence of normally toxic levels of H_3AsO_3. *AtNIP3;1* promoter activity is confined largely to the root, although not in the root tip (Xu et al. 2015). The overall conclusion is that AtNIP3;1 participates in H_3AsO_3 uptake and root-to-shoot translocation (Xu et al. 2015).

Studies based on a range of heterologous expression systems have demonstrated that members of all three functional NIP subclasses have the ability to channel uncharged As species. The outcome of expressing the rice (*OsNIP1;1*, *OsNIP2;1*, *OsNIP2;2*, and *OsNIP3;1*) and *A. thaliana* (*AtNIP1;1*, *AtNIP1;2*, *AtNIP5;1*, and *AtNIP7;1*) genes in frog oocytes is an increased influx of H_3AsO_3, moreover the expression of *AtNIP3;1*, *HvNIP1;2*, *HvNIP2;1*, *HvNIP2;2*, and *OsNIP3;3* in yeast enhances the cells' sensitivity to H_3AsO_3 providing additional evidence for the H_3AsO_3 permeabilities of NIPs (Fig. 2; Ma et al. 2008; Kamiya et al. 2009; Katsuhara et al. 2014; Xu et al. 2015).

A QTL mapping study in rice, based on a cross between the $H_2AsO_4^-$ tolerant cv. Bala and the sensitive cv. Azucena, was able to identify three genomic regions harboring genes determining the tolerance of the former cultivar (Norton et al. 2008). Analysis of the progeny suggested that an individual needs only to inherit any two of the three tolerance loci from cv. Bala for it to be tolerant. One of the three QTL regions harbored two genes which were differentially regulated when the plants were exposed to As stress: one encodes an aminoacylase and the other is *OsNIP4;1*; both are more actively transcribed in the tolerant parent (Norton et al. 2008). The latter gene is particularly significant as NIPs are implicated in the transport of H_3AsO_3 into the root. However, the heterologous expression of *OsNIP4;1* – unlike that of other NIPs – in an As-sensitive yeast cell line does not increase their H_3AsO_3 sensitivity (Katsuhara et al. 2014). The mechanistic basis of OsNIP4;1 on H_3AsO_3 tolerance remains to be determined.

OsNIP3;1, required for the uptake and translocation of H_3BO_3 (Hanaoka et al. 2014), also transports H_3AsO_3 when expressed in frog oocytes (Ma et al. 2008). *OsNIP3;1* is downregulated in response to an elevated supply of $H_3AsO_3/H_2AsO_3^-$ but not of $H_2AsO_4^-$ (Chakrabarty et al. 2009). As-responsive downregulation may help to lower the level of OsNIP3;1-mediated As root uptake under B-deficient conditions. All acquired information on As transport mechanisms controlling As fluxes into and within plants, particularly to edible plant parts such as rice grains, is highly valuable for the development of breeding strategies or the engineering of minimal-As-accumulating plants.

Two independent analyses have failed to identify any QTL linked to either *OsNIP2;1* or *OsNIP3;1* associated with the grain content of either $H_2AsO_3^-$/ $H_2AsO_4^-$ or $C_2H_7AsO_2$ (Kuramata et al. 2013; Norton et al. 2014). However, one QTL region (harboring *OsNIP2;2*) has been identified as contributing to the methylated As content of the grain (Kuramata et al. 2013). When tested at the seedling stage, both the shoot and root As contents in an *OsNIP2;2* knockout line are indistinguishable from those recorded in the wild type (Ma et al. 2008). Since *OsNIP2;2* is expressed in the node below the panicle after the onset of grain filling (Yamaji and Ma 2009), it has been suggested that differences in the grain $C_2H_7AsO_2$ content are due to a genotype-dependent transport efficiency and/or expression of *OsNIP2;2* (Kuramata et al. 2013). Carey et al. (Carey et al. 2010; Carey et al. 2011) have shown that $C_2H_7AsO_2$ is highly mobile in the panicle vascular system and is readily translocated into the grain. Whether OsNIP2;2 is permeable to either CH_5AsO_3 or $C_2H_7AsO_2$ remains to be shown.

So far, the indication is that the toxic metalloid As (both in its reduced and uncharged forms) transport in plants is handled largely by NIPs. Whether NIP-mediated H_3AsO_3 transport is simply an adventitious nonphysiological side activity, as a consequence of the compound's structural similarity to that of certain other essential metalloid nutrient substrates, or whether it has evolved as a genetically or physiologically implemented detoxification strategy along the lines of the GLPs in microbes, still remains to be resolved. Given that plants are sessile, it may well be that, in addition to their efflux activity from the root, the involvement of NIPs in As cell-to-cell translocation adds to the final compartmentalization of As-phytochelatin complexes in vacuoles of specific As-tolerant cell types (Moore et al. 2011) and/or an ability to protect As-sensitive cells. The latter two hypothesized roles of NIPs may also be supported by the observation that when AtNIP1;2- and AtNIP5;1-mediated As transport are disrupted *in planta*, the level of H_3AsO_3 tolerance is not increased, even though the tissue As content is markedly lowered (Kamiya et al. 2009). These findings indicate that H_3AsO_3 tolerance cannot be solely explained by a decreased As content in plants. The importance of gaining a better understanding of the regulation and mode of As transport has practical importance, as it will guide breeding strategies to selectively route As fluxes to targeted locations within or outside of crop plants depending on the objectives (i.e., accumulation, enrichment in, or exclusion from certain tissues) and to generate crop varieties that take up little or no As or at least do not translocate it to the edible parts of the plant.

4.4 NIP-Mediated Transport of Antimony

Trivalent and pentavalent Sb species have no known physiological role for plants, rather they are toxic (Kamiya and Fujiwara 2009). Homologous and heterologous expression systems have been used to show that various NIPs (Bienert et al. 2008a, b; Kamiya and Fujiwara 2009) and mammalian and microbial GLPs (reviewed by Maciaszczyk-Dziubinska et al. 2012) facilitate the movement of trivalent uncharged Sb species. The expression of *AtNIP5;1*, *AtNIP6;1* and *AtNIP7;1*, *LjNIP5;1* and *LjNIP6;1*, and *OsNIP3;2* and *OsNIP2;1* in a metalloid-tolerant yeast mutant abolishes the tolerance when the transformants were exposed to $C_8H_4K_2O_{12}Sb_2$ (potassium antimonyl tartrate) (Bienert et al. 2008a, b). The two independent *AtNIP1;1* T-DNA insertion mutants mentioned above in the context of tolerance to H_3AsO_3 are also able to both maintain root growth in the presence of toxic levels of $C_8H_4K_2O_{12}Sb_2$ and limit the accumulation of Sb (Kamiya and Fujiwara 2009). As the knockout of other NIPs (such as AtNIP1;2 and AtNIP5;1) expressed in the root do not reduce Sb sensitivity, it is likely that AtNIP1;1 is responsible, at least in part, for regulating and mediating the entry of Sb (Kamiya and Fujiwara 2009). Thus, NIPs belonging to each of the three functional subgroups NIP-I (AtNIP1;1), NIP-II (AtNIP5;1, AtNIP6;1, AtNIP7;1, LjNIP5;1, LjNIP6;1, and OsNIP3;2), and NIP-III (OsNIP2;1) facilitate the transport of Sb across plant membranes (Fig. 2). The Sb concentrations used in yeast and *A. thaliana* toxicity assays (up to 100 μM) do not occur in natural soils (Bienert et al. 2008a; Kamiya and Fujiwara 2009). Nevertheless, localized pollution associated with certain industrial activity has led to heavy loading with Sb_2O_3, so knowledge of Sb transport mechanisms is of relevance in the context of phytoremediation measures based on either Sb hyperaccumulators or on crop plants able to restrict the quantity of Sb translocated to edible parts. The likelihood is that the involvement of NIPs in the transport of trivalent Sb is an adventitious feature of these channels, which are presumed to have evolved as a means of transporting metalloids of physiological significance such as boric acid or silicic acid. Various microbial GLPs have also proven to be Sb permeable (reviewed by Maciaszczyk-Dziubinska et al. 2012; Zangi and Filella 2012; Mukhopadhyay et al. 2014; Mandal et al. 2014), even though there is no known biological requirement for this element. The nonspecificity of AQP/GLP channels is exploited in some cases in order to infiltrate curative drugs into parasitic or abnormal cells (notably cancerous cells). For example, Sb-containing drugs used to kill certain protozoan parasites are effectively taken up by the target organism via their AQP transport systems (Mandal et al. 2014). Two of the major drugs used to combat leishmaniasis are the pentavalent antimonials sodium stibogluconate (Pentostam) and meglumine antimoniate (Glucantime) (Mukhopadhyay et al. 2014). One of the five AQPs of *Leishmania major* (LmAQP1) is known to be involved in the as yet mechanistically non-understood uptake process of these drugs. Both experimentally induced and naturally occurring mutations in *LmAQP1* have been shown to reduce the uptake of Sb and hence increase the parasite's tolerance of the drugs (Mandal et al. 2014).

A similar scenario applies with respect to the arsenical drug melarsoprol, which enters the target cell via an AQP; drug resistance arises when the AQP is mutated to a form that hinders the free passage of the drug (Baker et al. 2012; Alsford et al. 2012). As pointed out in a recent review (Pommerrenig et al. 2015), it has been suggested that antimonous acid (H_3SbO_3) is the form of Sb generally permeating through NIPs and other GLPs when $C_8H_4K_2O_{12}Sb_2$ (antimony potassium tartrate) is provided as the source of Sb source in toxicity assays. This conclusion is largely based on the physicochemical similarity of H_3SbO_3 with H_3AsO_3 (Porquet and Filella 2007). Salts of H_3SbO_3 formally exist. In water, they form a gelatinous precipitate, which is formed by antimony trioxide (Sb_2O_3 * H_2O) which is itself potentially formed by $C_8H_4K_2O_{12}Sb_2$. However, the uncharged H_3SbO_3 is suggested to be metastable and, thus, does not occur in nature in significant quantities (Vink 1996). Some doubt remains therefore as to the form of Sb that permeates AQPs. Therefore, scientific efforts should be initiated to assess which Sb species permeates AQPs.

4.5 NIP-Mediated Transport of Germanium

Due to the absence of any known biological function and the rarity of Ge in most soils, the permeability of certain NIPs to this element is again likely a serendipitous effect of the structural similarity of Ge compounds to those formed by other physiologically significant metalloids. The bioavailable forms of Ge are the polar tetrahedral ortho-acid (H_4GeO_4) and the nonpolar, planar meta-acid form (H_2GeO_3), the chemical properties of which resemble, respectively, H_4SiO_4 and H_3BO_3 (Fig. 1). Neither of these forms has been exhaustively quantified in natural soils, the rhizosphere, or within plant tissue. The element is present in many silicate minerals in quantities of up to a few ppm; an estimate of the mean soil Ge concentration is 1.6 mg kg^{-1} (Rosenberg 2009). The dissociation behavior of germanic acid (pK_{a1}=9) resembles that of H_3BO_3 and H_4SiO_4, suggesting that at physiological pH, the prevalent form is non-charged and therefore capable of being transported by NIPs (Fig. 1). It has been long assumed that the uptake and translocation properties of Ge are similar to those shown by Si (Nikolic et al. 2007, Takahashi et al. 1976a, b). Plants containing high amounts of Si (particularly grasses) tend to be more sensitive to excess Ge than those containing little Si (Nikolic et al. 2007). The Ge concentration in soil-grown plants ranges from 0.01 mg kg^{-1} (*Magnoliopsida* species) to 1 mg kg^{-1} (*Poaceae* species), reflecting the more effective H_4SiO_4 transporter machinery present in grasses (Ma and Yamaji 2015), which comprises the NIP-IIIs and the Lsi2-type efflux transporters. The former facilitate the passive transport of Si across the plasma membrane between the apoplast/soil solution and plant cells down concentration gradients (Ma et al. 2006), while the latter are responsible for the efflux of Si from the cell (Ma et al. 2007).

Long before the discovery of Si and Ge transporters (Ma et al. 2006, 2007) and the molecular basis for the dual transport functions of NIPs and Lsi2-type transporters was described (reviewed by Ma and Yamaji 2015), existing knowledge of the chemical similarity between Si and Ge hydroxylated compounds was exploited in the use of Ge as an Si analog in toxicity screens (Ma et al. 2002; Nikolic et al. 2007). This form of screen was used to identify the rice *lsi* mutants (Ma et al. 2002). Subsequent mapping approaches identified the underlying responsible NIP aquaporin (*OsNIP2;1* and *OsNIP2;2*) and Lsi2-type transporter (*OsLsi2* and *OsLsi3*) genes (Ma et al. 2006; Ma et al. 2007; Yamaji and Ma 2009; Yamaji et al. 2015). The radioactive ^{68}Ge isotope and the non-radioactive isotopes in the form of germanic oxide (GeO_2) are frequently used as chemical tracer analogs for studying Si transport features of certain NIPs *in planta* as well as in the *S. cerevisiae* and *Xenopus laevis* frog oocyte heterologous expression systems (Ma et al. 2006; Nikolic et al. 2007; Schnurbusch et al. 2010; Mitani-Ueno et al. 2011; Gu et al. 2012; Hayes et al. 2013; Bárzana et al. 2014). A genome-wide association mapping study in rice has shown that some Ge sensitive loci coincide with known QTL underlying Si or As accumulation, but none map in the vicinity of either *OsNIP2;1* or *OsNIP2;2* (Talukdar et al. 2015). A QTL associated with Ge sensitivity lies within 200 Kbp of *OsLsi2*. *OsNIP4;1* (Os01g02190) is located within the genomic region of the detected loci. *OsNIP4;1* is strongly expressed in the inflorescence and particularly in the anthers (Liu et al. 2009). However, substrate selectivity data are not available, making it difficult to interpret its function with respect to Ge tolerance. The chemical similarities between the nonpolar, planar H_3BO_3 and H_2GeO_3 have prompted Hayes et al. (Hayes et al. 2013) to use Ge treatment as a surrogate for the effect of B toxicity on barley and wheat. A barley cultivar showing a mild reaction to the presence of GeO_2 is also tolerant to high levels of B; the underlying basis for B tolerance is a very low transcript abundance of *HvNIP2;1*, the gene implicated as encoding a B and Si transporter (Schnurbusch et al. 2010).

In summary, the nonspecific selectivity of NIP-IIIs being permeable to Si, B, and Ge represents a valuable feature, allowing to use Ge as a suitable tracer in science to mimic and characterize Si and B transport processes or to screen graminaceous crop populations for altered functions of NIP-III channels and related proteins

(Hayes et al. 2013). Ge is an important element for the semiconductor industry. However, unlike most metalloids and metals, it is not generally found in concentrated form in nature, so it has been suggested that plant accumulators could be exploited to extract it from contaminated but also agricultural soils. In this context, NIPs could potentially be engineered to increase the efficiency of the extraction process, allowing Ge to be recovered from biomass grown for the purpose of phytomining. Ge could then be extracted from, e.g., plant digestates of bioenergy crops or from straw or other not used plant residuals as a second add-on "yield" value.

4.6 *NIP-Mediated Transport of Selenium*

Se is essential in the human and animal diet, but is not essential for plant growth. The biologically active form of Se is the derived amino acid selenocysteine, which is inserted into bacterial, archeal, and eukaryotic mRNA by a specific tRNA. Because of the lower reduction potential of selenocysteine compared to cysteine itself, this compound has an important role in the catalytic sites of glutathione peroxidases and thioredoxin reductases, which act as protectants against oxidative stress (Lobanov et al. 2009). Vegetables and fruits represent the major source of dietary Se. The content of Se within plant tissue is rather low, presumably because it has no benefit for the plant; nevertheless, the element is readily taken up from the soil (Pilon-Smits and Quinn 2010). Thus, a suggested strategy to counteract Se deficiency in the diet is Se biofortification of staple crops, which would require the selection of Se accumulators or effective translocators of Se into the edible part of the plant.

The most prominent forms of soil Se are selenite ($HSeO_3^-$) and selenate ($HSeO_4^-$), with the latter predominating in well-aerated soils. The similar structure and pK_a values of selenate and sulfate result in the former being recognized and transported by sulfate transporters (Sors et al. 2005). The cross talk between selenate and sulfur metabolism makes this transport system unfavorable in the context of biofortification, as modifications to sulfur transport may have detrimental effects on a range of important traits, thereby outweighing any advantages of enhanced Se accumulation (Bienert and Chaumont 2013). H_2SeO_3 is a diprotic weak acid with pk_{a1} and pk_{a2} vales of 2.57 and 6.6, respectively, so that at physiological pHs it exists predominantly in the form of both $HSeO_3^-$ and SeO_3^{2-} (Fig. 1). Phosphate transporters (such as rice OsPT2) have been implicated in the active uptake of $HSeO_3^-$ into the root (Zhang et al. 2014). Under acidic conditions, selenous acid (H_2SeO_3) predominates (Fig. 1). The standard AQP inhibitors $HgCl_2$ and $AgNO_3$ both inhibit the uptake of H_2SeO_3 into the rice and maize root (Zhang et al. 2012; Zhang et al. 2010). Supplying $HSeO_3^-$ in a kinetic study of Se uptake into the maize root has shown that, when grown in an acidic (pH 3) medium, uptake is mostly in the form H_2SeO_3 (Zhang et al. 2010). Se uptake kinetics follow a linear trend which may suggest that the limiting step is a channel-mediated transport mechanism.

The first plant H_2SeO_3 transporter to be identified was OsNIP2;1 (Zhao et al. 2010); when grown in the presence of $HSeO_3^-$, the loss-of-function mutant *Osnip2;1* accumulates significantly less Se in its shoot and xylem sap than does the wild type. In contrast, the mutant and the wild type accumulate an equal amount of Se when grown on a medium supplemented with $HSeO_4^-$. Further experiments have revealed that H_2SeO_3 is most likely the Se form transported by OsNIP2;1 (Zhao et al. 2010a). The ability of OsNIP2;1 to transport Se has been further demonstrated by heterologously expressing it in yeast (Zhao et al. 2010a). NIPs may be involved in the intercellular transport of Se as well as in its uptake. Once $HSeO_4^-$ is taken up, it is reduced to $HSeO_3^-$ in both the chloroplast and the cytoplasm, before being further reduced to the Se^{2-} ion and hence incorporated into selenocysteine or selenomethionine; these amino acids can be nonspecifically incorporated into proteins instead of cysteine, leading to toxicity (Pilon-Smits and Quinn 2010). Still unresolved is whether (1) NIP isoforms of plant species other than rice are permeable to Se, (2) the permeability of NIPs to H_2SeO_3 is a feature of only the H_4SiO_4-permeable NIP-III isoforms present in both *Liliopsida* and *Magnoliopsida* species, and (3) the engineering of *NIP*s could represent viable means of directing Se flux in staple crops.

5 PIP-Mediated Metalloid Transport in Plants

On the basis of their sequence, the PIPs are the most homogeneous of the plant AQPs and also the most numerous (Anderberg et al. 2012). Two PIP subgroups (PIP1 and PIP2) are recognized and share a sequence identity above 50 %. The PIP1s have a longer N terminal and a shorter C terminal domain than the PIP2s, as well as having a shorter extracellular loop A (Chaumont et al. 2001). *PIP1* and *PIP2* genes behave differently when heterologously expressed in frog oocytes: in general, only PIP2s are able to induce a significant level of transmembrane water movement (Fetter et al. 2004; see chapter "Heteromerization of Plant Aquaporins"). When a *PIP1*/*PIP2* pair cloned from several section *Liliopsida* and *Magnoliopsida* species is co-expressed in frog oocytes, their products interact to modify their trafficking into and/or stability within the host membrane, thereby cooperating to synergistically increase water permeability (see chapter "Heteromerization of Plant Aquaporins"). A combination of physiological and molecular genetic evidence indicates that PIP water channels are highly important for the plant's water homeostasis (Maurel et al. 2015; Chaumont and Tyerman 2014). A small number of PIPs have been shown to be permeable to molecules other than water, including H_2O_2 and urea (reviewed by Maurel et al. 2015), and of note in the context of this chapter, they also transport uncharged metalloid species.

5.1 PIP-Mediated Transport of Boron

Direct evidence for the involvement of PIPs in B transport is fragmentary. Maize ZmPIP1;1 was the first plant AQP shown to have the capacity to transport H_3BO_3: the heterologous expression of *ZmPIP1;1* in frog oocytes results in a 30 % increase in B permeability over that achieved in control oocytes or those expressing ZmPIP3 (renamed ZmPIP2;5), AtNLM1, or EcGlpF (Dordas et al. 2000). The H_3BO_3 permeability of plasma membranes isolated from squash (*Cucurbita pepo*) vesicles is partially inhibited by the AQP inhibitors $HgCl_2$ and phloretin and is reversibly rescued by treatment with 2-mercaptoethanol. As mentioned earlier, this sort of compound-dependent on-off transport behavior is indicative of AQP-mediated transport. Dordas et al. (2000) have suggested that some H_3BO_3 enters the plant cell via passive diffusion through the plasma membrane lipid bilayer, while the rest is transported through PIP1 channels. Therewith this study provided the first experimental indication that plant AQPs are involved in metalloid transport and particularly in B transport. Subsequently, it has been shown that transferring either maize plants or transgenic tobacco plants overexpressing GFP:ZmPIP1 to a B-deficient medium for about 1 h results in the rapid disappearance of ZmPIP1 channels from the root apex cell plasma membrane (Goldbach et al. 2002). The implication is that the *ZmPIP1* product cannot be directly involved in B uptake under B-deficient conditions, since otherwise its upregulation would have been expected, as is the case for *AtNIP5;1* (Takano et al. 2006). Instead, the removal of B-permeable proteins from the plasma membrane may serve to prevent an undesirable loss of B from the root. A possible hypothesis is that the B permeability shown by certain PIPs only functions when the supply of B is non-limiting; alternatively, it may be that the removal of PIPs from the plasma membrane is independent of any potential H_3BO_3 channeling activity associated with these membrane pores. Based on yeast toxicity growth assays, H_3BO_3 permeability has also been inferred for the grapevine PIP isoforms VvTnPIP1;4 and VvTnPIP2;3 (Sabir et al. 2014).

The barley HvPIP1;3 and HvPIP1;4 resemble ZmPIP1;1 at the sequence level, and localize to the plasma membrane in both heterologous and native expression systems, in contrast to many PIP1s derived from other species (see chapter "Heteromerization of Plant Aquaporins"). The B permeability of these PIP1s has been investigated using a yeast toxicity growth assay (Fitzpatrick and Reid 2009). Both proteins increase the sensitivity of the yeast cells to exogenously supplied B, and an analysis of the cellular B content has confirmed that both are capable of mediating the uptake of B (Fitzpatrick and Reid 2009). The quantitative response of these HvPIP1s to a variation in the external concentration of B is unclear, since the transcription of their genes is unresponsive to the B nutritional status of the plant. In contrast, the transcription of both *OsPIP2;4* and *OsPIP2;7* does respond to the rice plant B nutritional status: they are downregulated in the shoot and strongly upregulated in the root when the external concentration of B is raised (Kumar et al. 2014). The heterologous expression of *OsPIP2;4* and *OsPIP2;7* in a yeast mutant frequently used to assess As permeability results in an increased sensitivity to B and in a significantly higher accumulation of B. When these proteins are constitutively expressed in *A. thaliana*, the plants produce more biomass and longer roots when being exposed to high levels of B but do not accumulate either more or less B. However, a short-term kinetic uptake assay has suggested that the stems and roots of the *OsPIP2*-expressing plants contain more B than do those of the wild type (Kumar et al. 2014). While the outcomes of heterologous expression clearly imply that certain PIPs are permeable to B, it remains to be demonstrated that the observed differences in B content of plants derive from a capacity of the PIPs to transport B, rather than reflecting a secondary effect of an AQP function unrelated to B transport. For example, a PIP-mediated change in the flux of water will alter the plant water status and hence its transpiration rate. As the transport of B within the plant depends strongly on the volume of the transpiration stream, an altered tissue B status can occur independently of active B uptake. The failure to measure tissue Ca in the above study is unfortunate, since B and Ca share a similar mobility through the xylem and distribution within the plant. A critical experiment would be to demonstrate whether or not plants experiencing a dissimilar B transport and PIP protein amount are also differentiated with respect to transpiration rate. Why ZmPIP1;3/PIP1;4 and ZmPIP2;2 are impermeable to B despite sharing a high level of sequence similarity with ZmPIP1;1 remains a puzzle (Bárzana et al. 2014). In brief, the assumption is that certain PIP1 and PIP2 isoforms possess residual permeability to H_3BO_3 sufficient to facilitate its transmembrane transport when expressed in a heterologous expression context; however, irrefutable evidence for their participation in B transport in plants is still lacking.

5.2 PIP-Mediated Transport of Arsenic

To date, the only claim that PIPs can be permeable to H_3AsO_3 was made by Mosa et al. (Mosa et al. 2012), who were able to demonstrate the downregulation of *OsPIP1;2*, *OsPIP1;3*, *OsPIP2;4*, *OsPIP2;6*, and *OsPIP2;7* in the root and shoot in response to $H_2AsO_3^-$ treatment. The heterologous expression of OsPIP2;4, *OsPIP*2;6, and *OsPIP*2;7 in frog oocytes caused increased As uptake, and the constitutive expression of *OsPIP2;4*, *OsPIP2;6*, and *OsPIP2;7* in *A. thaliana* results in an enhancement to the plant's tolerance toward $H_2AsO_3^-$, in contrast to the expectation that the transgene products should have increased the uptake of As (Mosa et al. 2012). The transgenic plants, however, show no evidence of an increased accumulation of As in either their shoot or their root.

The responsiveness of *PIP*s to As stress is a feature displayed by a number of plant species. As in rice, the abundance of five *Brassica juncea PIP1* and eight *PIP2* transcripts is reduced by exposing the plants to $H_3AsO_3/H_2AsO_3^-$ stress (Srivastava et al. 2013). Whether the observed variation was influenced, even in part, by diurnal cycling (which is known to affect *PIP* expression, see review by Heinen et al. 2009) cannot be ascertained. A subsequent whole genome transcriptome profiling of *B. juncea* subjected to $H_3AsO_3/H_2AsO_4^-$ stress has identified *PIP1;1* and *PIP2;2* as both being significantly downregulated by the stress (Srivastava et al. 2015). The stress also decreases the tissue water content of the plants, which inhibits seedling growth; at the same time increases are induced with respect to the production of reactive oxygen species, the extent of lipid peroxidation and in the level of root oxidation (Srivastava et al. 2013). Given that reactive oxygen species act to downregulate *PIP2* in the root (Hooijmaijers et al. 2012) and to drive the internalization of plasma membrane-localized PIPs (Wudick et al. 2015), it has yet to be resolved whether the altered state of *PIP* transcription is a direct effect of the As stress or whether it is rather a secondary effect, generated, for example, by a raised level of reactive oxygen species.

While direct uptake assays in heterologous expression systems provide a line of evidence suggesting the permeability of specific PIPs to metalloids, it remains puzzling why orthologous isoforms, despite their sharing a high degree of overall sequence homology and being 100 % identical in the regions of the protein known to determine selectivity (the NPA motifs and the ar/R selectivity filters) and reach the plasma membrane in the heterologous expression systems, are nevertheless impermeable to As and other metalloids.

6 TIP-Mediated Metalloid Transport in Plants

The TIPs are localized in the tonoplast (the vacuolar membrane). Vacuolar subtypes are characterized by a specific set of TIP isoforms dependent on the developmental stage of the plant and the cell differentiation status (Jauh et al. 1999). The TIPs make an important contribution to cellular osmoregulation, turgor, osmo-sensing, cell growth, and vacuolar differentiation, thanks to their capacity to transport water across the tonoplast (reviewed in Maurel et al. 2015). The various TIP subgroups are highly variable with respect to sequence, especially within their ar/R selectivity filter, resulting in a broad substrate spectrum, including urea (Liu et al. 2003, Soto et al. 2008), NH_3 (Jahn et al. 2004; Loqué et al. 2005), glycerol (Gerbeau et al. 1999; Li et al. 2008), H_2O_2 (Bienert et al. 2007) and various metalloids (as discussed below). It has been suggested that these transport functions are additive to the water transport function.

6.1 TIP-Mediated Transport of Boron

The heterologous expression of maize *ZmTIP1;2* in yeast increases the host cells' sensitivity to the presence of H_3BO_3 in the growth medium and increases H_3BO_3 flux in an iso-osmotic swelling assay when being expressed in frog oocytes (Bárzana et al. 2014). No attempt has been made so far to test whether this increased B permeability can be explained by a rise in the passive transmembrane diffusion of H_3BO_3 through the lipid bilayer induced by an increased rate of water transport. The substrate selectivity of the grapevine TIPs VvTnTIP1;1 and VvTnTIP2;2 has been assessed by expressing them in yeast, and both proteins strongly induce the cells' sensitivity to externally supplied B (Sabir et al. 2014). The potential physiological significance of these vacuolar-localized proteins to plant B homeostasis has not been investigated, either in conditions of B under- or oversupply. The *A. thaliana* pollen-specific gene *AtTIP5;1* appears to be induced by B stress, and its ectopic expression in the rest of the plant significantly increases the level of the plant tolerance to normally toxic levels of B (Pang et al. 2010). The interpretation of these outcomes might be that the plant is able to sequester B into the vacuoles when B is oversupplied. While the pollen specificity of *AtTIP5;1* has been ascribed to the high demand for B during pollen germination and pollen tube growth, the way in which AtTIP5;1 affects the transport of B within the pollen remains to be demonstrated. There is no convincing molecular or physiological evidence as yet for the involvement of TIPs in B homeostasis.

A QTL mapping approach targeting B efficiency in *A. thaliana* has been described by Zeng et al. (Zeng et al. 2008). The focus was on a trait referred to as a "B efficiency coefficient" (BEC), defined as the ratio between the seed yield of a given genotype grown under limiting B conditions and its seed yield when grown under non-limiting conditions. Five QTL have been identified, of which three – including the largest effect one named *AtBE1–2* – map within the same genomic region as a QTL for seed yield under limiting B conditions. The *AtBE1–2* harboring region also contains the *BOR1* homolog *BOR5* (*At*1*g*74810) and *AtTIP3;1* (*At*1*g*73190), while the *AtBE2* region contains *AtTIP4;1* (*At*2*g*25810). The implication is that at least two TIPs may well contribute to B efficiency, although as yet neither *TIP* gene product has been associated with B homeostasis. No *NIP* gene maps within any of QTL regions associated with either BEC or seed yield under limiting B conditions. Transcription profiling of contrasting B deficiency-tolerant citrus rootstocks has revealed that again a *TIP4;1* gene variant is substantially upregulated within the first 24 h of exposure to B deficiency but only in the tolerant genotype (Zhou et al. 2015). The significance of B to vacuolar function (if any) and the B storage capacity of different vacuole types remain obscure.

6.2 TIP-Mediated Transport of Arsenic

The As hyperaccumulator fern species *Pteris vittata* tolerates high concentrations of As in the growth substrate. The species reduces $H_2AsO_4^-$ to $H_2AsO_3^-$, which is then moved into the lamina of its fronds, where it is stored as free $H_3AsO_3/H_2AsO_3^-$. Few of the proteins contributing to these transport processes have yet been described. Indriolo et al. (Indriolo et al. 2010) have isolated the genes *PvACR3* and *PvACR3;1*, which encode proteins similar to the active ACR3 $H_2AsO_3^-$ efflux permease present in yeast. Like its yeast ortholog, PvACR3 actively transports As and localizes it to the vacuolar membrane in the gametophyte, where it is presumably detoxified. He et al. (He et al. 2015) have transformed a *P. vittata* cDNA library into yeast in an attempt to identify further As transporting proteins via a functional complementation assay. The screen has revealed *PvTIP4;1* gene, which encodes a protein permeable to $H_3AsO_3/H_2AsO_3^-$. Within its native species, *PvTIP4;1* transcription is largely confined to the roots. Unlike other TIP family members, PvTIP4;1 localizes to the plasma membrane rather than to the tonoplast. The capacity of PvTIP4;1 to transport As has been explored in both yeast and *A. thaliana*. Its heterologous expression in yeast results in an increased sensitivity to externally supplied $H_2AsO_3^-$ and in an increased uptake of As; furthermore, the mutation of the cysteine residue in the R3 position of its ar/R selectivity filter abolishes its ability to transport As (He et al. 2015). The constitutive expression of *PvTIP4;1* in *A. thaliana* boosts the accumulation of As and causes $H_2AsO_3^-$ sensitivity.

The conclusion is that certain TIPs are As permeable and that As sequestration is probably adopted for physiological As detoxification. Evidence, albeit indirect, showing that some TIPs can influence membrane permeability to metalloids has arisen from a study of the hydrangea (*Hydrangea macrophylla*) TIP1 HmPALT1, which, when heterologously expressed in yeast, facilitates the transmembrane diffusion of a not determined form of the Al^{3+} ion (Negishi et al. 2012). The form of Al transported across the tonoplast may be aluminum hydroxide (H_3AlO_3), an uncharged compound which shares some physicochemical similarities to certain AQP-channeled metalloid species.

7 XIP-Mediated Metalloid Transport in Plants

7.1 *XIP-Mediated Transport of Boron*

The plant and fungal AQP subfamily denoted as XIPs was first discovered by Danielson and Johanson (2008). While XIPs occur in many sections, *Magnoliopsida* species, the *Brassicaceae* spp. (including *A. thaliana*), and *Poaceae* lack any *XIPs* (Abascal et al. 2014). It is possible that other AQP isoforms have adopted the function of XIPs in these taxa. Based on the nature of their selectivity filter, the XIPs resemble the NIPs more closely than they do either the TIPs or the PIPs (Bienert et al. 2011). Their absence from both *A. thaliana* and rice, the two leading model plant species, reasons that little is known of their physiological role in plants. Initial studies support the notion that XIPs are not highly permeable to water, but favor larger uncharged solutes (Bienert et al. 2011; Lopez et al. 2012). The expression of six *Solanaceae XIPs* (*NtXIP1;1α* and *NtXIP1;1β*, *StXIP1;1α* and *StXIP1;1β*, *SlXIP1;1α* and *SlXIP1;1β*) in yeast results in an increased sensitivity to externally supplied H_3BO_3 (Bienert et al. 2011), suggesting the permeability of XIPs to H_3BO_3. The evidence supports the idea that the XIPs contribute to metalloid transport in plants, but this suggestion needs experimental confirmation. Whether XIPs facilitate the transport of other metalloids such as H_3AsO_3 or H_4SiO_4 remains to be seen.

8 Outlook

Given the rarity of At, Po, and Te and the lack of any biological significance for any of these metalloids in most organisms, any potential AQP-mediated transport associated with them is unlikely to be of any biological importance (Pommerrenig et al. 2015). At present, whether uncharged forms of these trace elements are transported *in planta* by AQPs is unknown. A number of challenges and open questions associated with plant AQP-mediated metalloid transport need to be addressed to complement the present knowledge. These are: (1) Which plant AQPs are permeable to which metalloid(s)? (2) Which metalloid-permeable AQPs are physiologically and actively involved in metalloid metabolism or response reactions? (3) How are plant AQPs regulated at the transcriptional and posttranslational level in response to metalloid exposure? (4) How do plant AQPs cooperatively orchestrate the transport of a given metalloid in one plant species? (5) What sequence motifs determine the metalloid selectivity of an AQP? (6) How can the ability of AQPs to transport and modify metalloid level and distribution be exploited to generate plants showing tolerance to either a high or a low level of metalloid? The answers to these questions will bear on the potential of plants to be exploited for certain agricultural conditions, for phytoremediation, for phytomining, or for biofortification. Finally, it will be interesting to analyze in an evolutionary and ecophysiological context when and where the ability of plant AQPs to channel metalloids was transformed into a main channel function.

Acknowledgments This work was supported by an Emmy Noether grant 1668/1-1 from the Deutsche Forschungsgemeinschaft. We thank all scientists for uncovering the exciting roles of AQPs in metalloid transport homeostasis. We apologize to all authors whose contributions to these research areas could not have been mentioned.

References

Abascal F, Irisarri I, Zardoya R (2014) Diversity and evolution of membrane intrinsic proteins. Biochim Biophys Acta 1840:1468–1481

Agre P, Kozono D (2003) Aquaporin water channels: molecular mechanisms for human diseases. FEBS Lett 555:72–78

Ahmadpour D, Geijer C, Tamás MJ, Lindkvist-Petersson K, Hohmann S (2014) Yeast reveals unexpected roles and regulatory features of aquaporins and aquaglyceroporins. Biochim Biophys Acta 1840:1482–1491

Alsford S, Eckert S, Baker N, Glover L, Sanchez-Flores A, Leung KF, Turner DJ, Field MC, Berriman M, Horn D (2012) High-throughput decoding of antitrypanosomal drug efficacy and resistance. Nature 482:232–236

Anderberg HI, Danielson JÅ, Johanson U (2011) Algal MIPs, high diversity and conserved motifs. BMC Evol Biol 21(11):110

Anderberg HI, Kjellbom P, Johanson U (2012) Annotation of *Selaginella moellendorffii* major intrinsic proteins and the evolution of the protein family in terrestrial plants. Front Plant Sci 20(3):33

Baker N, Glovera L, Munday JC, Aguinaga Andrés D, Barrett MP, de Koning HP, Horn (2012) Aquaglyceroporin 2 controls susceptibility to melarsoprol and pentamidine in African trypanosomes. Proc Natl Acad Sci U S A 109:10996–11101

Bárzana G, Aroca R, Bienert GP, Chaumont F, Ruiz-Lozano JM (2014) New insights into the regulation of aquaporins by the arbuscular mycorrhizal symbiosis in maize plants under drought stress and possible implications for plant performance. Mol Plant-Microbe Interact 27:349–363

Bienert GP, Chaumont F (2011) Plant aquaporins: roles in water homeostasis, nutrition, and signalling processes. In: Geisler M (ed) Transporters and pumps in plant signaling, 1st edn. Springer Publishers, Berlin-Heidelberg, pp 3–36

Bienert GP, Chaumont F (2013) Selenium and aquaporins. In: Kretsinger RH, Uversky VN, Permyakov EA (eds) Encyclopedia of metalloproteins. Springer, New York, pp 1891–1893
Bienert GP, Jahn TP (2010a) Major intrinsic proteins and arsenic transport in plants: new players and their potential roles. In: Jahn TP, Bienert GP (eds) MIPs and their role in the exchange of metalloids, advances in experimental medicine and biology, vol 679. Landes Bioscience Publishers, New York, pp 111–125
Bienert GP, Jahn TP (eds) (2010b) MIPs and their role in the exchange of metalloids, advances in experimental medicine and biology, vol 679. Landes Bioscience Publishers, New York
Bienert GP, Schjoerring JK, Jahn TP (2006) Membrane transport of hydrogen peroxide. Biochim Biophys Acta 1758:994–1003
Bienert GP, Møller AL, Kristiansen KA, Schulz A, Møller IM, Schjoerring JK, Jahn TP (2007) Specific aquaporins facilitate the diffusion of hydrogen peroxide across membranes. J Biol Chem 282:1183–1192
Bienert GP, Thorsen M, Schüssler MD, Nilsson HR, Wagner A, Tamás MJ, Jahn TP (2008a) A subgroup of plant aquaporins facilitate the bi-directional diffusion of $As(OH)_3$ and $Sb(OH)_3$ across membranes. BMC Biol 10(6):26
Bienert GP, Schüssler MD, Jahn TP (2008b) Metalloids: essential, beneficial or toxic? Major intrinsic proteins sort it out. Trends Biochem Sci 33:20–26
Bienert GP, Bienert MD, Jahn TP, Boutry M, Chaumont F (2011) *Solanaceae* XIPs are plasma membrane aquaporins that facilitate the transport of many uncharged substrates. Plant J 66:306–317
Bienert GP, Desguin B, Chaumont F, Hols P (2013) Channel-mediated lactic acid transport: a novel function for aquaglyceroporins in bacteria. Biochem J 454:559–570
Bogacki P, Peck DM, Nair RM, Howie J, Oldach KH (2013) Genetic analysis of tolerance to boron toxicity in the legume *Medicago truncatula*. BMC Plant Biol 27(13):54
Carbrey JM, Song L, Zhou Y, Yoshinaga M, Rojek A, Wang Y, Liu Y, Lujan HL, DiCarlo SE, Nielsen S, Rosen BP, Agre P, Mukhopadhyay R (2009) Reduced arsenic clearance and increased toxicity in aquaglyceroporin-9-null mice. Proc Natl Acad Sci U S A 106:15956–15960
Carey AM, Scheckel KG, Lombi E, Newville M, Choi Y, Norton GJ, Charnock JM, Feldmann J, Price AH, Meharg AA (2010) Grain unloading of arsenic species in rice. Plant Physiol 152:309–319
Carey AM, Norton GJ, Deacon C, Scheckel KG, Lombi E, Punshon T, Guerinot ML, Lanzirotti A, Newville M, Choi Y, Price AH, Meharg AA (2011) Phloem transport of arsenic species from flag leaf to grain during grain filling. New Phytol 192:87–98
Chakrabarty D, Trivedi PK, Misra P, Tiwari M, Shri M, Shukla D, Kumar S, Rai A, Pandey A, Nigam D, Tripathi RD, Tuli R (2009) Comparative transcriptome analysis of arsenate and arsenite stresses in rice seedlings. Chemosphere 74:688–702
Chaumont F, Tyerman SD (2014) Aquaporins: highly regulated channels controlling plant water relations. Plant Physiol 164:1600–1618
Chaumont F, Barrieu F, Wojcik E, Chrispeels MJ, Jung R (2001) Aquaporins constitute a large and highly divergent protein family in maize. Plant Physiol 125:1206–1215
Chen CH, Lewin J (1969) Silicon as a nutrient element for *Equisetum arvense*. Can J Bot 47:125–131
Chiba Y, Mitani N, Yamaji N, Ma JF (2009) HvLsi1 is a silicon influx transporter in barley. Plant J 57:810–818
Choi WG, Roberts DM (2007) *Arabidopsis* NIP2;1, a major intrinsic protein transporter of lactic acid induced by anoxic stress. J Biol Chem 282:24209–24218
Danielson JA, Johanson U (2008) Unexpected complexity of the aquaporin gene family in the moss *Physcomitrella patens*. BMC Plant Biol 22(8):45
Danielson JAH, Johanson U (2010) Phylogeny of major intrinsic proteins. In: Jahn TP, Bienert GP (eds) MIPs and their role in the exchange of metalloids, advances in experimental medicine and biology, vol 679. Landes Bioscience Publishers, New York, pp p19–p32
Dean RM, Rivers RL, Zeidel ML, Roberts DM (1999) Purification and functional reconstitution of soybean nodulin 26. An aquaporin with water and glycerol transport properties. Biochemistry 38:347–353

Deshmukh RK1, Vivancos J, Guérin V, Sonah H, Labbé C, Belzile F, Bélanger RR (2013) Identification and functional characterization of silicon transporters in soybean using comparative genomics of major intrinsic proteins in *Arabidopsis* and rice. Plant Mol Biol 83:303–315

Diehn TA, Pommerrenig B, Bernhardt N, Hartmann A, GP B (2015) Genome-wide identification of aquaporin encoding genes in *Brassica oleracea* and their phylogenetic sequence comparison to *Brassica* crops and *Arabidopsis*. Front Plant Sci 7(6):166

Dordas C, Brown PH (2000) Permeability of boric acid across lipid bilayers and factors affecting it. J Membr Biol 175:95–105

Dordas C, Chrispeels MJ, Brown PH (2000) Permeability and channel-mediated transport of boric acid across membrane vesicles isolated from squash roots. Plant Physiol 124:1349–1362

Durbak AR, Phillips KA, Pike S, O'Neill MA, Mares J, Gallavotti A, Malcomber ST, Gassmann W, McSteen P (2014) Transport of boron by the tassel-less1 aquaporin is critical for vegetative and reproductive development in maize. Plant Cell 26:2978–2995

Dynowski M, Schaaf G, Loque D, Moran O, Ludewig U (2008) Plant plasma membrane water channels conduct the signalling molecule H_2O_2. Biochem J 414:53–61

Epstein E (1994) The anomaly of silicon in plant biology. Proc Natl Acad Sci U S A 91:11–17

Fetter K, Van Wilder V, Moshelion M, Chaumont F (2004) Interactions between plasma membrane aquaporins modulate their water channel activity. Plant Cell 16:215–228

Finnegan PM, Chen W (2012) Arsenic toxicity: the effects on plant metabolism. Front Physiol 6(3):182

Fitzpatrick KL, Reid RJ (2009) The involvement of aquaglyceroporins in transport of boron in barley roots. Plant Cell Environ 32:1357–1365

Fortin MG, Morrison NA, Verma DP (1987) Nodulin-26, a peribacteroid membrane nodulin is expressed independently of the development of the peribacteroid compartment. Nucleic Acids Res 15:813–824

Funakawa H, Miwa K (2015) Synthesis of borate cross-linked rhamnogalacturonan II. Front Plant Sci 21(6):223

Gerbeau P, Güclü J, Ripoche P, Maurel C (1999) Aquaporin Nt-TIPa can account for the high permeability of tobacco cell vacuolar membrane to small neutral solutes. Plant J 18:577–587

Goldbach HE, Wimmer MA, Chaumont F, Matoh T, Volkmann D, Baluška F, Ruth Wingender R, Schulz M, Yu O (2002) Rapid responses of plants to Boron deprivation – where are the links between boron's primary role and secondary reactions? In: Goldbach HE, Brown PH, Rerkasem B, Thellier M, Wimmer MA, Bell RW (eds) Boron in plant and animal nutrition. Springer, New York, pp p167–p180

Grégoire C1, Rémus-Borel W, Vivancos J, Labbé C, Belzile F, Bélanger RR (2012) Discovery of a multigene family of aquaporin silicon transporters in the primitive plant Equisetum arvense. Plant J 72:320–330

Gu R, Chen X, Zhou Y, Yuan L (2012) Isolation and characterization of three maize aquaporin genes, ZmNIP2;1, ZmNIP2;4 and ZmTIP4;4 involved in urea transport. BMB Rep 45:96–101

Gupta AB, Sankararamakrishnan R (2009) Genome-wide analysis of major intrinsic proteins in the tree plant *Populus trichocarpa*: characterization of XIP subfamily of aquaporins from evolutionary perspective. BMC Plant Biol 9:134

Hanaoka H, Uraguchi S, Takano J, Tanaka M, Fujiwara T (2014) OsNIP3;1, a rice boric acid channel, regulates boron distribution and is essential for growth under boron-deficient conditions. Plant J 78:890–902

Hara-Chikuma M, Verkman AS (2006) Physiological roles of glycerol-transporting aquaporins: the aquaglyceroporins. Cell Mol Life Sci 63:1386–1392

Hayes JE, Pallotta M, Baumann U, Berger B, Langridge P, Sutton T (2013) Germanium as a tool to dissect boron toxicity effects in barley and wheat. Funct Plant Biol 40:618–627

He Z, Yan H, Chen Y, Shen H, Xu W, Zhang H, Shi L, Zhu YG, Ma M (2015) An aquaporin PvTIP4;1 from *Pteris vittata* may mediate arsenite uptake. New Phytol 209:746. doi:10.1111/nph.13637

Heinen RB, Ye Q, Chaumont F (2009) Role of aquaporins in leaf physiology. J Exp Bot 60:2971–2985

Herrera M, Hong NJ, Garvin JL (2006) Aquaporin-1 transports NO across cell membranes. Hypertension 48:157–164

Hooijmaijers C1, Rhee JY, Kwak KJ, Chung GC, Horie T, Katsuhara M, Kang H (2012) Hydrogen peroxide permeability of plasma membrane aquaporins of *Arabidopsis thaliana*. J Plant Res 125:147–153

Hu W, Hou X, Huang C, Yan Y, Tie W, Ding Z, Wei Y, Liu J, Miao H, Lu Z, Li M, Xu B, Jin Z (2015) Genome-wide identification and expression analyses of aquaporin gene family during development and abiotic stress in banana. Int J Mol Sci 16:19728–19751

Indriolo E, Na G, Ellis D, Salt DE, Banks JA (2010) A vacuolar arsenite transporter necessary for arsenic tolerance in the arsenic hyperaccumulating fern *Pteris vittata* is missing in flowering plants. Plant Cell 22:2045–2057

Isayenkov SV, Maathuis FJ (2008) The *Arabidopsis thaliana* aquaglyceroporin AtNIP7;1 is a pathway for arsenite uptake. FEBS Lett 582:1625–1628

Jahn TP, Møller AL, Zeuthen T, Holm LM, Klaerke DA, Mohsin B, Kühlbrandt W, Schjoerring JK (2004) Aquaporin homologues in plants and mammals transport ammonia. FEBS Lett 574:31–36

Jauh GY, Phillips TE, Rogers JC (1999) Tonoplast intrinsic protein isoforms as markers for vacuolar functions. Plant Cell 11:1867–1882

Johanson U, Karlsson M, Johansson I, Gustavsson S, Sjövall S, Fraysse L, Weig AR, Kjellbom P (2001) The complete set of genes encoding major intrinsic proteins in *Arabidopsis* provides a framework for a new nomenclature for major intrinsic proteins in plants. Plant Physiol 126:1358–1369

Kamiya T, Fujiwara T (2009) *Arabidopsis* NIP1;1 transports antimonite and determines antimonite sensitivity. Plant Cell Physiol 50:1977–1981

Kamiya T, Tanaka M, Mitani N, Ma JF, Maeshima M, Fujiwara T (2009) NIP1;1, an aquaporin homolog, determines the arsenite sensitivity of *Arabidopsis thaliana*. J Biol Chem 284:2114–2120

Katsuhara M, Sasano S, Horie T, Matsuhmoto T, Rhee J, Shibasaka M (2014) Functional and molecular characteristics of barley and rice NIP aquaporins transporting water, hydrogen peroxide and arsenite. Plant Biotechnol 31:213–219

Khabudaev KV, Petrova DP, Grachev MA, Likhoshway YV (2014) A new subfamily LIP of the major intrinsic proteins. BMC Genomics 4(15):173

Kumar K, Mosa KA, Chhikara S, Musante C, White JC, Dhankher OP (2014) Two rice plasma membrane intrinsic proteins, OsPIP2;4 and OsPIP2;7, are involved in transport and providing tolerance to boron toxicity. Planta 239:187–198

Kuramata M, Abe T, Kawasaki A, Ebana K, Shibaya T, Yano M, Ishikawa S (2013) Genetic diversity of arsenic accumulation in rice and QTL analysis of methylated arsenic in rice grains. Rice 11(6):3

Laforenza U, Bottino C, Gastaldi G (2015) Mammalian aquaglyceroporin function in metabolism. Biochim Biophys Acta 1858:1–11

Leonard A, Holloway B, Guo M, Rupe M, Yu G, Beatty M, Zastrow-Hayes G, Meeley R, Llaca V, Butler K, Stefani T, Jaqueth J, Li B (2014) Tassel-less1 encodes a boron channel protein required for inflorescence development in maize. Plant Cell Physiol 55:1044–1054

Li GW, Peng YH, Yu X, Zhang MH, Cai WM, Sun WN, Su WA (2008) Transport functions and expression analysis of vacuolar membrane aquaporins in response to various stresses in rice. J Plant Physiol 165:1879–1888

Li RY, Ago Y, Liu WJ, Mitani N, Feldmann J, McGrath SP, Ma JF, Zhao FJ (2009) The rice aquaporin Lsi1 mediates uptake of methylated arsenic species. Plant Physiol 50:2071–2080

Li N, Wang J, Song WY (2015) Arsenic uptake and translocation in plants. Plant Sci. doi:10.1093/pcp/pcv143

Liu LH, Ludewig U, Gassert B, Frommer WB, von Wirén N (2003) Urea transport by nitrogen-regulated tonoplast intrinsic proteins in Arabidopsis. Plant Physiol 133:1220–1228

Liu Q, Wang H, Zhang Z, Wu J, Feng Y, Zhu Z (2009) Divergence in function and expression of the NOD26-like intrinsic proteins in plants. BMC Genomics 15(10):313

Liu K, Liu LL, Ren YL, Wang ZQ, Zhou KN, Liu X, Wang D, Zheng M, Cheng ZJ, Lin QB, Wang JL, Wu FQ, Zhang X, Guo XP, Wang CM, Zhai HQ, Jiang L, Wan JM (2015) Dwarf and tiller-

enhancing 1 regulates growth and development by influencing boron uptake in boron limited conditions in rice. Plant Sci 236:18–28
Lobanov AV, Hatfield DL, Gladyshev VN (2009) Eukaryotic selenoproteins and selenoproteomes. Biochim Biophys Acta 1790:1424–1428
Lombi E, Holm PE (2010) Metalloids, soil chemistry and the environment. In: Jahn TP, Bienert GP (eds) MIPs and their role in the exchange of metalloids, advances in experimental medicine and biology, vol 679. Landes Bioscience Publishers, New York, pp p33–p41
Lopez D, Bronner G, Brunel N, Auguin D, Bourgerie S, Brignolas F, Carpin S, Tournaire-Roux C, Maurel C, Fumanal B, Martin F, Sakr S, Label P, Julien JL, Gousset-Dupont A, Venisse JS (2012) Insights into *Populus XIP aquaporins*: evolutionary expansion, protein functionality, and environmental regulation. J Exp Bot 63:2217–2230
Loqué D, Ludewig U, Yuan L, von Wirén N (2005) Tonoplast intrinsic proteins AtTIP2;1 and AtTIP2;3 facilitate NH_3 transport into the vacuole. Plant Physiol 137:671–680
Luyten K, Albertyn J, Skibbe WF, Prior BA, Ramos J, Thevelein JM, Hohmann S (1995) Fps1, a yeast member of the MIP family of channel proteins, is a facilitator for glycerol uptake and efflux and is inactive under osmotic stress. EMBO J 14:1360–1371
Ma JF, Yamaji N (2015) A cooperative system of silicon transport in plants. Trends Plant Sci 20:435–442
Ma JF, Goto S, Tamai K, Ichii M, Wu GF (2002) A rice mutant defective in Si uptake. Plant Physiol 130:2111–2117
Ma JF, Takahashi E (2002) Soil, Fertilizer, and Plant Silicon Research in Japan. In: Elsevier Science, Amsterdam
Ma JF, Tamai K, Yamaji N, Mitani N, Konishi S, Katsuhara M et al (2006) A silicon transporter in rice. Nature 440:688–691
Ma JF, Yamaji N, Mitani N, Tamai K, Konishi S, Fujiwara T, Katsuhara M, Yano M (2007) An efflux transporter of silicon in rice. Nature 448:209–212
Ma JF, Yamaji N, Mitani N, Xu XY, Su YH, McGrath SP et al (2008) Transporters of arsenite in rice and their role in arsenic accumulation in rice grain. Proc Natl Acad Sci U S A 105: 9931–9935
Maciaszczyk-Dziubinska E, Wawrzycka D, Wysocki R (2012) Arsenic and antimony transporters in eukaryotes. Int J Mol Sci 13:3527–3548
Mandal G, Orta JF, Sharma M, Mukhopadhyay R (2014) *Trypanosomatid aquaporins*: roles in physiology and drug response. Diseases 2:3–23
Marschner H (2012) Marschner's mineral nutrition of higher plants. 3rd Edition from Petra Marschner, Academic Press, London
Martínez-Cuenca MR, Martínez-Alcántara B, Quiñones A, Ruiz M, Iglesias DJ, Primo-Millo E, Forner-Giner MÁ (2015) Physiological and molecular responses to excess Boron in citrus macrophylla W. PLoS One 10:e0134372
Maurel C, Boursiac Y, Luu DT, Santoni V, Shahzad Z, Verdoucq L (2015) Aquaporins in plants. Physiol Rev 95:1321–1358
Meharg AA, Jardine L (2003) Arsenite transport into paddy rice (*Oryza sativa*) roots. New Phytol 157:39–44
Meharg AA, Zhao F-J (eds) (2012) Arsenic and rice. Springer, Dordrecht
Mitani N, Ma JF (2005) Uptake system of silicon in different plant species. J Exp Bot 56:1255–1261
Mitani N, Yamaji N, Ma JF (2008) Characterization of substrate specificity of a rice silicon transporter, Lsi1. Pflugers Arch 456:679–686
Mitani N, Yamaji N, Ma JF (2009a) Identification of maize silicon influx transporters. Plant Cell Physiol 50:5–12
Mitani N, Chiba Y, Yamaji N, Ma JF (2009b) Identification and characterization of maize and barley Lsi2-like silicon efflux transporters reveals a distinct silicon uptake system from that in rice. Plant Cell 21:2133–2142
Mitani N, Yamaji N, Ago Y, Iwasaki K, Ma JF (2011) Isolation and functional characterization of an influx silicon transporter in two pumpkin cultivars contrasting in silicon accumulation. Plant J 66:231–240

Mitani-Ueno N, Yamaji N, Ma JF (2011) Silicon efflux transporters isolated from two pumpkin cultivars contrasting in Si uptake. Plant Signal Behav 6:991–994

Mollapour M, Piper PW (2007) Hog1 mitogen-activated protein kinase phosphorylation targets the yeast Fps1 aquaglyceroporin for endocytosis, thereby rendering cells resistant to acetic acid. Mol Cell Biol 27:6446–6456

Montpetit J, Vivancos J, Mitani-Ueno N, Yamaji N, Rémus-Borel W, Belzile F, Ma JF, Bélanger RR (2012) Cloning, functional characterization and heterologous expression of TaLsi1, a wheat silicon transporter gene. Plant Mol Biol 79:35–46

Moore KL, Schröder M, Wu Z, Martin BG, Hawes CR, McGrath SP, Hawkesford MJ, Feng Ma J, Zhao FJ, Grovenor CR (2011) High-resolution secondary ion mass spectrometry reveals the contrasting subcellular distribution of arsenic and silicon in rice roots. Plant Physiol 156:913–924

Mosa KA, Kumar K, Chhikara S, Mcdermott J, Liu Z, Musante C, White JC, Dhankher OP (2012) Members of rice plasma membrane intrinsic proteins subfamily are involved in arsenite permeability and tolerance in plants. Transgenic Res 21:1265–1277

Mukhopadhyay R, Bhattacharjee H, Rosen BP (2014) Aquaglyceroporins: generalized metalloid channels. Biochim Biophys Acta 1840:1583–1591

Murata K, Mitsuoka K, Hirai T, Walz T, Agre P, Heymann JB et al (2000) Structural determinants of water permeation through aquaporin-1. Nature 407:599–605

Nable RO, Bañuelos GS, Paull JG (1997) Boron toxicity. Plant Soil 193:181–198

Nagarajan Y, Rongala J, Luang S, Singh A, Shadiac N, Hayes J, Sutton T, Gilliham M, Tyerman SD, McPhee G, Voelcker NH, Mertens HDT, Kirby N, Lee J-G, Yingling YG, Hrmova M (2015) Na^+-dependent anion transport by a barley efflux protein revealed through an integrative platform. Plant Cell 28:202–218

Nakagawa Y, Hanaoka H, Kobayashi M, Miyoshi K, Miwa K, Fujiwara T (2007) Cell-type specificity of the expression of OsBOR1, a rice efflux boron transporter gene, is regulated in response to boron availability for efficient boron uptake and xylem loading. Plant Cell 19:2624–2635

Negishi T, Oshima K, Hattori M, Kanai M, Mano S, Nishimura M, Yoshida K (2012) Tonoplast- and plasma membrane-localized aquaporin-family transporters in blue hydrangea sepals of aluminum hyperaccumulating plant. PLoS One 7:e43189

Nikolic M, Nikolic N, Liang Y, Kirkby EA, Römheld V (2007) Germanium-68 as an adequate tracer for silicon transport in plants. Characterization of silicon uptake in different crop species. Plant Physiol 143:495–503

Norton GJ, Nigar M, Williams PN, Dasgupta T, Meharg AA, Price AH (2008) Rice-arsenate interactions in hydroponics: a three-gene model for tolerance. J Exp Bot 59:2277–2284

Norton GJ, Douglas A, Lahner B, Yakubova E, Guerinot ML, Pinson SR, Tarpley L, Eizenga GC, McGrath SP, Zhao FJ, Islam MR, Islam S, Duan G, Zhu Y, Salt DE, Meharg AA, Price AH (2014) Genome wide association mapping of grain arsenic, copper, molybdenum and zinc in rice (*Oryza sativa L.*) grown at four international field sites. PLoS One 9:e89685

O'Neill MA, Eberhard S, Albersheim P, Darvill AG (2001) Requirement of borate cross-linking of cell wall rhamnogalacturonan II for *Arabidopsis* growth. Science 294:846–849

Pang Y, Li L, Ren F, Lu P, Wei P, Cai J, Xin L, Zhang J, Chen J, Wang X (2010) Overexpression of the tonoplast aquaporin AtTIP5;1 conferred tolerance to boron toxicity in *Arabidopsis*. J Genet Genomics 37:389–397

Park W, Scheffler BE, Bauer PJ, Campbell BT (2010) Identification of the family of aquaporin genes and their expression in upland cotton (*Gossypium hirsutum L.*). BMC Plant Biol 10:142

Parker MD, Boron WF (2013) The divergence, actions, roles, and relatives of sodium-coupled bicarbonate transporters. Physiol Rev 93:803–959

Pérez S, Rodríguez-Carvajal MA, Doco T (2003) A complex plant cell wall polysaccharide: rhamnogalacturonan II. A structure in quest of a function. Biochimie 85:109–121

Pilon-Smits EAH, Quinn CF (2010) Selenium Metabolism in Plants. In: Hell R, Mendel RR (eds) Cell biology of metals and nutrients, 225 plant cell monographs 17. Springer, Berlin-Heidelberg

Pommerrenig B, Diehn TA, Bienert GP (2015) Metalloido-porins: essentiality of nodulin 26-like intrinsic proteins in metalloid transport. Plant Sci 238:212–227

Porquet A, Filella M (2007) Structural evidence of the similarity of $Sb(OH)_3$ and $As(OH)_3$ with glycerol: implications for their uptake. Chem Res Toxicol 20:1269–1276

Reuscher S, Akiyama M, Mori C, Aoki K, Shibata D, Shiratake K (2013) Genome-wide identification and expression analysis of aquaporins in tomato. PLoS One 8:e79052

Richey DP, Lin EC (1972) Importance of facilitated diffusion for effective utilization of glycerol by *Escherichia coli*. J Bacteriol 112:784–790

Rosen BP, Tamas MJ (2010) Arsenic transport in prokaryotes and eucaryotic microbes. In: Jahn TP, Bienert GP (eds) MIPs and their role in the exchange of metalloids, advances in experimental medicine and biology, vol 679. Landes Bioscience Publishers, New York, pp p47–p56

Rosenberg E (2009) Germanium: environmental occurrence, importance and speciation. Rev Environ Sci Biotechnol 8:29–57

Sabir F, Leandro MJ, Martins AP, Loureiro-Dias MC, Moura TF, Soveral G, Prista C (2014) Exploring three PIPs and three TIPs of grapevine for transport of water and atypical substrates through heterologous expression in aqy-null yeast. PLoS One 9:e102087

Sade N, Vinocur BJ, Diber A, Shatil A, Ronen G, Nissan H et al (2009) Improving plant stress tolerance and yield production: is the tonoplast aquaporin SlTIP2;2 a key to isohydric to anisohydric conversion? New Phytol 181:651–661

Sakurai J, Ishikawa F, Yamaguchi T, Uemura M, Maeshima M (2005) Identification of 33 rice aquaporin genes and analysis of their expression and function. Plant Cell Physiol 46: 1568–1577

Sakurai G, Satake A, Yamaji N, Mitani-Ueno N, Yokozawa M, Feugier FG, Ma JF (2015) In silico simulation modeling reveals the importance of the Casparian strip for efficient silicon uptake in rice roots. Plant Cell Physiol 56:631–639

Schnurbusch T, Hayes J, Hrmova M, Baumann U, Ramesh SA, Tyerman SD, Langridge P, Sutton T (2010) Boron toxicity tolerance in barley through reduced expression of the multifunctional aquaporin HvNIP2;1. Plant Physiol 153:1706–1715

Shatil-Cohen A, Moshelion M (2012) Smart pipes: the bundle sheath role as xylem-mesophyll barrier. Plant Signal Behav 7:1088–1091

Shinkai Y, Sumi D, Toyama T, Kaji T, Kumagai Y (2009) Role of aquaporin 9 in cellular accumulation of arsenic and its cytotoxicity in primary mouse hepatocytes. Toxicol Appl Pharmacol 237:232–236

Sors TG, Allis DR, Salt DE (2005) Selenium uptake, translocation, assimilation and metabolic fate in plants. Photosynth Res 86:373–389

Soto G, Alleva K, Mazzella MA, Amodeo G, Muschietti JP (2008) AtTIP1;3 and AtTIP5;1, the only highly expressed Arabidopsis pollen-specific aquaporins, transport water and urea. FEBS Lett 582:4077–4082

Srivastava S, Srivastava AK, Suprasanna P, D'Souza SF (2013) Quantitative real-time expression profiling of aquaporin-isoforms and growth response of *Brassica juncea* under arsenite stress. Mol Biol Rep 40:2879–2886

Srivastava S, Srivastava AK, Sablok G, Deshpande TU, Suprasanna P (2015) Transcriptomics profiling of Indian mustard (*Brassica juncea*) under arsenate stress identifies key candidate genes and regulatory pathways. Front Plant Sci 6:646

Takahashi E, Matsumoto H, Syo S, Myake Y (1976a) Difference in the mode of germanium uptake between silicophile plants and non-silicophile plants: comparative studies on the silica nutrition in plants (part 3). J Sci Soil Manure Jpn 47:217–288

Takahashi E, Syo S, Myake Y (1976b) Effect of germanium on the growth of plants with special reference to the silicon nutrition: comparative studies on the silica nutrition in plants (part 1). J Sci Soil Manure Jpn 47:183–190

Takano J, Noguchi K, Yasumori M, Kobayashi M, Gajdos Z, Miwa K, Hayashi H, Yoneyama T, Fujiwara T (2002) *Arabidopsis* boron transporter for xylem loading. Nature 420:337–340

Takano J, Wada M, Ludewig U, Schaaf G, von Wirén N, Fujiwara T (2006) The Arabidopsis major intrinsic protein NIP5;1 is essential for efficient boron uptake and plant development under boron limitation. Plant Cell 18:1498–1509

Takano J, Miwa K, Fujiwara T (2008) Boron transport mechanisms: collaboration of channels and transporters. Trends Plant Sci 13:451–457

Takano J, Tanaka M, Toyoda A, Miwa K, Kasai K, Fuji K, Onouchi H, Naito S, Fujiwara T (2010) Polar localization and degradation of Arabidopsis boron transporters through distinct trafficking pathways. Proc Natl Acad Sci U S A 107:5220–5225

Talukdar P, Douglas A, Price AH, Norton GJ (2015) Biallelic and genome wide association mapping of germanium tolerant loci in rice (*Oryza sativa L.*). PLoS One 10:e0137577

Tanaka M, Wallace IS, Takano J, Roberts DM, Fujiwara T (2008) NIP6;1 is a boric acid channel for preferential transport of boron to growing shoot tissues in *Arabidopsis*. Plant Cell 20:2860–2875

Tanaka M, Takano J, Chiba Y, Lombardo F, Ogasawara Y, Onouchi H, Naito S, Fujiwara T (2011) Boron-dependent degradation of NIP5;1 mRNA for acclimation to excess boron conditions in *Arabidopsis*. Plant Cell 23:3547–3559

Thorsen M, Jacobson T, Vooijs R, Navarrete C, Bliek T, Schat H, Tamás MJ (2012) Glutathione serves an extracellular defence function to decrease arsenite accumulation and toxicity in yeast. Mol Microbiol 84:1177–1188

Tsukaguchi H, Shayakul C, Berger UV, Mackenzie B, Devidas S, Guggino WB, van Hoek AN, Hediger MA (1998) Molecular characterization of a broad selectivity neutral solute channel. J Biol Chem 273:24737–24743

Venkatesh J, Yu JW, Park SW (2013) Genome-wide analysis and expression profiling of the *Solanum tuberosum aquaporins*. Plant Physiol Biochem 73:392–404

Vink BW (1996) Stability relations of antimony and arsenic compounds in the light of revised and extended Eh-pH diagrams. Chem Geol 130:21–30

Wallace IS, Roberts DM (2004) Homology modeling of representative subfamilies of *Arabidopsis* major intrinsic proteins. Classification based on the aromatic/arginine selectivity filter. Plant Physiol 135:1059–1068

Wang HS, Yu C, Fan PP, Bao BF, Li T, Zhu ZJ (2015) Identification of two cucumber putative silicon transporter genes in *Cucumis sativus*. J Plant Growth Regul 34:332–338

Warrington K (1923) The effect of boric acid and borax on broad bean and certain other plants. Ann Bot 37:629–672

Wu B, Song J, Beitz E (2010) Novel channel enzyme fusion proteins confer arsenate resistance. J Biol Chem 285:40081–40087

Wudick MM, Li X, Valentini V, Geldner N, Chory J, Lin J, Maurel C, Luu DT (2015) Subcellular redistribution of root aquaporins induced by hydrogen peroxide. Mol Plant 8:1103–1114

Xu W, Dai W, Yan H, Li S, Shen H, Chen Y, Xu H, Sun Y, He Z, Ma M (2015) *Arabidopsis* NIP3;1 plays an important role in arsenic uptake and root-to-shoot translocation under arsenite stress conditions. Mol Plant 8:722–733

Yamaji N, Mitatni N, Ma JF (2008) A transporter regulating silicon distribution in rice shoots. Plant Cell 20:1381–1389

Yamaji N, Ma JF (2009) A transporter at the node responsible for intravascular transfer of silicon in rice. Plant Cell 21:2878–2883

Yamaji N, Chiba Y, Mitani-Ueno N, Ma JF (2012) Functional characterization of a silicon transporter gene implicated in silicon distribution in barley. Plant Physiol 160:1491–1497

Yamaji N, Sakurai G, Mitani-Ueno N, Ma JF (2015) Orchestration of three transporters and distinct vascular structures in node for intervascular transfer of silicon in rice. Proc Natl Acad Sci U S A 112:11401–11406

Yang HC, Cheng J, Finan TM, Rosen BP, Bhattacharjee H (2005) Novel pathway for arsenic detoxification in the legume symbiont *Sinorhizobium meliloti*. J Bacteriol 187:6991–6997

Yool AJ, Campbell EM (2012) Structure, function and translational relevance of aquaporin dual water and ion channels. Mol Asp Med 33:553–561

Zangi R, Filella M (2012) Transport routes of metalloids into and out of the cell: a review of the current knowledge. Chem Biol Interact 197:47–57

Zeng C, Han Y, Shi L, Peng L, Wang Y, Xu F, Meng J (2008) Genetic analysis of the physiological responses to low boron stress in *Arabidopsis thaliana*. Plant Cell Environ 31:112–122

Zhang L, Yu F, Shi W, Li Y, Miao Z (2010) Physiological characteristics of selenite uptake by maize roots in response to different pH levels. J Soil Sci Plant Nutr 173:417–422

Zhang H, Feng X, Zhu J, Sapkota A, Meng B, Yao H, Qin H, Larssen T (2012) Selenium in soil inhibits mercury uptake and translocation in rice (*Oryza sativa L.*). Environ Sci Technol 46:10040–10046

Zhang DY, Ali Z, Wang CB, Xu L, Yi JX, Xu ZL et al (2013) Genome-wide sequence characterization and expression analysis of major intrinsic proteins in soybean (*Glycine max L.*). PLoS One 8:e56312

Zhang L, Hu B, Li W, Che R, Deng K, Li H, Yu F, Ling H, Li Y, Chu C (2014) OsPT2, a phosphate transporter, is involved in the active uptake of selenite in rice. New Phytol 201:1183–1191

Zhao XQ, Mitani N, Yamaji N, Shen RF, Ma JF (2010a) Involvement of silicon influx transporter OsNIP2;1 in selenite uptake in rice. Plant Physiol 153:1871–1877

Zhao FJ, Ago Y, Mitani N, Li RY, Su YH, Yamaji N, McGrath SP, Ma JF (2010b) The role of the rice aquaporin Lsi1 in arsenite efflux from roots. New Phytol 186:392–399

Zhou GF, Liu YZ, Sheng O, Wei QJ, Yang CQ, Peng SA (2015) Transcription profiles of boron-deficiency-responsive genes in citrus rootstock root by suppression subtractive hybridization and cDNA microarray. Front Plant Sci 28(5):795

Zou Z, Gong J, Huang Q, Mo Y, Yang L, Xie G (2015) Gene structures, evolution, classification and expression profiles of the aquaporin gene family in castor bean (*Ricinus communis* L.). PLoS One 10(10):e0141022

Plant Aquaporins and Mycorrhizae: Their Regulation and Involvement in Plant Physiology and Performance

J.M. Ruiz-Lozano and R. Aroca

Abstract The establishment of a mycorrhizal symbiosis can change plant aquaporin gene expression and protein accumulation. However, the regulation of plant aquaporins seems to be dependent on the plant and fungal species involved in the symbiosis. The implications of such regulation on plant water relations, plant physiology and plant performance under optimal or stressful conditions have been the subject of intensive investigation in the last years. Results from different studies suggest that mycorrhizal symbioses act on host plant aquaporins and alter both plant water relations and plant physiology in order to cope better with stressful environmental conditions such as drought. The fungal aquaporins have been related to water transport in the fungal mycelium and in the internal exchange membranes at the symbiotic interface. Indeed, it is generally observed that mycorrhizal plants exhibit higher osmotic and hydrostatic root hydraulic conductance under drought stress conditions. Moreover, mycorrhizal plants also grow to a greater extent than non-mycorrhizal plants under drought conditions, indicating that the changes induced by the symbiosis on plant aquaporins contribute to enhance the plant tolerance to drought. These effects are likely to be the result of the combined action of different aquaporins regulated by the mycorrhizal symbiosis (including PIPs, TIPs, NIPs and SIPs), influencing the transport of water and, most probably, also of other solutes of physiological importance for the plant under drought stress conditions.

1 Introduction

The term mycorrhiza applies to a mutualistic symbiosis between roots of most higher plants and a group of soil fungi belonging to the phyla Glomeromycota, Basidiomycota or Ascomycota (Varma 2008). There are several types of mycorrhizal symbiosis, according to the plant and fungus involved and the morphological

J.M. Ruiz-Lozano (✉) • R. Aroca
Departamento de Microbiología del Suelo y Sistemas Simbióticos, Estación Experimental del Zaidín (CSIC), Profesor Albareda n° 1, 18008 Granada, Spain
e-mail: juanmanuel.ruiz@eez.csic.es

F. Chaumont, S.D. Tyerman (eds.), *Plant Aquaporins*, Signaling and Communication in Plants, DOI 10.1007/978-3-319-49395-4_15

characteristics of the structures formed during the association. The most abundant are the arbuscular mycorrhizal (AM) symbiosis, which are formed by almost 80 % of terrestrial plants, including many important agricultural species. The fungi involved develop specialized structures called arbuscules inside the root cells, where there is an exchange of nutrients between both symbionts (Genre et al. 2005). Another type of abundant mycorrhiza is the ectomycorrhizal symbiosis, involving an important number of tree species (Wang and Ding 2013). In ectomycorrhizae, the fungi form a Hartig net, where resources are exchanged with the host plant root (Agerer 2001). By the mycorrhizal association, the plant receives soil nutrients (especially phosphorus) and water, while the fungus receives a protected ecological niche and plant-derived carbon compounds for its nutrition (Varma 2008).

During the establishment of a mycorrhizal symbioses (especially during the AM symbiosis), plant root cells undergo extensive morphological alterations in order to accommodate the presence of an endophytic symbiont, with most of these changes concerning vacuolar or cytoplasmic membrane systems. In fact, during the formation of the AM symbiosis, the plant plasma membrane extends to form a novel periarbuscular membrane, which closely surrounds the fungal hyphae resulting in an estimated three- to tenfold increase in the outer plant cell surface area (Genre et al. 2005, 2008). In addition, both AM and ectomycorrhizal symbioses have been shown to alter root hydraulic properties (Muhsin and Zwiazek 2002b; Khalvati et al. 2005; Lehto and Zwiazek 2011; Bárzana et al. 2012). Thus, Krajinski et al. (2000) already hypothesized a variation of expression affecting genes that encode membrane-associated proteins, and it is not surprising that the establishment of a mycorrhiza can change plant aquaporin gene expression and protein accumulation. The implications that such regulation has on plant water relations, plant physiology and plant performance under optimal or stressful conditions have been the subject of intensive research.

2 Arbuscular Mycorrhizal Symbiosis and Plant Aquaporins

Research on regulation of host plant aquaporins by AM symbiosis is relatively recent. The first report on the modulation of aquaporin genes by the AM symbiosis was provided by Roussel et al. (1997) followed by Krajinski et al. (2000), who found mycorrhiza-induced expression of TIP (tonoplast intrinsic protein) aquaporins in parsley and *Medicago truncatula*, respectively. Downregulation of host plant aquaporins was described by Ouziad et al. (2006), who showed a decrease in the expression of PIP (plasma membrane intrinsic protein) and TIP aquaporins by mycorrhizal colonization in tomato plants. Uehlein et al. (2007) found PIP and NIP (nodulin26-like intrinsic protein) aquaporin genes from *Medicago truncatula* that were also induced by mycorrhization. These authors related the mycorrhiza-induced change in expression of the two genes with physiological changes in the plant roots, i.e. the symbiotic exchange processes located at the periarbuscular membrane (Uehlein et al. 2007). In any case, in the studies mentioned above, plants were cultivated under optimal conditions, and the question of host plant aquaporin regulation under water deficit conditions remained unresolved. Thus, Porcel et al. (2006) cloned genes encoding PIPs from soybean and lettuce, and their expression pattern was studied in AM and non-AM plants cultivated under well-watered or drought stress conditions. The starting hypothesis in this study was that if AM fungi can transfer water to the root of the host plants, the plant should increase its permeability for water and aquaporin genes should be upregulated in order to allow a higher rate of transcellular water flow. However, the results obtained showed that the PIP genes studied were downregulated in both plant species under drought stress and that such downregulation was even more severe in plants colonized by *G. mosseae* than in non-AM plants (Porcel et al. 2006). The expression of *GmPIP2* gene from soybean was analysed in a time course, and it was found that in AM plants the downregulation of *GmPIP2* occurred before than in non-AM plants. It was proposed that such an effect of the AM symbiosis advancing the downregulation of *GmPIP2* gene may have physiological importance to cope with drought stress. In fact, according to Aharon et al. (2003) and Jang et al. (2007), the overexpression of PIP aquaporins in transgenic tobacco and *Arabidopsis* improves plant vigour under favourable growth conditions, but the overexpression of such PIP genes had a negative effect during drought stress, causing fast wilting. Hence, the decreased expression of plasma membrane aquaporin genes during drought stress in roots of AM plants can be a regulatory mechanism to limit water loss from cells (Barrieu et al. 1999; Porcel et al. 2006).

The expression of four PIP aquaporin genes in roots from *Phaseolus vulgaris* was analysed in mycorrhizal and non-mycorrhizal plants subjected to drought, cold or salinity in order to illustrate the complexity of the response of aquaporin genes to AM fungi (Aroca et al. 2007). Three of these PIP genes showed differential regulation by AM symbiosis under the specific conditions of each stress applied. In fact, *PvPIP1;1* expression was slightly inhibited by *G. intraradices* inoculation under drought stress conditions, while non-mycorrhizal plants did not change its expression pattern. Cold stress inhibited *PvPIP1;1* expression similarly in AM and non-AM plants. Finally, salinity raised expression of *PvPIP1;1* in both groups of plants, but the enhancement was considerably higher in AM plants. The expression of *PvPIP1;2* was inhibited by the three stresses in the same way in AM and non-AM plants. In contrast, *PvPIP1;3* expression showed important differences in AM and non-AM plants according to the stress imposed. This gene was clearly induced in non-AM plants under drought stress but inhibited in AM plants. Under salinity the expression of this gene was also induced in both groups of plants, especially in AM. Under cold stress the behaviour was the opposite since it was inhibited in non-AM plants and induced in AM. The expression of *PvPIP2;1* was induced in non-AM plants under drought stress but was downregulated in AM plants. The response of *PvPIP2;1* expression to cold stress was not significant for any of the two plant groups and, again, the gene was considerably upregulated under salinity, especially in AM plants. Thus, the expression of each PIP gene analysed responded differently to each stress, and this response also depended on the AM fungal presence. These results point to the possibility that each PIP gene analysed could have a different function and regulation by AM symbiosis under the specific conditions of each stress studied (Aroca et al. 2007).

When root hydraulic conductivity (*Lp*) was measured in *P. vulgaris* plants, it was found that the regulation of root hydraulic properties by AM symbiosis was strongly correlated with the amount of PIP2 protein and its phosphorylation state, resulting in enhanced *Lp* values under drought, cold and salinity stresses in AM plants (Aroca et al. 2007).

Giovannetti et al. (2012) focused on two putative aquaporin genes, *LjNIP1* and *LjXIP1*, which were found to be upregulated in a transcriptomic analysis performed on roots of *Lotus japonicus* colonized by the AM fungus *Gigaspora margarita*. Using a laser microdissection approach, they demonstrated that *LjNIP1* was specifically expressed in arbuscule-containing cells, whereas *LjXIP1* transcripts were present in both non-colonized cortical cells from mycorrhizal roots and in cortical cells from non-mycorrhizal roots. The potential role of *LjXIP1* remains to be elucidated. In contrast, functional experiments with yeast protoplasts demonstrated that *LjNIP1* can transport water. On the basis of these functional results for *LjNIP1*, of its localization in the inner membrane system of arbusculated cells and of expression timing, it was proposed that LjNIP1 protein was potentially involved, directly or indirectly, in cell turgor regulation, in facilitating colonized cell adaptation to osmotic stresses and/or in the actual transfer of water from the fungus to the plant (Giovannetti et al. 2012).

Recently, Liu et al. (2014) conducted an experiment in which AM and non-AM rice plants were subjected to different temperature and exogenous trehalose treatments. Trehalose was used since it has been shown to act as important abiotic stress protectant and as a signalling molecule. Thus, authors of this study wanted to elucidate if trehalose might stimulate the expression of fungal and plant aquaporin genes and if trehalose might also stimulate AM fungal and rice root water uptake. The results showed that AM fungal inoculation enhanced rice root water uptake at both normal and low temperatures. At low temperature, AM rice plants showed higher expression levels for several plant PIPs than non-AM rice plants. Application of exogenous trehalose demonstrated that trehalose could regulate AM fungal and rice water absorption by inducing the expression of several *OsPIPs* and a fungal aquaporin gene. It was concluded that one of the mechanisms by which AM fungi improve plant resistance to low temperature was a fungal-enhanced trehalose accumulation in rice, which could act as a signal inducing fungal and host plant aquaporins expression that then maintained better water relations in mycorrhizal plants at low temperatures (Liu et al. 2014).

2.1 Arbuscular Mycorrhizal Fungal Aquaporins

The AM fungi also have aquaporin genes. Aroca et al. (2009) cloned the first aquaporin from an AM fungus (*GintAQP1*). Although the functionality of this aquaporin could not be demonstrated, authors found evidence supporting the idea that fungal aquaporins could compensate the downregulation of host plant aquaporins caused by drought. Also, they found that *GintAQP1* expression was upregulated in parts of the mycelium that were not osmotically stressed by NaCl, while other parts of the mycelium were stressed. This suggests possible communication between unstressed and stressed mycelium. More recently, Li et al. (2013) have described two functional aquaporins (*GintAQPF1* and *GintAQPF2*) from the same AM fungus (*Rhizophagus intraradices*). GintAQPF1 was localized in the plasma membrane, whereas GintAQPF2 was localized both in plasma and intracellular membranes. Both aquaporins could transport water, as shown by heterologous expression in yeast protoplasts, and the expression of the two genes in arbuscule-enriched cortical cells and extraradical mycelia of maize roots was enhanced significantly under drought stress. Thus, the two AM fungal aquaporins were related to water transport in the extraradical mycelium and in the periarbuscular membrane (Li et al. 2013). In any case, future research is needed to understand the role of AM fungal aquaporins for the fungus or for the symbiosis, under optimal and water deficit conditions.

2.2 *Regulation of Aquaporins by the AM Symbiosis Under Drought Stress Conditions and Influence on the Transport of Water and Other Solutes of Physiological Importance*

Although many aquaporins are highly selective for water, the selectivity filters of plant aquaporins show a high divergence (Sui et al. 2001), suggesting wide functional diversity for these proteins (Bansal and Sankararamakrishnan 2007) (See chapter "Structural Basis of the Permeation Function of Plant Aquaporins"). Thus, it has become increasingly clear that some aquaporins do not exhibit a strict specificity for water and can transport other small neutral molecules such as urea (Liu et al. 2003), ammonia (Loque et al. 2005), carbon dioxide (CO_2) (Uehlein et al. 2003), boric acid (Mitani et al. 2008), hydrogen peroxide (H_2O_2) (Bienert et al. 2007), silicic acid (Ma and Yamaji 2006) and some other molecules with physiological significance (Bienert et al. 2008), highlighting the paramount relevance of aquaporins for plant physiology.

The function and regulation of aquaporins have been intensively integrated to explain the remarkable hydraulic properties of plants. However, the identification of aquaporin substrates other than water, as mentioned above, has opened the possibilities for the involvement of aquaporins in many other processes of physiological significance for plants (Chaumont and Tyerman 2014; Li et al. 2014). In the mycorrhizal symbiosis, the importance of aquaporins for both water and nutrient exchange was recognized by Maurel and Plassard (2011) and supported recently by results obtained with AM maize plants (Bárzana et al. 2014), elaborated upon below.

The first report of involvement of a host plant aquaporin in the functioning of AM symbiosis under drought stress by a mechanism unrelated to water transport was provided by Porcel et al. (2005). They studied the effects of reduced expression of the PIP aquaporin-encoding gene NtAQP1 in mycorrhizal NtAQP1-antisense tobacco plants under well-watered and drought stress conditions. The study aimed at elucidating whether or not the impairment in NtAQP1 gene expression affected the AM fungal colonization pattern, as well as to find out if such impairment had any effect on the symbiotic efficiency of two AM fungi. The reduction of NtAQP1 expression had no effect on root colonization, suggesting that either NtAQP1 function is irrelevant for the process of root colonization or that the impairment in NtAQP1 gene expression had been compensated for by changes in the abundance or the activity of other aquaporin isoforms. In contrast, when Porcel et al. (2005) measured the symbiotic efficiency of the two AM fungi (in terms of plant biomass production), they observed that under drought stress, mycorrhizal wild-type plants grew faster than mycorrhizal NtAQP1 antisense plants. This indicates that the symbiotic efficiency of both AM fungi was greater in wild-type than in antisense plants and that the transport meditated by NtAQP1 seems to be important for the efficiency of the symbiosis under drought stress conditions (Porcel et al. 2005). This was related to the fact that NtAQP1 allows CO_2 passage and is involved in plant growth promotion (Uehlein et al. 2003).

Recently, Bárzana et al. (2014) conducted an investigation aimed at elucidating in which way the AM symbiosis modulates the expression of the whole set of aquaporin genes present in maize, both under optimal water conditions and under different drought stress scenarios. An additional objective was to characterize those aquaporins showing regulation by the AM symbiosis, in order to shed further light on the molecules that could be involved in the mycorrhizal responses to drought. The AM symbiosis regulated the expression of a wide number of aquaporin genes in the host plant (16 genes out of 36 existing in maize), comprising members of the different aquaporin subfamilies (Bárzana et al. 2014). Several of these AM-regulated aquaporins (ZmPIP1;3, ZmPIP2;2, ZmTIP1;1, ZmTIP1;2, ZmNIP1;1, ZmNIP2;1 and ZmNIP2;2) were functionally characterized in heterologous expression systems with *Xenopus laevis* oocytes and by yeast complementation. It was shown that they can transport water, but also other molecules of physiological importance for plant performance under both normal and stress conditions (glycerol, urea, ammonia, boric acid, silicon or hydrogen peroxide). The regulation of these genes depended on the watering conditions and on the severity of the drought stress imposed. The different aquaporin regulation patterns suggest that under short-term drought conditions, the AM symbiosis may stimulate further the physiological processes in which these aquaporins are involved, but when the drought becomes sustained, the AM symbiosis restricts most of the processes in which these aquaporins participate (Fig. 1).

AM maize plants maintained higher values of root water flux than non-AM plants. These effects have been related to the increased absorbing surface caused by fungal hyphae growing in the soil, combined with the ability of the fungus to take up water from soil pores inaccessible to roots (Marulanda et al. 2003; Allen 2009; Ruth et al. 2011). Thus, under such conditions, mycelial water uptake from soil pores inaccessible to roots and its transference from AM fungal hyphae to plant cells can be critical to improve water supply to the plant, potentially increasing flow via cell-to-cell and apoplastic pathways (Bárzana et al. 2012). In this sense, all the maize PIP2s regulated by the AM symbiosis showed capacity to transport water, especially ZmPIP2;2, which had a particularly high intrinsic water transport capacity. It has also been proposed that TIPs may provide a quick way for cellular osmotic balance by controlling the exchange of water between vacuole and cytosol (Forrest and Bhave 2007), playing an important role under osmotic stress conditions (Katsuhara et al. 2008). Thus, since TIPs regulate the exchange of water between vacuole and cytosol, they may also have an influence on root water flux by affecting exchange of water between transcellular and symplastic water pathways. In the study with maize, *ZmTIP1;1* and *ZmTIP1;2* were highly expressed in all treatments and in the oocyte system both exhibited a high capacity for water transport. Therefore, regulation of PIP and TIP aquaporins was proposed to be a key factor in regulation of plant water transport by AM symbiosis (Fig. 1, blue arrows).

Some maize aquaporins, including ZmNIP1;1, ZmTIP4;1 and ZmTIP4;2, were shown to have the capacity to transport glycerol. The physiological function of glycerol in plants remains unclear, while its utilization is well known in fungi and bacteria (Dietz et al. 2011). However, a study has shown a transfer of glycerol from host plants to pathogenic fungi (Wei et al. 2004), and Gustavsson et al. (2005) suggested that export of plant-derived glycerol may be also important for symbiotic fungi. Thus, ZmNIP1;1, ZmTIP4;1 and ZmTIP4;2 may be important for the AM symbiosis or for the plant-fungus interaction under sustained drought stress (Fig. 1, green arrows). This would explain why these aquaporins were so finely regulated by the AM symbiosis (Bárzana et al. 2014).

Nitrogen is one of the most important nutrients for all living organisms, being needed for the synthesis of compounds essential for growth. The ammonium ion (NH_4^+) and its conjugated base ammonia (NH_3) are the potential primary sources of N (besides NO_3^-) in plant nitrogen nutrition. Transport of urea and NH_3/NH_4^+ into the vacuole would avoid their toxicity in the cytoplasm and/or allow storage of N (Wang et al. 2008), and whenever required as an N-source, these compounds could be remobilized by a passive, low-affinity transport pathway, such as that provided by TIPs (Liu et al. 2003). In maize several TIPs have been shown to transport these compounds (Liu et al. 2003; Loque et al. 2005), including ZmTIP1;1 and ZmTIP1;2, which were regulated by the AM symbiosis (Bárzana et al. 2014). In the AM symbiosis, ammonium is suggested to be the major nitrogen compound transferred to the host plant, with urea playing a role as an intermediate solute (Govindarajulu et al. 2005; Tian et al. 2010; Perez-Tienda et al. 2011). This suggests that these aquaporins could be involved in the fungus-based nitrogen nutrition of the host plants (Fig. 1, orange arrows), as was also proposed for ectomycorrhizal fungi (Dietz et al. 2011).

Boron and silicon are metalloids with key structural functions in plant cells (see chapter "Plant Aquaporins and Metalloids"): boron cross-links the pectin fraction of cell walls and polymers of hydrated silica-gel are important for the physical strength of plant cells, especially in monocots like maize (Miwa et al. 2009). Most of the aquaporins regulated by the AM symbiosis have the capacity to transport boron, and

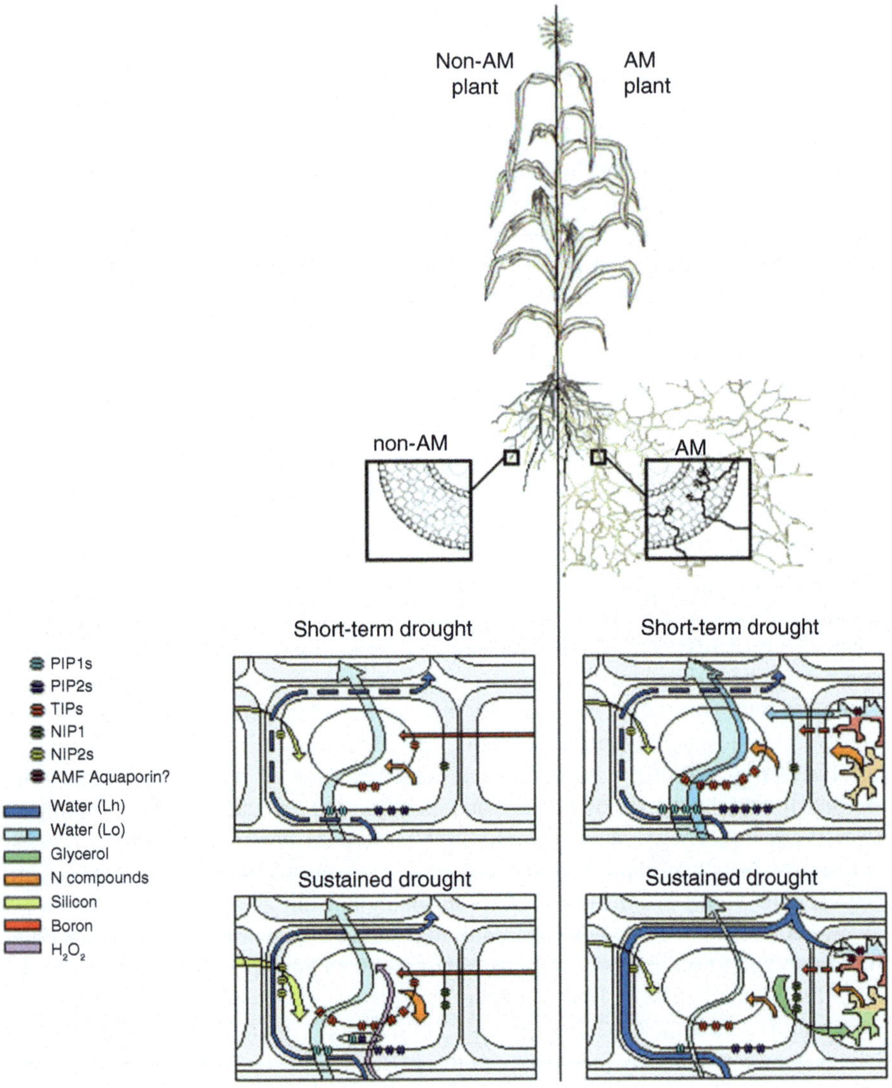

ZmNIP1;1 and ZmNIP2;2 could transport silicon (Bárzana et al. 2014). Thus, the regulation of these aquaporins by the AM symbiosis could have structural functions in maize plants under abiotic stress conditions, including drought (Fig. 1, red and yellow arrows).

Hydrogen peroxide (H_2O_2) is one of the most abundant reactive oxygen species (ROS) continuously produced in the metabolism of aerobic organisms. At low concentrations, it acts as a signal molecule controlling different essential processes in plants during normal growth and development, but it also functions as a defensive signal molecule against various abiotic and biotic stresses (Miller et al.

Fig. 1 Hypothetical model presenting the role of maize aquaporins in the regulation by the AM symbiosis of plant physiology and performance under different growing conditions. Under short-term drought, the expression of most PIPs in AM plants was kept high or even increased. Also, AM fungal hyphae increased the absorbing surface in soil. This, combined with the fungal ability to take up water from soil pores inaccessible to roots, allowed AM plants to maintain higher *Lp* values than non-AM plants (as shown in *blue arrows*). AM fungal aquaporins could be involved in the release of water from hyphae to plant. TIPs may also have an influence on *Lp* by affecting exchanges of water between transcellular and symplastic water pathways. The aquaporins regulated by the AM symbiosis can transport a variety of compounds of physiological importance for the plant, including glycerol (as shown in *green arrows*), that may be important for the plant-fungus interaction under sustained drought stress conditions. Moreover, nitrogen compounds (shown in *orange arrows*) provided by the root and the AMF may be translocated through ZmTIP1;1 and ZmTIP1;2 into the vacuole and stored. Under sustained drought, a remobilization of N stored in the vacuole may be necessary in non-AM plants, which upregulated ZmTIP1;1 and ZmTIP1;2. The B requirements (shown in *red arrows*) in non-AM plants may be guaranteed by the aquaporins that can transport B. In AM plants, the AM fungus may provide directly B to the host plant (shown in *dashed red arrows*), and plant aquaporins involved in B transport are downregulated in order to avoid toxicity due to an excess of B. In plants, Si can mechanically impede penetration by fungi and, thereby, a diminution of Si uptake (shown in *yellow arrows*) in mycorrhizal plants can be expected. Thus, mycorrhization reduced the expression of ZmNIP2;1 and ZmNIP2;2 (transporting Si). In non-AM plants, ZmNIP2;2 increased its expression under sustained drought conditions, probably to avoid plant lodging. TIP1s (for instance, ZmTIP1;1) could play a key role in the detoxification of excess H_2O_2 generated under stress conditions. Additionally, the mobilization of H_2O_2 via aquaporins could serve as a regulatory mechanism for membrane internalization of plant PIPs (as shown in non-AM roots under sustained drought), with subsequent decrease of water transport

2010). Several aquaporins regulated by the AM symbiosis could transport H_2O_2, especially ZmTIP1;1 (Bárzana et al. 2014). Bienert et al. (2007) proposed that TIP1s could play a key role as an additional mechanism for the detoxification of excess H_2O_2 generated under stress conditions (Fig. 1, purple arrows). This idea fits with the high gene expression and protein content maintained for ZmTIP1;1 under drought stress conditions. Additionally, the transport and mobilization of H_2O_2 via aquaporins could serve as a regulatory mechanism for membrane internalization of plant PIPs (Boursiac et al. 2008), with subsequent effects on water transport in AM plants.

3 Ectomycorrhizal Symbiosis and Aquaporins

Most land trees and shrubs establish a symbiosis with ectomycorrhizal (EM) fungi. In the EM symbiosis, the fungal hyphae do not penetrate living cells of the root, but instead they form a mantle around the roots and a net between epidermal and cortical cells called the Hartig net (Barea et al. 2011). Similar to AM fungi, EM fungi also receive carbon compounds from the host plant and they provide to the host plant mineral nutrients and water (Guehl and Garbaye 1990; García et al. 2011). However, EM fungi are able to grow without the presence of host roots, since they are capable of obtaining carbon compounds from external sources due to hydrolytic enzyme activities (Tedersoo et al. 2012; Moore et al. 2015).

As for AM fungi, EM fungi also increase abiotic stress tolerance of the host plant, including higher tolerance to drought and salt stress (Yi et al. 2008; Kipfer et al. 2012), both of which can cause dehydration of plant tissues (Aroca et al. 2012). In this part of the chapter, it will be first described how EM symbiosis modifies plant water relations and more precisely root water uptake. Later, the effects of EM symbiosis on plant aquaporin regulation will be summarized, and finally the regulation of the EM fungal aquaporins under different environmental conditions will be mentioned. Moreover, the regulation of aquaporins and water relations in the ectendomycorrhizal (EDM) symbiosis will be briefly described. In this symbiosis, the fungus forms a mantle and a Hartig net but also forms intracellular structures, although the cell walls of both partners remain unaltered (Dexheimer and Pargney 1991).

3.1 Water Relations in EM Plants

EM symbioses enhance drought and salinity tolerance of the host plants (Yi et al. 2008; Kipfer et al. 2012), and also regulate differentially leaf transpiration rate (E) and root water uptake, changing leaf water status. Under non-stressful conditions, Xu et al. (2015) found that white spruce (*Picea glauca*) trees inoculated with the EM fungus *Laccaria bicolor* had higher E and leaf water potential than non-inoculated trees. The inoculated plants also presented higher root water uptake capacity in terms of root hydraulic conductance normalized on a root volume basis. However, Calvo-Polanco and Zwiazek (2011) did not find any differences in E or root hydraulic conductance between trees (jack pine or white spruce) inoculated or not with the EM fungus *Suillus tomentosus* under optimal growth conditions. So, the response of E and root hydraulic conductance to EM symbiosis may be dependent on the combination of tree-fungus species. In fact Muhsin and Zwiazek (2002a) found previously the same results as Xu et al. (2015) in white spruce trees but using *Hebeloma crustuliniforme* as the EM fungus. Similarly, the fungus *L. bicolor* only increased root hydraulic conductance and E in white spruce trees, but not in *Ulmus americana*, *Populus tremuloides* or *Betula papyrifera* trees (Table 1). From Table 1, it can be inferred that under optimal growth conditions, EM symbioses rarely modify E (2 cases out of 11), although E was never decreased. On the other hand, increased root hydraulic conductance is not always matched with increased E. Unfortunately, in most cases, leaf water status was not determined, so it is difficult to correlate changes in root hydraulic conductance and E by EM symbiosis with changes in leaf water status.

Under osmotic stress (drought or salinity), the behaviour of root hydraulic conductance and E in EM plants is also species dependent. Thus, the EM fungus *H. crustuliniforme* increased root hydraulic conductance in white spruce and aspen (*P. tremuloides*) trees, but not in *B. papyrifera* trees under salt stress (Table 2). Most

Table 1 Effects of EM symbiosis on leaf transpiration rate (E), root hydraulic conductivity (*Lp*), plant growth and leaf water status (LWS) of trees growing under optimal growth conditions

Plant species	Fungal species	E	Lp	Growth	LWS	Source
Picea glauca	*Laccaria bicolor*	↑	↑	→	↑	Xu et al. (2015)
P. glauca	*Suillus tomentosus*	→	→	→	?	Calvo-Polanco and Zwiazek (2011)
Pinus banksiana	*S. tomentosus*	→	→	↑	?	Calvo-Polanco and Zwiazek (2011)
Ulmus americana	*Hebeloma crustuliniforme*	→	↑	→	?	Calvo-Polanco et al. (2009)
U. americana	*L. bicolor*	→	→	→	?	Calvo-Polanco et al. (2009)
Populus balsamífera	*H. crustuliniforme*	→	↑	→	→	Siemens and Zwiazek (2008)
P. tremuloides	*L. bicolor*	→	→	↑	?	Yi et al. (2008)
P. tremuloides	*H. crustuliniforme*	→	→	→	?	Yi et al. (2008)
Betula papyrifera	*L. bicolor*	→	→	→	?	Yi et al. (2008)
B. papyrifera	*H. crustuliniforme*	→	→	→	?	Yi et al. (2008)
P. glauca	*H. crustuliniforme*	↑	↑	↑	?	Muhsin and Zwiazek (2002a)

The direction of the arrows indicates the direction of the change. ↑, increase; ↓, decrease; →, no change; ?, non-recorded

interesting is that the same EM fungus *H. crustuliniforme* had no effect on root hydraulic conductance in aspen trees under optimal growth conditions, but increased it under salt stress (Yi et al. 2008; Tables 1 and 2). Again, there is no relationship between changes in root hydraulic conductance and E caused by EM symbiosis under osmotic stress (Table 2). Also, the growth promotion caused by EM symbiosis under osmotic stress was not always related to changes in root hydraulic conductance or E (Table 2), indicating that other mechanisms are behind the better performance of EM plants under osmotic stress, such as better antioxidative mechanisms (Alvarez et al. 2009), synthesis of specific proteins linked to EM symbiosis (Kraj and Grad 2013) or better nutritional status (Danielsen and Polle 2014).

3.2 Root Water Uptake in EM Plants

Although root hydraulic conductance is one factor that determines root water uptake (U), root hydraulic conductance and U do not always correlate (Aroca et al. 2001; Doussan et al. 2006), potentially because root hydraulic conductance is typically determined in detached root systems (Li and Liu 2010) and the influence of leaf transpiration is lost. Vandeleur et al. (2014) found that shoot topping reduced root hydraulic conductance from 30 to 60 % depending on the plant species. Moreover, root hydraulic conductance is frequently measured in roots removed from their surrounded soil, and the soil hydraulic conductance factor is missing as well as the

Table 2 Effects of EM symbiosis on leaf transpiration rate (E), root hydraulic conductivity (*Lp*), plant growth and leaf water status (LWS) of trees growing under stressful conditions

Plant species	Fungal species	Stress	E	Lp	Growth	LWS	Source
Picea glauca	*S. tomentosus*	10 % Polyethylene glycol	→	→	→	?	Calvo-Polanco and Zwiazek (2011)
P. glauca	*S. tomentosus*	60 mM NaCl	→	→	→	?	Calvo-Polanco and Zwiazek (2011)
Pinus banksiana	*S. tomentosus*	10 % Polyethylene glycol	↓	→	→	?	Calvo-Polanco and Zwiazek (2011)
P. banksiana	*S. tomentosus*	60 mM NaCl	→	→	↑	?	Calvo-Polanco and Zwiazek (2011)
Populus x canescens	*Paxillus involutus*	Stopping watering	→	↑	↑	↑	Beniwal et al. (2010)
Populus tremuloides	*Laccaria bicolor*	25 mM NaCl	→	→	↑	?	Yi et al. (2008)
P. tremuloides	*Hebeloma crustuliniforme*	25 mM NaCl	→	↑	→	?	Yi et al. (2008)
Betula papyrifera	*L. bicolor*	25 mM NaCl	→	→	→	?	Yi et al. (2008)
B. papyrifera	*H. crustuliniforme*	25 mM NaCl	→	→	→	?	Yi et al. (2008)
P. glauca	*H. crustuliniforme*	25 mM NaCl	↑	↑	↑	?	Muhsin and Zwiazek (2002a)

The direction of the arrows indicates the direction of the change. ↑, increase; ↓, decrease; →, no change; ?, non-recorded

potential hydraulic lift mechanism (the water that is absorbed by deeper roots and released into upper soil layers) (Doussan et al. 2006). Therefore, besides root hydraulic conductance, U determinations are essential to really ascertain if EM symbiosis actually enhances capacity of host roots to take up water. Bogeat-Triboulot et al. (2004) found an increase of U in *Pinus pinaster* trees inoculated with the EM fungus *Hebeloma cylindrosporum*. At the same time, EM trees had also higher root hydraulic conductance normalized on a root area basis and had an elevated amount of soil adhering to the roots, perhaps facilitating root water absorption. In this case, the root hydraulic conductance was determined by the high pressure flow meter (HPFM) technique, in which the roots remain in the soil and not disturbed. However, in a previous study, Colpaert and Van Assche (1993) found a negative effect on U by EM inoculation in *Pinus sylvestris* trees, but accompanied by a reduction in plant growth, which indirectly may decrease U, because aerial parts demanded less water. Measuring U in intact trees is necessary to understand the effects of EM on plant water relations as a whole.

The transport of water from root epidermal cells to root xylem vessels can follow three paths, namely, apoplastic, symplastic and transcellular (see chapter "Aquaporins and Root Water Uptake"). Since symplastic and transcellular paths cannot be discriminated empirically, the sum of both is called the cell-to-cell path. The apoplastic path corresponds to water circulating through cell walls, and cell-to-cell path corresponds to water flowing by the plasmodesmata and crossing cell membranes (plasma membrane and/or tonoplast) (Steudle and Peterson 1998). Since EM fungal mycelia penetrate root apoplastic spaces, EM symbiosis may modify the proportion of water flowing through each root pathway.

In earlier studies, Behrmann and Heyser (1992) found no evidence supporting that EM symbiosis enhanced the transport capacity of the apoplastic path using different apoplastic tracers. However their results could be limited to the EM association studied (*Pinus sylvestris-Suillus bovinus* association). In contrast, Muhsin and Zwiazek (2002b) found evidence supporting that EM symbiosis enhanced root hydraulic conductance by increasing the amount of water flowing via the apoplastic path. This conclusion was based on studies of the inhibition of root hydraulic conductance by $HgCl_2$, since root hydraulic conductance of EM roots presented less inhibition than that of non-mycorrhizal plants. Anyway, these results could change depending on the pH of the soil solution, since the differences in the proportion of apoplastic water flow between EM and non-EM plants depend on the root medium pH, being more pronounced at extreme pH (Siemens and Zwiazek 2011). Vesk et al. (2000) found that EM sheaths could be very impermeable to water, so under these circumstances the cell-to-cell pathway should predominate, since apoplastic access to water will be limited by fungal structures. In this sense, Lee et al. (2010) and Xu et al. (2015) found that hydraulic conductivity increased for root cortical cells of EM trees compared to non-EM trees, indicating that the cell-to-cell path could be enhanced by the EM symbiosis. All the above results should be taken with caution, since most probably it depends on the species involved in the EM symbiosis.

3.3 *Aquaporin Regulation by EM Symbioses*

Since root hydraulic conductance is governed in part by aquaporins, and more precisely the cell-to-cell path (Maurel et al. 2008), EM symbioses may regulate aquaporin expression and activity of the host trees. The activity of aquaporins has been estimated by using different chemicals that inhibit aquaporins, which could also affect fungal aquaporin activity. As mentioned above, Muhsin and Zwiazek (2002b) found that aquaporin activity was downregulated by the symbiosis between *Ulmus americana* and *Hebeloma crustuliniforme* partners using $HgCl_2$ as an inhibitor of aquaporin activity, indicating that flow via the apoplastic path may have increased. However, using the same aquaporin inhibitor, no enhancement of aquaporin activity in the symbiosis between *Pinus banksiana* and *Suillus tomentosus* partners was observed (Lee et al. 2010). However the use of $HgCl_2$ could have side effects that can interfere with the root hydraulic conductance measurements. Aquaporin activity is enhanced by phosphorylation of different serine residues (Johansson et al. 1998; Azad et al. 2008) (see also chapter "Plant Aquaporin Posttranslational Regulation"). However, the phosphorylation state of aquaporins in EM plants has not been assayed yet, although it has been determined in AM plants (Aroca et al. 2007; Calvo-Polanco et al. 2014; Bárzana et al. 2015).

The influence of EM symbioses on the expression of specific aquaporins of the host plant has been studied. Downregulation of the expression of two PIP genes out of eight by the symbiosis of the EM fungus *Laccaria bicolor* in *Picea glauca* roots has been reported (Xu et al. 2015). However, in this study EM symbiosis enhanced root hydraulic conductance (at both whole root and cell levels), so it is possible that such enhanced root hydraulic conductance could be mediated by aquaporins of the EM fungus or by an increase in water flow through the apoplastic path. In addition, post-translational modification could take place (see chapter "Plant Aquaporin Posttranslational Regulation") and enhance aquaporin activity in EM roots. Two other studies reported an upregulation in the expression of several PIP genes in EM roots. Tarkka et al. (2013) found an increase in the expression of five PIPs and one SIP (small and basic intrinsic protein) aquaporin genes in *Quercus robur* roots inoculated with the EM fungus *Piloderma croceum* using RNA-seq technique, but no measurement of root hydraulic conductance was performed. Marjanovic et al. (2005) found also an increase in the expression of one PIP (*PttPIP2;5*) gene in poplar (*Populus tremula* x *tremuloides*) roots inoculated with the EM fungus *Amanita muscaria*. Interestingly, this PIP gene had high capacity of transporting water when expressing in *Xenopus laevis* oocytes, and their expression correlated with higher root hydraulic conductance in EM trees. Obviously, more work is needed in order to establish the function of plant aquaporins in the regulation of root hydraulic conductance by EM symbiosis. These studies should include the use of plants with altered levels in the expression of specific aquaporins.

As commented above, EM fungi have their own aquaporins as do AM fungi. Dietz et al. (2011) found seven aquaporin genes in the *L. bicolor* genome, three of them having a high water and ammonia transport capacity when expressed in *Xenopus laevis* oocytes. Most recently, Nehls and Dietz (2014), analysing the genome of 480 fungal species, found around 50 putative orthodox aquaporin genes accounting for several EM fungal species. These findings point to the potential role of EM fungal aquaporins in the water relations of host plants. Hence, when white spruce trees were inoculated with *L. bicolor* strains overexpressing JQ585595 *L. bicolor* aquaporin, they showed enhanced root hydraulic conductance under optimal growth temperatures, but lower under suboptimal temperatures (Xu et al. 2015). Moreover, it has been found that one *L. bicolor* aquaporin (*LbAQP1*) is essential in the formation of the Hartig net in trembling aspen trees (Navarro-Ródenas et al. 2015) and this was related to the capacity of this aquaporin to transport NO, H_2O_2 and CO_2 that could act as signalling molecules. Obviously, more studies are necessary in other EM fungal species to understand the possible role of fungal aquaporins in EM symbiosis.

3.4 Aquaporins and Ectendomycorrhizal (EDM) Symbiosis

Ectendomycorrhizal (EDM) symbiosis also may increase plant host tolerance to environmental stresses, although this symbiosis has been less studied. Morte et al. (2010) found that symbiosis with the fungus *Terfezia claveryi* (desert truffle) was essential for the survival of *Helianthemum almeriense* plants under arid conditions. In a previous greenhouse study, *T. claveryi*-inoculated plants showed higher leaf water potential, transpiration rate and net photosynthesis rate than non-inoculated plants (Morte et al. 2000). Similar results were found by Turgeman et al. (2011) in the symbiosis between *Helianthemum sessiliflorum* and *Terfezia boudieri*.

Regarding root hydraulic properties modified by EDM symbiosis, Siemens and Zwiazek (2008) found no effect of *Wilcoxina mikolae* inoculation on root hydraulic conductance of balsam poplar (*Populus balsamifera*) trees. Unfortunately no other studies examining the effect of EDM symbiosis on root hydraulic properties have been published to date. Nevertheless, Navarro-Ródenas et al. (2013) found that *T. claveryi* inoculation modified the expression of *H. almeriensis* aquaporins, although a defined trend was not observed, since each aquaporin responded differently. As for the other mycorrhizal fungi, EDM fungi also possess their own aquaporins. Navarro-Ródenas et al. (2012) cloned one aquaporin from the EDM fungus *T. claveryi*. This aquaporin was able to transport water and CO_2 when expressed in *Saccharomyces cerevisiae*, and its expression responded to an osmotic stress applied *in vitro*.

4 Conclusions

Based on the reviewed literature, we propose that AM symbioses act on host plant aquaporins in a concerted manner to alter both plant water relations and physiology and allowing the plant to cope better with stressful conditions (Fig. 1). In support of this idea, it is generally observed that AM plants exhibit higher root hydraulic conductance under drought stress conditions. Moreover, AM plants also grow more than non-AM plants under drought conditions, indicating that, apart from the improved P nutrition, all these changes induced by the AM symbiosis on plant aquaporins contributed to the enhanced plant tolerance to drought (Bárzana et al. 2014). These effects are likely the result of the combined action of the different aquaporins regulated by the AM symbioses (including PIPs, TIPs, NIPs and SIPs), influencing the transport of water and, most probably, also of signalling molecules and other solutes of physiological importance for the plant under drought stress conditions. In

EM symbioses the regulation of plant aquaporins seems to be highly dependent on the plant and fungal species involved in the symbiosis. Moreover, the regulation of root hydraulic conductivity has been related also to the activity of the EM fungal aquaporins and with increases in water flow via the apoplastic pathway.

5 Perspectives

There is a broad consensus that it is necessary to analyse the diversity of plant aquaporin isoforms, of their substrates and their cellular localizations in order to understand their physiological functions with respect to whole plant hydraulics, plant development, nutrient acquisition and plant responses to various environmental stresses (Gomes et al. 2009; Li et al. 2014). The main objective of future research should be to identify those aquaporin isoforms regulated by the AM symbiosis having a key influence on the capacity of roots to transport water or the capacity for the transport *in planta* of other solutes. Moreover, the role of fungal aquaporins during formation and functioning of AM symbioses requires future investigation. For that, the localization of these aquaporins in the fungal mycelium is of great interest.

More integrative physiological studies are also needed to understand the role of aquaporins in EM and EDM symbioses. These studies should include measurements of the rate of whole root water uptake normalized to root surface area and plant water status. Also, the phosphorylation state of aquaporins in roots of EM and EDM plants should be determined, as well as the use of plants with altered expression levels of specific aquaporins. Although the study of *L. bicolor* aquaporins has increased our knowledge about the role of fungal aquaporins in the EM symbiosis, the study of aquaporins from other EM (EDM) fungi is necessary.

Acknowledgements This work is part of a MINECO-FEDER project (AGL2014-53126-R) and a project financed by Junta de Andalucía (P11-CVI-7107).

References

Agerer R (2001) Exploration types of ectomycorrhizae – a proposal to classify ectomycorrhizal mycelial systems according to their patterns of differentiation and putative ecological importance. Mycorrhiza 11:107–114

Aharon R, Shahak Y, Wininger S, Bendov R, Kapulnik Y, Galili G (2003) Overexpression of a plasma membrane aquaporins in transgenic tobacco improves plant vigour under favourable growth conditions but not under drought or salt stress. Plant Cell 15:439–447

Allen MF (2009) Bidirectional water flows through the soil-fungal-plant mycorrhizal continuum. New Phytol 182:290–293

Alvarez M, Huygens D, Fernandez C, Gacitúa Y, Olivares E, Saavedra I, Alberdi M, Valenzuela E (2009) Effect of ectomycorrhizal colonization and drought on reactive oxygen species metabolism of *Nothofagus dombeyi* roots. Tree Physiol 29:1047–1057

Aroca R, Tognoni F, Irigoyen JJ, Sánchez-Díaz M, Pardossi A (2001) Different root low temperature response of two maize genotypes differing in chilling sensitivity. Plant Physiol Biochem 39:1067–1073

Aroca R, Porcel R, Ruiz-Lozano JM (2007) How does arbuscular mycorrhizal symbiosis regulate root hydraulic properties and plasma membrane aquaporins in *Phaseolus vulgaris* under drought, cold or salinity stresses? New Phytol 173:808–816

Aroca R, Bago A, Sutka M, Paz JA, Cano C, Amodeo G, Ruiz-Lozano JM (2009) Expression analysis of the first arbuscular mycorrhizal fungi aquaporin described reveals concerted gene expression between salt-stressed and non-stressed mycelium. Mol Plant-Microbe Interact 22:1169–1178

Aroca R, Porcel R, Ruiz-Lozano JM (2012) Regulation of root water uptake under abiotic stress conditions. J Exp Bot 63:43–57

Azad AK, Katsuhara M, Sawa Y, Ishikawa T, Shibata H (2008) Characterization of four plasma membrane aquaporins in tulip petals: a putative homologous regulated by phosphorylation. Plant Cell Physiol 49:1196–1208

Bansal A, Sankararamakrishnan R (2007) Homology modeling of major intrinsic proteins in rice, maize and Arabidopsis: comparative of the selectivity filters. BMC Struct Biol 7:27–44

Barea JM, Palenzuela J, Cornejo P, Sánchez-Castro I, Navarro-Fernández C, López-García A, Estrada B, Azcón R, Ferrol N, Azcón-Aguilar C (2011) Ecological and functional roles of mycorrhizas in semi-arid ecosystems of Southeast Spain. J Arid Environ 75:1292–1301

Barrieu F, Marty-Mazars D, Thomas D, Chaumont F, Charbonnier M, Marty F (1999) Desiccation and osmotic stress increase the abundance of mRNA of the tonoplast aquaporin BobTIP26-1 in cauliflower cells. Planta 209:77–86

Bárzana G, Aroca R, Paz JA, Chaumont F, Martinez-Ballesta MC, Carvajal M, Ruiz-Lozano JM (2012) Arbuscular mycorrhizal symbiosis increases relative apoplastic water flow in roots of the host plant under both well-watered and drought stress conditions. Ann Bot 109: 1009–1017

Bárzana G, Aroca R, Bienert P, Chaumont F, Ruiz-Lozano JM (2014) New insights into the regulation of aquaporins by the arbuscular mycorrhizal symbiosis in maize plants under drought stress and possible implications for plant performance. Mol Plant-Microbe Interact 27:349–363

Bárzana G, Aroca R, Ruiz-Lozano JM (2015) Localized and non-localized effects of arbuscular mycorrhizal symbiosis on accumulation of osmolytes and aquaporins and on antioxidant systems in maize plants subjected to total or partial root drying. Plant Cell Environ 38:1613–1627

Behrmann P, Heyser W (1992) Apoplastic transport through the fungal sheath of *Pinus sylvestris Suillus bovinus* ectomycorrhizae. Bot Acta 105:427–434

Beniwal RS, Langenfeld-Heyser R, Polle A (2010) Ectomycorrhiza and hydrogel protect hybrid poplar from water deficit and unravel plastic responses of xylem anatomy. Environ Exp Bot 69:189–197

Bienert GP, Møller ALB, Kristiansen KA, Schulz A, Møller IM, Schjoerring JK, Jahn TP (2007) Specific aquaporins facilitate the diffusion of hydrogen peroxide across membranes. J Biol Chem 282:1183–1192

Bienert GP, Schüssler MD, Jahn TP (2008) Metalloids essential, beneficial or toxic? Major intrinsic proteins sort it out. Trends Biochem Sci 33:21–26

Bogeat-Triboulot MB, Bartoli F, Garbaye J, Marmeisse R, Tagu D (2004) Fungal ectomycorrhizal community and drought affect root hydraulic properties and soil adherence to roots of *Pinus pinaster* seedlings. Plant Soil 267:213–223

Boursiac Y, Boudet J, Postaire O, Luu D-T, Tournaire-Roux C, Maurel C (2008) Stimulus-induced downregulation of root water transport involves reactive oxygen species-activated cell signalling and plasma membrane intrinsic protein internalizatization. Plant J 56:207–218

Calvo-Polanco M, Zwiazek JJ (2011) Role of osmotic stress in ion accumulation and physiological responses of mycorrhizal white spruce (*Picea glauca*) and jack pine (*Pinus banksiana*) to soil fluoride and NaCl. Acta Physiol Plant 33:1365–1373

Calvo-Polanco M, Jones MD, Zwiazek JJ (2009) Effects of pH on NaCl tolerance of American elm (*Ulmus americana*) seedlings inoculated with *Hebeloma crustuliniforme* and *Laccaria Bicolor*. Acta Physiol Plant 31:515–522

Calvo-Polanco M, Molina S, Zamarreño AM, García-Mina JM, Aroca R (2014) The symbiosis with the arbuscular mycorrhizal fungus *Rhizophagus irregularis* drives water transport in flooded tomato plants. Plant Cell Physiol 55:1017–1029

Chaumont F, Tyerman SD (2014) Aquaporins: highly regulated channels controlling plant water relations. Plant Physiol 164:1600–1618

Colpaert JV, Van Assche JA (1993) The effects of cadmium on ectomycorrhizal *Pinus sylvestris* L. New Phytol 123:325–333

Danielsen L, Polle A (2014) Poplar nutrition under drought as affected by ectomycorrhizal colonization. Environ Exp Bot 108:89–98

Dexheimer J, Pargney JC (1991) Comparative anatomy of the host-fungus interface in mycorrhizas. Experientia 47:312–321

Dietz S, von Bülow J, Beitz E, Nehls U (2011) The aquaporin gene family of the ectomycorrhizal fungus *Laccaria bicolor*: lessons for symbiotic functions. New Phytol 190:927–940

Doussan C, Pierret A, Garrigues E, Pages L (2006) Water uptake by plant roots: II – modelling of water transfer in the soil root-system with explicit account of flow within the root system – comparison with experiments. Plant Soil 283:99–117

Forrest KL, Bhave M (2007) Major intrinsic proteins (MIPs) in plants: a complex gene family with impact on plant phenotype. Funct Integr Genomics 7:263–289

García AN, Arias SPB, Morte A, Sánchez-Blanco MJ (2011) Effects of nursey preconditioning though mycorrhizal inoculation and drought in *Arbutus unedo* L. plants. Mycorrhiza 21: 53–64

Genre A, Chabaud M, Timmers T, Bonfante P, Barker DG (2005) Arbuscular mycorrhizal fungi elicit a novel intracellular apparatus in *Medicago truncatula* root epidermal cells before infection. Plant Cell 17:3489–3499

Genre A, Chabaud M, Faccio A, Barker DG, Bonfante P (2008) Prepenetration apparatus assembly precedes and predicts the colonization patterns of arbuscular mycorrhizal fungi within the root cortex of both *Medicago truncatula* and *Daucus carota*. Plant Cell 20:1407–1420

Giovannetti M, Balestrini R, Volpe V, Guether M, Straub D, Costa A, Ludewig U, Bonfante P (2012) Two putative-aquaporin genes are differentially expressed during arbuscular mycorrhizal symbiosis in *Lotus japonicus*. BMC Plant Biol 12:186–199

Gomes D, Agasse A, Thiébaud P, Delrot S, Gerós H, Chaumont F (2009) Aquaporins are multifunctional water and solute transporters highly divergent in living organisms. Biochim Biophys Acta 1788:1213–1228

Govindarajulu M, Pfeffer PE, Jin H, Abubaker J, Douds DD, Allen JW, Bücking H, Lammers PJ, Shachar-Hill Y (2005) Nitrogen transfer in the arbuscular mycorrhizal symbiosis. Nature 435:819–823

Guehl JM, Garbaye J (1990) The effects of ectomycorrhizal status on carbon dioxide assimilation capacity, water-use efficiency and response to transplanting in seedlings of *Pseudotsuga menziesii* (Mirb) Franco. Ann For Sci 21:551–563

Gustavsson S, Lebrun A-S, Nordén K, Chaumont F, Johanson U (2005) A novel plant major intrinsic protein in *Physcomitrella patens* most similar to bacterial glycerol channels. Plant Physiol 139:287–295

Jang JY, Lee SH, Rhee JY, Chung GC, Ahn SJ, Kang H (2007) Transgenic Arabidopsis and tobacco plants overexpressing an aquaporin respond differently to various abiotic stresses. Plant Mol Biol 64:621–632

Johansson I, Karlsson M, Shukla V, Chrispeels MJ, Larsson C, Kjellbom P (1998) Water transport activity of the plasma membrane aquaporin PM28A is regulated by phosphorylation. Plant Cell 10:451–459

Katsuhara M, Hanba YT, Shiratake K, Maeshima M (2008) Expanding roles of plant aquaporins in plasma membranes and cell organelles. Funct Plant Biol 35:1–14

Khalvati MA, Hu Y, Mozafar A, Schmidhalter U (2005) Quantification of water uptake by arbuscular mycorrhizal hyphae and its significance for leaf growth, water relations, and gas exchange of barley subjected to drought stress. Plant Biol 7:706–712

Kipfer T, Wohlgemuth T, van der Heiden MGA, Ghazoul J, Egli S (2012) Growth response of drought-stressed *Pinus sylvestris* seedlings to single- and multi-species inoculation with ectomycorrhizal fungi. PLoS ONE 7:e35275

Kraj W, Grad B (2013) Seasonal dynamics of photosynthetic pigments, protein and carbohydrate contents in *Pinus sylvestris* L. seedlings inoculated with *Hebeloma crustuliniforme* and *Laccaria bicolor*. J Plant Nutr 36:633–650

Krajinski F, Biela A, Schubert D, Gianinazzi-Pearson V, Kaldenhoff R, Franken P (2000) Arbuscular mycorrhiza development regulates the mRNA abundance of *Mtaqp1* encoding a mercury-insensitive aquaporin of *Medicago truncatula*. Planta 211:85–90

Lee SH, Calvo-Polanco M, Chung GC, Zwiazek JJ (2010) Role of aquaporins in root water transport of ectomycorrhizal jack pine (*Pinus banksiana*) seedlings exposed to NaCl and fluoride. Plant Cell Environ 33:769–780

Lehto T, Zwiazek JJ (2011) Ectomycorrhizas and water relations of trees: a review. Mycorrhiza 21:71–90

Li QM, Liu BB (2010) Comparison of three methods for determination of root hydraulic conductivity of maize (*Zea mays* L.) roots. Agric Sci China 9:1438–1447

Li T, Hu Y-J, Hao Z-P, Li H, Wang Y-S, Chen B-D (2013) First cloning and characterization of two functional aquaporin genes from an arbuscular mycorrhizal fungus *Glomus intraradices*. New Phytol 197:617–630

Li G, Santoni V, Maurel C (2014) Plant aquaporins: roles in plant physiology. Biochim Biophys Acta 1840:1574–1582

Liu LH, Ludewig U, Gassert B, Frommer WB, Von Wirén N (2003) Urea transport by nitrogen-regulated tonoplast intrinsic proteins in Arabidopsis. Plant Physiol 133:1220–1228

Liu Z, Ma L, He X, Tian C (2014) Water strategy of mycorrhizal rice at low temperature through the regulation of PIP aquaporins with the involvement of trehalose. Appl Soil Ecol 84:185–191

Loque D, Ludewig U, Yuan L, Von Wirén N (2005) Tonoplast intrinsic proteins AtTIP2;1 and AtTIP2;3 facilitate NH_3 transport into the vacuole. Plant Physiol 137:671–680

Ma JF, Yamaji N (2006) Silicon uptake and accumulation in higher plants. Trends Plant Sci 11:392–397

Marjanovic Z, Uehlein N, Kaldenhoff R, Zwiazek JJ, Weiss M, Hampp R, Nehls U (2005) Aquaporins in poplar: what a difference a symbiont makes! Planta 222:258–268

Marulanda A, Azcón R, Ruiz-Lozano JM (2003) Contribution of six arbuscular mycorrhizal fungal isolates to water uptake by Lactuca sativa L. plants under drought stress. Physiologia Plantarum 119:526–533

Maurel C, Plassard C (2011) Aquaporins. For more than water at the plant fungus interface. New Phytol 190:815–817

Maurel C, Verdoucq L, Luu DT, Santoni V (2008) Plant aquaporins: membrane channels with multiple integrated functions. Annu Rev Plant Biol 59:595–624

Miller G, Suzuki N, Ciftci-Yilmaz S, Mittler R (2010) Reactive oxygen species homeostasis and signaling during drought and salinity stress. Plant Cell Environ 33:453–467

Mitani N, Yamaji N, Ma JF (2008) Characterization of substrate specificity of a rice silicon transporter, Lsi1. Eur J Physiol 456:679–686

Miwa K, Kamiya T, Fujiwara T (2009) Homeostasis of the structurally important micronutrients, B and Si. Curr Opin Plant Biol 12:307–311

Moore JAM, Jiang J, Post WM, Classen AT (2015) Decomposition by ectomycorrhizal fungi alters soil carbon storage in a simulation model. Ecosphere 6:29

Morte A, Lovisolo C, Schubert A (2000) Effect of drought stress on growth and water relations of the mycorrhizal association *Helianthemum almeriense-Terfezia claveryi*. Mycorrhiza 10:115–119

Morte A, Navarro-Ródenas A, Nicolás E (2010) Physiological parameters of desert truffle mycorrhizal *Helianthemum almeriense* plants cultivated in orchards under water deficit conditions. Symbiosis 52:133–139

Muhsin TM, Zwiazek JJ (2002a) Colonization with *Hebeloma crustuliniforme* increases water conductance and limits shoots sodium uptake in white spruce (*Picea glauca*) seedlings. Plant Soil 238:217–225

Muhsin TM, Zwiazek JJ (2002b) Ectomycorrhizas increase apoplastic water transport and root hydraulic conductivity in *Ulmus Americana* seedlings. New Phytol 153:153–158

Navarro-Ródenas A, Ruiz-Lozano JM, Kaldenhoff R, Morte A (2012) The aquaporin *TcAQP1* of the desert truffle *Terfezia claveryi* is a membrane pore water and CO_2 transport. Mol Plant-Microbe Interact 25:259–266

Navarro-Ródenas A, Bárzana G, Nicolás E, Carra A, Schubert A, Morte A (2013) Expression analysis of aquaporins from desert truffle mycorrhizal symbiosis reveals a fine-tuned regulation under drought. Mol Plant-Microbe Interact 26:1068–1078

Navarro-Ródenas A, Xu H, Kemppainen M, Pardo AG, Zwiazek JJ (2015) *Laccaria bicolor* aquaporin *LbAQP1* is required for Hartig net development in trembling aspen (*Populus tremuloides*). Plant Cell Environ (In press). doi:10.1111/pce.12552

Nehls U, Dietz S (2014) Fungal aquaporins: cellular functions and ecophysiological perspectives. Appl Microbiol Biotechnol 98:8835–8851

Ouziad F, Wilde P, Schmelzer E, Hildebrandt U, Bothe H (2006) Analysis of expression of aquaporins and Na^+/H^+ transporters in tomato colonized by arbuscular mycorrhizal fungi and affected by salt stress. Environ Exp Bot 57:177–186

Pérez-Tienda J, Testillano PS, Balestrini R, Valentina Fiorilli V, Azcón-Aguilar C, Ferrol N (2011) GintAMT2, a new member of the ammonium transporter family in the arbuscular mycorrhizal fungus *Glomus intraradices*. Fungal Genet Biol 48:1044–1055

Porcel R, Gómez M, Kaldenhoff R, Ruiz-Lozano JM (2005) Impairment of *NtAQP1* gene expression in tobacco plants does not affect root colonization pattern by arbuscular mycorrhizal fungi but decreases their symbiotic efficiency under drought. Mycorrhiza 15:417–423

Porcel R, Aroca R, Azcón R, Ruiz-Lozano JM (2006) PIP aquaporin gene expression in arbuscular mycorrhizal *Glycine max* and *Lactuca sativa* plants in relation to drought stress tolerance. Plant Mol Biol 60:389–404

Roussel H, Bruns S, Gianinazzi-Pearson V, Hahlbrock K, Franken P (1997) Induction of a membrane intrinsic protein-encoding mRNA in arbuscular mycorrhiza and elicitor-stimulated cell suspension cultures of parsley. Plant Sci 126:203–210

Ruth B, Khalvati M, Schmidhalter U (2011) Quantification of mycorrhizal water uptake via high-resolution on-line water content sensors. Plant Soil 342:459–468

Siemens JA, Zwiazek JJ (2008) Root hydraulic properties and growth of balsam poplar (*Populus balsamifera*) mycorrhizal with *Hebeloma crustuliniforme* and *Wilcoxia mikolae* var. mikolae. Mycorrhiza 18:393–401

Siemens JA, Zwiazek JJ (2011) *Hebeloma crustuliniforme* modifies root hydraulic responses of trembling aspen (*Populist tremuloides*) seedlings to changes in external pH. Plant Soil 345:247–256

Steudle E, Peterson CA (1998) How does water get through roots? J Exp Bot 49:775–788

Sui H, Han BG, Lee JK, Walian P, Jap BK (2001) Structural basis of water-specific transport through AQP1 water channel. Nature 414:872–878

Tarkka MT, Hermann S, Wubet T, Feldhahh L, Recht S, Kurth F, Mailander S, Bonn M, Neef M, Angay O, Bacht M, Graf M, Maboreke H, Fleischmann F, Grams TEE, Ruess L, Schadler M, Brandl R, Scheu S, Scherey SD, Grosse I, Buscot F (2013) OakContigDF159.1, a reference library for studying differential gene expression in *Quercus robur* during controlled biotic interactions: use for quantitative transcriptomic profiling of oak roots in ectomycorrhizal symbiosis. New Phytol 199:529–540

Tedersoo L, Naadel T, Bahram M, Pritsch K, Buegger F, Leal M, Koljalg U, Poldmaa K (2012) Enzymatic activities and stable isotope patterns of ectomycorrhizal fungi in relation to phylogeny and exploration types in an afrotropical rain forest. New Phytol 195:832–843

Tian C, Kasiborski B, Koul R, Lammers PJ, Bucking H, Shachar-Hill Y (2010) Regulation of the nitrogen transfer pathway in the arbuscular mycorrhizal symbiosis: gene characterization and the coordination of expression with nitrogen flux. Plant Physiol 153:1175–1187

Turgeman T, Ben Asher J, Roth-Bejerano N, Kagan-Zur V, Kapulnik Y, Sitrit Y (2011) Mycorrhizal association between the desert truffle *Terfezia boudieri* and *Helianthemum sessiliflorum* alters plant physiology and fitness. Mycorrhiza 21:623–630

Uehlein N, Lovisolo C, Siefritz F, Kaldenhoff R (2003) The tobacco aquaporin NtAQP1 is a membrane CO_2 pore with physiological functions. Nature 425:734–737

Uehlein N, Fileschi K, Eckert M, Bienert GP, Bertl A, Kaldenhoff R (2007) Arbuscular mycorrhizal symbiosis and plant aquaporin expression. Phytochemistry 68:122–129

Vandeleur RK, Sullivan W, Athman A, Jordans C, Gilliham M, Kaiser BN, Tyerman SD (2014) Rapid shoot-to-root signaling regulates root hydraulic conductance via aquaporins. Plant Cell Environ 37:520–538

Varma A (2008) Mycorrhiza. State of the art, genetics and molecular biology, eco-function, biotechnology, eco-physiology, structure and systematics, 3rd edn. Springer, Berlin

Vesk PA, Ashford AE, Markovina AL, Allaway WG (2000) Apoplasmic barriers and their significance in the exodermis and sheath of *Eucalyptus pilularis-Pisolithus tinctorious* ectomycorrhizas. New Phytol 145:333–346

Wang Y, Ding GJ (2013) Influence of ectomycorrhiza on nutrient absorption of *Pinus massoniana* seedlings under water stress. For Res 26:227–233

Wang W-H, Köhler B, Cao F-Q, Liu LH (2008) Molecular and physiological aspects of urea transport in higher plants. Plant Sci 175:467–477

Wei Y, Shen W, Dauk M, Wang F, Selvaraj G, Zou J (2004) Targeted gene disruption of glycerol-3-phosphate dehydrogenase in *Colletotrichum gloeosporioides* reveals evidence that glycerol is a significant transferred nutrient from host plant to fungal pathogen. J Biol Chem 279:429–435

Xu H, Kemppainen M, El Kayal W, Lee SH, Pardo AG, Cooke JEK, Zwiazek JJ (2015) Overexpression of *Laccaria bicolor* aquaporin JQ585595 alters root water transport properties in ectomycorrhizal white spruce (*Picea glauca*) seedlings. New Phytol 205:757–770

Yi H, Calvo-Polanco M, MacKinnon MD, Zwiazek JJ (2008) Responses of ectomycorrhizal *Populus tremuloides* and *Betula papyrifera* seedlings to salinity. Environ Exp Bot 62:357–363

The manufacturer's authorised representative in the EU is Springer Nature Customer Service Centre GmbH, Europaplatz 3, 69115 Heidelberg, Germany. If you have any concerns regarding our products, please contact ProductSafety@springernature.com

Printed and bound by CPI Group (UK) Ltd, Croydon, CR0 4YY
01/10/2026
02225869-0004